Metal–Organic Frameworks

Metal–Organic Frameworks

Materials Modeling towards Potential Engineering Applications

edited by

Jianwen Jiang

CRC Press
Taylor & Francis Group
Boca Raton London New York

CRC Press is an imprint of the
Taylor & Francis Group, an **informa** business

Metal–Organic Frameworks: Materials Modeling towards Potential Engineering Applications

First published 2015 Pan Stanford Publishing Pte. Ltd.

Published 2022 by CRC Press
Taylor & Francis Group
6000 Broken Sound Parkway NW, Suite 300
Boca Raton, FL 33487-2742

ISBN 13: 978-981-4613-45-3 (hbk)
ISBN 13: 978-981-4613-46-0 (ebk)

Visit the Taylor & Francis Web site at
http://www.taylorandfrancis.com

and the CRC Press Web site at
http://www.crcpress.com

British Library Cataloguing-in-Publication Data
A catalogue record for this book is available from the British Library.

Contents

Foreword

Thanks to the diligent work of Omar Yaghi and other creative synthetic chemists, it is now possible to synthesize a variety of metal–organic frameworks (MOFs) that show much promise for gas storage (e.g., methane and hydrogen) and for gas separations, especially those that separate climate-warming carbon dioxide from nitrogen as desired for stack gases from power plants prior to release to the atmosphere.

The possible number of different MOFs is very large. In a finite time, it is not feasible to synthesize even a small fraction of these possible structures. To guide experimentalists toward those structures that are likely to be most useful, theoretical or computational methods are needed. Fortunately, such methods have been under active development for some time. It is therefore now appropriate to assemble a set of articles that describe these methods to indicate their present and future utility.

The worldwide community of chemists and chemical engineers is grateful to the authors of these assembled articles and especially to Jianwen Jiang for his diligent service as editor of this timely volume.

John Prausnitz
University of California, Berkeley, USA

Preface

Metal–organic frameworks (MOFs), also called porous coordination polymers or porous coordination networks, have emerged as a special class of nanoporous materials. Constituted by the enormous choices of metal clusters and organic linkers, MOFs possess a wide range of surface area, pore volume, and functionality; thus they have been considered versatile materials for storage, separation, catalysis, etc. Over the past decade, thousands of MOFs have been synthesized, characterized and tested. The number of MOFs is constantly growing and in principle unlimited. It is, therefore, practically infeasible to experimentally test and select appropriate MOFs from infinite candidates for applications.

With rapidly evolving computational resources, atomistic/molecular modeling has become an indispensable tool in materials science and engineering. Sophisticated modeling at a microscopic level provides wealthy insights that are otherwise experimentally inaccessible and elucidates underlying physics from the bottom up. Furthermore, modeling can secure the fundamental interpretation of experimental observations and guide the rational selection and design of materials.

This book compiles a broad collection of recent modeling studies in the field of MOFs toward potential engineering applications. It contains 13 chapters and is categorized into 5 different topics:

- In Chapter 1, Caroline Mellot-Draznieks, Ben Slater, and Raimondas Galvelis give an overview of computational approaches for the crystal structure prediction of MOFs. They highlight the simulation-informed principles to design and characterize new MOFs and more systematically understand structure–property relationships.

- Adsorption phenomenon and associated properties of MOFs are summarized in Chapters 2–4. While David Dubbeldam and Krista S. Walton discuss the classical molecular simulations, Michael Fischer and Michael Fröba, as well as Lukáš Grajciar, Miroslav Rubeš, Ota Bludský, and Petr Nachtigall, review theoretical approaches, particularly density functional theory, to describe adsorption on coordinatively unsaturated metal sites in MOFs.
- Chapters 5 and 6 are devoted to both adsorption and diffusion in MOFs. George K. Papadopoulos presents statistical mechanics–based modeling for sorbate equilibria and transport in MOFs. On the other hand, Giovanni Garberoglio highlights the importance of quantum effects to accurately describe hydrogen behavior in MOFs.
- Numerous modeling studies of MOFs are focused on separation, as summarized in a series of chapters. In Chapter 7, Qingyuan Yang, Dahuan Liu, and Chongli Zhong review both adsorption- and membrane-based separation processes for CO_2-related systems, olefin/paraffin mixtures, and other gas mixtures. Seda Keskin discusses, in Chapter 8, adsorption-based separation for typical CO_2-containing gas mixtures in MOFs and membrane-based CO_2 separation in MOFs and mixed-matrix membranes. In Chapter 9, Carlos Nieto-Draghi and Javier Pérez-Pellitero survey the modeling of gas separation in zeolitic imidazolate frameworks. Alternatively, Alexandre F. P. Ferreira, Ana M. Ribeiro, João C. Santos, Marta C. Campo, and Alírio E. Rodrigues demonstrate the modeling of cyclic adsorptive separation in Chapter 10, with emphasis on pilot-scale modeling and process simulation. In Chapter 11, Anjaiah Nalaparaju and Jianwen Jiang review the simulation studies of ionic MOFs for gas adsorption and separation, water treatment, biofuel purification, and drug loading.
- Modeling studies of catalysis in MOF-based compounds are documented in Chapters 12 and 13. Qiuju Zhang and Liang Chen, as well as Remedios Cortese and Dario Duca, discuss the catalytic properties of MOFs and several typical chemical reactions catalyzed by MOFs.

From this book, one can clearly witness the increasingly important role of computational modeling in the characterization, screening, and design of MOFs for potential engineering applications. Microscopic insights from the bottom up are imperative to quantitatively unravel fundamental mechanisms, cohesively complement experiments, and effectively facilitate new development in this burgeoning field.

I wish to express my gratitude to the outstanding authors for their invaluable contribution to this book and the editorial team of Pan Stanford Publishing for its technical support.

Jianwen Jiang
Singapore

MOFs and a more systematic understanding of structure–property relationships. This is indeed a distinct subfield from the much larger field of MOF computational chemistry, whereby CSP aims to genuinely predict (existing or hypothetical) crystal structures at the atomic level without recourse to any experimental crystal structure data, that is, ab initio. While CSP has a longstanding history in the field of molecular solids such as pharmaceuticals [1] and more recently molecular cage materials [2], a number of recent reviews have highlighted how the field of CSP of periodic solid MOFs is still relatively young and presents significant challenges to methodologies. Nevertheless it has evolved dramatically over the last decade with major methodological developments in the area of inorganic solids and, later, in the area of MOFs [3, 4].

We will not cover the following specific topics, which were recently reviewed: force field development [5], simulation of guest-responsive flexible MOFs [6], and simulation of the adsorption properties of MOFs [7, 8] except when it involves the exploration of computationally generated hypothetical MOFs, and instead aim to give a general overview of the modeling of MOFs.

We will formally divide the predictive computational approaches of MOFs into six categories. Section 1.2 gives an overview of the various strategies developed so far for the genuine prediction of the crystal structure of MOFs, including large-scale computational methods. Section 1.3 discusses structure determination of MOFs, highlighting how computational approaches may be used in a predictive fashion when new MOFs are synthesized where their structures cannot be determined through the conventional structure solution methods from powder diffraction data. In Section 1.4 the computational modification of MOFs, typically through virtual ligand functionalization of MOFs or topologies, is addressed. Such approaches appeared only recently and have been adopted with the aim to develop more systematic functional and topological screening for MOFs for targeted adsorption properties. Quantitative structure–property relationship (QSPR) methods, beginning to emerge from the literature using a database of hypothetical MOFs or virtually modified ones, will also be presented. Section 1.5 presents one of most speculative and heuristic use of CPS in the computational design and evaluation of new families of MOFs and their subsequent successful targeted synthetic trials, exemplified in the very active area of polyoxometalate metal–organic frameworks (POMOFs). Section 1.6 deals with the exploration of the energy landscape of MOFs, focusing, on the one hand, on zeolitic imidazolate frameworks (ZIFs) and, on the other hand, on the understanding of the thermodynamics underlying MOFs making use of chiral ligands. Finally, in Section 1.7 we present the prediction of intrinsic physical properties of MOFs such as mechanical properties of MOFs or the estimation of electronic band gaps, a rapidly emerging area. Although applied to existing structures (therefore not belonging to predictive approaches but rather traditional modeling approaches), these types of calculations may now be readily applied to the screening of hypothetical MOFs, opening up opportunities for the computationally assisted fine-tuning of properties. A short, final section considers the first studies at the time of writing related to the computational studies of defects in MOFs.

1.2 Crystal Structure Prediction

1.2.1 *Introduction*

Let us briefly define what is meant by CSP in the much larger context of computational simulations of solids. The now-routine computational chemistry of solids, using force field methods or electronic structure techniques, may be used to identify the equilibrium structure of a predefined solid, starting from its known experimental structure or an alternative approximate structural model. By contrast, genuine structure prediction involves a minimal amount of empirical knowledge, for example, leaving the atomic coordinates, the cell parameters, and even the symmetry and space group undetermined, while fixing the chemical composition by specifying the targeted atomic content (and stoichiometry). Genuine CSP should therefore result in a list of candidate periodic crystal structures or networks that are compatible with a list of predefined targeted (topological, structural, or chemical) features and that might be further refined using either force field or electronic structure calculations.

Turning to the range of methodologies required to generate a hypothetical crystal structure from scratch, they are well reviewed in Refs. [3, 4]. In short, one needs procedures that are capable of, first, efficiently exploring the configurational space defined by the atoms or building blocks in a cell and identifying the regions of interest and, second, quantifying the quality of the predicted crystal structures with respect to the predefined target features. A number of strategies for predicting crystal structures of periodic nets have been developed. Many methods rely on global optimization methods, such as simulated annealing (SA) [9], Monte Carlo basin hoping (MCBH) [10], and genetic algorithms (GAs) [11], which involve a "cost function" over which minima are searched for. In SA methods, temperature is used in conjunction with a Monte Carlo (or molecular dynamics) scheme for allowing the continuous change of ions, atoms, or building block configurations to overcome barriers between local minima. Depending on the SA scheme, the generated crystal structures may include the global minimum or a list of metastable structures (local minima). In an evolutionary (or genetic) algorithm, Darwinian evolution is mimicked where initially random configurations on the energy hypersurface tend to form a population that may survive and persist through crossovers or mutations. Another class of methods consists of topological enumeration, or tiling, used with great success in the prediction of zeolites [12, 13]. Also, the metadynamics approach [14] applied to the pressure-induced crystal structure transformations of SiO_2 [15] allowed the exploration of the Gibbs free energy as a function of the cell parameters by history-dependent dynamics to avoid already visited minima. Not all methods mentioned above have been used today to genuinely predict MOF periodic nets. For example, the powerful metadynamics method has not been applied to the structure prediction of MOFs at the time of writing, to our knowledge. Below, we will detail strategies developed so far for the CSP of MOFs.

1.2.2 *The AASBU Method*

The very first computational approach for predicting crystal structures of MOFs starting from the knowledge of their constituent building blocks used SA, referred to as the automated assembly of secondary building unit (AASBU) method (Fig. 1.1) [16]. Originally developed for inorganic framework solids [16–18], it was further extended to the structure prediction of MOFs [19].

The key feature of the method lies in the combination of pre-defined building units and empirical "gluing" rules for assembling them. The exploration of their autoassembly in 3D space uses a sequence of SA and energy minimization steps. The inorganic unit is modeled by a rigid body, $(M_xL_y)^{inorg}$ (M: metal centers; L: ligand atoms). The organic unit, typically here a di- or tri-carboxylate molecule, $(C_nL_m)^{org}$, may be treated as a flexible body using a generic force field like *cvff*. The rules that control the assembly of the organic and inorganic units are encapsulated in a "cost" function consisting of a Lennard–Jones-like term plus a force field that favors "sticky atom" pairs. The $L^{org} \cdots L^{inorg}$ "sticky atom" pairs are those that form the organic–inorganic linkages, parameterized as a highly attractive potential well with a minimum at very short $L^{org} \cdots L^{inorg}$ separation.

This potential is used as a means to "glue" organic and inorganic units together. Similarly, $L^{inorg} \cdots L^{inorg}$ "sticky atom" pairs are those that form inorganic–inorganic clustering. A repulsive potential between $L^{org} \cdots L^{org}$ pairs avoids overlapping or connections of organic molecules to one another. The validity of the method was addressed with the simulation of existing and prototypic hybrid frameworks, such as Materials of the Institute Lavoisier (MIL)-53, HKUST-1, and MOF-5, as shown in Fig. 1.1. All these structures were predicted successfully using various procedures, either using independent organic and inorganic units or using a large hybrid building block like in the case of HKUST-1. As a typical output, the MIL-53 topology was produced in space group *Cc* among a list of 54 candidates, with the final high symmetry, *Imma*, in good agreement with that of the experimental structure, *Imcm*, the other structures consisting of a series of not-yet-synthesized topologies ranging from lamellar to 3D extended structures. One advantage of the AASBU method is to perform a topology-oriented search for crystal structures (assembling individual building units or whole cages) able to highlight topological relationships within groups of related materials, generating expected topologies together with new ones.

Any building block can combine with any other building block, provided that the geometry and chemical composition local to the point of connection (marked by purple crosses in Fig. 1.2b) is the same as in crystallographically determined structures. Building blocks are combined stepwise, and when an atomic overlap occurs at a particular step, a different building block is chosen or a different connection site, until all possibilities are exhausted. While the total number of steps in each generation process can vary, there are always three steps when, instead of adding a building block, a periodic boundary is imposed by connecting any two building blocks (steps 2 and 4 in Fig. 1.2b). When no more building blocks can be added, the crystal generation procedure is complete. Note that no force field (or quantum mechanical) energy minimizations are involved since the pieces are connected according to the geometric rules that govern how the building blocks are connected in existing MOFs. By construction, this approach can generate a very large number of structures per second per CPU, and because it is enumerative, an entire portion of space of hypothetical MOFs can be systematically explored. It is important to point out that this approach simultaneously explores conformational as well as compositional space, with no explicit constraints in terms of topology or symmetries of the resulting structures. Still, no attempt has been included in the work to predict which conformer is lower in energy, leaving the question of the viability and relative stabilities of the hypothetical MOFs generated open. The code is in development and may be informally referred to as a "crystal generator" or a "MOF generator." This approach has been further elegantly applied for the large-scale screening of existing and hypothetical MOFs for specific adsorption/separation processes (see Section 1.4).

1.3 Using Structure Prediction to Solve the Structures of MOFs

The comprehensive determination of the crystal structures of MOFs is a crucial prerequisite in order to rationalize both their properties and their synthesis. When single crystals of sufficient size or quality may not be easily obtained, structure determination from powder data is still one of the most efficient approaches but still is a critical and time-consuming step. Yet, it is frequent that when a new MOF solid is synthesized, laboratory X-ray powder diffraction (XRPD) reveals such a complex pattern that the structure solution is abandoned or not even undertaken. In general, crystal structures tend to become more and more complex to solve, for example, involving larger cell volumes or chemical compositions, leading to the absence of atomic contrast, rendering the process of structure solution nontrivial and even more challenging. Computational model building methods combined with geometry optimizations, used independently and thus ahead of the exploitation of the X-ray patterns of the targeted compounds, have proved to be particularly useful for proposing initial structural models suited for further Rietveld refinement [21].

1.3.1 *The Decoration Strategy*

A most illustrative example is that of the structure prediction of the two highly complex MIL-100 [22] and MIL-101 [23] crystal structures, chromium benzene-1,3,5-tricarboxylate (BTC) and chromium benzene-1,4-dicarboxylate (BDC), respectively. In both cases, laboratory XRPD revealed complex patterns with an important overlap of peaks at low angles, ruling out the use of ab initio structure solution methods. A computational structure prediction approach was undertaken. More details on the strategy adopted to predict these two crystal structures may be found in Ref. [24].

In a first step, an examination of the chemical composition of both compounds allowed the identification of the constitutive building blocks among a list of candidate ones, a hybrid supertetrahedron (ST) in both cases (see Fig. 1.3a). In each ST, the four vertices are occupied by a trimer of corner-sharing octahedra, the organic linker occupying the faces (BTC) or the edges (BDC) of the ST. In a second step, drawing an analogy between the inter-ST linkages expected in MIL-100/101 with the Si–O–Si linkages typical of zeolites (see Fig. 1.3c), it was inferred that the assembly of STs should lead to zeotype structures.

Subsequent steps consisted of constructing candidate structural models by "decorating" upper-scale versions of zeolite frameworks,

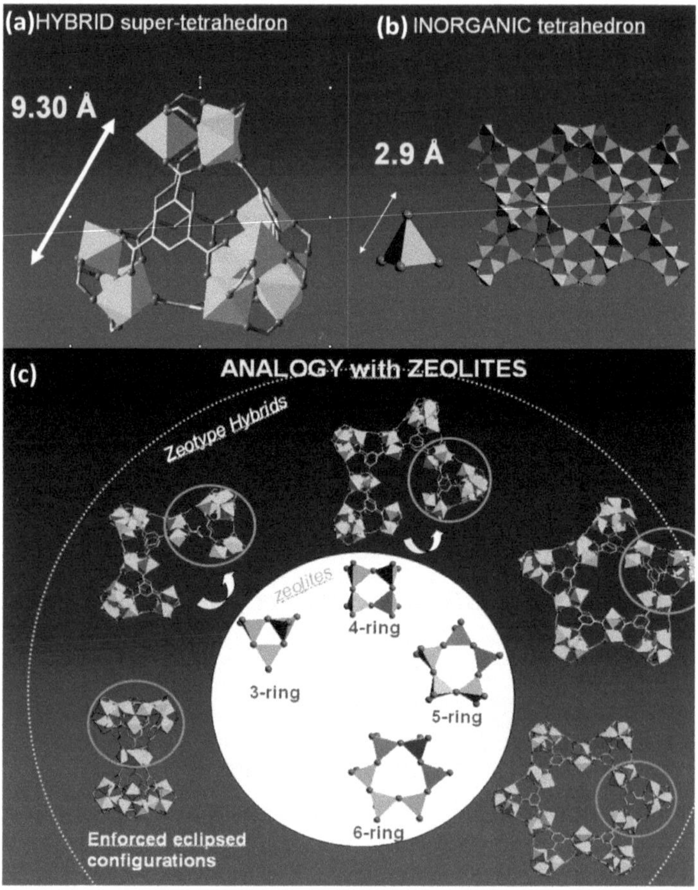

Figure 1.3 Comparison between (a) the preconstructed hybrid supertetrahedron (ST) in MIL-100 and (b) the corner sharing of SiO_4 tetrahedra in zeolitic silicates. (c) Analogy between zeolitic rings (from three to six rings) and the rings that may be possibly made from ST–ST linkages. Only five rings and six rings may be formed. Adapted from Ref. [24], copyright 2005, with permission from Elsevier.

as reported with their three-letter code from the IZA database [25], that is, substituting each SiO_4 tetrahedron by the required hybrid ST in a corner-sharing fashion. It resulted in only three polymorphs possessing exclusively five rings and six rings with the MTN, MEP, and unknown hexagonal (HEX) topologies (Fig. 1.4), each

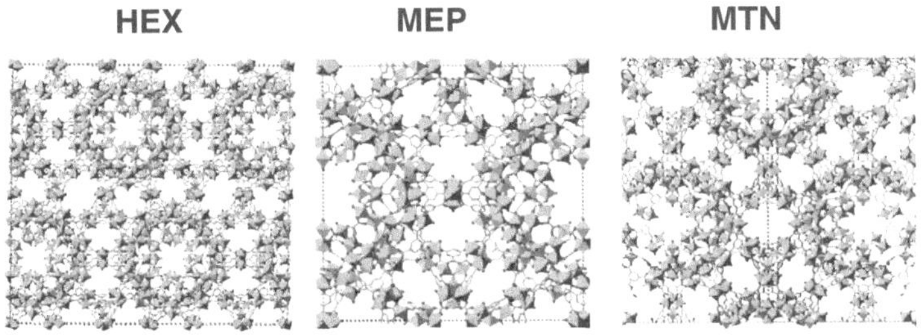

Figure 1.4 The three predicted zeotypic MOFs structures made of ST–ST linkages, all three corresponding to possible polymorphs of MIL-100. Adapted from Ref. [22].

of which was subsequently geometry-optimized using the generic *uff* force field. Finally, for each compound, the correct structure was identified by comparing the simulated patterns of the three possible polymorphs to the observed one, leading to the conclusion that both compounds are MTN-type structures (interestingly found as the energetically favored ones).

The predicted model structures of MIL-100 and MIL-101 have unusually a large cell dimension (over 370,000 Å^3 and 700,000 Å^3, respectively) and were directly used for their further structural refinement using synchrotron data. The cases of MIL-100 and MIL-101 remain exceptional, since some prior knowledge on the constitutive building blocks together with their connectivity mode is necessary. Still, these two MOFs structure are the most illustrative cases of a close interplay between the two fields of structure prediction and structure determination. Another attractive possibility of the decoration strategy is the production of new hypothetical unknown topologies [26].

More recently, such "decoration strategies" of networks were extensively used for exploring the energy landscape of whole families of existing or hypothetical MOFs, such as POMOFs and ZIFs, as developed in more detail in Sections 1.5 and 1.6.

1.3.2 *Ligand Replacement Strategy in Flexible MOFs*

Ligand replacement strategies are especially useful to computationally assist the structure determination of MOFs when topological relationships are identified between a new MOF structure (yet unsolved) and an existing MOF for which the crystal structure has already been determined experimentally, either with single crystals or with powder diffraction data.

The ligand replacement strategy was used with great success for the computational structure solution of the isoreticular MIL-88A-B-C-D series [27]. Starting with the experimentally known crystal structure of iron fumarate-MIL-88A, the structures of the three as-synthesized isoreticular frameworks were derived by replacing the fumarate ligand of the parent MIL-88A structure with larger ligands, that is, terephthalate (MIL-88B), 2,6-naphthalenedicarboxylate (MIL-88C), and 4,4'-biphenyldicarboxylate (MIL-88D) (Fig. 1.5). Further lattice energy minimizations were required to allow the relaxation of the ligand itself and that of the cell needed to accommodate the larger ligand, using the experimental cell parameters as target values while retaining the maximum authorized symmetry. The resulting models were then directly used to perform the Rietveld refinements on laboratory powder diffraction data. The same ligand replacement strategy was used to solve the crystal structure of aluminum-naphtalate MIL-69 ($Al(OH)(O_2C$-$C_{10}H_6$-$CO_2)H_2O$), assuming that a similar framework topology to that of the aluminum terephthalate MIL-53 might occur in the naphtalate, MIL-69 [28].

Also, the full range of the reversible swelling of the MIL-88A-B-C-D series [29, 30] and MIL-89 [31] upon dehydration or adsorption of various organic solvents could be structurally followed in the absence of exploitable (i.e., proper for Rietveld refinement) powder data. The full path of their structural transformations was followed applying energy minimizations initially to the closed form and taking the sequence of experimental cell parameters upon swelling as target values.

In a similar spirit, the crystal structure of the breathing Co(BDP) (BDP = 1,4-benzene-dipyrozolate) solid was determined using the cell parameters obtained via the successful indexing of its diffraction powder pattern while using the atomic coordinates of the topologically related MIL-53(Fe) crystal structure, replacing the BDC linker of the parent structure with the BDP linker [32]. Further force field and density functional theory (DFT) calculations were

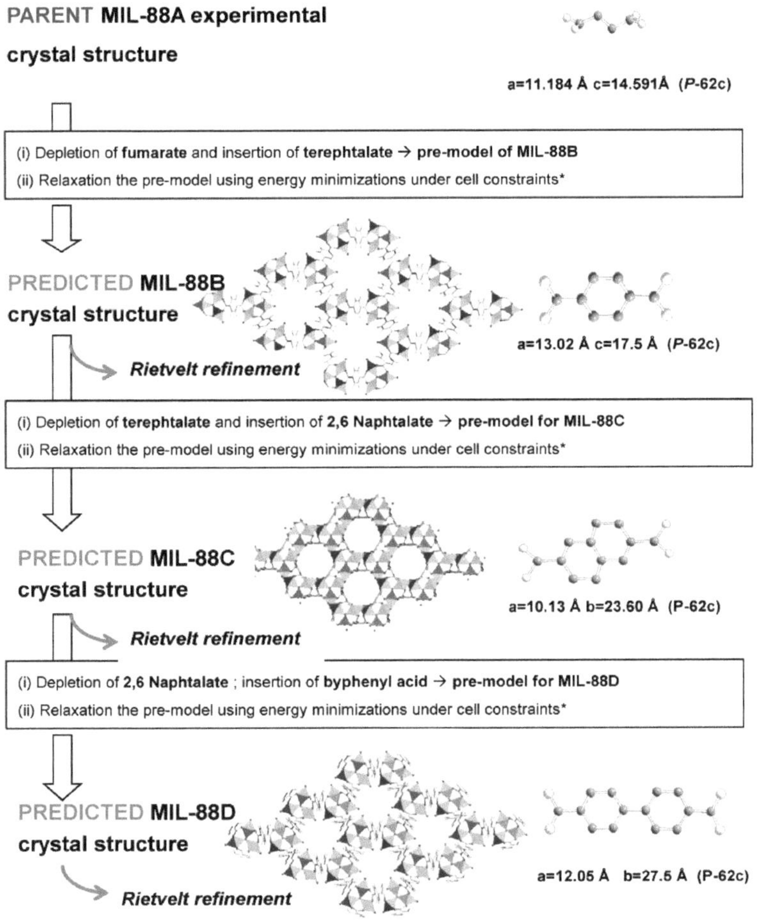

Figure 1.5 Scheme for the ligand replacement strategy used for the computational structure elucidation of MIL-88A-B-C-D series of compounds. The method consists of gradually replacing the organic ligand in the MIL-88A "parent" crystal structure solved from XRPD by another of larger size and then minimizing the lattice energy of the resulting model. The converged models may be used for further Rietveld refinement, when possible. Adapted from Ref. [27].

used to geometry optimize the resulting flexible structure, providing plausible models in view of the experimental diffraction patterns of the flexible Co(BDP) under various N_2 pressures.

1.4 Computational Screening of Hypothetical MOFs through Ligand Functionalization

There is a very large amount of literature exploring through simulations the adsorption/separation performances of MOFs, which will not be considered here, as a very large number of references and reviews may be consulted [33–35]. Despite the now routine use of simulations for predicting gas adsorption isotherms or selectivities, we lack a generic tool for predicting the performances of existing or hypothetical MOF structures for a targeted application, probably because of the lack of general correlation principles. With MOFs, computational approaches are particularly adapted to evaluate the performances of hypothetical structures and explore the impact of topologies of linkers through computationally modified MOFs, yielding guidance on the potential interest of targeted topologies or ligands ahead of experimental trials.

In that spirit, Torrisi et al. [36, 37] proposed MIL-53-type solids in which the linker was computationally functionalized with various chemical groups (i.e., OH–, COOH–, NH_2–, and CH_4–) with the aim to improve CO_2 capture. On the basis of the experimental crystallographic structure of MIL-53(Al), the models were constructed and geometry-optimized using periodic DFT calculations in order to optimize the position of the ligand substituents. The affinity of the model structures for CO_2 was further evaluated by computing binding energies or adsorption enthalpies through a series of DFT (cluster and periodic models) and grand canonical Monte Carlo (GCMC) simulations (Fig. 1.6). In line with our previous cluster DFT calculations [38, 39], the results point to OH– and COOH– functional groups as the best candidates among those studied in terms of capacity and CO_2/CH_4 selectivity, evidencing the crucial role of polar groups in CO_2 capture and the particularly penalizing impact of bulky nonpolar groups. Interestingly, these computational findings regarding the impact of such functional groups were confirmed

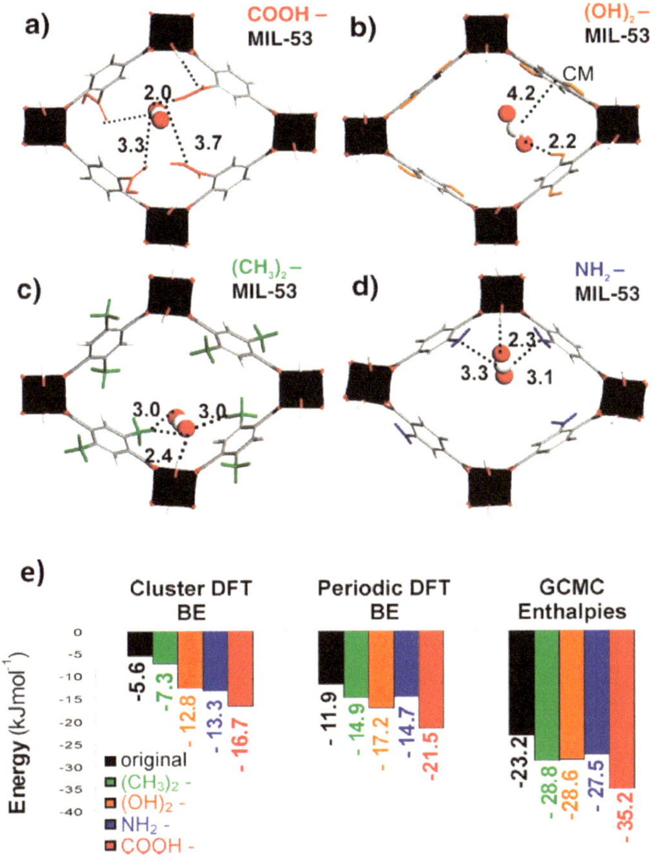

Figure 1.6 Top: DFT-calculated most stable adsorption site for CO_2 in the computed hypothetical model of MIL-53(Al^{3+}) in its open form functionalized with (a) NH_2– (blue), (b) $(OH)_2$– (orange), (c) –COOH– (red), and (d) $(CH_3)_2$– (green) groups. Adapted from Ref. [37]. Bottom: Histograms of CO_2 binding energies (BEs) from cluster and periodic DFT calculations and adsorption enthalpies calculated by GCMC simulations. Reprinted with permission from Ref. [36]. Copyright 2010 American Chemical Society.

experimentally a year later in the area of nanoporous organic polymer for CO_2 sorption [40].

ZIFs constitute another large family with attractive features for optimizing adsorption capacity and selectivity to capture CO_2 from gas mixtures by controlling the pore size and its chemistry through

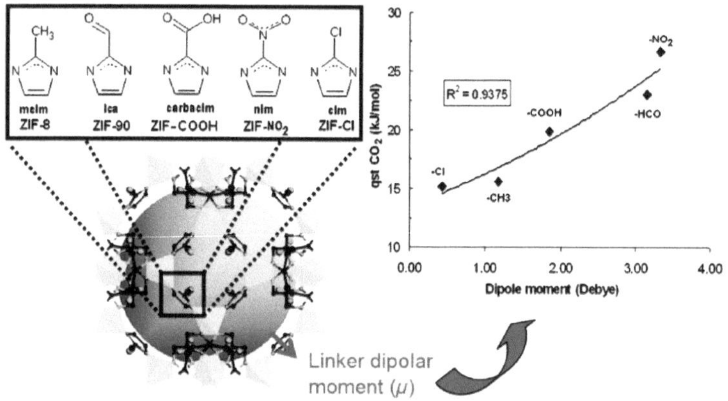

Figure 1.7 Linkers of the sodalite-ZIF solids computationally functionalized at the −2 position (left) and the expected CO_2 isosteric heat of adsorption in the corresponding functionalized ZIFs (right). Reprinted with permission from Ref. [41]. Copyright 2011 American Chemical Society.

the choice of a functionalized linker. Nieto-Draghi et al. [41] studied a series of computationally functionalized ZIF-8 models for the enhancement of CO_2 uptake and highlighted the dipole of the linker as the main criterion for enhanced affinity. Virtually increasing the dipole moment from 0.3 D up to 8.3 D in the sodalite-type ZIF, the authors find an exponential relationship between the ligand dipole moment and the computed isosteric heat of adsorption of CO_2 (Fig. 1.7).

The above approaches have been rapidly followed by more systematic and large-scale QSPR studies—directly inspired from the quantitative structure–activity relationship (QSAR) methods largely used in drug design—enabling us to correlate selected characteristics of the host–guest system represented by a series of descriptors with a target property. A typical QSPR approach was proposed by Amrouche et al. [42] to compute the isosteric heat of adsorption of various polar and nonpolar (Ar, CH_4, C_2H_6, N_2, O_2, CO_2, H_2S, SO_2, H_2O, CH_3CN) molecules in a large variety of functionalized sodalite-ZIFs structures (Fig. 1.8). Typically, the main descriptors include not only the dipolar and quadrupolar moments of the organic linker constitutive of the ZIF but also the pore mean curvature, the number of functional groups present on the

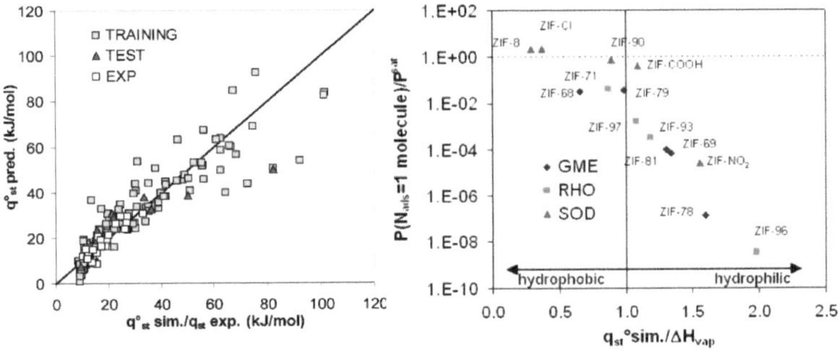

Figure 1.8 (Left) Comparison of the predicted and experimental isosteric heat of adsorption of various gases in a range of ZIFs where the prediction is obtained using the QSPR model. (Right) Correlative approach for the hydrophobic/hydrophilic nature of the different ZIFs studied. P^{sat} and $P(N_{\mathrm{ads}} = 1)$ are the saturation pressure and the pressure at which one molecule of water is adsorbed per unit cell, respectively. The ratio $q_{\mathrm{st}}/\Delta H_{\mathrm{vap}}$ is an estimate for the relative affinity of a water molecule to approach a solid surface embedded in liquid water. Reproduced with permission from Ref. [42].

linker, the dipolar moment of the adsorbed gas, and its atmospheric boiling temperature, finally proposing an equation able to predict the isosteric heat of adsorption for ZIFs, together with a simple mean of estimating the hydrophobic/hydrophylic nature of the ZIF.

In the same spirit, the derivation of QSAR models was used to identify the key factors that drive the caffeine uptake in a series of MIL-88 solids, such as the polarity, polarizability, and H-donor capacity of the organic linker introducing variously hydrophilic and acidic functional groups (–Br, –F, $-CH_3$, $-NH_2$, $-NO_2$, –OH, $-CF_3$) [43].

On a much larger scale, the screening approach proposed above by Wilmer et al. was successfully developed to find structure–property relationships of MOF topologies for methane adsorption [20], as well as xenon/krypton [44] and CO_2/CH_4 and CO_2/N_2 separations [45]. These large-scale results exhibit sharp correlations between structural (e.g., pore size, surface area, and pore volume) and chemical features, with absorbent evaluation criteria such as CO_2working capacity, CO_2 uptake, and heat of adsorption of CO_2/N_2 selectivity (Fig. 1.9).

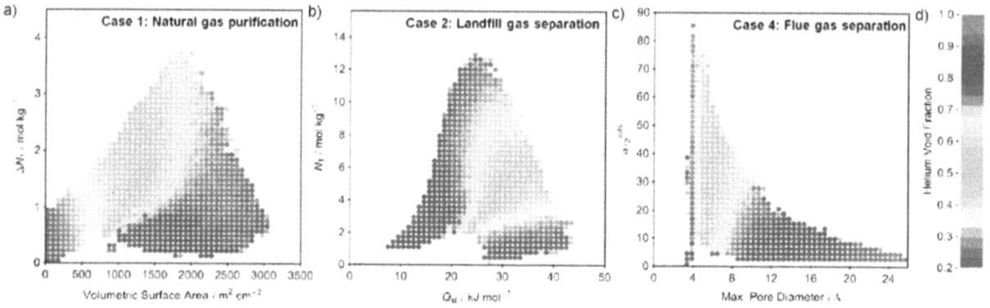

Figure 1.9 Structure–property relationships obtained from simulated adsorption of CO_2, CH_4, and N_2 over 130,000 hypothetical MOFs. Correlations are visible between (a) CO_2 working capacity (ΔN_1) and surface area, (b) CO_2 uptake (N_1) at 2.5 bars and CO_2 heat of adsorption (Q_{st}), and (c) selectivity over N_2 (α^{ads}) and maximum pore size. Each plot is constructed with filled circles over a 50×50 grid if more than 25 structures exist within the region. The color of each circle represents the average (d) helium void fraction of all structures in that plot region. Reproduced with permission from Ref. [45].

These recent studies point toward the potential of correlative approaches and especially QSPR approaches to predict the thermodynamic properties of MOFs. Ultimately the goal of such approaches might be to find the "perfect" structure for a target separation process, with the short-term benefit (i.e., even before a target structure is identified) to obtain very clear structure–property relationships over a wide region of the phase space, guiding synthetic chemists to be able to leverage them.

1.5 Inventing New Solids: Computational Design of Polyoxometalate Organic Frameworks

1.5.1 *Current Context*

Polyoxometalates (POMs) consist of a large family of soluble anionic metal clusters of d-block transition metals in high oxidation states (W^{VI}, $Mo^{V,VI}$, $V^{IV,V}$) with a diversity of structures [46] possessing magnetic, redox, and catalytic properties [47]. Besides their assembly into redox-active inorganic extended frameworks, [48] POMs were only recently regarded as possible building blocks for the design of POM-based MOFs, so-called POMOFs [49]. The combination of the extensive electrochemical activity and the diverse coordination modes in POMs, with the properties of MOFs, is obviously one attractive route toward the discovery of new multifunctional materials. This is motivated by the creation of redox-active porous solids, provided that the electrochemical activity of POMs may be retained once incorporated in the insoluble MOFs.

The current approach to produce POM-based MOFs is to consider the POM as an isolated inorganic building block, the inter-POM connectivity being achieved through a POM-M-L-M-POM scheme, where M is a transition metal and L is a multidendate organic linker, such as a dicarboxylate or a diphosphonate [50]. On the basis of this approach, a new class of POM-based MOFs has emerged, the so-called POMOFs (note that, confusingly, this name has also been used for fully inorganic POM-based frameworks as an abbreviation of POM open frameworks) built from the connection of POMs to one another through bridging organic linkers. Yet, although the number of hydrothermally synthesized POMOFs has continuously increased, 3D frameworks are scarce and those obtained by rational design strategy even more. We will highlight in this section how simulations play a role in this field.

1.5.2 *Computational Design and Evaluation of Hypothetical POMOFs*

The general philosophy for synthesizing new extended MOFs has been the replacement of small building blocks with larger ones, using the reticular chemistry concepts developed by Yaghi et al. [51] Computationally this time, it was expected to be able to identify key features of particular topologies that may lend themselves to the design of POMOFs, predict their crystal structures, and minimize their lattice energies through geometry optimization so as to identify viable candidates further amenable to synthetic trials.

Exploiting the analogy with zeolites (Fig. 1.10) and a decoration strategy, it was shown in 2005 that it was possible to construct and geometry-optimize a hypothetical 3D POMOF possessing a tetrahedral zeotype net converging toward a viable crystal model [52]. Willing to extend this initial work, the aim was to explore computationally the relative energies of a large number of

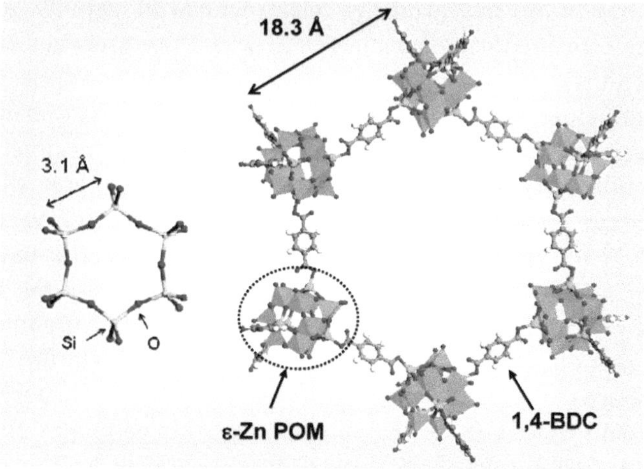

Figure 1.10 Schematic analogy between a six-ring in a zeolitic silicate and an six-ring in a Z-POMOF. Reproduced with permission from Ref. [53]. Copyright 2009 American Chemical Society.

hypothetical POMOFs possessing tetrahedral nets (typified by dense silica polymorphs and zeotypes) and possibly synthesize a new family of 3D POMOFs [53]. The choice to explore tetrahedral nets of POMOFs was driven by the extraordinary corpus of available structures of zeolites, with over 180 unique zeotype topologies described experimentally [54], supported by the fact that a number of MOFs had already been synthesized with zeolite topologies, either fortuitously or rationally, as illustrated by the spectacular discovery of ZIFs [55].

Having zeotypes in mind, the design strategy chosen above requires the search for a suitable cationic POM, together with a suitable bidendate organic linker. For the POM, the authors turned to the recently developed family of compounds based on the ε-Keggin isomer of the reduced $\{PMo_{12}O_{40}\}$ anion [56] and chose its Zn(II) derivative, denoted by ε-Zn, due to the likelihood of the external Zn coordination sites to adopt a tetrahedral arrangement and to the ability of Zn-ε-Keggin to react with bifunctional linkers. Regarding the linker, we chose the rigid BDC to form the targeted POM-Zn-L-Zn-POM linkages, as BDC is a common linker in MOFs. With this combination, the Zn-ε-Keggin cation possesses the ideal tetrahedral shape in which four Zn cations are exposed in a regular tetrahedral arrangement—in a similar fashion as oxygen atoms in SiO_4 tetrahedra. Thus, it may replace the "Si" site in a parent SiO_2 net, while the rigid anionic ligand may mimic the oxygen atom in the tetrahedral net. A selection of the 21 known uninodal (i.e., possessing one tetrahedral T site per asymmetric unit) hypothetical Z-POMOFs were constructed using a decoration strategy, that is, formally replacing the Si-O-Si links by POM-BDC-POM or POM-im-POM links using zeotypes from the IZA database. Also, dense structure types known as the most thermodynamically favored silicates (α-quartz [QUA], cristobalite [CRI], and coesite [COE]) were constructed as Z-POMOFs (Fig. 1.11). The general code TOBUNPOROUS [57] was written to automatically (i) replace each SiO_4 tetrahedron with an ε-Keggin building block in the appropriate orientation, (ii) replace each bridging oxygen atom by a linker, and (iii) scale up the unit cell dimensions from a zeolite to a POMOF appropriately. Finally, further lattice energy minimizations used the generic *uff* force field plus rigid bodies on the ε-Keggin and the linker, while authorizing POM-linker (Metal-O) bond relaxation, yielding a final ranking of computed lattice energies of all hypothetical POMOFs.

As in silicates, the dense QUA-, CRI-, and COE-like Z-POMOFs are the most stable ones (Fig. 1.12). Yet, unlike silica polymorphs, it is not the QUA that is most stable but rather the CRI-like (or diamond) topology. It is noteworthy that diamond-like MOFs have been reported as prevalent in terms of topologies (more than 70% of synthesized MOFs possess diamond-like nets), yielding credit to these predictions. Also, it is apparent that the most stable structures are those where the orbital overlap between the metal centers of the POM and the oxygen atoms of the carboxylic groups of the linker is maximized, making the crystal structures with the narrowest distribution of Zn–Zn distances the most stable ones and reducing even further the range of acceptable POM-linker-POM connections to the CRI-, BTC-, and SOD-type Z-POMOFs. Finally, the simulations were followed by numerous synthesis attempts of POMOFs made of ε-Keggin and BDC, leading to the first Z-POMOF-1 structure, $[NBu_4]_3[PMo_{12}O_{36}(OH)_4Zn_4(BDC)_2]\cdot 2H_2O$, which was found to possess the predicted most stable CRI topology,

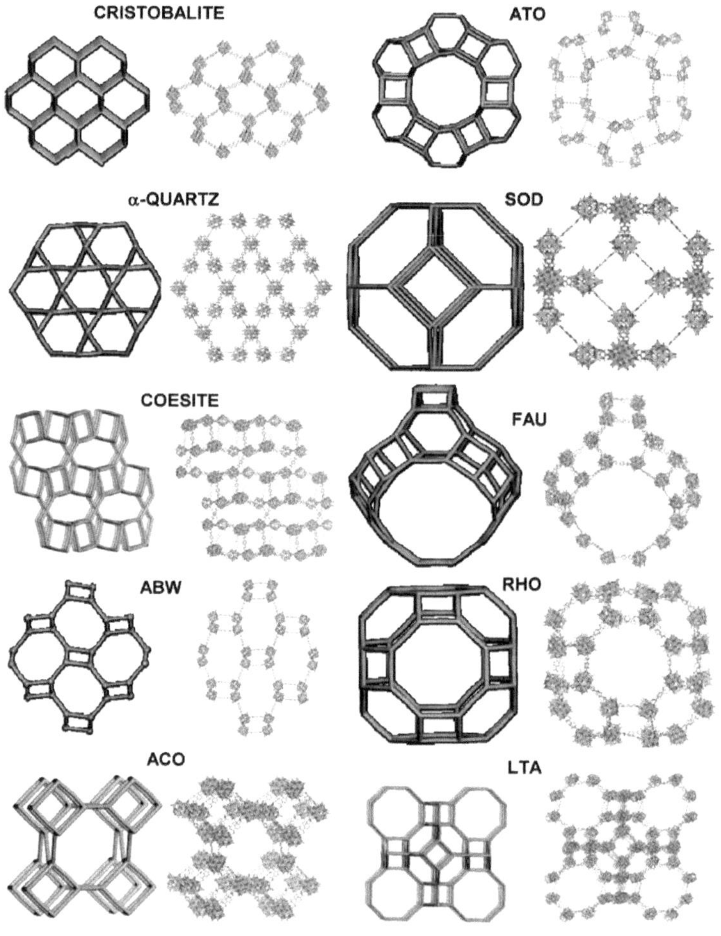

Figure 1.11 Hypothetical crystal structures of Z-POMOFs made of ε-Keggin ions and BDC linkers and possessing zeotype topologies. A subselection of uninodal together with multimodal nets is shown. Adapted from Ref. [53]. Copyright 2009 American Chemical Society.

synthesized in a triply interpenetrated version, while exhibiting high electrocatalytic activity for the reduction of bromates. Incidentally, this concurrence between the predicted most stable structure and the synthesized one suggests that a thermodynamic rather than kinetic control may be at play in Z-POMOF-1 synthesis. No other Z-POMOF has been synthesized to date; however, BTC and SOD

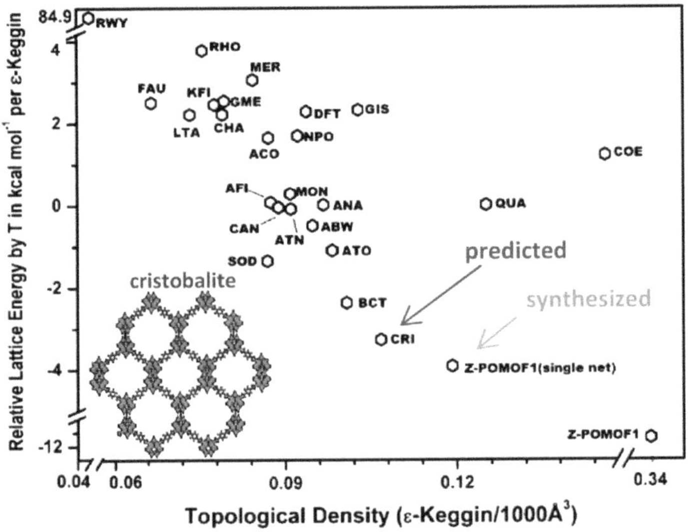

Figure 1.12 Relative lattice energies of hypothetical Z-POMOFs with density. The cristobalite-like Z-POMOF is predicted as the most stable structure among those studied. For comparison the lattice energy of the single net and triply interpenetrated nets of the experimental cristobalite-like Z-POMOF1 is shown for comparison, quantifying the further stabilization emanating from the interpenetration of nets. Reproduced with permission from Ref. [53]. Copyright 2009 American Chemical Society.

appear as likely synthetic targets. Also, the geometry optimization of the interpenetrated structure Z-POMOF-1 provided a quantification of the net-net interactions (~7 kcal/mol), indicating the added stabilization achieved through the formation of the more dense and interpenetrated structure.

Using imidazolate as the linker, a similar computational approach predicted the zni- and CRI-like structures as the most stable ones among the selection of topologies studied (Fig. 1.13) [58]. Disappointingly, the attempts to synthesize imidazole POMOFs materials did not lead to 3D structures but rather to 2D or molecular structures so far [58, 59]. Still, the molecular structure obtained with benzimidazole (bim) with the $\varepsilon(\mathrm{bim})_4$ compound is indeed promising in terms of rational design and synthesis since

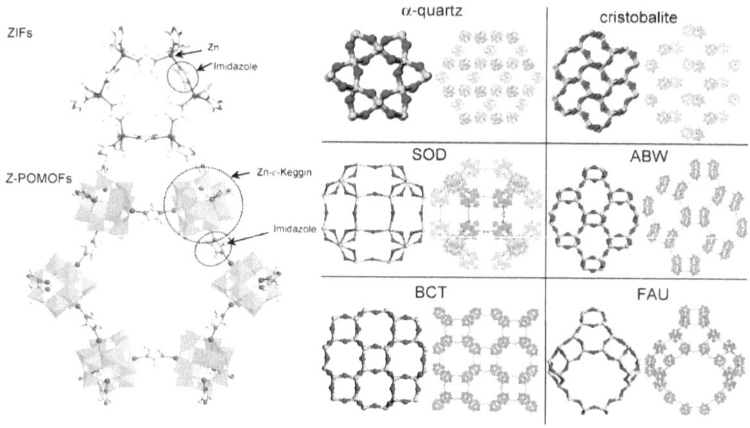

Figure 1.13 (Left) Analogy between ZIFs and the construction of Z-POMOFs from imidazolate and ε-Keggin ions. (Right) Selection of hypothetical Z-POMOFs constructed from dense silicates and zeotypes. Reproduced with permission from Ref. [58] from the PCCP Owner Societies.

it possesses the targeted tetrahedral building block required to propagate 3D a CRI-type structure.

1.5.3 *Thermodynamic Control and Template Effect?*

Using 1,3,5-benzene tricarboxylate linkers and tetrabutyl ammonium (TBA) counterions in hydrothermal conditions, a series of novel Zn-ε-Keggin-based POMOFs possessing electrocatalytic efficiency for the hydrogen evolution reaction were synthesized [60]. It includes $\varepsilon(\mathrm{trim})_{4/3}$, a new 3D open framework adopting a novel (3,4)-connected net, ofp [61]. At first, it was surprising that the "default" ctn-(I-43d) and bor-like (P-43m) nets, already synthesized as covalent organic frameworks [62], are not the observed outcomes, although they possess more symmetric and simple topological features (one crystallographic tetrahedral T site and one crystallographic triangular site) than the observed opf net (I-42m, 2 tetrahedral sites, 1 triangular site).

Force field–based calculations were used to evaluate the relative lattice energy of the ofp net in comparison with ctn- and bor-like polymorphs and find that the stability of the more complex ofp net

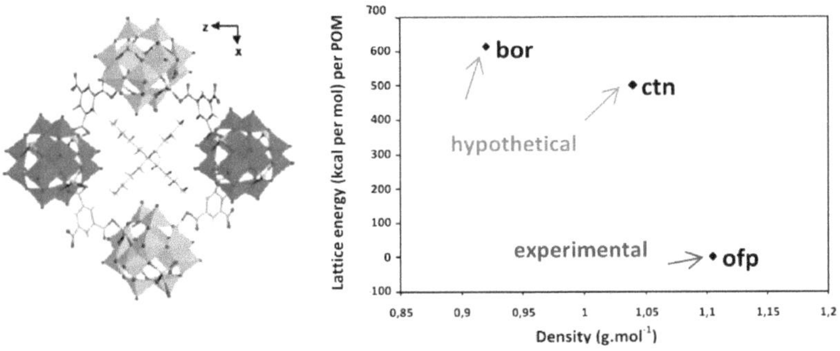

Figure 1.14 (Left) View of a four-membered ring in the experimental $\varepsilon(\text{trim})_{4/3}$ with a docked and energy-minimized TBA molecule. (Right) Variation of lattice energies with density for the three template-free $\varepsilon(\text{trim})_{4/3}$ polymorphs, showing the experimentally observed **opf**-like net as more stable than the two hypothetical **bor**- and **ctn**-like polymorphs. Adapted from Ref. [60].

is directly related to its greatest density (Fig. 1.14). In line with the previous study of Z-POMOF-1, the calculations in conjunction with the experimentally observed outcomes suggest that the successfully synthesized POMOFs so far were obtained under thermodynamic control.

It was further considered that the absence of the default bor- and ctn-like topologies could point toward the key role of the TBA template molecules, in the absence of which no crystallization occurs. Docking of TBA molecules in the porous volume of $\varepsilon(\text{trim})_{4/3}$ was undertaken to locate the TBA molecules, the position of which could not be determined from single crystal diffraction. Four preferred sites for TBA were found, with in total 16 molecules located among the 18 ones of the as-synthesized $\varepsilon(\text{trim})_{4/3}$. The various strengths of TBA–POMOF interactions suggest that TBA molecules play either a structure directing role or a space-filling role, depending on their location in the channels. A subset of TBA molecules is found to occupy the center of four-membered rings exhibiting the strongest short-range host–guest interactions among all docked TBA molecules, suggesting that they might stabilize the formation of four-membered rings of interconnected POMs.

1.6 Exploring the Energy Landscapes of MOFs

1.6.1 *Importance of Dispersive Forces in MOFs*

As a preliminary and general comment, it is worth underlying that the impact of dispersive forces on the energy landscape of MOFs can be pivotal. For example, Walker et al. [63] studied the archetypal Al-MIL53 material, which shows a remarkable property of "breathing" in the absence of any guest molecule. Liu et al. [64] reported that guest-free Al-MIL53 transforms with increasing temperatures from a narrow-pore (np) material to a large-pore (lp) material with a change in the cell volume of 864 to 1419 cubic Angstroms. Such bistability is at the time of writing still relatively rare, but the importance of this property and flexibility of MOFs in general has been the subject of a number of reviews, for example [65]. The Walker et al. [63] study revealed that the denser np phase had a lower enthalpy than the lp phase and that the dispersive interaction was crucial for stabilizing the np phase over the lp phase. This study prompted the question of the driving force that gives rise to the lp phase. It was further shown that the vibrational entropy of the lp phase is higher so that the Helmholtz free energy of the lp phase becomes more favorable than that of the np phase at elevated temperatures around room temperature, in agreement with experiments.

1.6.2 *Energy Landscape of Zeolitic Imidazolate Frameworks*

ZIFs constitute an important and well-documented subclass of MOFs, where tetrahedrally coordinated cations (Zn, Co, Fe [66], and recently Li and B [67]) are linked by imidazole derivatives to create topologies that resemble those of silica polymorphs, including zeolites. Access to new ZIF topologies has been partly achieved, for example, in the Zn-based family, through the judicious choice of linkers. Still, the factors that determine which topologies will form as ZIFs, and which will not, is an important question. The energy landscape of MOFs is increasingly well studied by both experiment and theory, revealing surprising properties of these materials. For example, Li et al. [68] identified negative linear compressibility in a zinc formate material, which was shown to be facilitated by cooperative tilting of the formate moieties and shrinkage of the metal–oxygen bonds. Until recently, investigation of the energy landscape of MOF materials was chiefly being conducted using simulation approaches. In that respect, a number of computational predictions using DFT methods have been reported recently on likely the energy landscape of ZIFs, exploring hypothetical topologies and comparing them to existing ones in various ways.

Leoni et al. [69] eported around 30 topologically different tetrahedral nets of $Zn(imid)_2$. They developed a sophisticated topological study relying on scores such as the topological density (accumulation of vertices around a reference point), the genus (a measure of the complexity of the structure and its connectedness), and the tiling scheme. For example, it is found that the most stable $Zn(imid)_2$ polymorph is the very compact **zni** topology also observed in a number of experimental structures [67]. Alternatively, the diamond-like (or CRI) network is specifically studied as a favorable candidate topology [70], with a complex energy landscape comprising more than 30 different conformations contrasting with the scarce occurrence of this topology among ZIFs. Still, it is noteworthy that these DFT studies did not include dispersive interactions corrections.

In 2009, willing to compare the energy landcape of ZIFs to that of zeolites Mellot-Draznieks et al. [71] calculated the cohesive energies of a range of known and hypothetical ZIF materials (based on zinc and imidazole) using the DFT approach (Fig. 1.15). In particular, these authors focused on exploring the relation between the energy of a given system and its density. The objective of the work was to establish whether there was a simple inverse relationship between thermodynamic stability and density, as had been found in siliceous zeolite materials [72]. Hypothetical ZIFs based on known zeolite topologies (www.iza-structure.org) were built and optimized, and the stability of all phases was measured relative to the **zni** topology, which is the densest zinc imidazole structure. It was found that there was, indeed, a general inverse relationship between stability and density, remarkably similar to that of siliceous zeolites particularly given the approximate tenfold

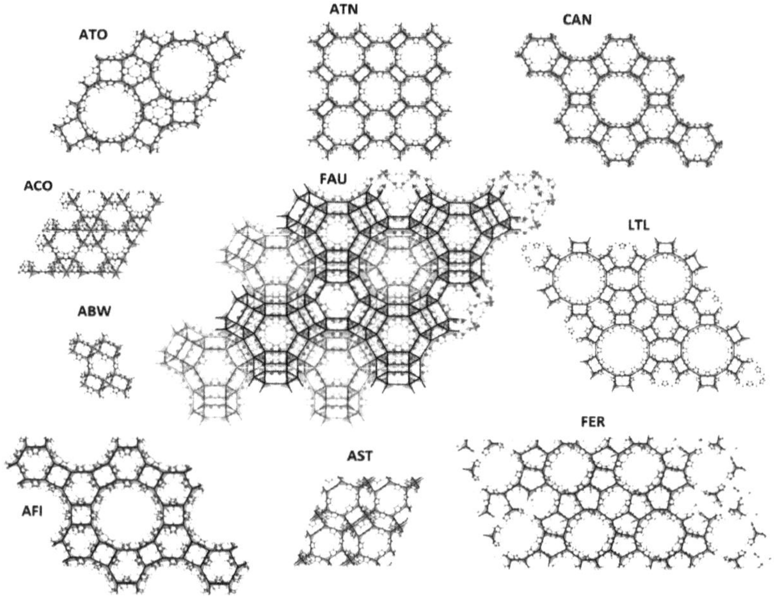

Figure 1.15 Simulated energy-minimized crystal structures of hypothetical ZIFs Reproduced from Ref. [71] with permission from the Royal Chemical Society.

difference in density, where a hypothetical **LTL** ZIF—the least stable structure considered—was found to be ~23 kJ/mol less stable than the **zni** analogue. ZIF-8 was found to be ~13 kJ/mol less stable than **zni**, which is very similar to the 14 kJ/mol measured by Hughes et al. through calorimetry. Although dispersion corrections were not included either, the simulations find that the variation in the total energy of the three SOD structures (ZIF-7, ZIF-8, ZIF-9) ranges from 12 to 17.5 kJ/mol, directly reflecting the impact of the linker on the stabilization of a given topology. The use of substituted imidazole in ZIF-7 and ZIF-9 allows the synthesis of metastable and distorted SOD structures, where stabilizing interactions between substituents may compensate the energetic cost of cooperative framework distortions.

This study was then recently revisited using prescriptions of DFT that included van der Waals interactions [73] where the main result, as was anticipated in the initial study, was that the variance

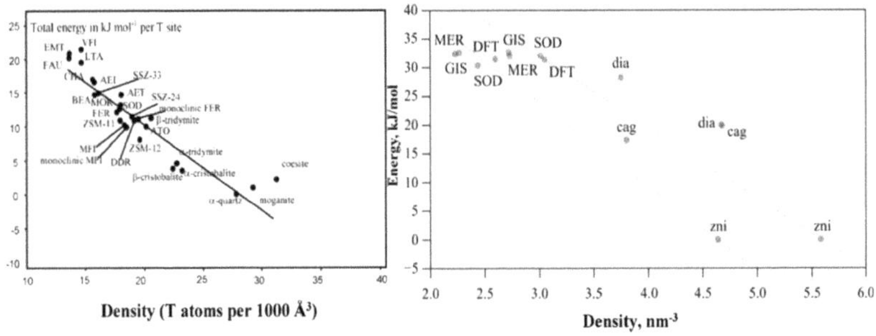

Figure 1.16 (Left) Calculated total energy of SiO_2 zeolites as a function of their framework density. Reproduced with permission from Ref. [72]. Copyright 1996 American Chemical Society. (Right) Relative DFT-D total lattice energies versus density on Zn-based (blue) and LiB-based (red) ZIFs. Reproduced with permission from Ref. [73].

in energy with density decreased substantially. It is found that the stabilization of dense ZIF structures relative to porous ones is even more pronounced due to stabilizing linker–linker dispersive interactions, with a 40 kJ/mol range rather than the previous 15 kJ/mol range found without dispersion corrections (Fig. 1.16). In addition it was found that, somewhat surprisingly, the landscape of zinc imidazole structures was very similar to analogous LiB frameworks.

The latter finding was surprising because the number of synthesized LiB frameworks is rather small in comparison to zinc-imidazole-based frameworks, yet their calculations suggest that the thermodynamic penalty to form metastable, lower-density materials is comparable to the numerous low-density ZIFs that have been successfully synthesized. This was interpreted as evidence that kinetic factors must present additional barriers to synthesis of porous LiB materials.

Also, the impact of the fluorinated linkers on the Zn-based ZIFs' energy landscape was recently reported by Leoni et al. [74] yet missing the inclusion of the above-mentioned dispersion corrections.

Overall, computational studies show without a doubt that a number of hitherto undiscovered topologies should be amenable to synthesis and that even the most open framework types might be obtained with appropriately substituted ligands. Interestingly, it has been shown through calorimetric [75] measurement that the cohesive energy of ZIF materials is not substantially reduced at very low density, and this has been attributed to dispersive (attractive van der Waals) interactions becoming less significant as the distance between the constituent units increases. Also, these experimental findings substantially confirm the above computational predictions, including the similar energy–density trends with respect to zeolites: ZIFs, like zeolites, are energetically metastable with respect to their dense components, suggesting that no immediate thermodynamic barrier precludes the development of highly porous materials.

1.6.3 *Understanding the Energy Landscape of Nonchiral and Chiral MOFs*

Chiral hybrid frameworks are of particular interest since they can readily be made from commercially available homochiral ligands and show promise in applications such as enantiomerically selective catalysis and separations [76]. As detailed below, computational tools can provide valuable information about the energetic contribution of different components of the crystal structure to the relative stabilities of hybrid frameworks.

In an initial work, Cheetham et al. [77] studied the formation of a series of zinc 4-cyclohexene-1,2-dicarboxylates, comparing the structures and stabilities of the *cis*- and *trans*-polymorphs. Starting with the *cis*-1,2-dicarboxylate of 4-cyclohexene, the reaction with zinc salts at 60°C yields the *cis*-hydrate **1**, while at 180°C it yields first the anhydrous *cis*-product **2** and then the *trans*-product **3**. Using DFT calculations, it is found that the *cis*-dihydrate **1** is more stable than the corresponding *cis*-anhydrous phase **2** and that the dehydration reaction is endothermic, confirming that **1** is indeed more stable at lower temperature, as observed. Also, the *trans*-structure **3** is found to be more stable than the polymorphic *cis*-structure **2**, consistent with its more dense structure, and suggesting that its formation is thermodynamically driven (Fig. 1.17). Such thermodynamic control is in line with early experimental and computational studies of metal diphosphonates [78].

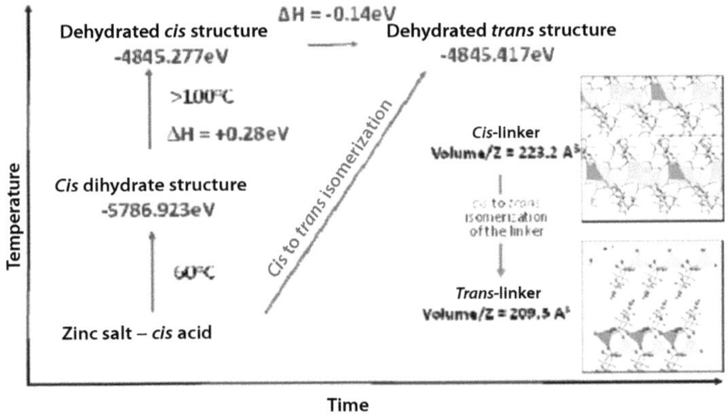

Figure 1.17 Scheme illustrating thermodynamic control during the hydrothermal synthesis of zinc cyclohexene-1,4-dicarboxylates according to Ref. [79]. By courtesy of A. K. Cheetham.

In a further study, the possibility of using *R*, *R*- and *S*, *S*-enantiomers of trans-1,2-dicarboxylate is also investigated, combining synthesis, calorimetry, and computational approaches [79]. The obtained racemic *R*, *R*–*S*, *S*- and chiral *R*, *R*-Zn-dicarboxylates possess fundamentally different topologies (Fig. 1.18), the first being constructed from four rings and the second of three rings of tetrahedrally coordinated zinc ions. Using this time the force field developed for Zn-based MOFs by Allendorf et al. [80], it is shown that the less dense racemic *R*, *R*–*S*, *S*-Zn-dicarboxylate structure is more stable than the chiral *R*, *R*-dicarboxylate phase, in excellent agreement with the calorimetric estimations of differences of enthalpies of solution for both compounds. The relative stabilities of the racemic and chiral phases is further rationalized in terms of the differences in the intralayer energy of the component single layers and the nonbonded interlayer interactions. On the one hand, the interlayer energy is stronger in the chiral *R*, *R*-structure than in the racemic *R*, *R*–*S*, *S*-structure. On the other hand, the intralayer energy of the chiral structure exceeds that of the racemic one, yielding a more favorable lattice energy for this structure despite its lower density.

On a larger scale, Cheetham et al. further studied the structurally diverse system of alkaline earth (Ca, Ba) [81] and Li [82] tartrates

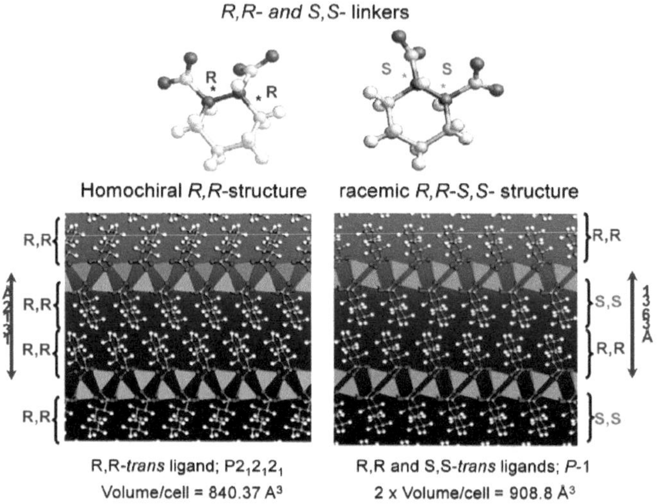

Figure 1.18 (Top) The trans-R, R-cyclohexane and trans-S, S-cyclohexane. dicarboxylic acids. (Bottom) Tdense chiral (left) and less dense racemic (right) structures. Illustration adapted from Ref. [79].

using L-, meso-, and D,L-tartaric acids (Fig. 1.19). The relative energetic stabilities obtained by DFT calculations and calorimetric methods were again in excellent agreement, shedding light on the complex behavior of these systems. It is shown that the thermodynamic stabilities of the metal tartrate phases, together with kinetic factors such as ligand isomerization, are key factors controlling the phase behavior. In the anhydrous lithium tartrate system, where only small electronic energy differences are computed between the various structures, the in-depth analysis of vibrational modes through zero-point vibrational energy is used to point out the structural details that are responsible for the energy difference between polymorphs.

1.7 Exploring Physical Properties of MOFs

1.7.1 *Estimation of Band Gaps*

By contrast with the considerable research efforts dedicated to adsorption/separation and catalysis applications for MOFs, research

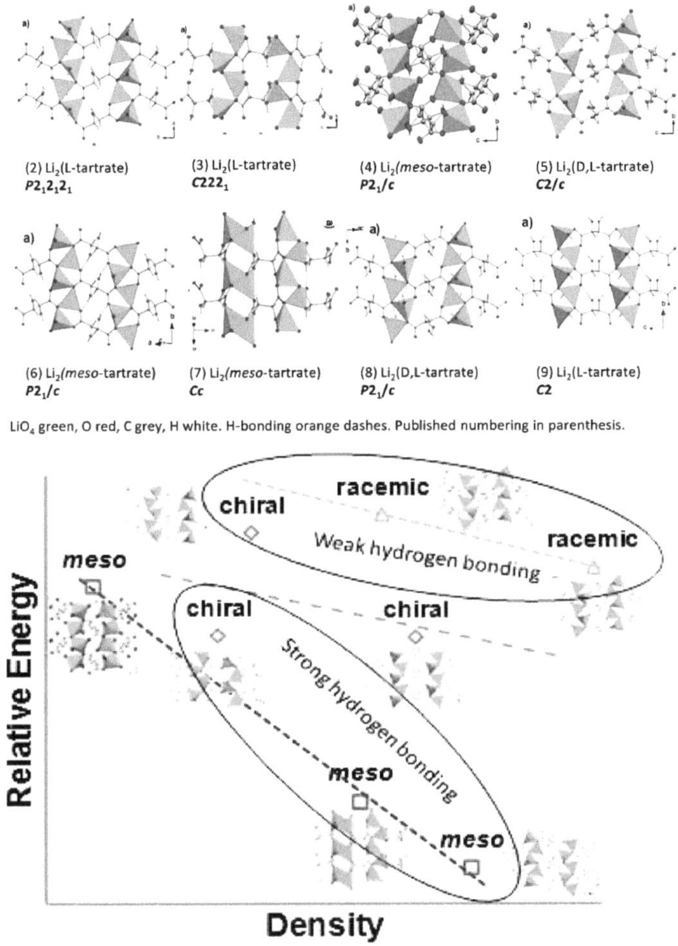

Figure 1.19 (Top) Eight anhydrous Li_2(tartrate)'s crystal structures. (Bottom) Anhydrous Li_2(tartrate)'s calculated energy versus density. Adapted from Ref. [82]. By courtesy of A. K. Cheetham.

focusing on the potential use of MOFs into electronic devices is more recent [83–85]. In addition, MOFs have recently proven to be wonderful structures for solar energy harvesting and photocatalysis [86], making it necessary to develop strategies to control their band structure.

In that respect, electronic structure calculations are seen as valuable routes to investigate the properties of electro- and photoactive frameworks. They are especially suited to study independently the two parts that can be tailored to alter the electronic band gap of MOFs, that is, the inorganic subnetwork on the one hand and the organic linker and its functionalization on the other hand. Still, experimental band gaps are larger than the calculated ones, consistent with the acknowledged fact that DFT usually underestimates the exited state energies (which may be partially corrected by the use of hybrid functionals such as B3LYP). Although accurate calculations of MOF band gaps remain challenging, in particular due to their very large unit cell, computational methods are useful to estimate trends in MOFs and understand the fundamentals of their physical properties.

Following the early work by Dovesi et al. [87] the semiconductor behavior of the archetypal MOF-5 has been both experimentally and computationally investigated [85]. It is acknowledged that the conjugated aromatic ring is responsible for the energy transfer to the inorganic subunit Zn_4O_{13}, which behaves like a quantum dot, the band gap being mainly driven by the energy differences between the highest occupied molecular orbitals (HOMOs) and lowest unoccupied molecular orbitals (LUMOs) of the linker [88]. Keeping the cubic topology of isoreticular metal–organic framework (IRMOF-1) with its $(OZn_4)^{6+}$ tetrahedral building blocks, the impact of a wide range of organic linkers on the electronic properties were calculated. The HOMO–LUMO energy difference is dominated by the C sp^2 states of the organic linker, with solids classified as semiconductors or insulators with band gaps between 1.0 eV and 5.5 eV, predicting that longer linkers might offer narrower band gaps. Further experimental studies on a number of these linkers have later confirmed these predictions [89].

Turning to the impact of the metal clusters on MOFs' band gaps, Cabrera et al. [90] theoretically explored in an early study the impact of the nature of the transition metal in IRMOF-1, concluding that the band gap remains unchanged upon replacement of Zn with Be, Mg, Ca, and Cd. By contrast, varying virtually the cobalt-to-zinc ion substitution ratio in the same Zn-based IRMOF-1 structure [91], it is expected that band gaps ranging from semiconducting to metallic

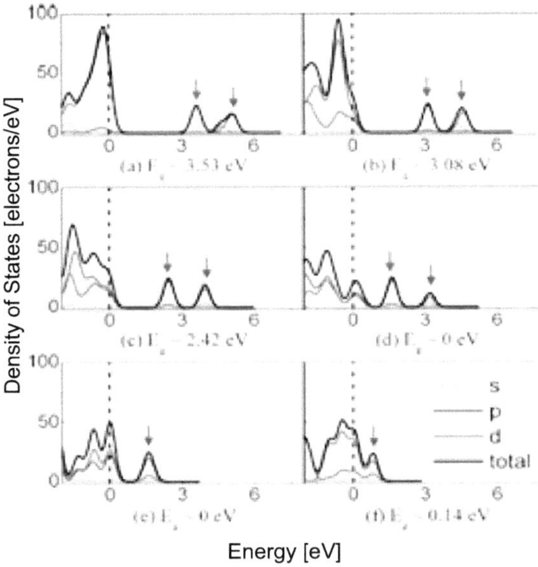

Figure 1.20 Partial density of states of IRMOF-1 (a) and cobalt-substituted IRMOF-1 with 6.25% (b), 12.5% (c), 25% (d), 50 (e), and 100% (f). Replacing this time Zn by Co in the metal clusters, the overlap between the Co and Zn *d* orbitals and the O and C *p* orbitals may be increased, resulting in a decrease of the band gap. Reproduced from Ref. [91] with permission from the PCCP Owner Societies.

states could be achieved (Fig. 1.20). Alternatively, by comparing the computed band gaps of three existing MOFs that contain the same biphenyl dicarboxylate ligands (namely, IRMOF-9, Zn5-BPDC, and CPO-7), Liu et al. [92] conclude that increasing the size of the Zn-based metal cluster leads to a red shift in band gaps.

A combined experimental and theoretical study of the complex UiO-66 (based on $Zr_6O_4(OH)_4$ clusters and terephtalate) MOF was also reported [93]. It is shown that in both hydroxylated and dehydroxylated structures, the carbon atoms of the organic linker are those involved in the modulation of the band gap, while Zr atoms do not play any role, while pinpointing different behaviors between UiO-66 and MOF-5 with respect to those of their corresponding oxides, ZrO_2 and ZnO.

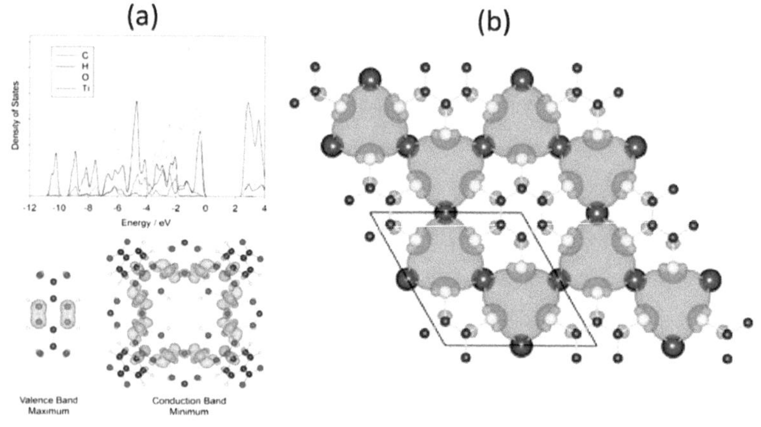

Figure 1.21 (a) Electronic density of states and density isosurfaces associated with the band edge states in MIL-125. Adapted from Ref. [95]. (b) Electron carrier wavefunction in PbC_6S_6. Adapted from Ref. [96].

An interesting discovery arises in the photochromism of a framework material built from TiO_2 octahedrons, MIL-125 [94], where an explanation for the striking white-to-black color change on UV excitation was provided on the basis of DFT calculations by Walsh and Catlow [95]. By examining the electronic and defect structure of this hybrid titanium system, they propose the origin of its remarkable photochromic properties (Fig. 1.21a). Band gap excitations (charge transfer from the organic to inorganic units) were found to have sufficient energy to create lattice vacancies, which results in the formation of gaseous O_2 coupled with the reversible reduction of Ti(IV) to photoactive Ti(III) ions. Importantly, the analysis of the electronic and defect structure of MIL-125 reveals a spatial separation of charges (electrons versus holes) between the organic and inorganic subnetworks, as a lower-energy process than in bulk TiO_2 and that of the band gap, explaining the facile color change of MIL-125 under UV light. The photostimulated chemical reduction of this prototypical material suggests immense potential developments of photochemical applications through the tailoring the MOFs and their band gap, a still emerging area. Similar techniques have been used by Walsh [96] to study the origin and nature of electrical charge carriers in Pb hexathiolate hybrid materials. As shown in Fig. 1.21b, the electron wave functions associated with conduction in the material involve the organic and inorganic network, making this class of material a truly hybrid semiconductor.

Obviously, it would be interesting to tailor the band structure of MOFs in order to achieve visible light absorption. In that purpose, it has been recently reported through a computational study that the band gap of MOF-5 (3.53 eV) could be modified to absorb visible light by substituting the oxygen anion of the metal oxide core ZnO_4, with sulfur anions (Zn_4S; 3.30 eV) or selenium anions (Zn_4Se; 2.91 eV) that allow a shift of the Fermi level due to their electron-rich properties [97].

1.7.2 *Mechanical Properties*

Elastic properties describe the response of materials to external loads and strains (Fig. 1.22). Having in mind that the main targeted applications of ZIFs are gas storage/separation and catalysis, it is crucial to know what mechanical loads such hybrid frameworks might withstand during industrial processes and to understand structure–mechanical property relationships. An extensive review by Tan et al. [98] examined the mechanical properties of MOFs by considering their Young's modulus, Poisson's ratio, and bulk and

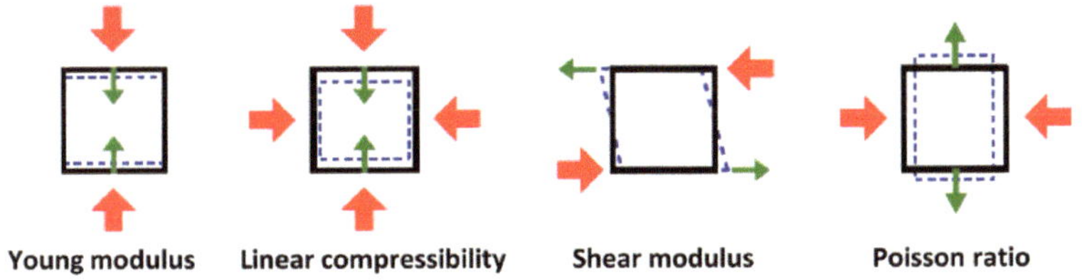

Figure 1.22 Scheme showing the various directional elastic properties from left to right: Young's modulus, E, characteristic of the uniaxial stiffness along a selected crystallographic direction; the linear compressibility, β, characteristic of the response to an isostatic compression; the shear modulus, G, characteristic of the resistance to shear; and Poisson's ratio, the ratio of transverse and axial strain in response to a uniaxial stress. Red arrows: stress applied; green arrows: resulting strain measured; blue: resulting deformation.

shear moduli, highlighting an important lack of understanding about their fundamental mechanical characteristics.

The computational way to estimate elastic properties usually includes the finite-difference method. In short, the unit cell of the crystal structure is slightly strained incrementally, and the stress tensor is computed after relaxing atomic positions. The corresponding compliance tensor component may then be extracted from strain–stress curves. Elastic properties depend on the second derivatives of energy with respect to atomic positions and cell vectors, thus requiring a very accurate level of theory.

Following the early first-principles study of the prototypical framework MOF-5 by Dovesi et al. [87] the elastic constants of MOF-5 were computationally studied by several authors using force fields [99–101] and DFT [102–105] approaches. It is worth noting that the discrepancies between DFT studies highlight the relative dependence of the results upon the particular DFT method used.

In conjunction with experimental investigations, the elastic properties of a couple of ZIFs, namely, the dense zinc-imidazolate **zni** framework [106] and the prototypical porous ($Zn(mIm)_2$) ZIF-8 [107], have been studied at the DFT level at the athermal approximation (0 K). The complete elasticity profile of ZIF-8 reveals its extremely low shear modulus. In turn, ab initio calculations provided excellent estimations over a wide range of mechanical properties such as elastic constants and Young's and shear moduli, discrepancies being attributed to the neglect of thermal effect and zero-point motions. Among others, the calculations reveal that the shear deformation in ZIF-8 is dominated by cooperative bending of the N-Zn-N angles in the ZnN_4 tetrahedra, together with the Zn-mIm-Zn bond angles, in line with previous experimental/computational findings on the relative compliance of ZnN_4 in the dense **zni** $Zn(Im)_2$ structure [106].

While the mechanical properties of the relatively rigid (i.e., nonbreathing) MOFs have largely explored theoretically, theoretical investigations on the mechanical properties of "breathing" solids such as MIL-53 are still scarce, probably due to their highly complex stimuli-responsive energy landscape.

Recently, Coudert et al. [108] performed ab initio calculations of the mechanical properties of a series of highly flexible MOFs,

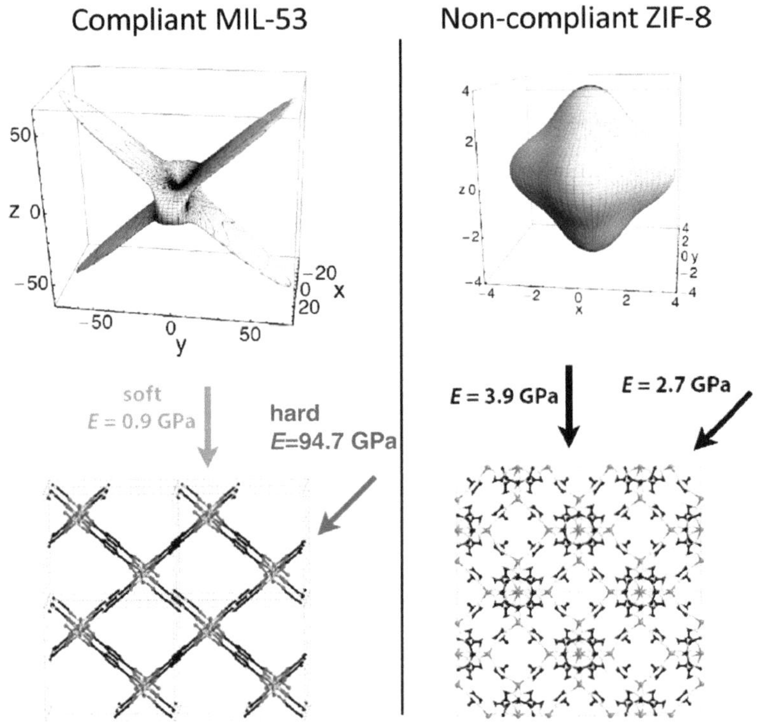

Figure 1.23 Comparison of the directional Young's modulus for the large-pore form of Al-MIL-53 (left) and for ZIF-8 (right). Adapted from Ref. [108]. By courtesy of F.-X. Coudert.

including (Al, Ga,) MIL-53 and V-MIL-47 and comparing them to those computed for nonbreathing MOFs such as ZIF-8. Among other conclusions, the Young's modulus of the lp form of MIL-53 is strongly anisotropic, exhibiting expected resistance to compression along the inorganic Al(OH) chains (y direction) and along the organic linkers (two directions in the xy plane), together with directions of a very low Young's modulus corresponding to the well-known breathing mode of deformation of MIL-53 (Fig. 1.23, left). These features contract with those of the less compliant ZIF-8 (Fig. 1.23, right) Also this type of solid is found to possess shear modulus anisotropy, which is significantly higher than in other MOFs, together with extremely high negative linear compressibilities.

Overall, it is apparent that the elastic behavior of MOFs, especially those exhibiting breathing properties, is extremely complex, leaving a key role to play in theoretical investigations. Since theoretical predictions of the elastic properties of several MOFs are rapidly developing, their systematic validation through experimental studies is more pressing. The full extent of the elastic region in MOFs, together with the exploration of their behavior outside the elastic region, remains to be explored in more depth. The impact of framework dynamics and temperature still needs to be investigated computationally.

1.8 Modelling Defects in MOFs

The design of functionalized MOFs (by linker functionalization, for example) is an elegant way for tuning the properties of the solid. This was performed extensively for catalytic purposes. Unexpectedly, a very high catalytic activity of ZIF-8 (nonfunctionalized) was found in acid-base-catalyzed reactions [109, 110], in particular in the transesterification of vegetable oil. Ab initio calculations were performed [109, 111] and concluded that active sites are located at defects and at the external surface of ZIF-8. The structure and hydroxylation/carbonation level of the latter were calculated as a function of operating conditions (Fig. 1.24a). A strongly frustrated

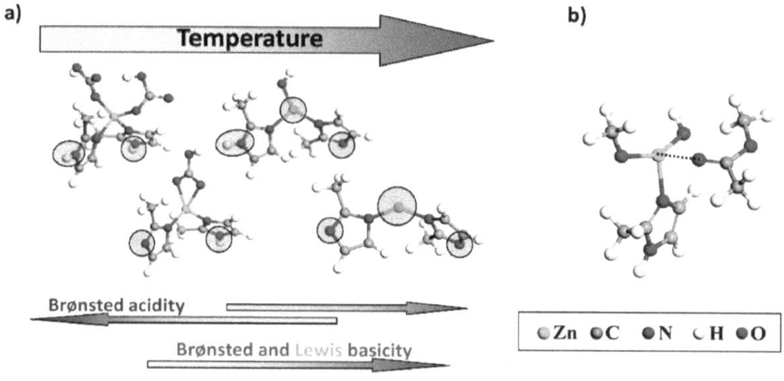

Figure 1.24 Modeling defects in ZIF-8. Aadapted from Ref. [111]. By courtesy of C. Chizallet.

environment of zinc (ZnII) was shown to gather the stability and reactivity conditions for being an efficient active site in the reaction (Fig. 1.24b). This study opens new perspectives in the optimization of the activity of the catalyst by promoting the presence of defects or a high surface-to-bulk ratio.

1.9 Concluding Remarks

In this chapter we have briefly described advances towards ab initio prediction of crystal structures of MOFs and the important role that modeling has had in identifying hybrid frameworks materials, which in many cases could not be solved by traditional diffraction techniques. Also, the review has highlighted the use of simulation-informed design principles to target and characterize new hybrid materials, as illustrated in the field of POM open frameworks. The rich and vast landscape of hybrid materials can be further navigated with the aid of modeling by screening materials, identifying desirable (and undesirable) properties. The ability to probe known and large numbers of hypothetical hybrid materials offers exciting prospects for mining databases and establishing informatics-based procedures to help the goal of designing a material with tailored properties and functionalities. In that respect, the use of QSPR techniques incorporating both experimental and hypothetical structures is emerging as a valuable route. The review has also shown how state-of-the-art methods have yielded important insights into the fundamental physical properties of materials, such as compressibility and response to temperature/pressure, and the electronic properties of emerging materials. One area that has received less focus but is important and remains elusive to study experimentally is crystal defects, including surfaces/interfaces, and point defects, such as vacancies. Because of the difficulty in studying and characterizing defects in this class of materials, there is substantial potential for modeling to make an important contribution to the elucidation of the effects of defects on chemical reactivity (e.g., in catalysis and gas storage) and physical properties (such as stiffness or porosity). With no doubt, modeling will continue to provide pivotal contributions to the development and direction of the field of hybrid materials.

References

1. Price, S. L. (2008). From crystal structure prediction to polymorph prediction: interpreting the crystal energy landscape, *Phys. Chem. Chem. Phys.*, **10**, 1996–2009.
2. Jones, J. T. A., Hasell, T., Wu, X. F., Bacsa, J., Jelfs, K. E., Schmidtmann, M., Chong, S. Y., Adams, D. J., Trewin, A., Schiffman, F., Cora, F., Slater, B., Steiner, A., Day, G. M., Cooper, A. I. (2011). Modular and predictable assembly of porous organic molecular crystals, *Nature*, **474**, 367–371.
3. Mellot-Draznieks, C. (2007). Role of computer simulations in structure prediction and structure determination: from molecular compounds to hybrid frameworks, *J. Mater. Chem.*, **17**, 4348–4358.
4. Woodley, S. M., Catlow, R. (2008) Crystal structure prediction from first principles, *Nat. Mater.*, **7**, 937–946.
5. Tafipolsky, M., Amirjalayer, S., Schmid, R. (2010). Atomistic theoretical models for nanoporous hybrid materials, *Microporous Mesoporous Mater.*, **129**, 304–318.
6. Coudert, F.-X., Boutin, A., Jeffroy, M., Mellot-Draznieks, C., Fuchs, A. H. (2009). Thermodynamic method and models to study flexible metal-organic frameworks, *ChemPhysChem*, **12**, 247–258.
7. Düren, T., Bae, Y.-S., Snurr, R. Q. (2009). Using molecular simulation to characterize metal-organic frameworks for adsorption applications, *Chem. Soc. Rev.*, **38**, 1237–1247.
8. Sumida, K., Rogov, D. L., Masom, J. A., McDolnald, T. M., Bloch, E. D., Herm, Z. R., Bae, T. H., Long, J. R. (2012). Carbon dioxide capture in metal-organic frameworks, *Chem. Rev.*, **112**, 724–781.
9. Kirkpatrick, S., Gellat, J. C. D., Vecchi, M. P. (1983). Optimization by simulated annealing, *Science*, **220**, 671–680.
10. Wales, D. J., Scheraga, H. A. (1999). Review: chemistry. Global optimization of clusters, crystals, and biomolecules, *Science*, **285**, 1368–1372.
11. Coley, D. A. (1999). *An Introduction to Genetic Algorithms for Scientists and Engineers* (World Scientific).
12. Foster, M. D, Simperler, A., Bell, R. G., Friedichs, O. D., Paz, F. A. A., Klinowski, J. (2004). Chemically feasible hypothetical crystalline networks, *Nat. Mater.*, **3**, 234–238.
13. O'Keeffe, M. (2008). Three-periodic nets and tilings: regular and related infinite polyhedral, *Acta Crystallogr. A*, **64**, 425–429.

14. Martoňak, R., Laio, A., Parrinello, M. (2003). Predicting crystal structures: the Parrinello-Rahman method revisited, *Phys. Rev. Lett.*, **90**, 075503.

15. Martoňak, R., Donadio, D., Oganov, A. R., Parrinello, M. (2006). Crystal structure transformations in SiO_2 from classical and ab initio metadynamics, *Nat. Mater.*, **5**, 623–626.

16. Mellot-Draznieks, C., Newsam, J. M., Gorman, A. M., Freeman, C. M., Férey, G. (2000). De novo prediction of inorganic structures developed through automated assembly of secondary building units (AASBU method), *Angew. Chem. Int. Ed.*, **39**, 2270–2275.

17. Mellot-Draznieks, C., Férey, G., Schön, C., Cancarevic, Z., Jansen, M. (2002). Computational design and prediction of interesting not-yet-synthesized structures of inorganics materials using building unit concepts, *Chem. Eur. J.*, **8**, 4102–4113.

18. Mellot-Draznieks, C., Girard, S., Férey, G. (2002). Novel inorganic frameworks constructed from double-four-ring (D4R) units: computational design structures and lattice energies of silicate, aluminophosphate and gallophosphate candidates, *J. Am. Chem. Soc.*, **124**, 15326–15335.

19. Mellot-Draznieks, C., Dutour, J., Férey, G. (2004). Hybrid organic–inorganic frameworks: routes for computational design and structure prediction, *Angew. Chem., Int. Ed.*, **43**, 6290–6296.

20. Wilmer, C. E., Leaf, M., Lee, C. Y., Farha, O. K., Hauser, B. G., Hupp, J. T., Snurr, R. Q. (2012). Large-scale screening of hypothetical metal-organic frameworks, *Nat. Chem.* **4**, 83–89.

21. Férey, G., Mellot-Draznieks, C., Serre, C., Millange, F. (2005). Crystallized frameworks with giant pores: are there limits to the possible? *Acc. Chem. Res.*, **38**, 217–225.

22. Férey, G., Serre, C., Mellot-Draznieks, C., Millange, F., Surblé, S., Dutour, J., Margiolaki, I. (2004). A hybrid solid with giant pores prepared by a combination of targeted chemistry, simulation and powder diffraction, *Angew. Chem., Int. Ed.*, **43**, 6296–6301.

23. Férey, G., Mellot-Draznieks, C., Serre, C., Millange, F., Dutour, J., Surblé, S., Margiolaki, I. (2005). A chromium terephthalate-based solid with unusually large pore volumes and surface area, *Science*, **309**, 2040–2042.

24. Mellot-Draznieks, C., Férey, G. (2005). Assembling molecular species into 3D frameworks: computational design and structure solution of hybrid materials, *Prog. Solid State Chem.*, **33**, 187–197.

25. Baerlocher, C., McCusker, L. B. Database of Zeolitic Structures, http://www.iza-strucure.org/databases.
26. Mellot-Draznieks, C., Dutour, J., Férey, G. (2004). Computational design of hybrid frameworks: structure and energetic of two $Me_3OF_3\{-O_2C-C_6H_6-CO_2-\}_3$ metal-dicarboxylate polymorphs, MIL-hypo-1 and MIL-hypo-2, *Z. Anorg. Allg. Chem.*, **630**, 2599–2604.
27. Surblé, S., Serre, C., Mellot-Draznieks, C., Millange, F., Férey, G. (2006). A new isoreticular class of metal-organic-frameworks with the MIL-88 topology, *Chem. Commun.*, **3**, 284–286.
28. Loiseau, T., Mellot-Draznieks, C., Muguerra, H., Férey, G., Haouas, M., Taulelle, F. (2005). Hydrothermal synthesis and crystal structure of a new three-dimansional aluminum-organic framework MIL-69 with 2,6-naphthalenedicarboxylate (ndc), Al(OH)(ndc) H_2O, *C. R. Chim.*, **8**, 765–772.
29. Mellot-Draznieks, C., Serre, C., Surblé, S., Audebrand. N., Férey, G. (2005). Very large swelling in hybrid frameworks: a combined computational and powder diffraction study, *J. Am. Chem. Soc.*, **127**, 16273–16278.
30. Serre, C., Mellot-Draznieks, C., Surblé, S., Audebrand, N., Filinchuk, Y., Férey, G. (2007). Role of solvent-host interactions that lead to very large swelling of hybrid frameworks, *Science*, **315**, 1828–1831.
31. Serre, C., Surblé, S., Mellot-Draznieks, C., Filinchuk, Y., Férey, G. (2008). Evidence of flexibility in the nanoporous iron(III) carboxylate MIL-89, *Dalton Trans.*, **40**, 5462–5464.
32. Salles, F., Maurin, G., Serre, C., Llewellyn, P. L., Knöfel, C., Choi, H. J., Filinchuk, Y., Oliviero, L., Vimont, A., Long, J. R., Ferey, G. (2010). Multistep N_2 breathing in the metal-organic framework Co(1,2-benzenedipyrazolate), *J. Am. Chem. Soc.*, **132**, 13782–13788.
33. Babarao, R., Jiang, J. W. (2008). Molecular screening of metal-organic frameworks for CO_2 storage, *Langmuir*, **12**, 6270–6278.
34. Li, J. R., Kuppler, R. J., Zhou, H. C. (2009). Selective gas adsorption and separation in metal-organic frameworks, *Chem. Rev. Soc.*, **38**, 1477–1504.
35. Getman, R. B., Bae, Y. S., Wilmer, C. E., Snurr, R. Q. (2012). Review and analysis of molecular simulations of methane, hydrogen, and acetylene storage in metal-organic-framworks, *Chem. Rev.*, **112**, 703–723.
36. Torrisi, A., Bell, R., Mellot-Draznieks, C. (2010). Functionalized MOFs for enhances CO_2 capture, *Cryst. Growth Design*, **10**, 2839–2841.

37. Torrisi, A., Bell, R., Mellot-Draznieks, C. (2013). Predicting the impact of functionalized ligands on CO_2 adsorption in MOFs: a combined DFT and grand canonical Monte Carlo study, *Microporous Mesoporous Mater.*, **168**, 225–238.

38. Torrisi, A., Mellot-Draznieks, C., Bell, R. G. (2010). Impact of ligands on CO_2 adsorption in metal-organic frameworks: first principles study of the interaction of CO_2 with functionalized benzenes. II. Effect of polar and acidic substituents, *J. Chem. Phys.*, **132**, 044705.

39. Torrisi, A., Mellot-Draznieks, C., Bell, R. G. (2009). Impact of ligands on CO_2 adsorption in metal-organic frameworks: first principles study of the interaction of CO_2 with functionalized benzenes. I. Inductive effects on the aromatic ring, *J. Chem. Phys.*, **130**, 194703.

40. Dawson, R., Adams, D. J., Cooper, A. I. (2011). Chemical tuning of CO_2 sorption in robust nanoporous organic polymers, *Chem. Sci.*, **2**, 1173–1177.

41. Amrouche, H., Aguado, S., Perez-Pellitero, J., Chizallet, C., Siperstein, F., Farrusseng, D., Bats, N., Nieto-Draghi, C. (2011). Experimental and computational study of functionality impact on sodalite-zeolitic imidazolate frameworks for CO_2 separation, *J. Phys. Chem. C*, **115**, 16425–16432.

42. Amrouche, H., Creton, B., Siperstein, F., Nieto-Draghi, C. (2012). Prediction of thermodynamic properties of adsorbed gases in zeolitic imidazolate frameworks, *RSC Adv.*, **2**, 6028–6035.

43. Gaudin, C., Cunha, D., Ivanoff, E., Horcajada, P., Cheve, G., Yasri, A., Loget, O., Serre, C. (2012). A quantitative structure activity relationship approach to probe the influence of the functionalization on the drug encapsulation of porous metal-organic frameworks, *Microporous Mesoporous Mater.*, **157**, 124–130.

44. Sikora, B. Wilmer, C. E, Greenfield M. L., Snurr R. Q. (2012). Thermodynamic analysis of Xe/Kr selectivity in over 137,000 hypothetical metal-organic frameworks, *Chem. Sci.*, **7**, 2217–2223.

45. Wilmer, C E., Farha, O. K., Bae, Y.-S., Hupp, J. T., Snurr, R. Q. (2012). Structure-property relationships of porous materials for carbons dioxide separation and capture, *Energy Environ. Sci.*, **12**, 9849–9856.

46. Dolbecq, A., Dumas, E., Mayer, C. R., Mialane, P. (2010). Hybrid organic-inorganic polyoxometalate compounds: from structural diversity to applications, *Chem. Rev.*, **110**, 6009–6048.

47. Special issue on polyoxometalates (2007). *J. Mol. Catal. A*, **1–2**, 1–242.

48. Mitchell, S. G., Streb, C., Miras, H. N., Boyd, T., Long, D. L., Cronin, L. (2010). Face-directed self-assembly of an electronically active Archimedean polyoxometalate architecture, *Nat. Chem.*, **2**, 308–312.

49. Zheng, S. T., Zhang, J., Yang, G. Y. (2009). Designed synthesis of POM-organic frameworks from {Ni(6)PW(9)} building blocks under hydrothermal conditions, *Angew. Chem., Int. Ed.*, **47**, 3909–3913.

50. Dolbecq, A., Mialane, P., Sécheresse, F., Keita, B., Nadjo, L. (2012). Functionalized polyoxometalates with covalently linked bisphosphonates, N-donor or carboxylate ligands: from electrocatalytic to optical properties, *Chem. Commun.*, **48**, 8299–8316.

51. Ockwig, N. W., Delgado-Friedrichs, O., O'Keeffe, M., Yaghi, O. M. (2005). Reticular chemistry: occurrence and taxonomy of nets and grammar for the design of frameworks, *Chem. Res.*, **38**, 176–182.

52. Dolbecq, A., Mellot-Draznieks, C., Mialane, P., Marrot, J., Férey, G., Sécheresse, F. (2005). Hybrid 2D and 3D frameworks based on epsilon-Keggin polyoxometallates: experiment and simulation, *Eur. J. Inorg. Chem.*, **15**, 3009–3018.

53. Rodriguez-Albelo, L. M., Ruiz-Salvador, A. R., Sampieri, A., Lewis, D. W., Gomez, A., Nohra, B., Mialane, P., Marrot, J., Sécheresse, F., Mellot-Draznieks, C., Biboum, R. N., Keita, B., Nadjo, L., Dolbecq, A. (2009). Zeolitic polyoxometalate-based metal-organic frameworks (Z-POMOFs): computational evaluation of hypothetical polymorphs and the successful targeted synthesis of the redox-active Z-POMOF1, *J. Am. Chem. Soc.*, **131**, 16078–16087.

54. International Zeolite Association, http://www.iza-structure.org/databases.

55. Park, K. S., Ni, Z., Cote, A. P., Choi, J. Y., Huang, R. D., Uribe-Romo, F. J., Chae, H. K., O'Keeffe, M., Yaghi, O. M. (2006). Exceptional chemical and thermal stability of zeolitic imidazolate frameworks, *Proc. Natl. Acad. Sci. U S A*, **103**, 10186–10191.

56. Mialane, P., Dolbecq, A., Lisnard, L., Mallard, A., Marrot, J., Scheresse, F. (2002). [Epsilon-$PMo_{12}O_{36}(OH)_4\{La(H_2O)_4\}_4$](5): the first epsilon-$PMo_{12}O_{40}$ Keggin ion and its association with the two-electron-reduced alpha-$PMo_{12}O_{40}$ isomer, *Angew. Chem., Int. Ed.*, **2002**(41), 2398–2401.

57. Ruiz-Salvador, A. R., Gomez, A., TOBUNPOROUS code. Code available from authors.

58. Rodriguez-Albelo, L. M., Ruiz-Salvador, A. R., Lewis, D. W., Gomez, A., Mialane, P., Marrot, J., Dolbecq, A., Sampieri, A., Mellot-Draznieks, C.

(2010). Zeolitic polyoxometalate-based metal-organic frameworks (Z-POMOFs) with imidazole ligands and ε-Keggin ions as building blocks, computational evaluation of hypothetical polymorphs and a synthesis approach, *Phys. Chem. Chem. Phys.*, **12**, 8632–8640.

59. Rodriguez-Albelo, L. M., Rousseau, G., Mialane, P., Marrot, J., Mellot-Draznieks, C., Ruiz-Salvador, A. R., Li, S., Lui, R., Zhang, G., Keita, B., Dolbecq, A. (2012). ε-Keggin-based coordination networks: synthesis, structure and application toward green synthesis of polyoxometalate@graphene hybrids, *Dalton Trans.*, **41**, 9989–9999.
60. Nohra, B., Moll, H. E., Albelo, M. R., Mialane, P., Marrot, J., Mellot-Draznieks, C., O'Keeffe, M., Biboum, R. N., lemaire, J., Keita, B., Nadjo, L., Dolbecq, A. (2011). Polyoxometalate-based metal organic framework (POMOFs): structural trends, energetics and high electrocatalytical efficiency for hydrogen evolution reaction, *J. Am. Chem. Soc.*, **133**, 13363–13374.
61. We refer here to the RCSR database (http://rcsr.anu.edu.au) developed by M. O'Keeffe designed for easily classifying nets, giving each net its own three-letter designation. The new "ofp" net refers to open-framework polyoxometalate. See also O'Keeffe, M., Peskov, M. A., Ramsden, S. J., Yaghi, O. M. (2008). The reticular chemistry structure resource (RCSR) database of, and symbols for, crystal nets, *Acc. Chem. Res.*, **41**, 1782–1789.
62. El-Kaderi, H. M., Hunt, J. R., Mendoza-Cortes, J. L., Cote, A. P., Taylor, R. E., O'Keeffe, M., Yaghi, O. M. (2007). Designed synthesis of 3D covalent organic frameworks, *Science*, **316**, 268–272.
63. Walker, A. M., Civalleri, B., Slater, B., Mellot-Draznieks, C., Cora, F., Zicovich-Wilson, C. M., Roman-Perez, G., Soler, J. M., Gale, J. D. (2010). Flexibility in a metal-organic framework material controlled by weak dispersion forces: the bistability of MIL–53, *Angew. Chem., Int. Ed.*, **49**, 7501–7503.
64. Liu, Y., Her, J.-H., Dailly, A., Ramirez-Cuesta, A. J., Neumann, D. A., Brown, C. M. (2008). Reversible structural transition in MIL-53 with large temperature hysteresis, *J. Am. Chem. Soc.*, **130**, 11813–11818.
65. Férey, G., Serre, C. (2009). Large breathing effects in three-dimensional porous hybrid matter: facts, analyses, rules and consequences, *Chem. Soc. Rev.*, **38**, 1380–1399.
66. Phan, A., Doonan, C. J., Uribe-Romo, F. J., Knobler, C. B., O'Keeffe, M., Yaghi, O. M. (2010). Synthesis, structure and carbon dioxide capture properties of zeolitic imidazolate frameworks, *Acc. Chem. Res.*, **43**, 58–67.

67. Zhang, J., Wu, T., Zhou, C., Chen, S. M., Feng, P. Y., and Bu, X. H. (2009). Zeolitic boron imidazolate frameworks, *Angew. Chem., Int. Ed.*, **48**, 2542–2545.

68. Li, W., Probert, M. R., Kosa, M., Bennet, T. D., Thirumurugan, A., Burwoord, R. P., Parinello, M., Howard, J. A. K., Cheetham, A. K. (2012). Negative linear compressibility of a metal-organic framework, *J. Am. Chem. Soc.*, 134, 11940–11943.

69. Baburin, I. A., Leoni, S., Seifert, G. (2008). Enumeration of not-yet-synthesized zeolitic zinc imidazolate MOF networks: a topological and DFT approach, *J. Phys. Chem. B*, **112**, 9437–9443.

70. Baburin, I. A., Leoni, S. (2009). Modelling polymorphs of metal-organic frameworks: a systematic study of diamondoid zinc imidazolates, *Cryst. Eng. Commun.*, **12**, 2809–2816.

71. Lewis, D. L., Ruiz-Salvador, R. A., Gomez, A., Rodriguez-Albelo, Coudert, F. X., Slater, B., Cheetham, A. K., Mellot-Draznieks, C. M. (2009). Zeolitic imidazole frameworks: structural and energetics trends compared with their zeolitic analogues, *Cryst. Eng. Commun.*, **11**, 2272–2276.

72. Henson, N. J., Cheetham, A. K., Gale, J. D. (1996). Computational studies of aluminum phosphate polymorphs, *Chem. Mater.*, **8**, 664–670.

73. Galvelis, R., Slater, B., Cheetham, A. K., Mellot-Draznieks, C. M. (2011). Comparison of the stability of zinc and lithium-boron zeolitic imidazole frameworks, *Cryst. Eng. Commun.*, **14**, 374–378.

74. Baburin, I. A., Leoni, S. (2012). The energy landscape of zeolitic imidazolate frameworks (ZIFs): towards quantifying the presence of substituents on the imidazole ring, *J. Mater. Chem.*, **22**, 10152–10154.

75. Hugues, J. T., Bennett, T. D., Cheetham, A. K., Navrotsky, A. (2013). Thermochemistry of zeolitic imidazolate frameworks of varying porosity, *J. Am. Chem. Soc.*, **135**, 598–601.

76. Ma, L. Q., Abney, C., Lin, W. B. (2009). Enantioselective catalysis with homochiral metal-organic frameworks, *Chem. Soc. Rev.*, **38**, 1248–1256.

77. Lee, C., Mellot-Draznieks, C., Slater, B., Wu, G., Harrison, W. T. A., Rao, C. N. R., Cheetham, A. K. (2006). Thermodynamic and kinetic factors in the hydrothermal synthesis of hybrid frameworks: zinc 4-cyclohexene-1,2-dicarboxylates, *Chem. Commun.*, 2687–2689.

78. Harvey, H. G., Slater, B., Attfield, M. P. (2004). Rational design of the pore system within the framework aluminum alkylenediphosphonate series, *Chem.-Eur. J.*, **10**, 3270–3278.

79. Bailey, A. J., Lee, C., Feller, R. K., Orton, J. B., Mellot-Draznieks, C., Slater, B., Harrison, W. T. A., Simoncic, P., Navrotsky, A., Grossel, M. C., Cheetham, A. K. (2008). Comparison of chiral and racemic forms of zinc cyclohexane trans-1,2-dicarboxylate frameworks: a structural, computational, and calorimetric study, *Angew. Chem., Int. Ed.*, **47**, 8634–8637.

80. Greathouse, J. A., Allendorf, M. D. (2006). Interaction of water with MOF-5 simulated by molecular dynamics, *J. Am. Chem. Soc.*, **128**, 10678–10679.

81. Appelhans, L. N., Kosa, M., Radha, A. V., Simoncic, P., Navrotsky, A., Parrinello, M., Cheetham, A. K. (2009). Phase selection and energetics in chiral alkaline earth tartrates and their racemic and meso analogues: synthetic, structural, computational and calorimetric studies, *J. Am. Chem. Soc.*, **131**, 15375–15386.

82. Yeung, H. H.-M., Kosa, M., Parrinello, M., Forster, P. M., Cheetham, A. K. (2010). Structural diversity and energetics in anhydrous lithium tartrates: experimental and computational studies of novel chiral polymorphs and their racemic and meso analogues, *Cryst. Growth Design*, **11**, 221–230.

83. Allendorf, M. D., Schwartzberg, A., Stavila, V., Talin A. A. (2011), A roadmap to implementing metal-organic frameworks in electronic devices: challenges and critical directions, *Chem. Eur. J.*, **17**, 11372–11388.

84. Cheetham, A. K., Rao, C. N. R. (2007). There is room in the middle, *Science*, **318**, 58–59; Rao, C. N. R., Cheetham, A. K., Thirumurugan, A. (2008). Hybrid inorganic-organic materials: a new family in condensed matter physics, *J. Phys.: Condens. Matter*, **20**, 083202.

85. Silva, C. Gomes, Corma, A., Garcia, H. (2010). Metal-organic frameworks as semiconductors, *J. Mater. Chem.*, **20**, 3141–3156.

86. Wang, C., Xie, Z. G., de Krafft, K. E., Lin, W. L. (2011). Doping metal organic frameworks for water oxidation, carbom dioxide reduction and organic photocatalysis, *J. Am. Chem. Soc.*, **133**, 13445–13454.

87. Civalleri, B., Napoli, F., Noël, Y., Roetti, C., Dovesi, R. (2006). Ab initio prediction of materials properties with CRYSTAL: MOF-5 as a case study, *Cryst. Eng. Commun.*, **8**, 364–371.

88. Kuc, A., Enyashin, A., Seifert, G. (2007). Metal-organic frameworks: structural, energetic, electronic, and mechanical properties, *J. Phys. Chem. B*, **111**, 8179–8186.

89. Gascon, J., Hernandez-Alonso, M. D., Almeida, A. R., van Klink, G. P. M., Kapteijn, F., Mul, G. (2008). Isoreticular MOFs as efficient photocatalyst with tunable band gap: an operando FTIR study of the photoinduced oxidation of propylene, *ChemSusChem*, **1**, 981–983.
90. Cabrera-Fuentes, M., Nicholson, D. M., Sumpter, B. G., Widom, M. (2005). Electronic structure and properties of isoreticular metal-organic-frameworks: the case of M-IRMOF1 (M= Zn, Cd, Be, Mg and Ca), *J. Chem. Phys.*, **123**, 124713.
91. Choi, J. H., Choi, Y. J., Lee, J. W., Shin, W. H., Kang, J. K. (2009). Tunability of electronic band gaps from semiconducting to metallic states via tailoring Zn ions in MOFs with Co ions, *Phys. Chem. Chem. Phys.*, **11**, 628–631.
92. Lin, C.-K., Zhao, D., Gao, W.-Y., Yang, Z., Ye, J., Xu, T., Ge, Q., Ma, S., Liu, D.-J. (2012), Tunability of band gaps in metal-organic frameworks, *Inorg. Chem.*, **51**, 9039–9044.
93. Valenzano, L., Civalleri, B., Chavan, S., Bordiga, S., Nilsen, M. H., Jakobsen, S., Lillerud, K. P., Lamberti, C. (2011). Disclosing the complex structure of UiO-66 metal organic framework: a synergic combination of experiment and theory, *Chem. Mater.*, **23**, 1700–1718.
94. Dan-Hardi, M., Serre, C., Frot, T., Rozes, L., Maurin, G., Sanchez, C. and Férey, G. (2009). A new photoactive crystalline highly porous titatium(IV) dicarboxylate, *J. Am. Chem. Soc.*, **131**, 10857.
95. Walsh, A. and Catlow, C. R. A. (2010). Photostimulated reduction processes in a titania hybrid metal-organic framework, *ChemPhys Chem*, **11**, 2341–2344.
96. Walsh, A. (2011). Effects of reduced dimensionality on the electronic structure and defect chemistry of semiconducting hybrid organic-inorganic PbS solids, *Proc. R. Soc. A: Math., Phys. Eng. Sci.*, **467**, 1970–1985; Walsh, A. (2010). Defect processes in a PbS metal organic framework: a quantum-confined hybrid semiconductor, *J. Phys. Chem. Lett.*, **1**, 1284–1287.
97. Choi, J. H., Jeon, H. J., Choi, K. M., Kang, J. K. (2012). Metal-organic frameworks for visible light absorption via anion substitution, *J. Mater. Chem.*, **22**, 10144–10147.
98. Tan, J. C., Cheetham, A. K. (2011). Mechanical properties of hybrid inorganic-organic framework materials: establishing fundamental structure-property relationships, *Chem. Soc. Rev.* **40**, 1059–1080.
99. Tafipolsky, M., Schmid, R. (2009). Systematic first principles parameterization of force fields for metal-organic frameworks using a genetic algorithm approach, *J. Phys. Chem. B*, **113**, 1341–1352.

100. Han, S. S., Goddard, W. A. (2007) Metal-organic frameworks provide large negative thermal expansion behavior, *J. Phys. Chem. C*, **111**, 15185–15191.

101. Greathouse, J. A., Allendorf, M. D. (2008). Force field validation for molecular dynamics simulations of IRMOF-1 and other isoreticular zinc carboxylate coordination polymers, *J. Phys. Chem. C*, **112**, 5795–5802.

102. Bahr, D. F., Reid, J. A., Mook, W. M., Bauer, C. A., Stumpf, R., Skulan, A. J., Moody, N. R., Simmons, B. A., Shindel, M. M., Allendorf, M. D. (2007). Mechanical properties of cubic zinc carboxylate IRMOF-1 metal-organic framework crystals, *Phys. Rev. B*, **76**, 184106.

103. Samanta, A., Furuta, T., Li. J. (2006). Theoretical assessment of the elastic constants and hydrogen storage capacity of some metal-organic framework materials, *J. Chem. Phys.*, **125**, 084714.

104. Mattesini, M., Soler, J. M., Yndurain, F. (2006). Ab initio study of metal-organic framework-5 Zn_4O(1,4-benzenedicarboxylate)$_3$: an assessment of mechanical and spectroscopic properties, *Phys. Rev. B*, **73**, 094111.

105. Zhou, W., Yildirim, T. (2006). Lattice dynamics of metal-organic frameworks: neutron inelastic scattering and first-principles calculations, *Phys. Rev. B*, **74**, 180301.

106. Bennett, T. D., Tan, J.-C., Moggach, S. A., Galvelis, R., Mellot-Draznieks, C., Reisner, B. A., Thirumurugan, A., Allan, D. R., Cheetham, A. K. (2010). Mechanical properties of dense zeolitic imidazolate frameworks (ZIFs): a high-pressure X-ray diffraction, nanoindentation and computational study of the zinc framework Zn(Im)$_2$, and its lithium-boron analogue, LiB(Im)$_4$, *Chem. Eur. J.*, **16**, 10684–10690.

107. Tan, J.-C., Civalleri, B., Lin, C.-C., Valenzano, L., Galvelis, R., Chen, P.-F., Bennett, T. D., Mellot-Draznieks, C., Zicovich-Wilson, C. M., Cheetham, A. K. (2012). Exceptionally low shear modulus in a prototypical imidazole-based metal-organic framework, *Phys. Rev. Lett.*, **108**, 095502.

108. Ortiz, A. U., Boutin, A., Fuchs, A. H., Coudert F. X. (2012). Anisotropic elastic properties of flexible metal-organic frameworks: how soft are soft porous crystals? *Phys. Rev. Lett.*, **109**, 195502.

109. Chizallet, C., Lazare, S., Bazer-Bachi, D., Bonnier, F., Lecocq, V., Soyer, E., Quoineaud, A. A., Bats, N. (2010). Catalysis of transesterification by a nonfunctionalized metal-organic framework: acido-basicity at the external surface of ZIF-8 probed by FTIR and ab initio calculations, *J. Am. Chem. Soc.*, **132**, 12365–12377.

110. Tran, U. P. N., Le, K. K. A., Phan, N. T. S. (2011). Expanding applications of metal–organic frameworks: zeolite imidazolate framework ZIF-8 as an efficient heterogeneous catalyst for the Knoevenagel reaction, *ACS Catal.*, **1**, 120–127.

111. Chizallet, C., Bats, N. (2010). External surface of zeolite imidazolate frameworks viewed ab initio: multifunctionality at the organic-inorganic interface, *J. Phys. Chem. Lett.*, **1**, 349–353.

Chapter 2

On the Application of Classical Molecular Simulations of Adsorption in Metal–Organic Frameworks

David Dubbeldam[a] and Krista S. Walton[b]

[a] *Van't Hoff Institute for Molecular Sciences, University of Amsterdam, Science Park 904, 1098XH, Amsterdam, The Netherlands*

[b] *School of Chemical & Biomolecular Engineering, Georgia Institute of Technology, 311 Ferst Dr. NW, Atlanta, GA 30332-0100, USA*

d.dubbeldam@uva.nl, krista.walton@chbe.gatech.edu

2.1 Introduction

Ordered crystalline porous materials offer the potential for selective adsorption by exploiting differences in molecular configurations. Zeolites are readily available, very stable, and cheap. A zeolite should have the right combination of high adsorption selectivity, combined with an adequate capacity for use in traditionally used fixed-bed devices. Recently, new classes of nanoporous materials have been designed that have good stability, high void volumes, and well-defined tailorable cavities of uniform size. Examples are metal–

Metal–Organic Frameworks: Materials Modeling towards Potential Engineering Applications
Edited by Jianwen Jiang

ISBN 978-981-4613-45-3 (Hardcover), 978-981-4613-46-0 (eBook)
www.panstanford.com

organic frameworks (MOFs) [Chui et al. (1999); Li et al. (1999); Eddaoudi et al. (2002); Yaghi et al. (2003); Kitagawa et al. (2004); Snurr et al. (2004); Mueller et al. (2006); Férey (2008); Robson (2008); Long and Yaghi (2009); Telfer (2010)], covalent organic frameworks (COFs) [Côté et al. (2005); El-Kaderi et al. (2007)], and zeolitic imidazolate frameworks (ZIFs) [Park et al. (2006)]. These novel materials possess almost unlimited structural variety because of the many combinations of building blocks that can be imagined. MOFs are synthesized under solvo- or hydrothermal conditions. During synthesis the building blocks self-assemble into crystalline materials that, after evacuation, can find applications in adsorption separations, air purification, gas storage, chemical sensing, and catalysis [Snurr et al. (2004); Mueller et al. (2006); Jiang (2012b)].

Each year the synthesis of several hundred new structures is reported, and this has created the urgent need to screen these efficiently. Molecular simulations have sufficiently advanced in both speed and accuracy to allow rapid evaluation of (hypothetical) structures for storage and/or separation devices [Wilmer et al. (2012); Krishna and van Baten (2011a); Wu et al. (2012a); Haldoupis et al. (2012)]. Screening poses the question Assuming certain requirements for an given industrial application, what nanoporous material would be most suitable? The overwhelming majority of studies for gas separation in MOFs have been focused on CO_2 capture [Keskin et al. (2010); D'Alessandro et al. (2010); Bae and Snurr (2011); Li et al. (2011); Férey et al. (2011); Jiang (2012a)]. In principle, the enormous number of MOFs with various combinations of organic linkers and metal cations creates an opportunity to tailor pore structure, size, and functionality. Significant problems, however, thwart attempts to discover or design such materials. The principles for designing (water-stable) MOFs are not yet well known [Low et al. (2009)]. Han et al. experimentally demonstrated the feasibility of efficiently screening MOF materials in parallel on the basis of adsorption capacity and adsorption and diffusion selectivities [Han et al. (2012)]. Similarly, many computational screening studies have appeared [Wilmer et al. (2012); Krishna and van Baten (2011a); Wu et al. (2012a); Lin et al. (2012)], as well as combined experimental/simulation work [Yazaydin et al. (2009)].

Regenerative adsorber systems can operate continuously by using:

(1) multiple fixed beds of an adsorbent;
(2) fluidized bed contactors with separate adsorption and desorption vessels; or
(3) rotary bed adsorbents that cycle continuously between adsorption and desorption operations.

Pressure swing adsorption (PSA) is an industrial separation technique at atmospheric temperature and high pressure [Ruthven (1984); Ruthven et al. (1994)]. In practice a batch setup with various vessels in parallel is used; some are in adsorption mode, while others are in regeneration mode. In pressure swing the regeneration is done at low pressure, while thermal swing adsorption uses high temperature. PSA is less sensitive to diffusional effects and largely dominated by adsorption. This is in contrast to membrane technology, where the difference in mobility of species leads to separation. The difficulty of screening lies in determining what "most suitable" means. The economics of PSA is crucially dependent on adsorption selectivity. High adsorption selectivity leads to smaller equipment volumes and lower capital and energy requirements. However, high selectivity means a strong affinity of the most adsorbing species and therefore higher regeneration costs. Besides selectivity, it is also the capacity of the material that determines the efficiency of a separation device [Krishna (2012)]. In pressure swing adsorbers, high adsorption capacities are desirable because they result in lower frequencies in the PSA regeneration cycles [Krishna and Long (2011)]. Other crucially important factors are cost of the material and stability (especially temperature and water stability) [Schoenecker et al. (2012)]; once the MOF is evacuated it is likely to become air and moisture sensitive [Caskey and Matzger (2009)]). A reliable screening methodology should take all these considerations into account.

In this book chapter we highlight simulation techniques that lie at the foundation of screening. A connection is made to how these properties are measured in experiment such that the reader hopefully will get a feel for the accuracy (and sometimes inaccuracy) of the computation of certain properties.

2.2 Geometric Properties

2.2.1 *Visualization of the Topology*

Data visualization is the process of displaying information in any sort of pictorial or graphical representation. Visualization Toolkit (VTK) is an open-source, freely available software system for 3D computer graphics, image processing, and scientific visualization [Schroeder et al. (1996)]. Visualization is an important tool to elucidate molecular mechanisms by connecting the macroscopic results (e.g., adsorption isotherms) to the microscopic molecular-level information.

Examining the atomic positions of a structure does not immediately lead to a thorough understanding of the structure. Snurr et al. [Snurr et al. (1993)], at the advent of computational zeolite research, already used visualization of energy contour plots and 3D density distributions of benzene in silicalite to obtain siting information. Details of the topology of the structure can be easily obtained using VTK to interpret and classify the structure. Dubbeldam et al. generated 3D energy landscapes using the free energy obtained from the Widom insertion method [Dubbeldam et al. (2003)]. The simulation cell is divided up into, for example, $150 \times 150 \times 150$ voxels. The adsorbate is randomly inserted millions of times, and the voxels corresponding to the atom positions of the adsorbate are updated with the current Boltzmann weight. The resulting dataset has regions with near-zero values that correspond to overlap with the structure. The ratio of the nonzero values to the total number of voxels is the void fraction. Multiplying by the volume of the unit cell we can compute the accessible pore volume. The near-zero values are rendered transparent. By also rendering the high values transparent we are able to visualize the "pore walls." The volume-rendered pictures are similar to isocontour surfaces but in general show less visual artifacts because the walls have a higher "thickness" (by defining an appropriate scalar to opacity function in VTK). Figure 2.1a shows the CHA-type structure. The unit cell is monoclinic, with $a = b = 13.5292$ Å, $c = 14.7483$ Å, and $\alpha = \beta = 90°$, $\gamma = 120°$. It is hard to make out the structure from the atomic positions. In Fig. 2.1b the pore energy landscape is

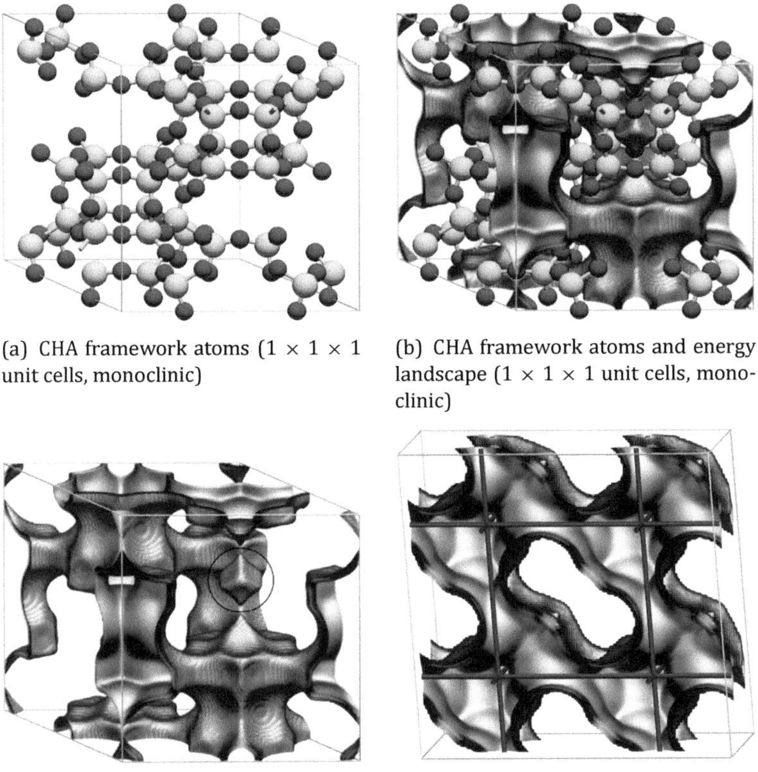

(a) CHA framework atoms (1 × 1 × 1 unit cells, monoclinic)

(b) CHA framework atoms and energy landscape (1 × 1 × 1 unit cells, monoclinic)

(c) CHA energy landscape (1 × 1 × 1 unit cells, monoclinic)

(d) CHA topology (2 × 2 × 2 primitive unit cells)

Figure 2.1 Analysis of the CHA zeolite topology. (a) Framework atoms, (b) framework atoms with the pore energy landscape, (c) energy landscape (a blocking pocket shown annotated with a black circle), and (d) lattice topology. The probe molecule is helium.

shown, and Fig. 2.1c shows only the pore energy landscape without the framework atoms. The latter is more clear and reveals the pore connectivity of the system. The channel system is now apparent, and any bottlenecks for diffusion are easy to identify. Note that it also immediately reveals "pockets" that are not accessible from the main channel systems. These should be removed and blocked during the simulations [Krishna and van Baten (2010a)]. Figure 2.1d shows the CHA channel system (with the pockets properly removed). The lattice ($\alpha = \beta = \gamma = 94.07°$) is close to a cubic lattice. This is not immediately apparent from Fig. 2.1a. The landscape shows no intracage barriers, and each cage has six windows connecting the cages to each other. The energy landscapes are crucial to understand the structure. They reflect the viewpoint of the probe molecule, and the pictures are different for each type of adsorbate.

2.2.2 *Studying Adsorption Sites*

Helium is a good probe molecule to identify possible adsorption sites of small molecules. Any surface location with a high concave curvature is a likely adsorption site, while convex surfaces are not. At a concave position the dispersion interaction with the wall is high. Adsorption siting information can be obtained from Monte Carlo (MC) or molecular dynamics (MD) simulations by keeping track of a 3D histogram of the atomic positions. Figure 2.2a shows the ITQ-29 zeolite (siliceous Linde Type A [LTA]). We have removed the framework itself and blocking pockets (the sodalite cages that are not accessible for larger adsorbates from the main channels) and only show the energy landscape. In Fig. 2.2b the topology is clear, the structure is cubic, and each cage is connected to six other cages. The pictures show a snapshot of methane at an average loading of eight molecules per cage at 298 K. Snapshot pictures are powerful tools, but one needs to examine many to have a feel for the behavior of the

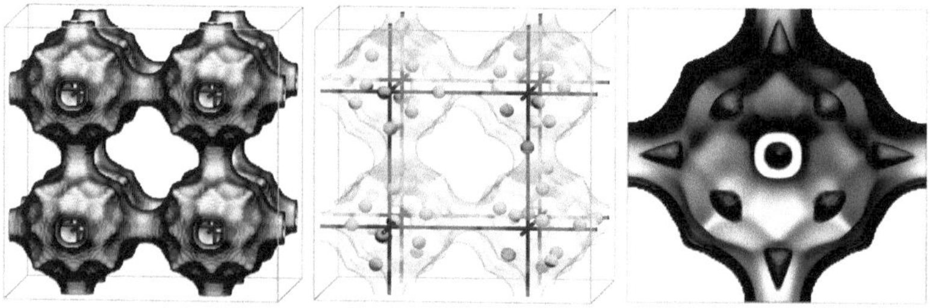

(a) ITQ-29 ($2 \times 2 \times 2$ uc) (b) ITQ-29 ($2 \times 2 \times 2$ uc) (c) ITQ-29 ($1 \times 1 \times 1$ uc)

Figure 2.2 ITQ-29 topology (siliceous LTA, cubic). (a) Energy landscape of $2 \times 2 \times 2$ unit cells (unit cell edge is 11.8671 Å), (b) snapshot at 298 K and eight methane molecules per cage, and (c) density of methane inside a cage at 298 K and eight molecules per cage.

system. By keeping track of the 3D histogram we can make "density plots" (Fig. 2.2c). Because the unit cells are the same, the picture is the same for each unit cell, and we can focus on just one. There are six sites identified in front of the eight-ring windows, and eight sites in the corners of the cage. Note that one could predict these sites from the energy landscapes by identifying the locations on the surface that are highly concave (the corners in the cage) or sites that maximize dispersion interactions from different wall directions (the location in front of the window).

For convenience we show information on the Mobil Five (MFI)-type zeolite in Fig. 2.3, which will be needed in a later section. MFI is the most studied zeolite system. The orthorhombic unit cell with edge lengths $a = 20.022$ Å, $b = 19.899$ Å, and $c = 13.383$ Å [van Koningsveld et al. (1987)] is shown in Fig. 2.3a. There are two straight and two zigzag channels per unit cell and four intersections per unit cell. To visualize molecules inside the structure the pore walls can be rendered transparent. Figure 2.3b shows a snapshot of hexane (nC6) molecules; Fig. 2.3c shows a snapshot of 2,3-dimethylbutane (2,3-DMB). Molecules are subjected to the minimum image convention rule, and this determines the amount of unit cells (for MFI the minimum number of unit cells is $2 \times 2 \times 2$). Snapshots are useful to detect differences in adsorption sites of the various species. Figure 2.3d is the average of many snapshots, and this density is the same for all unit cells. Therefore the density is very convenient to obtain the siting information at the unit cell level.

With VTK a 3D dataset can be manipulated with ease, and VTK offers advanced computational techniques like tricubic interpolation and contouring algorithms. It is straightforward to "slice" the energy or density landscape and visualize the slice by a contour or color plot. The minimum energies can be found, free energy bottlenecks can be identified, and one can perform computations of pore volumes and surface areas. Sarkisov and Harrison [Sarkisov and Harrison (2011)] developed a geometric percolation algorithm to visualize pathways for rigid spherical and nonspherical molecules as a probe. This type of tool is also able to detect whether the porous network in a material under consideration is fully accessible to a probe of a particular shape and size.

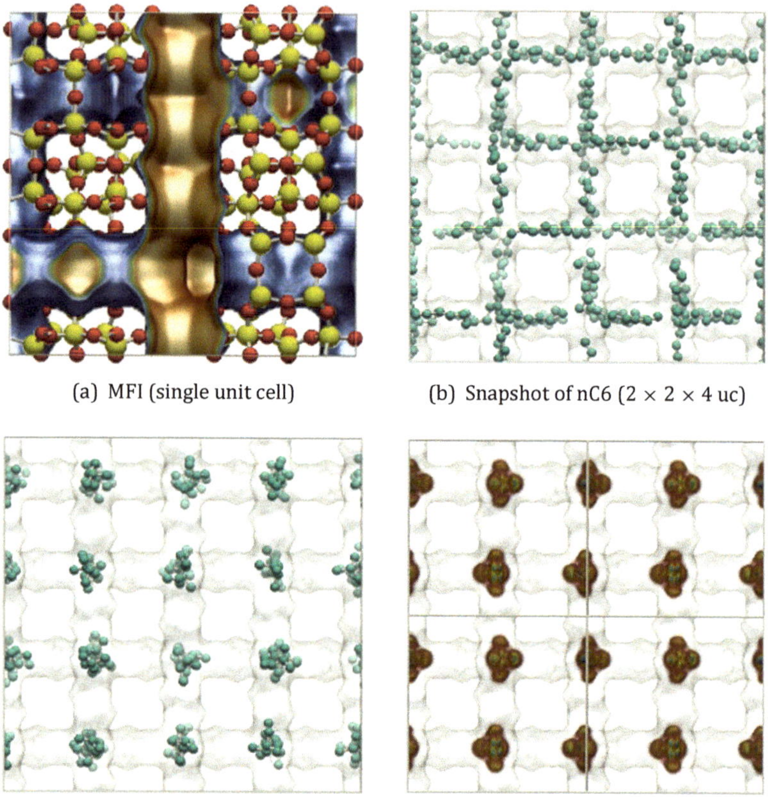

(a) MFI (single unit cell)

(b) Snapshot of nC6 (2 × 2 × 4 uc)

(c) Snapshot 2,3-DMB (2 × 2 × 4 uc)

(d) Density 2,3-DMB (2 × 2 × 4 uc)

Figure 2.3 Energy landscape of MFI. The MFI unit cell has edge lengths $a = 20.022$ Å, $b = 19.899$ Å, and $c = 13.383$ Å, with cell angles $\alpha = \beta = \gamma = 90°$. The MFI pore system consists of straight channels running in the c direction, which are connected via zigzag channels. About 29% of the structure is void. Color code: oxygen (red) and silicon (yellow). The snapshots and density plot are at 433 K and 100 kPa pressure.

2.2.3 *Void Fraction*

Knowledge of the density of porous materials is critical for full characterization of the adsorbents and for fixed-bed adsorber design studies. There are three densities that are relevant in this regard: bulk or bed density (ρ_b), particle density (ρ_p), and skeletal or solid density (ρ_s). These densities are related to one another through the external void fraction of packing (ϵ), the internal porosity or helium void fraction (ϵ_p), and the total packed bed voidage (ϵ_b) by the following equations [LeVan and Carta (2008)]:

$$\rho_b = \rho_p(1 - \epsilon) = \rho_s(1 - \epsilon_p)(1 - \epsilon) = \rho_s(1 - \epsilon_b) \tag{2.1}$$

The bulk density is the most straightforward to determine experimentally. One can simply fill a volume such as a graduated cylinder with a known mass of adsorbent powder or pellets to determine the mass of adsorbent per volume. The particle density is the adsorbent mass per volume occupied by the particle. This value is often available from commercial suppliers when purchasing adsorbents. However, it can also be measured when the particle shape and size are known, for example, cylindrical or spherical pellets, but the volume for irregular-shaped or very small particles must be estimated.

The skeletal or solid density, ρ_s, is the mass of the adsorbent per volume occupied by the particle but with the pores deducted. So, this density represents the "true density" of a porous material, as it was first referred to by McBain in 1932 [McBain (1932)]. Coolidge was the first to use the more descriptive "skeletal density" [Coolidge (1934)]. In recent years, it has been more common to refer to this property as the "helium density". This term derives from the experimental method of helium expansion that is used to determine such values. The two key assumptions in such experiments are that (1) helium is inert and adsorption is negligible and that (2) helium is small enough to penetrate nanopores. Determining ρ_s experimentally requires a simple apparatus with two known volumes separated by a valve. The first volume, V_1, has been loaded with a known mass of clean adsorbent and held under vacuum at constant temperature. The sample volume, V_2, is initially filled with helium and maintained at constant temperature and pressure. After equilibration, the valve is opened and helium is allowed to expand into V_1. The final temperature and pressure for both volumes are then used to calculate ρ_s from [Talu and Myers (2001); Neimark and Ravikovitch (1997); Malbrunot et al. (1997)]:

$$\rho_s = \frac{m}{\rho_{He1_1} V_1 - (\rho_{He2_0} - \rho_{He2_1}) V_2} \tag{2.2}$$

where ρ_{He1} and ρ_{He2} are the helium densities in volumes 1 and 2, respectively, at the initial condition (0) and final condition (1).

Talu and Myers [Talu and Myers (2001)] proposed a simulation methodology that mimics the experimental procedure. For consistency with experiment, the helium void fraction ξ is determined by probing the framework with a nonadsorbing helium molecule using the Widom particle insertion method:

$$\xi = \langle e^{-\beta U} \rangle \tag{2.3}$$

The available pore volume V_{pore} is simply

$$V_{\text{pore}} = \xi V \tag{2.4}$$

Computationally, one can also use the $r \rightarrow 0$ limit of the pore size distribution (see a later section) to evaluate the void fraction.

2.2.4 *Surface Areas*

Surface area is the most basic property of porous materials. Along with pore volume, surface area has become the main benchmark characterization method for any porous material. The most common experimental method for obtaining surface areas involves measuring nitrogen gas adsorption isotherm on the material at 77 K. The isotherm data are then fit by the Brunauer–Emmett–Teller (BET) model, and the surface area is then backed out of the BET parameters. There are a variety of commercial systems on the market for performing these types of measurements and automatically calculating the surface area using the BET equation [Brunauer et al. (1938)] or Langmuir model. The BET equation is given by

$$\frac{x}{v(1-x)} = \frac{1}{v_m c} + \frac{(c-1)x}{v_m c} \tag{2.5}$$

where $x = p/p_0$ is the ratio of the bulk pressure to vapor pressure, v is the volume of nitrogen adsorbed per gram of material at STP, v_m is the monolayer capacity, and c is related to the heat of adsorption. Adsorption data are plotted as the left-hand side of Eq. 2.5 versus x. A line is fit to the low pressure region and the slope and y intercept yield c and v_m, respectively. The surface area is then calculated from

$$A = v_m \sigma N_{\text{av}} \tag{2.6}$$

where σ_0 is the cross-sectional area of nitrogen at liquid density (16.2 Å) and N_{av} is Avogadro's number. There are several assumptions built into the BET model [Rouquerol et al. (2007); Parra et al. (1995); Adamson and Gast (1997)]. Adsorption is assumed to occur in a multilayer fashion, with the number of adsorbed layers being infinite at saturation. This means that there can be no appreciable overlap between monolayer and multilayer adsorption. In spite of these assumptions, the BET model has been shown to predict surface areas in good agreement with values obtained through geometric calculations from a model of the crystal structure.

Geometric surface areas can be calculated by using a simple MC integration technique in which a nitrogen probe (3.681 Å) molecule is rolled along the surface of the framework [Shrake and Rupley (1973); Duren et al. (2004, 2007); Sarkisov and Harrison (2011)]. The geometric surface area is calculated by randomly inserting the nitrogen probe around framework atoms and checking for overlap. The insertions can be efficiently performed by drawing random numbers on a sphere of radius σ. Another choice for the size of an atom is to take the Lennard–Jones (LJ) well depth $2^{1/6}\sigma$. If a probe size of 0 Å is used the so-called Connolly's surface area is computed [Connolly (1983)]. For sorption applications, molecule-accessible surface areas are physically more meaningful than Connolly surface areas [Duren et al. (2007)]. In addition to the geometric surface area, it is of course also equally possible to compute BET surface areas from simulated adsorption isotherms. MOFs provide an interesting set of materials to test the applicability of the BET model, in combination with molecular simulations, since MOFs are crystalline materials with the structure completely known from single-crystal x-ray diffraction studies.

Walton and Snurr [Walton and Snurr (2007)] examined the consistency of the BET model compared to the geometric calculation for several prototypical MOF structures with varying pore sizes. MOF-5 (isoreticular metal–organic framework [IRMOF]-1) and five isostructures, IRMOF-6, IRMOF-10, IRMOF-14, IRMOF-16, and IRMOF-18 (Fig. 2.4) [Eddaoudi et al. (2002); Li et al. (1999); Rowsell et al. (2004)]. Grand canonical Monte Carlo (GCMC) simulations were performed for nitrogen adsorption at 77 K for each MOF. The DREIDING force field [Mayo et al. (1990)] was used for each MOF

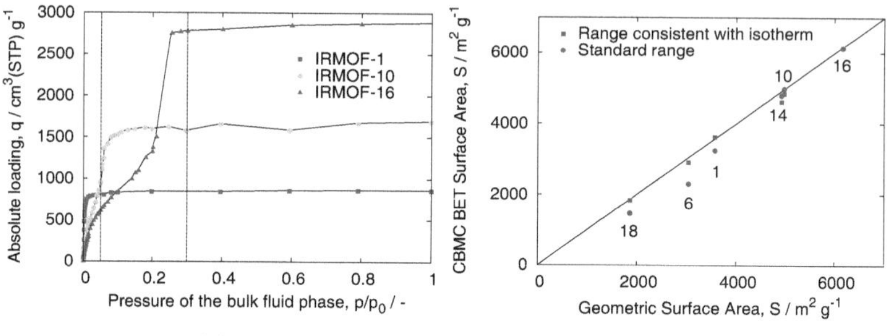

(a) (b)

Figure 2.4 BET surface areas in MOFs. (a) Adsorption isotherms for nitrogen in IRMOF-1, IRMOF-10, and IRMOF-16 at 77 K calculated from GCMC simulations; vertical lines denote the standard BET range. (b) Comparison of BET surface areas calculated from the standard range, BET surface areas calculated from the range identified through the consistency criteria, and geometric surface areas.

atom, and the TraPPE force field [Potoff and Siepmann (2001)] was used to model nitrogen. The BET model was then applied to the isotherm data to calculate a surface area. The plot of Eq. 2.5 produces a curve that is typically made up of three regions: concave at low pressure, linear at intermediate pressure, and convex to the x axis at high pressure. Choosing the proper linear region can be quite subjective, and the traditional BET range usually encompasses $0.05 < p/p_0 < 0.3$ with the assumption that monolayer formation occurs in this range. However, small-pore MOFs will reach saturation long before $p/p_0 = 0.05$, so the BET parameters would be meaningless. The range should be chosen on the basis of two major consistency criteria developed by Roquerol et al. [Rouquerol et al. (2007)]: (i) the selected pressure range should have values of $v(p_0 - p)$ increasing with p/p_0, and (ii) the y intercept of the linear region should yield a positive value for the c parameter. The simulated isotherms represent nitrogen adsorption in a perfect crystal, which could never be achieved by experiments. Thus, these results provide a more direct comparison for the accessible surface area calculation.

Figure 2.4(a) shows results for nitrogen adsorption simulations in IRMOF-1, IRMOF-10, and IRMOF-16 with the traditional BET range noted. Clearly, IRMOF-1 saturates below $p/p_0 = 0.05$.

This same issue is encountered with the other MOFs of similar pore diameter: IRMOF-6 and IRMOF-18. Figure 2.4(b) shows the accessible surface areas for each MOF compared with the BET surface areas calculated from the simulated isotherm using both the standard range and the range selected using the above consistency criteria. There is very good agreement between BET surface areas and the accessible surface area for the large-pore IRMOF-10, IRMOF-14, and IRMOF-16. However, the BET surface areas for the small-pore IRMOF-1, IRMOF-6, and IRMOF-18 using the standard pressure range exhibit significant deviations from the accessible surface area. These results provide compelling evidence of the importance of calculating the BET surface area from the proper region of the adsorption isotherm. Commercial "BET instruments" are typically set to automatically choose a fixed range for BET fitting. The operator must ensure that the range results in consistent BET model parameters. In addition, good agreement with the accessible surface area indicates that this is a valid method for characterization of materials for which the crystal structure is known. The calculation can be obtained quickly, and discrepancies with experimental BET surface areas can be used to determine if an experimental sample has been properly activated or if it possesses collapsed pores.

Figure 2.5 shows the surface area and accessible pore volumes as a function of the helium void fraction, computed using molecular simulation. MOFs and zeolites live in different worlds when it comes to surface areas and porosities. ZIFs have properties that lie in between zeolites and MOFs. The current record holder is NU-110, with a staggering 7400 m^2/g geometric surface area (computed from simulations) and 7100 m^2/g BET surface area obtained from experiment [Farha et al. (2012)]. Almost 93% of the NU-110 structure is void. Sarkisov established a theoretical limit of the accessible surface area in porous structures [Sarkisov (2012)]. For both crystalline and disordered structures the theoretical limit of the surface area is estimated at about 15,000 m^2/g. Farha et al. demonstrated computationally a new surface area ceiling for MOFs ($\approx$14600 m^2/g) [Farha et al. (2012)]. Both works give surface area limits that substantially exceeds what much of the MOF community had perceived to be a theoretical limit ($\approx$10500 m^2/g) [Schnobrich et al. (2010)]. The clear advantage of MOFs

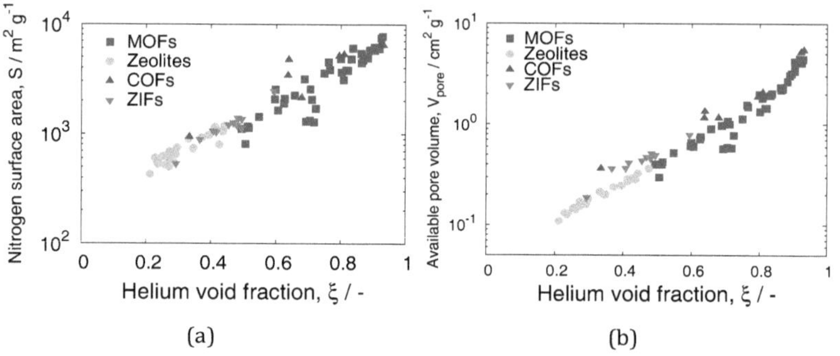

Figure 2.5 Typical geometric properties of MOFs, COFs, ZIFs, and zeolites. Data taken from Dubbeldam et al. (2012) and Farha et al. (2012). NU-110 has the highest geometric and BET surface area (7400 m^2/g and 7140 m^2/g, respectively) [Farha et al. (2012)], while COF-108 [El-Kaderi et al. (2007)] has the highest pore volume (5.46 cm^3/g) and void fraction (93.1%).

is their incredible pore volumes and surface areas. Almost every MOF framework atom is a wall atom, having interaction with the adsorbates. Zeolites have many framework atoms buried inside the walls. They have therefore relatively high framework densities and low surface areas. However, as we shall see, zeolites in general show high selectivities because the smaller channel dimensions allow for higher discrimination between adsorbing species. The question "zeolites vs. MOFs" translates to "selectivity vs. capacity". A current research area is therefore to find ways to combine the high selectivity property of zeolites with the large capacities of MOFs.

2.2.5 *Pore Size Distributions*

Pore size distributions (PSDs) are determined experimentally by applying one of several different models to nitrogen or argon adsorption isotherm data measured at 77 K. There are many reviews of this topic [Kaneko (1994); Do (2003)], so we only touch on a few important models here. The Horvath–Kawazoe (HK) method [Horvath and Kawazoe (1983)] is one of the most popular approaches for determining the pore size distribution of microporous structures. This method is based on statistical analysis of a fluid confined by a slit pore. The HK equation has since been extended for cylindrical [Saito and Foley (1991)] and spherical [Cheng and Yang (1994)] pores. The original HK model was developed on the basis of the assumption that the mean free energy change of an adsorbate molecule as it passes from bulk to adsorbed phase is related to the pressure at which pore filling occurs. Several assumptions are built into the method: (i) pore filling occurs in slit pores in a single step, (ii) no wall wetting or compression of the adsorbate occurs as bulk pressure increases, and (iii) the condensed-phase density is equal to the liquid density if the pore-filling pressure is less than the bulk gas pressure. These assumptions construct quite an ideal scenario; nevertheless, the HK method has been shown to reproduce PSDs obtained from more sophisticated models such as DFT, which treats the condensed phase in a more realistic manner. Techniques such as nonlocal density functional theory (NLDFT) [Lastoskie et al. (1993); Kluson and Scaife (2001)] and, more recently, quenched solid density functional theory (QSDFT) [Neimark et al. (2009)] have been incorporated into commercially available sorption systems for automatic calculation of PSDs from nitrogen or argon adsorption isotherms. QSDFT is a modification to NLDFT in which the effects of surface roughness and heterogeneity are accounted for explicitly.

To compute the PSD geometrically the method of Gelb and Gubbins can be used [Gelb and Gubbins (1999); Sarkisov and Harrison (2011)]. By determining the largest covering sphere for every point in the void volume, the cumulative pore volume curve is obtained. Let $V_{\mathrm{pore}}(r)$ be the volume of the void space "coverable" by spheres of radius r or smaller; a point x is in $V_{\mathrm{pore}}(r)$ only if we can construct a sphere of radius r that overlaps x and does not overlap any substrate atoms. This volume is equivalent to that enclosed by the pore's Connolly surface. $V_{\mathrm{pore}}(r)$ is a monotonically decreasing function of r and is easily compared with the "cumulative pore volume" curves often calculated in isotherm-based PSD methods. The derivative $-\mathrm{d}V_{\mathrm{pore}}(r)/\mathrm{d}r$ is the fraction of volume coverable by spheres of radius r but not by spheres of radius $r + \mathrm{d}r$ and is a direct definition of the PSD. The $V_{\mathrm{pore}}(r)$ function can be calculated by an MC volume integration. Alternatively, the PSD can be computed using a grid. The algorithm of Ban and Vlugt works as follows [Ban and Vlugt (2009)]:

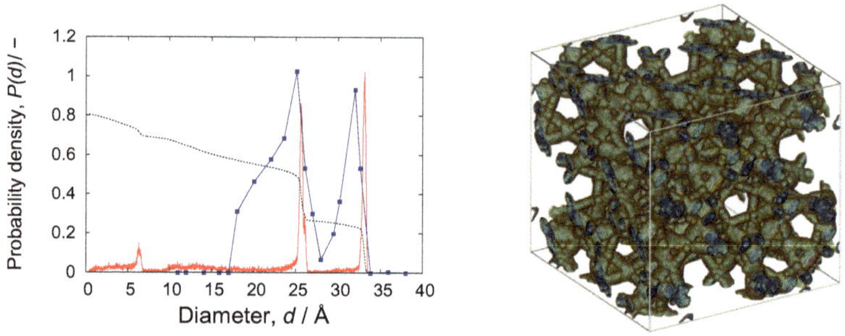

Figure 2.6 PSD of MIL-101. Blue line: experimental data [Anbia and Hoseini (2012)]; red line: computational PSD [Dubbeldam et al. (2012)]; green line: cumulative pore volume curve. The picture shows the MIL-101 unit cell (89 Å×89 Å×89 Å) containing 14,416 atoms.

(1) A 3D grid with a small spacing is constructed. Typically a grid size of 0.1 Å (in each direction) is used.
(2) A spherical test particle is positioned at a random position in the framework. The radius of this particle is chosen as the minimum distance between the center of the particle and any of the zeolite framework atoms, minus the radius of the closest framework atom.
(3) The diameter of the test particle is recorded for all grid points that are inside the test particle.
(4) This procedure is repeated many times.

PSD of Matériaux de l'Institut Lavoisier (MIL)-101 structure has one of the largest MOF unit cells, and solving the crystal structure was quite an achievement [Férey et al. (2005)]. Figure 2.6 compares the experimental PSD and the computational one. Qualitatively the structural features match well, and both show peaks around 23–25 Å and 32–33 Å. The experimental PSD in general shows wider peaks due to (i) the simulations using a rigid structure fixed at the crystallographic positions and (ii) the model assumptions to back out the experimental PSD. The computational PSD is the derivative of the cumulative pore volume curve. When properly normalized, the $d \rightarrow 0$ value is another way of obtaining the void fraction. To compare: the Widom method gives a helium void fraction of 0.809.

2.3 Single-Component Adsorption

2.3.1 *Adsorption*

There are two principal methods to measure the adsorption isotherm, volumetric and gravimetric [Keller and Staudt (2005); Condon (2006)]. In both methods the adsorbent is held at a constant temperature and the adsorptive pressure is increased step-wise. After increasing the pressure and holding it fixed again, the progress of the adsorption is monitored to determine when equilibrium is achieved. The re-equilibration time depends on the system and ranges from minutes to hours (and more than a week for dodecane in MFI). Volumetry is the oldest method to measure adsorption of gases in solids. The physics of volumetric gas adsorption experiments is simple: a given amount of sorptive gas is expanded into a vessel, which includes a sorbent sample and which initially has been evacuated. Upon expansion the sorptive gas is partly adsorbed on the (external and internal) surface of the sorbent material, partly remaining in the gas phase around the sorbent. By a mass balance, the amount of gas being adsorbed can be calculated if the void volume of the sorbent, that is, the volume that cannot be penetrated by the sorptive gas molecules is known—at least approximately. The principle of the gravimetric method is simpler than that of the volumetric method. For the gravimetric method, one simply brings in a pressure of the adsorbate and measures the mass gain of the sample. The isotherm is then simply mass gain versus pressure. For careful work, the gravimetric method also needs to be precalibrated. This calibration is for some small corrections: (a) the buoyancy of the sample and the balance equipment and (b) a molecular flow correction.

In adsorption simulation, the natural ensemble to use is the grand canonical ensemble (or μ, V, T ensemble). In this ensemble, the temperature T, the volume V, and the chemical potential μ are fixed. The equilibrium conditions are that the temperature and chemical potential of the gas inside and outside the adsorbent must be equal. At these conditions the average amount of molecules is computed and usually reported in units of, for example, molecules per unit cell, mole of adsorbate per kilogram of framework material, or milligram per gram.

2.3.2 Isotherm Shapes

The Langmuir-type isotherm remains the most widely used for practical applications [Yang (2003)]. Langmuir assumes that each adsorption site is independent and each site can be occupied by only a single molecule. The Langmuir isotherm for pure component adsorption is

$$q = q_{\mathrm{m}} \frac{BP}{1 + BP} \qquad \text{with} \quad B = \frac{\alpha}{\beta\sqrt{2\pi m k_{\mathrm{B}} T}} e^{-Q/RT} \tag{2.7}$$

where α is the sticking probability or accommodation coefficient for adsorption (upon a collision on the surface) and β is the rate constant for desorption (also called the Langmuir constant). At low pressures, the Langmuir isotherm reduces to a linear form, or Henry's law form:

$$q = K_{\mathrm{H}} P \qquad \text{and} \quad K_{\mathrm{H}} = B q_{\mathrm{m}} \tag{2.8}$$

where K_{H} is Henry's constant. Zeldowitsch assumed an exponentially decaying function of the site density to obtain the empirical Freundlich isotherm [Zeldowitsch (1934)]:

$$q = K_{\mathrm{H}} P^{1/n} \tag{2.9}$$

To avoid the indefinite increase in adsorption with pressure, the Langmuir–Freundlich isotherm was proposed [Yang (1987)]:

$$q = q_{\mathrm{m}} \frac{BP^{1/n}}{1 + BP^{1/n}} \tag{2.10}$$

The Langmuir and Langmuir–Freundlich isotherms are readily extended to n-component mixtures [Yang (2003)].

Even though the Langmuir isotherm is widely used, it is rare to find a Langmuir isotherm in microporous materials. Confinement results in a complex energy landscape with usually multiple adsorption sites that differ in energy. Linear, monobranched, and dibranched molecules in MFI are a prototypical system and studied in high detail [Vlugt et al. (1998, 1999); Vlugt (2000); Calero et al. (2001)]. MFI is a 3D zeolite system that consists of intersection channels; two linear channels per unit cell, two zigzag channels per unit cell, and four intersections per unit cell. Figure 2.7 shows the adsorption isotherms of 2,2-dimethylbutane (2,2-DMB) and 2,3-DMB. The molecules are bulky, too bulky for the channels, and they

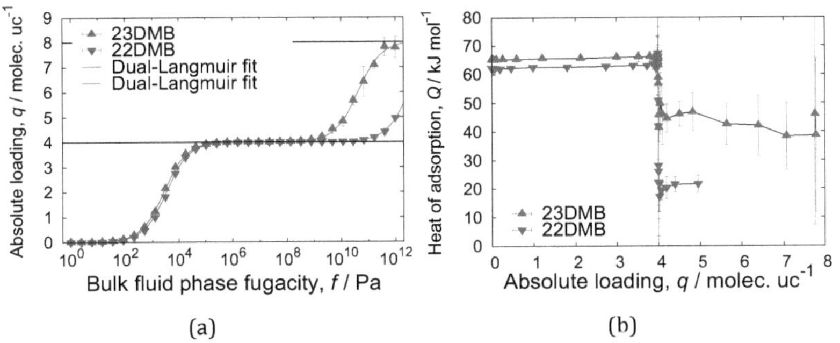

Figure 2.7 Adsorption isotherms and heats of adsorption of 2,2-dimethylbutane (2,2-DMB) and 2,3-dimethylbutane (2,3-DMB) in MFI at 433 K.

sit more snugly in the intersections. The molecules do not really interact with each other as they are spread over the intersections (the intersections are separated by the zigzag and linear channels). Up to four molecules per unit cell, the isotherm can be perfectly described by a Langmuir model [Krishna et al. (2002)]. However, it is possible for dibranched molecules to occupy a channel (after all, the molecules have to diffuse through the channels to reach an intersection), but it is energetically not favored until the pressure overcomes this energy penalty. At very high pressures, the molecules are pushed in the channels. There are another four sites in the linear and zigzag channels available, and the full isotherms can be accurately described by a dual Langmuir model with a maximum loading of eight molecules per unit cell (Fig. 2.7).

We can plot the heat of adsorption as a function of loading for the dibranched molecules (Fig. 2.7). Up to four molecules per unit cell, the heat of adsorption is found to be constant. This is not surprising: there is hardly any molecule–molecule interactions, and the heat represents the interaction of the molecule with the framework intersection atoms. After four molecules per unit a jump in the heat of adsorption is noticed: about 25 kJ/mol decrease for 2,3-DMB and 40 kJ/mol for 2,2-DMB. This energy difference is the energy penalty that needs to be overcome by the pressure increase. Clearly, the more bulky 2,2-DMB has a higher energy penalty to move in the smaller channels. Therefore 2,3-DMB fills up before 2,2-

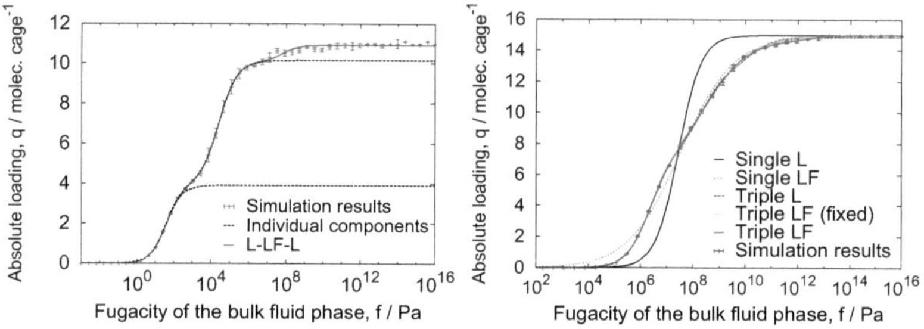

(a) Isobutane in MFI (b) Methane in LTA

Figure 2.8 Adsorption of isobutane in MFI zeolites and methane in ITQ-29 (siliceous LTA) at 298 K.

DMB, but both eventually will reach eight molecules per unit cell. Note that this (theoretical) analysis is possible because in molecular simulation we are not limited to a (practical) pressure range. In experiments it is difficult to reach pressures above 100 bar. The second part of the isotherms, the filling of the channels by the dibranched molecules, would be difficult to study. Moreover, we can find out where the molecules are (intersection or channel) and determine the energetics, location, and nature of the adsorption sites. Molecular simulation is therefore a great tool to obtain additional structure–property relations.

What happens when molecules interact and/or when adsorption sites are closer together? It turns out that this is by far the most common zeolite/MOF system. We show two examples in Fig. 2.8, isobutane in MFI and methane in LTA. Isobutane is also a branched molecule but smaller than 2,2-DMB and 2,3-DMB. Up to four molecules per unit cell it has identical behavior to the more bulky dibranched molecules; also isobutane initially is only found in the intersections. However, when the pressure is increased they have to occupy the linear and zigzag channels and another six molecules can be pushed in. At 1e8 Pascal, another molecule can be squeezed in. In fact, around 1e20 Pa or higher another molecule could be pushed in. This is partly due to modeling of atoms as "soft spheres" (as opposed to "hard spheres" that cannot be compressed). The isobutanes in the channels are interacting, and the system cannot be properly described with a triple Langmuir model, but one has to use a Langmuir–Freundlich term combined with a dual Langmuir model. The methane in the LTA system is more complicated; here the molecules are confined to a cage with three types of adsorption sites. The cages are arranged in a cubic lattice (each cage has six windows connecting them to the next cage). There are two sites found at infinite dilution, six sites in front of the window and eight sites in corners between the windows. Once these sites are filled up, at high loading, another site in the center of the cage can be filled (a packing effect). The cage is about 12 Å in diameter. This means that a molecule feels several adsorption sites and other molecules. The single- and multiple-Langmuir models are insufficient, and we need to resort to a triple Langmuir–Freundlich model to describe the system. The triple Langmuir–Freundlich model with all parameters free performs slightly better than using known information on the number and type of adsorption sites. This indicates that the correlations are difficult to capture theoretically and adsorption is definitely not simply an addition of all the individual adsorption sites. If the adsorption sites have a large difference in energy then first the most favorable sites are filled up before the next site type is filled. If the site types are close in adsorption energy than the site types are filled concurrently. The first filling type is captured by Langmuir; the latter is better described by Langmuir–Freundlich. Langmuir–Freundlich extends Langmuir to nonuniform surfaces [Yang (2003)]. The sequential filling is also enhanced at low temperatures and the concurrent filling at high temperatures. At sufficiently high temperature the concept of an adsorption site breaks down and the molecular density is very spread out. Figure 2.9 visualizes the density of a single methane molecule at 10 K, 77 K, and 298 K. At 10 K and 77 K the energy contour plots are plotted to show that the density at 10 K corresponds to the local potential energy minima. At 77 K also the corner sites come into play, which are higher in energy. At room temperature the density is spread out. It is instructive to compare the density at room temperature for a single methane molecule to the density for eight molecules per unit cell (see Fig. 2.2c). The filling depends on the potential energy landscape of the framework, on temperature, and on loading. The siting and filling of argon and nitrogen in IRMOF-1 have been studied in high detail in Ref. [Dubbeldam et al. (2007a)].

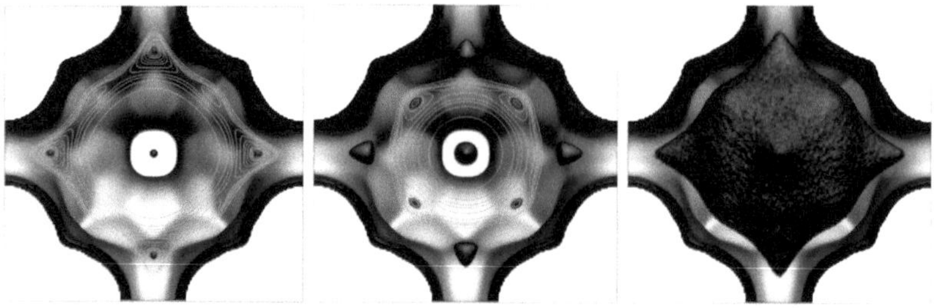

(a) 10K (b) 77K (c) 298K

Figure 2.9 Density plots of methane at infinite dilution at 10 K, 77 K, and 298 K, respectively. (a) Contour plot on a plane at fractional position $z = 1/2$. (b) Contour plot on a plane at fractional position $z = 1/3$.

2.3.3 *Maximum Loading*

In many theoretical descriptions fractional loading $\theta = q/q_{\rm m}$ is used (e.g., [Reed and Ehrlich (1981)]). This assumes the existence of "maximum loading." Except for lattice models, however, this concept is not easy to define. In simulation, but also in experiment, the absolute loading tends to increase with higher pressures. Therefore, from a practical point of view one can think of maximum loading as the loading at the highest pressure reachable by the experimental apparatus. An alternative could be to use an equation of state and compute the maximum loading as the liquid density available in $V_{\rm pore}$, but also this maximum is pressure and temperature dependent. A better-defined maximum loading can be obtained by examining the diffusion coefficients. At maximum loading there is a sudden order-in-magnitude decrease in self-and collective diffusivities when the diffusion becomes solid-like and effectively grinds to a halt. We note that this maximum loading is often higher than the maximum loadings as taken from isotherms at realistic pressures.

Still, sometimes one can extrapolate the isotherm to a plateau value. For the (small) cage-type confinement, the plateau value is an integer number of molecules per cage. For other structures with more open space the plateau value is a noninteger and temperature dependent because the fluid density is temperature dependent. This value can be computed accurately because $V_{\rm pore}$ is easy to compute and for many adsorbates the equation of state is known. The plateau value of an isotherm is therefore relatively easy to predict using molecular simulation. The simulations, however, assume a perfect crystal. It is for this reason that the isotherm from simulation can serve as a "target value" for experiments. It allows researchers to assess the quality of the synthesized crystals. There are many reasons why one expects the simulation values to be higher than in a real sample: (a) (partial) pore collapse, (b) leftover solvents from synthesis, (c) water adsorption, (d) imperfections and/or stacking faults during crystal growth, etc. But also the opposite can occur. If the experimental values are higher than the simulated isotherm, then (1) some key physics is missing in the model (e.g., polarization) or (2) the experiments also measure adsorption on the external surface of the crystal.

Some researchers have used the ratio of the experimental V_{pore} and the simulation value as a scaling value for the isotherm [Surble et al. (2006); Dubbeldam et al. (2007b)]. Usually good agreement is then observed. However, one can doubt whether positional information on the molecules is reliable, because the simulated and experimental systems differ and have different occupancy and energetics of the adsorption sites. The adsorption sites are shared with, for example, solvent molecules, and/or some adsorption sites are not accessible. Some simulations have included solvent molecules (DMF and water) in the model [Bae et al. (2009)] and observed a cancellation effect: the solvent molecules increase attraction, and the adsorbate loading is higher, but at the same time the solvent increases the crystal weight. Gravimetrically a cancellation effect is seen, but volumetrically the loading increases when solvents were included.

2.3.4 *Clustering of Molecules*

In Fig. 2.10 we show the behavior of CO_2 in IRMOF-1 as a function of density. With decreasing temperature, these isotherms exhibit dramatic steps with some similarities to those reported for CO_2 in other MOFs. Walton et al. presented a molecular model that predicts the inflections and steps in very good agreement with experiment and helps explain the adsorption mechanisms that are responsible

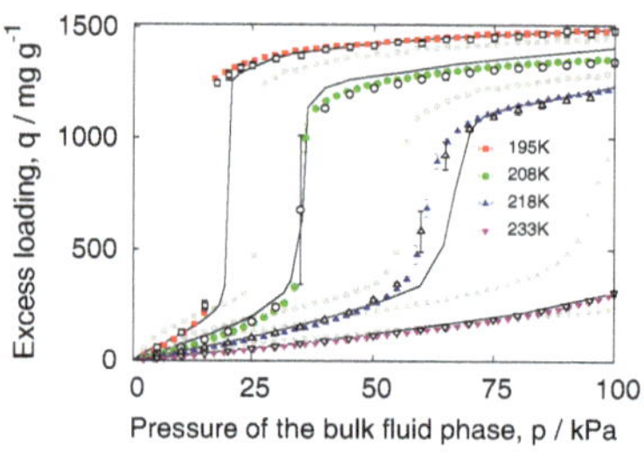

(a) Lines: experimental data, filled colored point: rigid-framework model [Dubbeldam et al. (2007c); Walton et al. (2008)] open symbols: flexible-framework model [Dubbeldam et al. (2007c)], gray-filled symbols: Dzubak-model [Dzubak et al. (2012)].

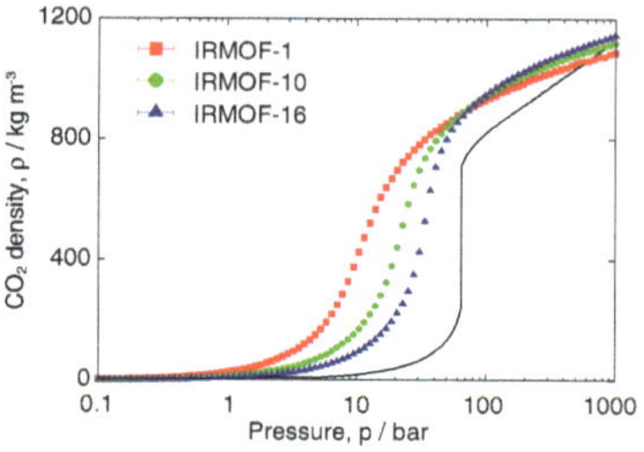

(b) Lines: NIST experimental database, filled symbols: simulations rigid-framework model [Dubbeldam et al. (2007c); Walton et al. (2008)].

Figure 2.10 CO_2 adsorption in IRMOF-1. (a) Step-like isotherms at low temperature and (b) CO_2 bulk behavior compared to increased confinement.

for this behavior [Walton et al. (2008)]. Analysis of the CO_2 positions reveals that, below the sharp rise in the isotherms of Fig. 2.10, molecules are mostly adsorbed near the corners of the MOF cavities. With increasing pressure, the pores fill, leading to a type V isotherm. Type V isotherms are relatively rare and are often accompanied by hysteresis. However, neither the experiments nor the simulations display hysteresis. To shed more light on the pore filling, Walton et al. plotted the density of CO_2 within the pore volume of IRMOF-1, IRMOF-10, and IRMOF-16 at 298 K. IRMOF-10 and IRMOF-16 possess approximately double and triple the pore volume per gram of IRMOF-1. The bulk density of CO_2 is also shown in the figure. From these results, we find that the pressure of the pore filling shifts toward the bulk condensation pressure with increasing pore size (IRMOF-1, IRMOF-10, and IRMOF-16).

From the Walton paper the following molecular picture emerges. Taking the CO_2 bulk-fluid behavior as a reference we see that relatively inert confinement (IRMOF-16) shifts the fluid density–pressure curve to lower pressures. For increasing confinement (IRMOF-10 and IRMOF-1) the curve shifts further to lower pressures and becomes more convoluted. This means that the original fluid behavior is retained, at least at low temperature. At low pressures, the molecules adsorb independently and they occupy the strongest adsorption sites. Usually these sites have stronger interactions than CO_2–CO_2 interactions. Once these sites are filled, CO_2 starts to "cluster" together and a rapid condensation-like filling occurs.

How and when the adsorbates cluster together are largely dependent on pressure and temperature. For a fluid, below the critical temperature T_c molecular clustering occurs, while above this threshold temperature clustering is absent. In the presence of a weakly interacting adsorbent, this behavior is only mildly modified and the critical temperature is still a good guideline. It is often stated that there is no vapor/liquid transition in microporous materials and that we can only speak of an adsorbed phase. This is not exactly true at the molecular level, because molecules can either move by themselves (similar to a gas phase) or move in a cluster with other molecules (similar to a fluid phase).

Clustering has a large influence on the adsorption and diffusion behavior of molecules [Krishna and van Baten (2010c,d, 2011c)]. Krishna and van Baten showed that stepped isotherm behavior is a characteristic for a guest molecule in many microporous hosts at $T < T_c$. The steepness of the isotherm is higher for open structures such as IRMOF-1 and COF-108 than when the molecules are strongly confined, for example, within the channels of MFI. The extent of clustering increases as T falls increasingly below T_c. The degree of clustering is loading dependent and approximately follows the trend of $1/\Gamma_i$ vs c_i, where $\Gamma_i \equiv \frac{\partial \ln f_i}{\partial \ln c_i} = \frac{f_i}{c_i}\frac{\partial f_i}{\partial c_i}$ is the thermodynamic factor and c_i is the loading or concentration within the pores. For those parts of the adsorption where $1/\Gamma_i > 1$ the slope of the isotherm $\frac{\partial c_i}{\partial f_i}$ exceeds the slope $\frac{f_i}{c_i}$ (the line through the origin to relevant point on the isotherm $\frac{f_i}{c_i}$). Under this condition, the self-diffusivity can sometimes exceed the transport diffusivity. The scenario $1/\Gamma_i > 1$ within micropores is a reflection of metastable equilibrium and is characterized by clustering of molecules. The greater the extent of clustering of molecules, the larger the value by which $1/\Gamma_i$ exceeds unity. Generally speaking, $1/\Gamma_i > 1$ occurs when $T < T_c$ and the pore concentrations are such that they lie within the binodal region of the bulk-fluid-phase equilibrium. Furthermore, the loading c_i should be lower than that at the liquid spinodal at that temperature. For a chosen guest at a specified T, $1/\Gamma_i$ tends to be higher for more open structures, such as IRMOF-1, and COFs with larger pore sizes. Conversely, for narrower pores the influence of the pore walls is higher, and this leads to a lower $1/\Gamma_i$ when compared at the same loading c_i. For strongly confined molecules, $1/\Gamma_i$ may never exceed unity, even at $T < T_c$. The clustering mechanism provides a rationale for earlier experimental studies for diffusion of alkanes in CuBTC, in which the Fick D_i was found to decrease with concentration in regions within which $1/\Gamma_i > 1.16$ [Chmelik et al. (2009)]. The temperatures in these experiments were also below the corresponding T_c of the alkanes. For adsorption of n-alkanes at any given temperature, clustering phenomena are of increasing importance with increasing chain length. The explanation for the type IV loading dependence of self-diffusivity of a variety of guest molecules in micro- and mesoporous materials, with pore sizes larger than 0.74 nm, can be traced to molecular clustering. Clustering of molecules is also to be expected when there is a strong degree of hydrogen bonding.

2.3.5 *Excess vs. Absolute Adsorption*

Both volumetric and gravimetric methods do not determine the actual amount of gas adsorbed but the so-called excess adsorption instead. The excess adsorption corresponds to the amount of material that occupies a certain volume element of the adsorption layer in excess compared to the amount of material that would be found in the same volume element if there was only nonadsorbed gas at the same pressure and temperature present. The gravimetric method fails because of the buoyancy of the adsorption layer in the surrounding gas, and the volumetric method fails because of the necessity to subtract the amount that was not adsorbed but left in the gas phase from the amount of gas that was added to the sample container [Mertens (2009)].

Statistical thermodynamic theories and molecular simulations of adsorption of gases on porous solids are formulated in the language of absolute thermodynamic variables; on the other hand, experimental measurements are reported as excess thermodynamic variables [Myers and Monson (2002)]. In Fig. 2.11 the computed

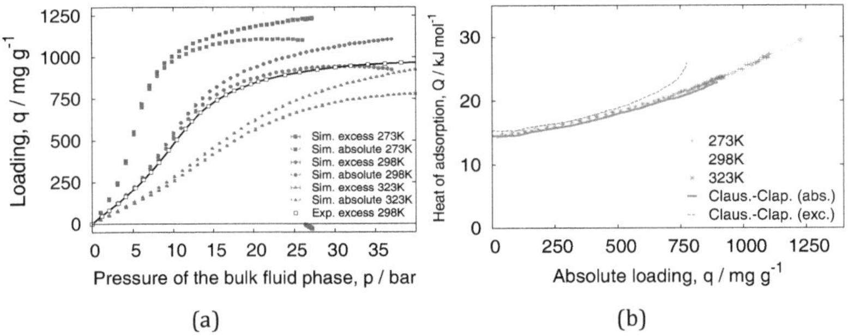

Figure 2.11 Absolute vs. excess adsorption. (a) Excess and absolute CO_2 isotherms in IRMOF-1 at 273 K, 298 K, and 323 K. (b) Comparison of heats of adsorption using direct computation in molecular simulation vs. the Clausius–Clapeyron equation using the absolute (correct way) and excess isotherms (wrong way).

isotherms for CO_2 at 273 K, 298 K, and 323 K are shown, both in absolute and in excess loadings. At low pressure (subatmospheric), the difference between absolute and excess adsorption is negligible. However, for higher pressures the difference will be more substantial. The loadings are plotted versus pressure, while in simulation the fugacity is used. The conversion and computation of the excess part are done using the Peng–Robinson equation of state. The pressure of 27 bar at 273 K corresponds to a fugacity of 50 bar (fugacity coefficient is 0.54). The absolute loadings are not affected by any gas–liquid transition, but the excess loadings are. At the vapor pressure the simulated excess loadings turn negative. The excess correction corresponds now to a liquid inside V_{pore}, which is much larger than the gas density.

The conversion from absolute to excess adsorption is

$$n_{\text{excess}} = n_{\text{absolute}} - \frac{pV_{\text{pore}}}{zRT} = n_{\text{absolute}} - \rho_{\text{pore}} V_{\text{pore}} \qquad (2.11)$$

where p is the pressure of the bulk fluid phase, R the gas constant, T the temperature, z the compressibility in the bulk fluid phase, and V_{pore} the pore volume. For consistency with experiment, the accessible pore volume $V_{\text{pore}} = \xi V$ has be determined once using a simulation of a single helium molecule at the reference conditions. Usually a reference temperature of 25°C (298 K) is chosen in experiment for the experimental determination of the helium void volume. Note that adsorption is always relative to this reference state, even though the real pore volume changes as a function of temperature. Both gravimetric and volumetric methods use helium as a reference gas to determine the buoyancy of the sample (gravimetric) or the void volume of the sample cell (volumetric). Helium is commonly considered as "nonadsorbing". Helium adsorption is in the μmol g^{-1} range as compared to the CO_2 sorption capacity, which is in the mmol g^{-1} range [Sakurovs et al. (2009)] and hence can be neglected. By definition and experimental practice, the volumetric as well as the gravimetric sorption experiments represent a differential measuring method with helium as a standard reference gas.

Isotherm fitting with fundamental models, such as Langmuir and Toth, should be done using absolute, not excess, loadings; this has been stressed by Myers and Monson [Myers and Monson (2002)] and Krishna [Krishna (2012)]. It is interesting to compare properties obtained from absolute isotherms to properties computed from uncorrected excess isotherms. We will do so for the heat of adsorption. Figure 2.11 shows the computed heats of adsorption at 273 K, 298 K, and 323 K using the fluctuation formula [Karavias and Myers (1991); Vlugt et al. (2008)]. The isotherms are fitted with cubic splines to numerically compute the heat using the Clausius–Clapeyron equation $Q = -R\left(\frac{\partial \ln p}{\partial 1/T}\right)_q$. The heats from the absolute isotherm match very well (the heats computed at different temperatures are almost identical), but at high loading the heats from excess isotherms show significant artifacts [Krishna (2012)]. The poorer the adsorption strength, the larger the difference between the absolute and excess loadings; the differences are particularly significant for poorly adsorbing gases such as H_2 and N_2 [Krishna (2012)].

Mertens et al. proposed to deconvolute the excess and absolute adsorption [Mertens (2009)]. Since in Eq. 2.11 the first term on the right is strictly monotonically increasing and the second one strictly monotonically decreasing, one can attempt to separate the two contribution by fitting the first term with strictly monotonically increasing fit functions as a function of pressure. The second term is always monotonically decreasing, but V_{pore} needs to be included as a fit parameter [Mertens (2009)]. Others have carried out a similar procedure in the case of gravimetric measurements using the so-called three-parameter isotherm. In contrast to the original meaning of the Langmuir function, its parameters will most likely not have a direct physical meaning.

2.3.6 *Taking Flexibility into Account or Not?*

Most work to date on zeolites and MOFs has been using the rigid-framework assumption (the frameworks are kept fixed). This allows for the use of grid interpolation techniques to compute host–adsorbate interactions very efficiently. To study the properties of the framework itself a flexible model is required of course, but the literature is divided on the issue of whether it is essential to include flexibility for adsorption and diffusion studies. A comparison between a flexible and a rigid model is not straightforward. For example, if one takes a flexible model for a structure and compares it to the rigid results, then differences are bound to show up. Importantly, this is due the fact that the flexible structure is *on average* not the same as the rigid one. Vlugt and Schenk avoided this pitfall explicitly; they modified the Demontis model for zeolites to take the bond lengths from the crystal data as input values for the bond reference values [Vlugt and Schenk (2002)]. That is, in the model the reference distances r_0 for O–O and Si–O are no longer constants and vary for different bonds of the same type. The minimum-energy structure (which is the crystal structure) is reproduced exactly when $T \rightarrow 0$ or $k/k_B \rightarrow \infty$, and the harmonic potentials describe the fluctuations around this equilibrium structure. At low loadings, the influence of the framework flexibility on the heat of adsorption and Henry coefficient are found to be rather small. For molecules such as isobutane and heptane with inflection behavior, the influence on the isotherms seems to be somewhat larger. Due to simplicity of the model, and the large range of k/k_B studied, the systems can be considered in some sense as worst-case scenarios.

Fritzsche et al. found little difference for diffusivities of methane in LTA [Fritzsche et al. (1998)]. The atomic positions for the runs with a rigid lattice were obtained by energy minimization. They attributed the influence of the lattice vibrations on diffusion found in Refs. [Demontis and Suffritti (1994a,b)] to the slight difference in the lattices used for runs with rigid and vibrating lattices, respectively. Garcia-Sanchez et al. studied the influence of different models ([Demontis et al. (1987, 1988)], [Nicholas et al. (1991)], [Hill and Sauer (1994, 1995)]) for flexibility in LTA with and without cations. The diffusion results depend very much on the model used [Garcia-Sanchez et al. (2007)]. The reason is that none of the current force fields are able to capture the average structure exactly, specifically the eight-ring window separating the cages. A small deviation of the window from the crystal structure leads to very different diffusivities. A "perfect" flexible model should, on average, have the same structure as the crystal structure from experiment (at the same temperature). Only then can one assess the true influence of flexibility on diffusion [Garcia-Sanchez et al. (2007)]. Krishna and van Baten even draw the unequivocal conclusion that lattice flexibility, per se, has no influence on the self-diffusivity of CH_4 in cage-type zeolites with eight-ring windows [Krishna and van Baten (2010b)]. They performed MD simulations with a rigid framework using the time-averaged values of the framework atoms positions obtained from the Hill–Sauer force field implementation. These rigid-framework simulations yield precisely the same diffusivity values as those obtained by Garcia-Sanchez et al. with a fully flexible framework. These results support the contention of Kopelevich and Chang that lattice vibrations in zeolites cannot drive diffusion [Kopelevich and Chang (2001)].

On the other hand, Leroy et al. use the Demontis model for silicalite and found that zeolite flexibility affects differently adsorbate transport properties according to alkane length and loading Self-diffusion was enhanced using a flexible silicalite for the lowest loadings and the shortest alkanes, namely, methane and *n*-butane, while no effect was observed for *n*-hexane and *n*-octane diffusion. The qualitative dependence of the self-diffusion with alkane length remains unchanged compared to rigid-framework simulations. Interestingly, the modified Demontis model [Vlugt and Schenk (2002)] was examined for methane and hexane, and for these little difference between flexible and rigid was observed. There is a clear-cut case where differences in diffusivities are to be expected, namely, when the molecules have the same dimensions as the channels. Examples are double-branched alkanes, cyclohexane, and benzene in silicalite. Indeed it has been found that the diffusivity of aromatics in silicalite changes by an order of magnitude if framework flexibility is taken into account [Forester and Smith (1997)].

So what about MOFs? Structural flexibility is a well-known property of MOFs [Fletcher et al. (2005)]. For example, MILs exhibit breathing [Serre et al. (2002)] and IRMOFs exhibit negative thermal expansion [Dubbeldam et al. (2007c)]. Also for MOFs little influence of flexibility on adsorption and diffusion has been found [Ford et al. (2009)]. The diffusion results for alkane as a function of chain length and of benzene were found to be in excellent agreement with nuclear magnetic resonance (NMR) experiments [Ford et al. (2012)]. The IRMOF-1 model of Dubbeldam et al. was used [Dubbeldam et al. (2007c)]. Figure 2.10 shows CO_2 isotherms in IRMOF-1 at four temperatures. The closed symbols are for a rigid lattice, and the open symbols are for a flexible simulation in the $\mu_1 N_2 pT$-ensemble [Wolf et al. (1993)]. Almost no influence can be detected, and even though IRMOF-1 can be visibly seen (in snapshots) to be moving quite a bit, the average effect is almost nearly identical to the rigid framework. Framework flexibility will be more important for structures that can change conformation under thermal conditions, such as IRMOF-7 [Amirjalayer and Schmid (2008)], breathing structures such as MILs, and structures with small pores or tight channels. Seehamart et al. [Seehamart et al. (2010)] found that framework flexibility has qualitative as well as quantitative effects on the diffusivity of ethane in the MOF Zn(tbip), which has relatively tight channels. Generally, little effect of framework flexibility has been found for zeolites and MOFs (see Ref. [Demontis and Suffritti (2009); Ford et al. (2009)] and references therein) unless the guest size approaches the framework pore size. For MOFs, flexibility is crucial when the framework undergoes large structural changes (e.g., breathing).

2.3.7 *Polarization*

Yazydin et al. [Yazaydin et al. (2009)] , Wilmer at al. [Wilmer and Snurr (2011)], and Krishna and Van Baten [Krishna and van Baten

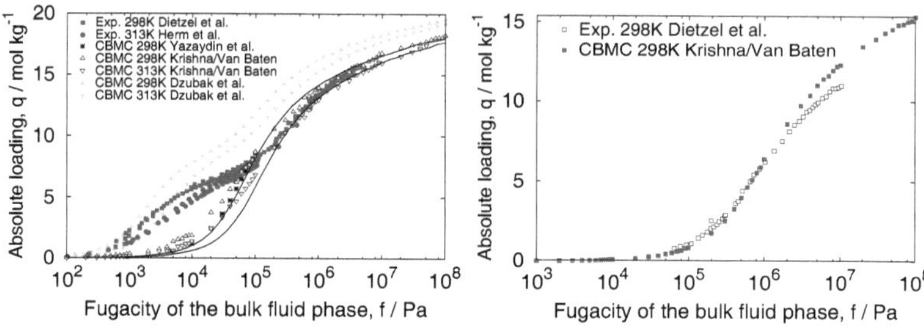

(a) CO_2 in Mg-MOF-74 (b) Methane in Mg-MOF-74

Figure 2.12 Failure of modeling. Simulations of CO_2 in Mg-MOF-74 compared to experiments.

(2011b)] pointed out the failure of configurational bias Monte Carlo (CBMC) modeling in the CO_2-Mg-MOF-74 system. Yazaydin et al. attributed this to orbital interactions between the CO_2 molecules and the open metal sites, which were not accounted for in their method. Figure 2.12, replotted from Ref. [Krishna and van Baten (2011b)], shows the comparison of CO_2 simulation and experiment. The experimental data is from two sources, Dietzel et al. [Dietzel et al. (2009)] and Herm et al. [Herm et al. (2011)]. Considering the small temperature difference, the experimental data agrees very well. However, the CBMC simulations are unable to reproduce the low-loading part of the CO_2 isotherm. Interestingly, for methane in Mg-MOF-74 the simulations and experiments agree well.

The high-loading parts of the isotherms match well, both for CO_2 and for methane. Therefore it is unlikely an experimental issue, but rather the simulations miss some key physics. It appears that a Lennard–Jones+Coulomb force field is unable to reproduce the experimental isotherm at low loading. The interactions of CO_2, water, CO, and alkenes with the metals of the MOF are complicated in nature, and future simulation methodology could focus on this system.

An initial attempt was recently published [Dzubak et al. (2012)]. The Buckingham dispersion functional form was chosen and fitted to representative Mg-MOF-74 clusters using the MP_2 level of theory. An $1/r^5$ term was added to obtain better agreement with the MP_2 potential energy surface. The Mg–CO_2 interaction was effectively increased, which increases the adsorption at low loading. In the same paper also an MP_2-calibrated model for the Zn-MOF-74 and IRMOF-1 structures was constructed. The MP_2-calibrated model does not reproduce the experimental data well (see Fig. 2.10, gray-filled symbols). The location of the inflections are missed, the CO_2 loading at low pressures and temperature is overpredicted, and the adsorption is underpredicted at higher pressures.

Unfortunately, the MP_2-calibrated model does not seem to solve the problem. By increasing the attraction around the open metal site not only the attraction with the first CO_2 molecule but also with all others CO_2 inside the cutoff range is increased. The sharpness of the experimental inflection suggests the effect is local near the open metal site. This suggests a polarization-type interaction, which can act locally at places where the electric field gradient is high (i.e., near the open metal site).

Wilmer et al. noted that metal atoms tend to be strongly charged, and so it may be that CO_2 molecules become polarized near metal atoms, which would result in a stronger interaction than could be accounted for in a fixed-charge model [Wilmer and Snurr (2011)]. This stronger interaction may account for the discrepancy between experimental and simulated adsorption measurements. Most likely, the Mg-MOF-74 CO_2 simulations could be improved by incorporating explicit polarization. This approach has previously successfully been applied to adsorption in cation-loaded zeolites [Lachet et al. (1997, 1998, 1999)]. In addition to conventional methods to include polarization, the charge equilibration method could also be used since it allows a dynamically re-equilibrate of charges in an MC simulation. Charge fluctuation models allow for charge transfer.

2.4 Mixture Adsorption Isotherms

2.4.1 *Experimental Measurements of Mixture Adsorption*

A volumetric adsorption apparatus is typically used for measuring multicomponent adsorption equilibrium data. The key to such a system lies in proper sampling and analysis of the gas-phase composition. In performing experiments with such an instrument, a known amount of adsorbate is injected from a sample cylinder into a closed loop and circulated through an adsorbent bed until equilibrium has been reached. An automatic sampling valve is used to sample the gas phase at prescribed intervals and inject the sample into a gas chromatograph (GC) for analysis. Equilibrium has been attained when there is no relative change in the gas-phase composition from one GC analysis to the next. At this point, the number of moles of adsorbate in the gas phase is calculated. The difference between the initial amount injected into the system and the amount left at equilibrium gives the amount adsorbed.

For pure-component adsorption measurements, repeating this procedure for different pressures at constant temperature allows for construction of the adsorption isotherm. Multicomponent measurements are not as straightforward, particularly if competition between the components for adsorption sites is great. Each measurement will yield a different set of adsorbed-phase and gas-phase compositions, making it impossible to obtain experimental data in a systematic fashion. For example, in previous work, Walton and LeVan [Walton and LeVan (2006); Walton et al. (2006)] measured multicomponent adsorption of CO and CO_2 in NaY zeolite. CO_2 was preferentially adsorbed for this system and was found to displace the adsorbed CO each time it was injected into the system. Thus, exploring the adsorption space of CO in the presence of CO_2 required carefully calculated experimental procedures. This was achieved by preloading the adsorbent with small loadings of CO_2 and then slowly increasing the amount of CO in the system. In general, multicomponent adsorption measurements are extremely difficult and technically demanding and provide a perfect example of where simulations and adsorption modeling can complement experiments. The most famous model for calculating mixture adsorption from pure-component adsorption isotherms is ideal adsorption solution theory (IAST).

2.4.2 *Ideal Adsorption Solution Theory*

It is possible to estimate the mixture loading from the pure components using the IAST of Myers and Prausnitz [Myers and Prausnitz (1965)]. Briefly, the basic equation of the IAST is the analogue of Raoult's law for vapor–liquid equilibrium, that is,

$$P_{y_i} = P_i^0(\pi)\,x_i, \qquad i = 1, 2, \ldots, n \tag{2.12}$$

where n is the number of components and x_i is the mole fraction in the adsorbed phase:

$$x_i = \frac{q_i}{q_1 + q_2 + \cdots + q_n} \tag{2.13}$$

where q_i is the loading expressed in mol kg^{-1} and $p_i^0\,(i\pi)$ is the pressure for the sorption of every pure component i, which yields the same spreading pressure π as for the mixture. The spreading pressure is defined by the Gibbs adsorption isotherm:

$$\frac{\pi A}{k_B T} = \rho \int_{p=0}^{p=p_i^0} \frac{q_i^0\,(p)}{p}\,\mathrm{d}p \tag{2.14}$$

where A is the surface area per m^3 of adsorbent, ρ the density of the crystallite expressed in kg m^{-3}, and $q_i^0\,(P)$ the pure component isotherm (emphasized by the superscript 0). The total amount adsorbed is obtained from

$$\begin{aligned} q_{\text{mix}} &\equiv q_1 + q_2 + \cdots + q_n \\ &= \frac{1}{\frac{x_1}{q_1^0(p_1^0)} + \frac{x_2}{q_2^0(p_2^0)} + \cdots + \frac{x_n}{q_n^0(p_n^0)}} \end{aligned} \tag{2.15}$$

The set of Eqs. 2.12, 2.13, 2.14, and 2.15 needs to be solved numerically to obtain the loadings of the individual components in the mixture.

Figure 2.13a shows the single-component isotherms of hexane isomers at 433 K in UiO-66. We will denote the order linear $\gg$ monobranched $\gg$ dibranched as "normal hierarchy," and the order dibranched $\gg$ monobranched $\gg$ linear as "reverse hierarchy." At low loading, the single-component isomers shows a reverse hierarchy. The heats of adsorption are consistent with this hierarchy. Note that at saturation the fluctuation formula for the heat of adsorption breaks down, and the error bar increases dramatically as saturation is approached. The single-component isotherms can be fitted using the Langmuir–Freundlich isotherms (Fig. 2.13c). With the fits the mixture can be predicted using the IAST and compared to the explicit simulations. Figure 2.13d shows that the IAST works very well for

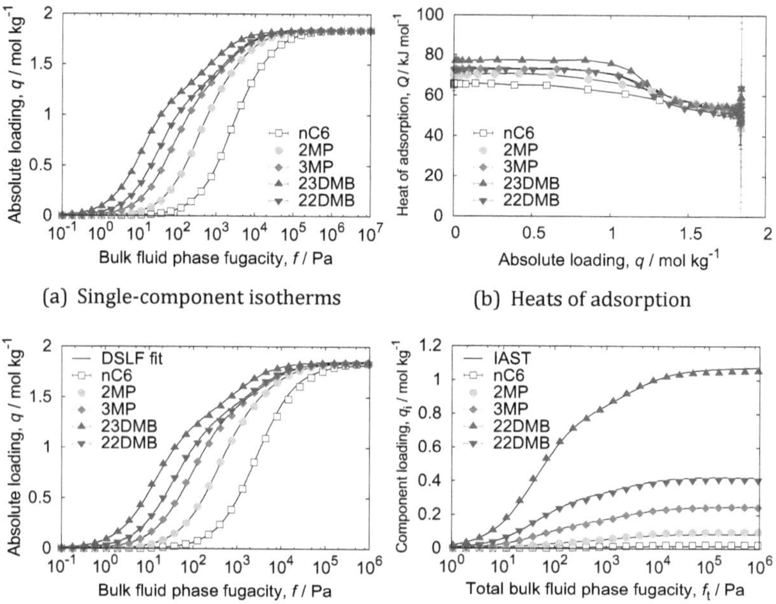

(a) Single-component isotherms

(b) Heats of adsorption

(c) Single-component isotherm fitting

(d) IAST-solution prediction and the mixture isotherms

Figure 2.13 Single-component isotherms of hexane isomers (hexane, 2-methylpentane, 3-methylpentane, 2,2-DMB, and 2,3-DMB), the heat of adsorption, the Langmuir–Freundlich fit to the pure-component isotherms, and the prediction of the IAST compared to the mixture simulation, for hexane isomers in UiO-66 at 433 K.

this system. The accuracy of IAST calculations has been established in several studies [Krishna et al. (2002); Krishna and Baur (2008)].

Measuring mixture isotherms experimentally is difficult. The IAST therefore provides a short cut to obtaining mixture information using only single-component isotherms as input. The IAST solution from CBMC single-component isotherms can directly be compared to CBMC mixture simulations and provide validation for using the IAST for a particular system. The IAST does not always work. The IAST assumes a homogeneous mixing of the different species and therefore can fail, for example, when molecular segregation occurs [Dubbeldam et al. (2008)]. The IAST incorporates activity coefficients in the adsorbed phase to account for the deviations from ideality. When activity coefficients are calculated the method is known as real adsorbed solution theory (RAST) [Steffan and Akgerman (2001); Ertoa et al. (2012)]. RAST overcomes some of the IAST limitations because it explicitly takes into account the nonideal behavior shown by many adsorption systems.

The main power of molecular simulation lies in the ability to obtain molecular-type information. A direct route is to examine the snapshots over a range of pressures [Vlugt et al. (1998)]. Figure 2.14 shows the pore energy landscape of the UiO-66 structure. The structure has two types of cages that alternate. From the figure we see that the double-branched molecule has a strong affinity for the smaller type cages. At low pressures almost no molecules are found in the larger cages. At high pressure also the larger cages start to fill up. The various pieces of information now can be combined: the heat of adsorption of 2,3-DMB is the highest because of the tight interaction with the smaller cages. Also the other types of molecule fit in but less snugly. At a loading of eight molecules per unit cell, when the smaller cages are filled up, inflection occurs. The heat of adsorption decreases precisely at the loading where the inflection occurs.

2.4.3 *Breakthrough Curves*

Various types of equipment have been developed for adsorption operations, including fixed beds, moving beds, fluidized beds, and stirred-tank contactors. Of these, fixed-bed adsorbers are most commonly applied; the adsorption area available per unit volume is greater in fixed beds than in most other configurations [Doran (2012)]. Adsorption in a fixed bed is carried out in a vertical cylindrical column filled with adsorbent particles. In chromatography, the dissolved component of a mixture that is to be separated in the chromatographic column is called "solute," and the "effluent" is the combination of the mobile phase and the solute exiting the column. The fluid containing the solute flows downward through the packed bed of adsorbents, usually at a constant flow rate. The saturation of the column is detected by monitoring the concentration of the solute in the fluid phase at the exit end of the column. The concentration profile in the fluid phase or the

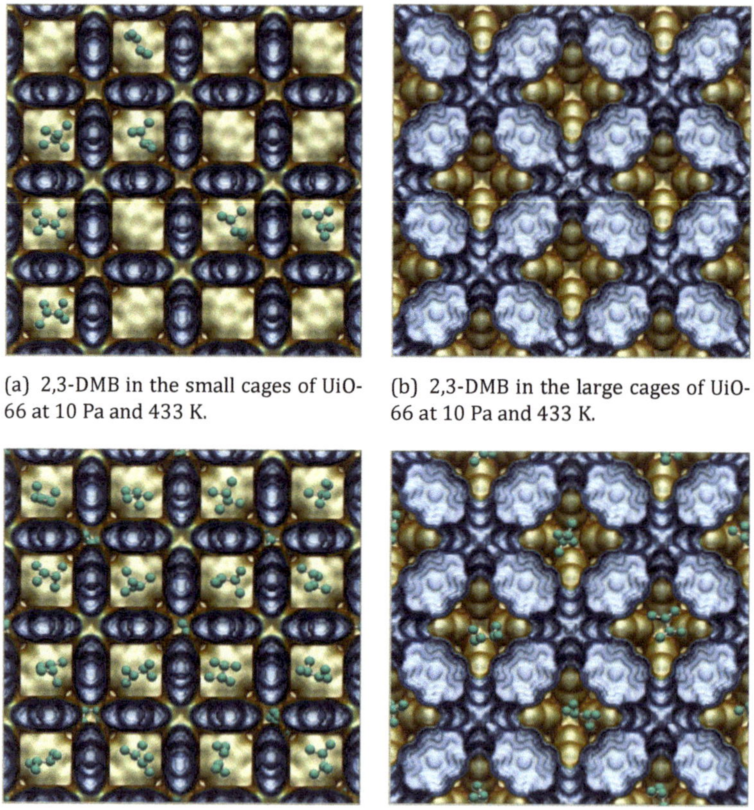

(a) 2,3-DMB in the small cages of UiO-66 at 10 Pa and 433 K.

(b) 2,3-DMB in the large cages of UiO-66 at 10 Pa and 433 K.

(c) 2,3-DMB in the small cages of UiO-66 at 10^5 Pa and 433 K.

(d) 2,3-DMB in the small cages of UiO-66 at 10^5 Pa and 433 K.

Figure 2.14 Snapshots of 2,3-DMB in UiO-66 ($2 \times 2 \times 2$ unit cells) at low and high pressure. The UiO-66 cubic unit cell has edge length $a = b = c = 20.7004$ Å. The UiO-66 pore system consists of two types of cages that alternate: four larger cages and eight smaller cages per unit cell. About 50% of the structure is void. Color code: zirconium (ochre), oxygen (red), carbon (cyan), and hydrogen (white).

variation of the solute concentration in the effluent as a function of time is called the "breakthough curve." The breakthrough curve is useful for indicating the saturation of the adsorbent bed, because the adsorbent is incapable of further adsorption. Once the break point is reached feed flow into the column is stopped. The adsorbed material is recovered by desorption or elution process (converse breakthrough process) and the adsorbent column is regenerated.

At the start of the process the topmost layer of the bed comes into contact with the mixture solution at concentration $c_{i,0}$. The fixed bed is initially devoid of adsorbates. The effluent coming out of the bottom of the bed is practically a solute-free carrier gas. The top of the bed adsorbs rapidly and gets saturated with adsorbates. Initially, the concentration in the fluid drops exponentially with the bed length. When the adsorbent near the inlet becomes nearly saturated, most of the mass transfer takes place further from the inlet, and the concentration profile then becomes S-shaped. Instead of plotting the adsorbate concentration, the concentration in the fluid phase in equilibrium with the adsorbent solid is usually plotted. The bulk of the adsorption takes place over a relatively narrow zone, the "adsorption zone," in which the concentration changes rapidly. The adsorption zone moves down through the bed with a rate that is usually much lower than the linear velocity of the solution in the bed. At the front of the adsorption zone molecules are adsorbed, while at the back they reach saturation. The moment the front of the adsorption zone reaches the end of the bed, the solute "breaks through" the bed. The solute now appears in the effluent. The concentration of the solute in the effluent rises rapidly as the adsorption zone passes through the bottom of the bed and reaches saturation (the entire adsorption zone leaves the bed). The slope of the breakthrough curve is a function of the rate of adsorption. The higher the adsorption rate, the steeper the breakthrough curve. If the adsorption process was instantaneous, the adsorption zone would be of an infinitesimal width, and the breakthrough line would be a vertical line when all the adsorbent solid was saturated. If a process targets purification of the solution, the operation is interrupted at or just before the breakthrough. The degree of column utilization can be improved by extending the column length, because the breakthrough curve can be shifted to the right without changing its shape. The shape of the breakthrough curve greatly influences the design and operation of fixed-bed adsorbers. The most efficient performance will be obtained when the shape of the isotherm is as sharp as possible.

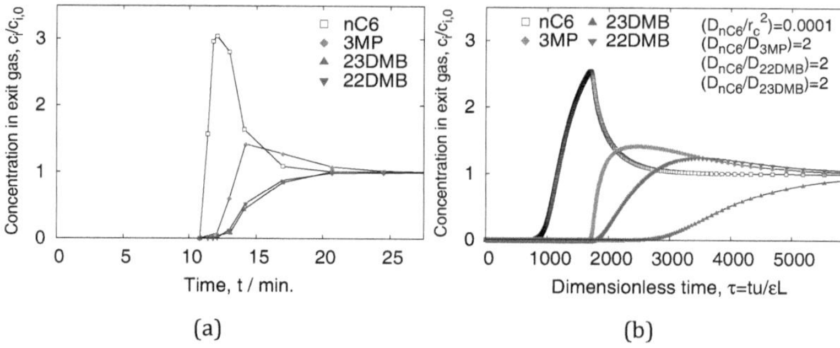

Figure 2.15 Breakthrough of hexane isomers in UiO-66. (a) Experimental breakthrough [Bárcia et al. (2011)] and (b) simulated breakthrough [Dubbeldam et al. (2012)].

The measurement of breakthrough curves is performed using a relatively simple apparatus. A typical system has three major components: flow controllers, a fixed bed with temperature control, and an instrument to measure the composition of the outlet stream. GCs and mass spectrometers (MSs) (sometimes together) are often used to monitor the outlet stream. An MS is more convenient because it can be configured to be constantly "on stream" such that the outlet is being continuously monitored; this is especially important in the case of fast and sharp breakthrough behavior. If the concentration front moves too quickly through the bed, a GC might only have time to analyze one sample before the entire front breaks through the bed. On the other hand, the MS cannot be used to analyze systems containing more than one species with similar molecular weights.

Recently, Bárcia et al. published results for alkane separation in UiO-66 [Bárcia et al. (2011)]. Figure 2.15 shows CBMC modeling and the comparison with experiments. The results are in qualitatively agreement. It should be noted that the experimental results are heavily influenced by diffusional effects, which tend to convolute the peaks. The breakthrough of *n*C6 appears first with an increase of concentration up to values significantly higher than the initial one, that is, the maximum value c/c_0 exceeds unity. This increase is due to continuous replacement of *n*-C6 in the adsorbent by the other more strongly adsorbing molecules. The dimensionless concentrations

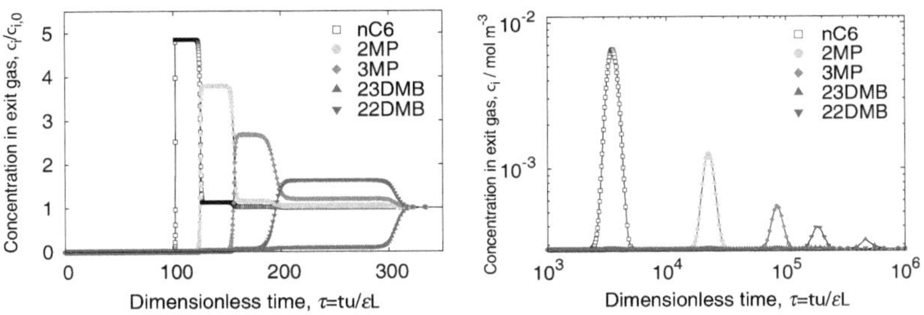

(a) $L = 0.3$ m; $\varepsilon = 0.4$; $u = 0.04$ m/s (at inlet); $c_{i0} = 20000/8.314/433$K

(b) $L = 1$ m; $\varepsilon = 0.4$; $u = 0.04$ m/s (superficial gas velocity); duration of pulse $= t = 10$ s (real time)

Figure 2.16 Simulated step and pulse breakthroughs of hexane isomers in UiO-66 at 433 K [Dubbeldam et al. (2012)].

exceeding unity is a well-known phenomenon in pressure swing adsorption (PSA) and are referred to as "overshoot" and "roll-up" [Van den Broeke and Krishna (1995); Krishna (2000)].

So far we discussed the "step" breakthrough where the concentration flow is maintained. Another common type of breakthrough is the "pulse" breakthrough. Here, a pulse of materials is injected for only a relatively short time (e.g., 10 seconds). The pulse breakthroughs are affected by both adsorption and desorption. If a material is capable of purification then the peaks corresponding to different components can be individually identified. Differences between species can always be amplified by making the column longer, and therefore materials need to be fairly evaluated for the same bed length, etc. Figure 2.16 shows the step and pulse breakthroughs for hexane isomers in UiO-66. The breakthrough curves and the mixture isotherms contain the same information, but the pulse curves are very suitable to visualize fractionation (all peaks are separated). The area under the peaks is the same for an equimolar mixture (the height of the peaks decreases).

The breakthrough calculations were performed using the methodology described in the paper by Krishna and Long [Krishna and Long (2011); Dubbeldam et al. (2012)]. In these breakthrough calculations the following set of assumptions was made:

(1) Intracrystalline diffusion. This is the controlling resistance; the macropore/mesopore diffusion resistance and diffusion in the gas phase are ignored.
(2) The plug flow of the n-component mixture is through the fixed bed.
(3) The bed is maintained under isothermal conditions at a fixed temperature.
(4) The pressure drop across the bed is considered to be negligible, that is, the total fugacity is maintained constant.

The partial pressures in the gas phase at any position and instant of time are obtained by solving the following set of partial differential equations for each of the species i in the gas mixture.

$$\frac{1}{RT}\varepsilon\frac{\partial p_i}{\partial t} = -\frac{1}{RT}\frac{\partial\left(up_i\right)}{\partial z} - \left(1-\varepsilon\right)\rho\frac{\partial q_i}{\partial t}; \qquad i = 1, 2, \ldots, 5 \tag{2.16}$$

In Eq. 2.16, t is the time, z is the distance along the adsorber, ρ is the framework density, ε is the bed voidage, and u is the superficial gas velocity. The variation of the superficial gas velocity along the length of the adsorber is properly accounted for. The molar loadings of the species i, q_i, at any position z and time t are determined from IAST calculations. Typically, the adsorber length L is divided into 100 slices. The number of slices is determined by checking that the obtained breakthrough results do not change on increasing it. In the breakthrough calculations, the pressure drop in the bed is considered to be negligible, and consequently the total pressure remains constant. As a consequence, the velocity u changes along the length of the adsorber due to adsorption into the porous materials. The velocity change is duly accounted for; the value of u at any slice in the adsorber is determined from a material balance across that slice. A combination of the discretized PDEs along with the algebraic IAST equilibrium model results in a set of differential algebraic equations (DAEs), which are solved using BESIRK [Kooijman and Taylor (1995)]. BESIRK is a sparse matrix solver based on the semi-implicit Runge–Kutta method originally developed by Michelsen [Michelsen (1976)], and extended with the Bulirsch–Stoer [Bulirsch and Stoer (1966)] extrapolation method. Use of BESIRK improves the numerical solution efficiency in solving the set of DAEs. The evaluation of the sparse Jacobian required in the numerical algorithm is largely based on analytic expressions [Krishna and Baur (2008)]. Further details of the adsorber model, along with the numerical procedures used, are provided in Refs. [Krishna and Baur (2008); Krishna and van Baten (2011b)]. Details of the numerical procedures used are published in Refs. [Krishna and Long (2011); Krishna and Baur (2008); Walton and LeVan (2003)].

To fairly compare different materials, the fractional voidage should be held constant ($\varepsilon = 0.4$ in [Dubbeldam et al. (2012)]). This implies the volume of adsorbents used is the same. The total mass of the corresponding adsorbents used is governed by the framework density. All the n-component mixture breakthrough calculations were carried out with the inlet gas maintained at fixed partial fugacities at the inlet. The breakthrough behaviors are usually conveniently presented in terms of the dimensionless outlet concentration, $\frac{c_i}{c_{i,0}} = \frac{f_i}{f_{i0}}$, representing the ratio of the partial fugacities at the outlet gas divided by the partial fugacities at the adsorber inlet, as a function of the dimensionless time, τ, obtained by dividing the actual time, t, by the contact time between the gas and the crystallites, $\varepsilon L/u$. For a given adsorbent, under chosen operating conditions, the breakthrough characteristics are uniquely defined by τ, allowing the results to be equally applicable to laboratory-scale equipment as well as to industrial-scale adsorbers. Experimental verification of the PSA breakthrough and pulse chromatographic simulation methodologies has been obtained by comparison with experiments [Bloch et al. (2012); He et al. (2012); Wu et al. (2012b)].

2.5 Screening

2.5.1 *Case Study: Hexane Isomers*

The separation of linear, monobranched, and dibranched isomers of alkanes is of significant importance in the petrochemical industry. For example, the dibranched alkanes in the 5–7 carbon number range are preferred components of high-octane gasoline and its selective removal from other isomers produced in an alkane isomerization reactor. This separation can be achieved by selective adsorption in ordered crystalline nanoporous materials, such as zeolites, MOFs, COFs, and ZIFs, by exploiting subtle differences on molecular configurations. Literally, several thousands of such materials have been synthesized, making the choice of adsorbent a daunting task. Dubbeldam et al. approach this task by resorting to molecular simulations carried out on a prescreened list of more than 100 nanoporous structures. Of all examined structures, only a handful show an ordered adsorption hierarchy [Dubbeldam et al. (2012)].

To provide insights into the origin of the differences in selectivities of various structures Dubbeldam et al. performed a systematic study of graphite slits and square channels of varying widths and carbon nanotubes of different diameters. Graphite slits have been artificially kept at a fixed distance, following the approach of Severson and Snurr [Severson and Snurr (2007)] and Krishna and van Baten [Krishna and van Baten (2011a)]. The channels have been created by intersecting the sheets orthogonally with other sheets (Fig. 2.17a). From the enthalpy of adsorption at large spacings and low loadings (i.e., this corresponds to the adsorption of a single molecule on a surface), plotted in Fig. 2.17b relative to nC6, we find that natural order of adsorption on a flat surface is the normal hierarchy. The enthalpies for monobranched molecules are higher than *n*C6, and the enthalpies of dibranches are even higher in energy. The normal hierarchy is therefore the order one would expect, and any modifications to the normal order are due to the specific confinement that favors specific molecules. This occurs at confinements around 6–7 Å, where the dibranched molecules fit most snugly because they interact strongly with multiple walls. The spacings of 5 Å and smaller deserve special attention. Here, we have very strong confinement and size exclusion. In Fig. 2.17c it is shown that the reverse hierarchies found are all in the 6–7 Å range. Below that range we can have high selectivity (as observed for ZIF-77) due to strong confinement and size exclusion, and above the 6–7 Å region we also find the normal hierarchy with lower selectivities. Note that Figs. 2.17c and 2.17d are qualitatively similar, and spacing and capacity go hand-in-hand. The selectivity is inversely related to spacing. ZIF-77 has a 2D channel system with larger main channels and smaller side channels. The larger main channels are size-excluding the dibranched molecules, while the

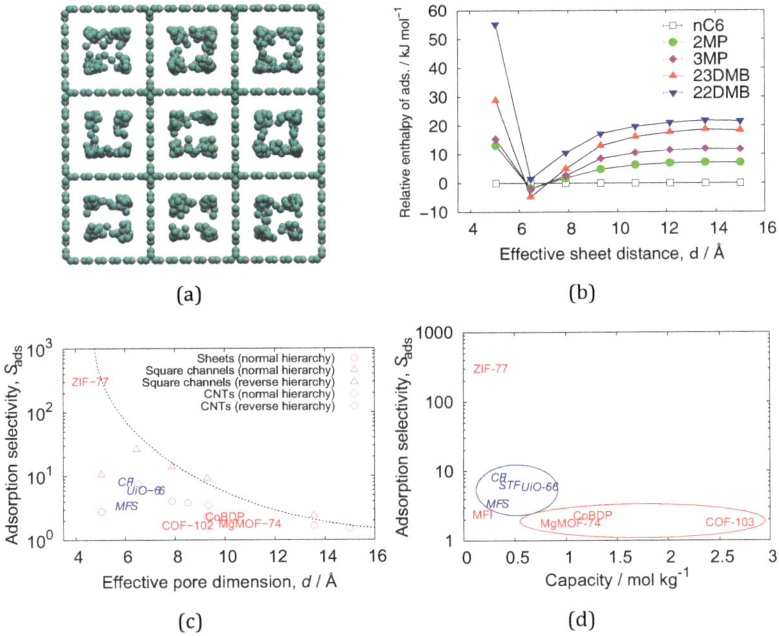

Figure 2.17 Hexane isomer selectivity at 433 K as a function of confinement. (a) Snapshot of hexane molecules in artificial carbon square channels made from carbon sheets; the spacing can be controlled. (b) Enthalpy of adsorption relative to nC6 for graphite square channels at infinite dilution. (c) Adsorption selectivity at 100 kPa as a function of spacing. (d) Adsorption selectivity at 100 kPa as a function of pore capacity. Red color denotes normal hierarchy, blue color denotes reverse hierarchy, and the dotted line shows the qualitative behavior of the selectivity.

smaller channels are size-excluding the monobranched molecules. Also of high interest is CoBDP, which combines good selectivity with a high pore volume. The CoBDP MOF structure was found using CBMC simulation by Krishna and van Baten as a potential alkane separation device [Krishna and van Baten (2011a)].

Dubbeldam et al. have rationalized the adsorption behavior of hexane (and heptane) isomers, and their screening study provides a thorough understanding. First, large-pore MOFs are not useful for alkane separations. Above a certain threshold (depending on the alkane chain length) the excess pore volume just goes to waste.

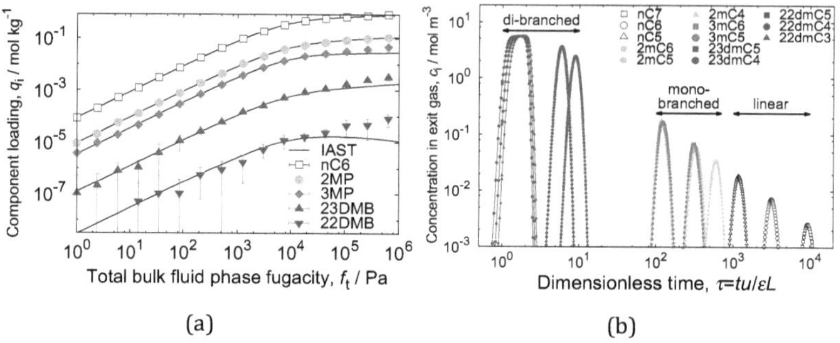

Figure 2.18 The ZIF-77 alkane separation device. (a) Hexane isomer mixture isotherm and (b) 13-component pentane/hexane/heptane isomer pulse breakthrough.

Below the threshold for selectivity, both branching hierarchies are possible, but the one corresponding to the largest pore volume (the normal hierarchy) is probably most efficient, also because adsorption and diffusion selectivity go hand-in-hand. To properly evaluate a structure one needs at least a five-component mixture with both the 2,2-dibranched *and* the 2,3-dibranched molecules included, because the 2,2-dibranched and 2,3-dibranched molecules have usually very different behaviors. The simulated breakthrough curves provide the best way of assessing the efficiency of a structure. The efficiency depends on (a) the selectivity and (b) the accessible pore volume. CoBDP provides a very good compromise between the two, but the selectivity of ZIF-77 is so high that this sieve fractionates to yield individual pure components. It can even do so over a range of chain lengths (see Fig. 2.18b) for the pulse breakthrough of a 13-component mixture of pentane/hexane/heptane isomers.

Two hierarchies are possible, normal and reverse. The reverse hierarchy, as found in UiO-66 and CFI, can only be found in strong confinement structures and will therefore correspond to structures with relatively small pore volumes. An important disadvantage of the reverse hierarchy is that the hierarchy of the diffusion coefficient is opposite the hierarchy of adsorption strengths. The reverse hierarchy structure does show higher selectivities, though, but much lower than ZIF-77 and MFI (at very high pressures). In contrast, the normal hierarchy of diffusivities would enhance the separation further. That implies the following, taking UiO-66 as the reference: (a) If fractionation is the target then a smaller structure would give better selectivities (if we compare the pulse breakthrough of UiO-66 vs. ZIF-77, then ZIF-77 is orders of magnitude better; compare the time axes), and (b) if we take an MOF with a slightly larger pore structure, then the selectivity is lower (and opposite) but compensated by a favorable diffusion hierarchy; this structure would have a larger capacity than UiO-66. We are now at a point where we can combine all of these concepts to select near-optimal structures (for alkanes) on the basis of this type of analysis. The methodology to reach these conclusions was computational large-scale screening.

2.5.2 *Screening Strategies*

Computed geometric properties such as pore volume, surface areas, and PSD, can be routinely calculated on the basis of the crystallographic data of structures. However, one cannot conclude from these values whether a structure will perform well, but one can use these values to decide that a structure will not be adequate. A structure can have a high geometrical surface area, but this does not mean all of this surface area is used. Moreover, it does not provide information on the adsorption strength. Binding energies and heats of adsorption do provide this type of information. Other types of information, such as selectivities at infinite dilution or at maximum loading, give hints but give no definitely proof of its performance as a separation device, because the shape of the isotherm is a very important factor. For storage/release applications a linear isotherm is preferred as this gives more control.

Lightweight screening on the basis of a few properties that can efficiently be computed allows for fast screening. On the other hand, some structures might be missed, and structures that were identified could turn out to be mediocre. Computers are fast enough nowadays to allow for a pipeline type of screening. First geometric properties are used to dismiss structures that are not porous enough. For the remainder the maximum loading is computed to assess the capacity. Structures with sufficient capacity are than evaluated using pure-component isotherms and mixture isotherms.

Structures that show decent selectivities are finally studied in detail using breakthrough simulations. The latter evaluates the efficiency of nanoporous materials in a real application setup. Finally, successful candidates should be examined closely in terms of adsorption sites and the mechanism in order to understand the origin of the separation. Hopefully, a pattern emerges from which design criteria could be deduced.

The unique separation capability of ZIF-77 would not have been unearthed using a conventional screening study based on, for example, PSDs. The reason is that the pores of ZIF-77 are highly noncylindrical. The PSD of ZIF-77 could be wrongly interpreted as that there are no pores larger than 4.5 Å in diameter. However, this is the *shortest* distance, and in the other direction the pore is much larger. In fact, linear, monobranched and 2,3-dibranched alkanes fit in. MD simulations can be used to compute the self-diffusion coefficients at 433 K and infinite dilution. The hierarchy of diffusivities is linear $>$ monobranched $>$ 2,3-dibranched $\ggg$ 2,2-dibranched. The 2,2-dibranched is too bulky to enter the pore system. It would be challenging to refine currently used conventional screening methodologies to be able to detect systems like ZIF-77. However, currently it is feasible to analyze over a 100 structures in full detail, and with current advances in computational efficiency, future screenings could handle many, many more structures.

However, alkanes are relatively inert and interact mainly via their shape and size. For alkanes, it is all about the molecule–wall distance. At small distances linear alkanes pack better than branched ones, but at intermediate distances a reversed hierarchy is observed, while at larger distances the linear is preferred again. The main challenge for the future is to extend screening methodology to much more complex systems. For CO_2, N_2, etc., charge–charge interaction is crucial and the adsorbate–adsorbate interactions are important. No clear recipe to select highly efficient CO_2 separation materials has been found to date.

Acknowledgments

This material is supported by the Netherlands Research Council for Chemical Sciences (NWO/CW) through a VIDI grant (David Dubbeldam) and the National Science Foundation CAREER Award 0969261 and PECASE Award ARO W911-10-0079 (Krista S. Walton). We thank R. Krishna for his fruitful discussion and for providing the IAST and breakthrough simulations, and T. J. H Vlugt, Aurora J. Cruz Cabeza, and Ariana Torres Knoop for comments on the manuscript.

References

Adamson, A., Gast, A. (1997). *Physical Chemistry of Surfaces*, 6th edn. (John Wiley & Sons, New York).

Amirjalayer, S., Schmid, R. (2008). Conformational isomerism in the isoreticular metal organic framework family: a force field investigation, *J. Phys. Chem. C*, **112**(38), 14980–14987.

Anbia, M., Hoseini, V. (2012). Development of MWCNT@MIL-101 hybrid composite with enhanced adsorption capacity for carbon dioxide, *Chem. Eng. J.*, **191**, 326–330.

Bae, Y., Dubbeldam, D., Nelson, A., Walton, K., Hupp, J., Snurr, R. (2009). Strategies for characterization of large-pore metal-organic frameworks by combined experimental and computational methods, *Chem. Mater.*, **21**, 4768–4777.

Bae, Y., Snurr, R. (2011). Development and evaluation of porous materials for carbon dioxide separation and capture, *Angew. Chem., Int. Ed.*, **50**(49), 11586–11596.

Ban, S., Vlugt, T. (2009). Zeolite microporosity studied by molecular simulation, *Mol. Simulat.*, **35**(12–13), 1105–1115.

Bárci, P., Guimarães, D., Mendes, P., Silva, J., Guillerm, V., Chevreau, H., Serre, C., Rodrigues, A. (2011). Reverse shape selectivity in the adsorption of hexane and xylene isomers in MOF UiO-66, *Microporous Mesoporous Mater.*, **139**(1–3), 67–73.

Bloch, E., Queen, W., Krishna, R., Zadrozny, J., Brown, C., Long, J. (2012). Hydrocarbon separations in a metal-organic framework with open iron(II) coordination sites, *Science*, **335**, 1606–1610.

Brunauer, S., Emmett, P., Teller, E. (1938). Adsorption of gases in multimolecular layers, *J. Am. Chem. Soc.*, **60**(2), 309–319.

Bulirsch, R., Stoer, J. (1966). Numerical treatment of ordinary differential equations by extrapolation methods, *Numer. Math.*, **8**(1), 1–14.

Calero, S., Smit, B., Krishna, R. (2001). Separation of linear, mono-methyl and di-methyl alkanes in the 5-7 carbon atom range by exploiting configurational entropy effects during sorption on silicalite-1, *Phys. Chem. Chem. Phys.*, **3**(19), 4390–4398.

Caskey, S., Matzger, A. (2009). Selected applications of metal-organic frameworks in sustainable energy technologies, *Mater. Matter*, **4**, 111–115.

Cheng, L., Yang, R. (1994). Improved Horvath-Kawazoe equations including spherical pore models for calculating micropore size distribution, *Chem. Eng. Sci.*, **49**(16), 2599–2609.

Chmelik, C., Karger, J., Wiebcke, M., Caro, J., van Baten, J., Krishna, R. (2009). Adsorption and diffusion of alkanes in CuBTC crystals investigated using infra-red microscopy and molecular simulations, *Microporous Mesoporous Mater.*, **117**(1–2), 22–32.

Chui, S.-Y., Lo, S.-F., Charmant, J., Orpen, A., Williams, I. (1999). A chemically functionalizable nanoporous material $[Cu_3(TMA)_2(H_2O)_3]_n$, *Science*, **283**(5405), 1148–1150.

Condon, J. (2006). *Surface Area and Porosity Determinations by Physisorption: Measurements and Theory* (Elsevier Science, Amsterdam).

Connolly, M. (1983). Solvent-accessible surfaces of proteins and nucleic-acids, *Science*, **221**(4612), 709–713.

Coolidge, A. (1934). Adsorption at high pressures. I, *J. Am. Chem. Soc.*, **56**(3), 554–561.

Côté, A. P., Benin, A. I., Ockwig, N. W., O'Keeffe, M., Matzger, A. J., Yaghi, O. M. (2005). Porous, crystalline, covalent organic frameworks, *Science*, **310**(5751), 1166–1170.

D'Alessandro, D., Smit, B., Long, J. (2010). Carbon dioxide capture: prospects for new materials, *Angew. Chem., Int. Ed.*, **49**(35), 6058–6082.

Demontis, P., Suffritti, G. (1994a). Molecular-dynamics investigation of the diffusion of methane in a cubic symmetry zeolite of type-ZK4, *Chem. Phys. Lett.*, **223**(4), 355–362.

Demontis, P., Suffritti, G. (1994b). Molecular-dynamics simulations of diffusion in a cubic symmetry zeolite, *Stud. Surf. Sci. Catal.*, **84**, 2107–2113.

Demontis, P., Suffritti, G. (2009). A comment on the flexibility of framework in molecular dynamics simulations of zeolites, *Micropor. Mesopor. Mat.*, **125**(1–2), 160–168.

Demontis, P., Suffritti, G., Quartieri, S., Fois, E., Gamba, A. (1987). Molecular dynamics studies on zeolites. II. A simple-model for silicates applied to anhydrous natrolite, *Zeolites*, **7**(6), 522–527.

Demontis, P., Suffritti, G., Quartieri, S., Fois, E., Gamba, A. (1988). Molecular dynamics on zeolites. III. Dehydrated zeolite-A, *J. Phys. Chem.*, **92**(4), 867871.

Dietzel, P., Besikiotis, V., Blom, R. (2009). Application of metal-organic frameworks with coordinatively unsaturated metal sites in storage and separation of methane and carbon dioxide, *J. Mater. Chem.*, **19**(39), 7362–7370.

Do, D. (2003). Pore characterization of carbonaceous materials by DFT and GCMC simulations: a review, *Adsorp. Sci. Tech.*, **21**(5), 389–423.

Doran, P. (2012). *Bioprocess Engineering Principles*, 2nd edn. (Academic Press, London).

Dubbeldam, D., Calero, S., Maesen, T., Smit, B. (2003). Understanding the window effect in zeolite catalysis, *Angew. Chem. Int. Ed.*, **42**(31), 3624–3626.

Dubbeldam, D., Frost, H., Walton, K., Snurr, R. (2007a). Molecular simulation of adsorption sites of light gases in the metal-organic framework IRMOF-1, *Fluid Phase Equilib.*, **261**(1), 152–161.

Dubbeldam, D., Frost, H., Walton, K., Snurr, R. (2007b). Molecular simulation of adsorption sites of light gases in the metal-organic framework IRMOF-1, *Fluid Phase Equilib.*, **261**(1–2), 152–161.

Dubbeldam, D., Galvin, C., Walton, K., Ellis, D., Snurr, R. (2008). Separation and molecular-level segregation of complex alkane mixtures in metal-organic frameworks, *J. Am. Chem. Soc.*, **130**(33), 10884–10885.

Dubbeldam, D., Krishna, R., Calero, S., Yazaydin, A. (2012). In silico screening of ordered crystalline nanoporous adsorbents for separation of alkane isomers, *Angew. Chem., Int. Ed.,* **51,** 11867–11871.

Dubbeldam, D., Walton, K., Ellis, D., Snurr, R. (2007). Exceptional negative thermal expansion in isoreticular metal-organic frameworks, *Angew. Chem., Int. Ed.,* **46**(24), 4496–4499.

Duren, T., Millange, F., Férey, G., Walton, K., Snurr, R. (2007). Calculating geometric surface areas as a characterization tool for metal-organic frameworks, *J. Phys. Chem. C*, **111**(42), 15350–15356.

Duren, T., Sarkisov, L., Yaghi, O., Snurr, R. (2004). Design of new materials for methane storage, *Langmuir*, **20**(7), 2683–2689.

Dzubak, A., Lin, L.-C., Kim, J., Swisher, J., Poloni, R., Maximoff, S., Smit, B., Gagliardi, L. (2012). Ab initio carbon capture in open-site metal-organic frameworks, *Nat. Chem.*, **4**, 810–816.

Eddaoudi, M., Kim, J., Rosi, N., Vodak, D., Wachter, J., O'Keeffe, M., Yaghi, O. (2002). Systematic design of pore size and functionality in isoreticular MOFs and their application in methane storage, *Science*, **295**(5554), 469–472.

El-Kaderi, H., Hunt, J., Mendoza-Cortés, J., Côté, A., Taylor, R., O'Keeffe, M., Yaghi, O. (2007). Designed synthesis of 3D covalent organic frameworks, *Science*, **316**(5822), 268–272.

Ertoa, A., Lancia, A., Musmarra, D. (2012). Real adsorbed solution theory model for competitive multicomponent liquid adsorption onto granular activated carbon, *Microporous Mesoporous Mater.*, **154**, 45–50.

Farha, O., Eryazici, I., Jeong, N., Hauser, B., Wilmer, C., Sarjeant, A., Snurr, R., Nguyen, S., Yazaydin, A., Hupp, J. (2012). Metal-organic framework materials with ultrahigh surface areas: is the sky the limit? *J. Am. Chem. Soc.*, **134**(36), 15016–15021.

Férey, G. (2008). Hybrid porous solids: past, present, future, *Chem. Soc. Rev.*, **37**(1), 191–214.

Férey, G., Mellot-Draznieks, C., Serre, C., Millange, F., Dutour, J., Surble, S., Margiolaki, I. (2005). A chromium terephthalate-based solid with unusually large pore volumes and surface area, *Science*, **309**, 2040–2042.

Férey, G., Serre, C., Devic, T., Maurin, G., Jobic, H., Llewellyn, P., de Weireld, G., Vimont, A., Daturi, M., Chang, J. (2011). Why hybrid porous solids capture greenhouse gases? *Chem. Soc. Rev.*, **40**(2), 550–562.

Fletcher, A. J., Thomas, K. M., Rosseinsky, M. J. (2005). Flexibility in metal-organic framework materials: Impact on sorption properties, *J. Solid State Chem.*, **178**(8), 2491–2510.

Ford, D., Dubbeldam, D., Snurr, R. (2009). The effect of framework flexibility on diffusion of small molecules in the metal-organic framework IRMOF-1, *Diffus. Fundam. III*, **11**, 1–8.

Ford, D., Dubbeldam, D., Snurr, R., Kunzel, V., Wehring, M., Stallmach, F., Karger, J., Muller, U. (2012). Self-diffusion of chain molecules in the metal-organic framework IRMOF-1: simulation and experiment, *J. Phys. Chem. Lett.*, **3**(7), 930–933.

Forester, T., Smith, W. (1997). Bluemoon simulations of benzene in silicalite-1—prediction of free energies and diffusion coefficients, *J. Chem. Soc. Faraday. T.*, **93**(17), 3249–3257.

Fritzsche, S., Wolfsberg, M., Haberlandt, R., Demontis, P., Suffritti, G., Tilocca, A. (1998). About the influence of lattice vibrations on the diffusion of methane in a cation-free LTA zeolite, *Chem. Phys. Lett.*, **296**(3–4), 253–258.

Garcia-Sanchez, A., Garcia-Perez, E., Dubbeldam, D., Krishna, R., Calero, S. (2007). A simulation study of alkanes in Linde Type A zeolites, *Adsorpt. Sci. Technol.*, **25**(6), 417–427.

Gelb, L., Gubbins, K. (1999). Pore size distributions in porous glasses: a computer simulation study, *Langmuir*, **15**(2), 305–308.

Haldoupis, E., Nair, S., Sholl, D. (2012). Finding MOFs for highly selective CO2/N2 adsorption using materials screening based on efficient assignment of atomic point charges, *J. Am. Chem. Soc.*, **134**(9), 4313–4323.

Han, S., Huang, Y., Watanabe, T., Dai, Y., Walton, K., Nair, S., Sholl, D., Meredith, J. (2012). High-throughput screening of metal-organic frameworks for CO2 separation, *Acs. Comb. Sci.*, **14**(4), 263–267.

He, Y., Krishna, R., Chen, B. (2012). Metal-organic frameworks with potential for energy-efficient adsorptive separation of light hydrocarbons, *Energy Environ. Sci.*, **5**, 9107–9120.

Herm, Z., Swisher, J., Smit, B., Krishna, R., Long, J. (2011). Metal-organic frameworks as adsorbents for hydrogen purification and precombustion carbon dioxide capture, *J. Am. Chem. Soc.*, **133**(15), 5664–5667.

Hill, J., Sauer, J. (1994). Molecular mechanics potential for silica and zeolite catalysts based on ab-initio calculations. I. Dense and microporous silica, *J. Phys. Chem.*, **98**(4), 1238–1244.

Hill, J., Sauer, J. (1995). Molecular mechanics potential for silica and zeolite catalysts based on ab-initio calculations. II. Aluminosilicates, *J. Phys. Chem.*, **99**(23), 9536–9550.

Horvath, G., Kawazoe, K. (1983). Method for the calculation of effective pore size distribution in molecular sieve carbon, *J. Chem. Eng. Japan*, **16**, 470–475.

Jiang, J. (2012a). Metal-organic frameworks for CO2 capture: what are learned from molecular simulations, in O. Ortiz and L. Ramirez (eds.), *Coordination Polymers and Metal Organic Frameworks* (Nova Science Publishers), 225–247.

Jiang, J. (2012b). Recent development of in silico molecular modeling for gas and liquid separations in metal-organic frameworks, *Curr. Opin. Chem. Eng.*, **1**(2), 138–144.

Kaneko, K. (1994). Determination of pore-size and pore-size distribution. 1. Adsorbents and catalysts, *J. Membr. Sci.*, **96**(1–2), 59–89.

Karavias, F., Myers, A. (1991). Isosteric heats of multicomponent adsorption–thermodynamics and computer-simulations, *Langmuir*, **7**, 3118–3126.

Keller, J., Staudt, R. (2005). *Gas Adsorption Equilibria: Experimental Methods and Adsorptive Isotherms* (Springer Verlag, Boston).

Keskin, S., van Heest, T., Sholl, D. (2010). Can metal-organic framework materials play a useful role in large-scale carbon dioxide separations? *ChemSusChem*, **3**(8), 879–891.

Kitagawa, S., Kitaura, R., Noro, S. (2004). Functional porous coordination polymers, *Angew. Chem., Int. Ed.*, **43**(18), 2334–2375.

Kluson, P., Scaife, S. (2001). Pore size distribution analysis of structure different microporous carbons—theoretical evaluation based on density functional theory and nitrogen and argon experimental adsorption isotherms at 77K, *Chem. Biochem. Eng. Q.*, **15**(3), 117–125.

Kooijman, H., Taylor, R. (1995). A dynamic nonequilibrium model for dynamic simulation of tray distillation-columns, *AIChE J.*, **41**(8), 1852–1863.

Kopelevich, D., Chang, H. (2001). Does lattice vibration drive diffusion in zeolites? *J. Chem. Phys.*, **114**(8), 3776–3789.

Krishna, R. (2000). Diffusion of binary mixtures in microporous materials: overshoot and roll-up phenomena, *Int. Comm. Heat Mass Transfer*, **27**(7), 893–902.

Krishna, R. (2012). Adsorptive separation of CO2/CH4/CO gas mixtures at high pressures, *Microporous Mesoporous Mater.*, **156**, 217–223.

Krishna, R., Baur, R. (2008). Modelling issues in zeolite based separation processes, *Sep. Purif. Technol.*, **33**(3), 213–253.

Krishna, R., Calero, S., Smit, B. (2002). Investigation of entropy effects during sorption of mixtures of alkanes in MFI zeolite, *Chem. Eng. J.*, **88**(1–3), 81–94.

Krishna, R., Long, J. (2011). Screening metal-organic frameworks by analysis of transient breakthrough of gas mixtures in a fixed bed adsorber, *J. Phys. Chem. C*, **115**(26), 12941–12950.

Krishna, R., van Baten, J. (2010a). Comment on "comparative molecular simulation study of CO2/N2 and CH4/N2 separation in zeolites and metal-organic frameworks", *Langmuir*, **26**(4), 2975–2978.

Krishna, R., van Baten, J. (2010b). Comment on "modeling adsorption and self-diffusion of methane in LTA zeolites: the influence of framework flexibility", *J. Phys. Chem. C*, **114**(41), 18017–18021.

Krishna, R., van Baten, J. (2010c). Highlighting a variety of unusual characteristics of adsorption and diffusion in microporous materials induced by clustering of guest molecules, *Langmuir*, **26**(11), 8450–8463.

Krishna, R., van Baten, J. (2010d). Investigating cluster formation in adsorption of CO2, CH4, and Ar in zeolites and metal organic frameworks at suberitical temperatures, *Langmuir*, **26**(6), 3981–3992.

Krishna, R., van Baten, J. (2011a). In silico screening of metal-organic frameworks in separation applications, *Phys. Chem. Chem. Phys.*, **13**(22), 10593–10616.

Krishna, R., van Baten, J. (2011b). Investigating the potential of MgMOF-74 membranes for CO2 capture, *J. Membrane Sci.*, **377**(1–2), 249–260.

Krishna, R., van Baten, J. (2011c). A rationalization of the type IV loading dependence in the Karger-Pfeifer classification of self-diffusivities, *Microporous Mesoporous Mater.*, **142**(2–3), 745–748.

Lachet, V., Boutin, A., Tavitian, B., Fuchs, A. (1997). Grand canonical Monte Carlo simulations of adsorption of mixtures of xylene molecules in faujasite zeolites, *Faraday Discuss.*, **106**, 307–323.

Lachet, V., Boutin, A., Tavitian, B., Fuchs, A. (1998). Computational study of p-xylene/m-xylene mixtures adsorbed in NaY zeolite, *J. Phys. Chem. B*, **102**(46), 9224–9233.

Lachet, V., Boutin, A., Tavitian, B., Fuchs, A. (1999). Molecular simulation of p-xylene and m-xylene adsorption in Y zeolites. Single components and binary mixtures study, *Langmuir*, **15**(25), 8678–8685.

Lastoskie, C., Gubbins, K., Quirke, N. (1993). Pore size distribution analysis of microporous carbons: a density functional theory approach, *J. Phys. Chem.*, **97**, 4786–4796.

LeVan, M. and Carta, G. (2008). Adsorption and ion exchange, in R. Perry and D. Green (eds.), *Perry's Chemical Engineers' Handbook* (McGraw-Hill, New York).

Li, H., Eddaoudi, M., O'Keeffe, M., Yaghi, O. (1999). Design and synthesis of an exceptionally stable and highly porous metal-organic framework, *Nature*, **402**(6759), 276–279.

Li, J., Ma, Y., McCarthy, M., Sculley, J., Yu, J., Jeong, H., Balbuena, P., Zhou, H. (2011). Carbon dioxide capture-related gas adsorption and separation in metal-organic frameworks, *Coordin. Chem. Rev.,* **255**(15–16), 1791–1823.

Lin, L.-C., Berger, A., Martin, R., Kim, J., Swisher, J., Jariwala, K., Rycroft, C., Bhown, A., Deem, M., Haranczyk, M., Smit, B. (2012). *In silico* screening of carbon-capture materials, *Nat. Mater.,* **11**(7), 633–641.

Long, J., Yaghi, O. (2009). The pervasive chemistry of metal-organic frameworks, *Chem. Soc. Rev.*, **38**(5), 1213–1214.

Low, J., Benin, A., Jakubczak, P., Abrahamian, J., Faheem, S., Willis, R. (2009). Virtual high throughput screening confirmed experimentally: porous coordination polymer hydration, *J. Am. Chem. Soc.*, **131**(43), 15834–15842.

Malbrunot, P., Vidal, D., Chahine, R., Bose, T. (1997). Adsorbent helium density measurement and its effect on adsorption isotherms at high pressure, *Langmuir,* **13**, 539–544.

Mayo, S., Olafson, B., Goddard, W. (1990). DREIDING: a generic force field for molecular simulations, *J. Phys. Chem.*, **94**(26), 8897–8909.

McBain, J. (1932). *The Sorption of Gases and Vapours by Solids* (George Routledge & Sons, London).

Mertens, F. (2009). Determination of absolute adsorption in highly ordered porous media, *Surf. Sc.*, **603**(10-12), 1979–1984.

Michelsen, M. (1976). An efficient general purpose method for the intergration of stiff ordinary differential equations, *AIChE J.*, **22**(3), 594–597.

Mueller, U., Schubert, M., Teich, F., Puetter, H., Schierle-Arndt, K., Pastre, J. (2006). Metal-organic frameworks—prospective industrial applications, *J. Mater. Chem.*, **16**(7), 626–636.

Myers, A., Monson, P. (2002). Adsorption in porous materials at high pressure: Theory and experiment, *Langmuir,* **18**(26), 10261–10273.

Myers, A., Prausnitz, J. (1965). Thermodynamics of mixed-gas adsorption, *AIChE J.*, **11**(1), 121.

Neimark, A., Lin, Y., Ravikovitch, P., Thommes, M. (2009). Quenched solid density functional theory and pore size analysis of micro-mesoporous carbons, *Carbon,* **47**(7), 1617–1628.

Neimark, A., Ravikovitch, P. (1997). Calibration of pore volume in adsorption experiments and theoretical models, *Langmuir,* **13**, 5148–5160.

Nicholas, J., Hopfinger, A., Trouw, F., Iton, L. (1991). Molecular modeling of zeolite structure. 2. Structure and dynamics of silica sodalite and silicate force-field, *J. Am. Chem. Soc.*, **113**(13), 4792–4800.

Park, K., Ni, Z., Côté, A., Choi, J., Huang, R., Uribe-Romo, F., Chae, H., O'Keeffe, M., Yaghi, O. (2006). Exceptional chemical and thermal stability of zeolitic imidazolate frameworks, *P. Natl. Acad. Sci. USA,* **103**(27), 10186–10191.

Parra, J., de Sousa, J., Bansal, R., Pis, J., Pajares, J. (1995). Characterization of activated carbons by the BET equation—an alternative approach, *Adsorpt. Sci. Technol.*, **12**, 51–65.

Potoff, J., Siepmann, J. (2001). Vapor-liquid equilibria of mixtures containing alkanes, carbon dioxide, and nitrogen, *AIChE J.*, **47**(7), 1676–1682.

Reed, D., Ehrlich, G. (1981). Surface diffusivity and the time correlation of concentration fluctuations, *Surf. Sci.*, **105**(2–3), 603–628.

Robson, R. (2008). Design and its limitations in the construction of bi- and poly-nuclear coordination complexes and coordination polymers (aka MOFs): a personal view, *Dalton Trans.*, **38**, 5113–5131.

Rouquerol, J., Llewellyn, P., Rouquerol, F. (2007). Is the BET equation applicable to microporous adsorbents? *Stud. Surf. Sci. Catal.,* **160**, 49–56.

Rowsell, J., Millward, A., Park, K., Yaghi, O. (2004). Hydrogen sorption in functionalized metal-organic frameworks, *J. Am. Chem. Soc.*, **126**, 5666–5667.

Ruthven, D. (1984). *Principles of Adsorption & Adsorption Processes* (John Wiley & Sons, New York).

Ruthven, D., Farooq, S. and Knaebel, K. (1994). *Pressure Swing Adsorption* (John Wiley & Sons, New Jersey).

Saito, A., Foley, H. (1991). Curvature and parametric sensitivity in models for adsorption in micropores, *AIChE J.*, **37**, 429–436.

Sakurovs, R., Day, S., Weir, R. (2009). Causes and consequences of errors in determining sorption capacity of coals for carbon dioxide at high pressure, *Int. J. Coal. Geol.*, **77**(1–2), 16–22.

Sarkisov, L. (2012). Accessible surface area of porous materials: Understanding theoretical limits, *Adv. Mat.*, **24**(23), 3130–3133.

Sarkisov, L., Harrison, A. (2011). Computational structure characterisation tools in application to ordered and disordered porous materials, *Mol. Phys.*, **37**(15), 1248–1257.

Schnobrich, J., Koh, K., Sura, K., Matzger, A. (2010). A framework for predicting surface areas in microporous coordination polymers, *Langmuir*, **26**(8), 5808–5814.

Schoenecker, P., Carson, C., Jasuja, H., Flemming, C., Walton, K. (2012). Effect of water adsorption on retention of structure and surface area of metal-organic frameworks, *Ind. Eng. Chem. Res.*, **51**(18), 6513–6519.

Schroeder, W., Martin, K., Lorensen, B. (1996). *The Visualization Toolkit: An Object-Oriented Approach to 3D Graphics*, 4th edn. (Kitware, Inc., Clifton Park, New York, USA).

Seehamart, K., Nanok, T., Karger, J., Chmelik, C., Krishna, R., Fritzsche, S. (2010). Investigating the reasons for the significant influence of lattice flexibility on self-diffusivity of ethane in Zn(tbip), *Microporous Mesoporous Mater.*, **130**(1–3), 92–96.

Serre, C., Millange, F., Thouvenot, C., Noguès, M., Marsolier, G., Louër, D., Férey, G. (2002). Very large breathing effect in the first nanoporous chromium(III)-based solids: MIL-53 or Cr(III)(OH) x [O(2)C-C(6)H(4)-CO(2)] x [HO(2)C-C(6)H(4)-CO(2)H](x) x H(2)O(y), *J. Am. Chem. Soc.*, **124**(45), 13519–13526.

Severson, B., Snurr, R. (2007). Monte Carlo simulation of n-alkane adsorption isotherms in carbon slit pores, *J. Chem. Phys.*, **126**(13), 134708.

Shrake, A., Rupley, J. (1973). Environment and exposure to solvent of protein atoms. Lysozyme and insulin, *J. Mol. Biol.*, **79**(2), 351–371.

Snurr, R., Bell, A., Theodorou, D. (1993). Prediction of adsorption of aromatic-hydrocarbons in silicalite from grand-canonical Monte-Carlo simulations with biased insertions, *J. Phys. Chem.*, **97**(51), 13742–13752.

Snurr, R., Hupp, J., Nguyen, S. (2004). Prospects for nanoporous metal-organic materials in advanced separations processes, *AIChE J.*, **50**(6), 1090–1095.

Steffan, D., Akgerman, A. (2001). Thermodynamic modeling of binary and ternary adsorption on silica gel, *AIChE J.*, **47**(5), 1234–1246.

Surble, S., Millange, F., Serre, C., Duren, T., Latroche, M., Bourrelly, S., Llewellyn, P., Férey, G. (2006). Molecular simulation of adsorption sites of light gases in the metal-organic framework IRMOF-1, *J. Am. Chem. Soc.*, **128**(46), 14889–14896.

Talu, O., Myers, A. (2001). Molecular simulation of adsorption: Gibbs dividing surface and comparison with experiment, *AIChE. J.*, **47**(5), 1160–1168.

Telfer, S. (2010). What are these things called MOFs? *Chem. N. Z.*, **74**, 9–14.

Van den Broeke, L., Krishna, R. (1995). Experimental verification of the Maxwell-Stefan theory for micropore diffusion, *Chem. Eng. Sci.*, **50**(16), 2507–2522.

van Koningsveld, H., van Bekkum, H., Jansen, J. (1987). On the location and disorder of the tetrapropylammonium (TPA) ion in zeolite ZSM-5 with improved framework accuracy, *Acta Cryst. B*, **43**, 127–132.

Vlugt, T. (2000). Adsorption and diffusion in zeolites: a computational study, Ph.D. thesis, University of Amsterdam.

Vlugt, T., Garcia-Perez, E., Dubbeldam, D., Ban, S., Calero, S. (2008). Computing the heat of adsorption using molecular simulations: the effect of strong coulombic interactions, *J. Chem. Theory. Comput.*, **4**(7), 1107–1118.

Vlugt, T., Krishna, R., Smit, B. (1999). Molecular simulations of adsorption isotherms for linear and branched alkanes and their mixtures in silicalite, *J. Phys. Chem. B*, **103**, 1102–1118.

Vlugt, T., Schenk, M. (2002). Influence of framework flexibility on the adsorption properties of hydrocarbons in the zeolite silicalite, *J. Phys. Chem. B*, **106**, 12757–12763.

Vlugt, T., Zhu, W., Kapteijn, F., Moulijn, J., Smit, B., Krishna, R. (1998). Adsorption of linear and branched alkanes in the silicalite-1, *J. Am. Chem. Soc.*, **120**(22), 5599–5600.

Walton, K., Abney, M., LeVan, M. (2006). CO2 adsorption in Y and X zeolites modified by alkali metal cation exchange, *Mircrop. Mesopor. Mat.*, **91**(1-3), 78–84.

Walton, K., LeVan, M. (2003). Consistency of energy and material balances for bidisperse particles in fixed-bed adsorption and related applications, *Ind. Eng. Chem. Res.*, **42**(26), 6938–6948.

Walton, K., LeVan, M. (2006). A novel adsorption cycle for CO2 recovery: experimental and theoretical investigations of a temperature swing compression process, *Sep. Sci. Tech.*, **41**(3), 485–500.

Walton, K., Millward, A., Dubbeldam, D., Frost, H., Low, J., Yaghi, O., Snurr, R. (2008). Understanding inflections and steps in carbon dioxide adsorption isotherms in metal-organic frameworks, *J. Am. Chem. Soc.*, **130**(2), 406–407.

Walton, K., Snurr, R. (2007). Applicability of the BET method for determining surface areas of metal-organic frameworks, *J. Am. Chem. Soc.*, **129**, 8552–8556.

Wilmer, C., Leaf, M., Lee, C., Farha, O., Hauser, B., Hupp, J., Snurr, R. (2012). Large-scale screening of hypothetical metal-organic frameworks, *Nat. Chem.*, **4**(2), 83–89.

Wilmer, C., Snurr, R. (2011). Towards rapid computational screening of metal-organic frameworks for carbon dioxide capture: calculation of framework charges via charge equilibration, *Chem. Eng. J.*, **171**(3), 775–781.

Wolf, R., Lee, M., Davis, R., Fay, P., Ray, J. (1993). Pressure-composition isotherms for palladium hydride, *Phys. Rev. B*, **48**(17), 12415–12418.

Wu, D., Wang, C., Liu, B., Liu, D., Yang, Q., Zhong, C. (2012a). Large-scale computational screening of metal-organic frameworks for CH_4/H_2 separation, *AIChE J.*, **58**(7), 2078–2084.

Wu, H., Yao, K., Zhu, Y., Li, B., Shi, Z., Krishna, R. (2012b). Cu-TDPAT, an rht-type dual-functional metal–organic framework offering significant potential for use in H2 and natural gas purification processes operating at high pressures, *J. Phys. Chem. C*, **116**, 16609–16618.

Yaghi, O., O'Keeffe, M., Ockwig, N., Chae, H., Eddaoudi, M., Kim, J. (2003). Reticular synthesis and the design of new materials, *Nature*, 423(6941), 705–714.

Yang, R. (1987). *Gas Separation by Adsorption Processes* (Imperial College Press, London).

Yang, R. (2003). *Adsorbents: Fundamentals and Applications* (John Wiley & Sons, Hoboken, New Jersey).

Yazaydin, A., Snurr, R., Park, T., Koh, K., Liu, J., LeVan, M., Benin, A., Jakubczak, P., Lanuza, M., Galloway, D., Low, J., Willis, R. (2009). Screening of metal-organic frameworks for carbon dioxide capture from flue gas using a combined experimental and modeling approach, *J. Am. Chem. Soc.*, **131**(51), 18198–18199.

Zeldowitsch, J. (1934). Adsorption site energy distribution, *URSS*, **1**, 961–973.

Chapter 3

Modeling the Adsorption of Small Molecules at Coordinatively Unsaturated Metal Sites: Density Functional Theory and Molecular Mechanics Approaches

Michael Fischer[a] **and Michael Fröba**[b]

[a] *Department of Chemistry, University College London, 20 Gordon Street, London WC1H 0AJ, United Kingdom*

[b] *Institute of Inorganic and Applied Chemistry, University of Hamburg, Martin-Luther-King-Platz 6, 20146 Hamburg, Germany*

m.fischer@ucl.ac.uk, froeba@chemie.uni-hamburg.de

The presence of coordinatively unsaturated metal sites is a particularly interesting feature of many MOFs. These sites can serve as preferential adsorption sites or catalytically active centers, thereby potentially enhancing the materials' performance in gas storage, gas separation, or catalytic applications. This chapter reviews the different theoretical approaches to model the adsorption of small molecules (molecules having no more than three nonhydrogen atoms) at these sites. After briefly presenting the most important experimental techniques that are used to investigate the localized adsorption at the metal sites, the first part of this review surveys computational studies using electronic structure

Metal–Organic Frameworks: Materials Modeling towards Potential Engineering Applications
Edited by Jianwen Jiang

ISBN 978-981-4613-45-3 (Hardcover), 978-981-4613-46-0 (eBook)
www.panstanford.com

methods, especially density functional theory (DFT). These methods are particularly useful to develop an atomistic understanding of the interactions governing adsorption, and the most important (sometimes conflicting) conclusions arising from DFT calculations are outlined. The second part discusses molecular mechanics calculations, primarily grand canonical Monte Carlo simulations. This method is predominantly employed to predict macroscopic quantities, such as adsorption isotherms. While calculations using standard parameters may provide qualitative insights, the shortcomings of literature force fields to model localized metal–adsorbate interactions are pointed out. In the last part, different approaches to overcome these deficiencies are discussed. It is shown how recent improvements that bridge the gap between electronic structure calculations and molecular mechanics methods enable a more realistic representation of localized interactions between metal centers and adsorbed molecules.

3.1 Introduction

One of the most intriguing features of porous metal–organic frameworks (MOFs) is their chemical versatility, that is, the possibility to synthesize ordered inorganic-organic hybrid materials with different metal centers, different linker molecules, and different (coordinating or noncoordinating) functional groups. A particularly interesting property for many applications in adsorption, separation, and catalysis is the presence of coordinatively unsaturated metal sites (also termed "open" or "accessible" metal sites) in many of these materials. These sites are typically coordinated by solvent molecules after the synthesis but can be made accessible through solvent removal (activation). As it will be outlined in the following, accessible metal sites have been evidenced to act as preferential adsorption sites for many adsorbate species. The main emphasis of this chapter will lie on theoretical approaches to assess the interaction of adsorbed molecules with unsaturated metal sites. Due to the vast amount of literature available, the discussion concentrates on "small" molecules, understood here as molecules having no more than three nonhydrogen atoms. Among these molecules, hydrogen, methane, and carbon dioxide are discussed in most detail, but other species such as carbon monoxide, water, and small hydrocarbons will also be included where appropriate.

The field of MOF synthesis started to thrive in the late 1990s, with groundbreaking work being performed by Yaghi and coworkers. In these early stages, researchers from the Yaghi group also discovered the first system with coordinatively unsaturated metal sites, an MOF termed Zn(bdc) (with bdc = 1,4-benzenedicarboxylate), reported in 1998 [1]. The following year, Williams and coworkers published their work on $Cu_3(BTC)_2$ (BTC = 1,3,5-benzentricarboxylate), also termed HKUST-1 [2]. Due to its facile synthesis, combined with high porosity, $Cu_3(BTC)_2$ has been widely studied for numerous applications, making it one of the most well-characterized MOFs. Therefore, this MOF will be used as a showcase example to outline the concept of coordinatively unsaturated metal sites here:

The inorganic building unit of $Cu_3(BTC)_2$ is the Cu_2 paddlewheel, visualized in Fig. 3.1. In this unit, which is also quite common in coordination chemistry, for example, in copper acetate and copper benzoate, each of the Cu centers is (almost) square-planar-coordinated by four oxygen atoms of the linker carboxylate groups. The fifth coordination site, perpendicular to the plane, is occupied by the other copper atom. While there is no metal–metal

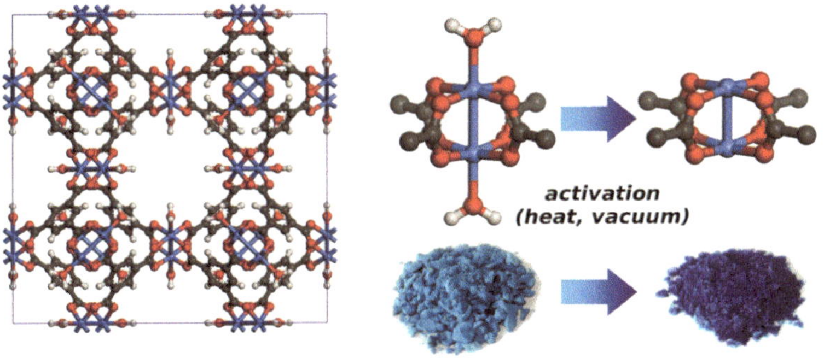

Figure 3.1 (Left) Structure of $Cu_3(BTC)_2$, including water molecules coordinated to the metal sites. Color scheme: gray = carbon, red = oxygen, white = hydrogen, and light blue = copper. (Right) View of the Cu_2 paddlewheel building unit before and after activation. The color change of an experimental sample is also shown.

bonding in the classical sense, the two copper atoms interact with each other through antiferromagnetic coupling [3]. The sixth coordination site, which points into the pore space of $Cu_3(BTC)_2$, is occupied by coordinating solvent molecules (e.g., water, ethanol, DMF) after the synthesis. Upon activation, that is, evacuation and heating to moderately elevated temperatures (with the optimal conditions depending on the type of solvent), the coordinated solvent molecules can be removed. This leads to the formation of coordinatively unsaturated copper sites, which can act as preferential adsorption sites for different molecules. For the example of $Cu_3(BTC)_2$, two other features are noteworthy: First, a significant color change from turquoise to dark blue is observed upon desolvation (Fig. 3.1). Second, the Cu–Cu distance is reduced from approximately 2.65 Å for an MOF with fully coordinated Cu_2 paddlewheels to about 2.50 Å for a desolvated system [4].

The Cu_2 paddlewheel is arguably the most frequent MOF building unit with accessible metal sites. In addition to $Cu_3(BTC)_2$, it is found in various other important systems, such as MOF-505 and the isoreticular MOFs of the NOTT series reported by Schröder and coworkers [5–7], several highly porous PCN materials synthesized in the Zhou group [8–10], as well as Cu(mip) (mip = 5-methylisophthalate), which is particularly interesting for catalytic applications [11]. M_2 paddlewheel building units have also been reported for other metals, such as zinc [12] and chromium [13]. While it is hardly possible to name all MOFs that have, or potentially have, coordinatively unsaturated metal sites, a few of the most important inorganic building units with accessible metal centers are listed below and visualized in Fig. 3.2:

- Triangular units consisting of three metal centers bridged by a common μ-oxo atom; these units are present—among others—in Matériaux de l'Institut Lavoisier (MIL)-100/MIL-101 [14, 15] and in an indium MOF with *soc* topology [16].
- Square-planar M_4Cl units that constitute various MOFs reported by Long's group [17–19].
- Chains of five-coordinated metal centers that are connected by multidentate linkers to build up a framework with

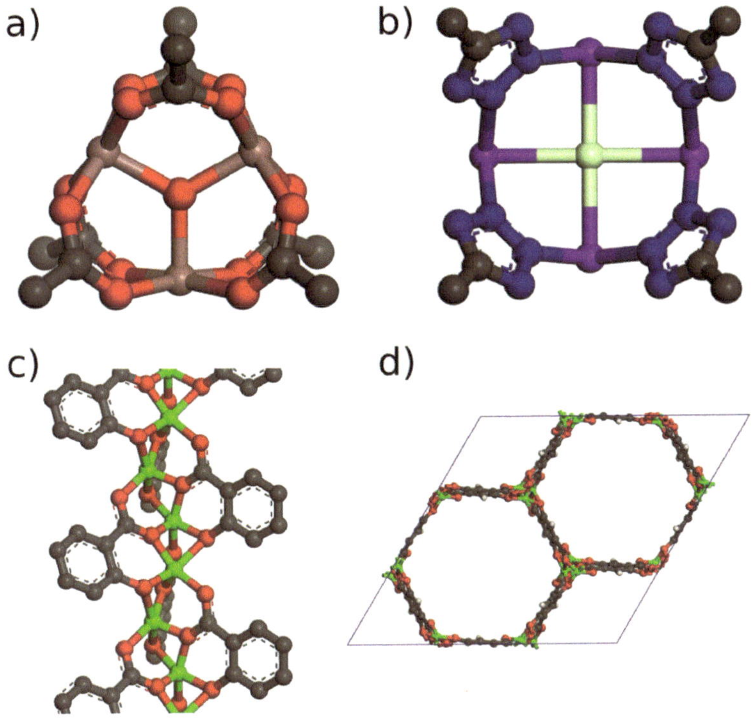

Figure 3.2 Examples of building units with coordinatively unsaturated metal sites. (a) Triangular In_3O cluster from the structure of In-*soc*-MOF. (b) Square-planar Mn_4Cl cluster from the structure of $(Mn_4Cl)_3(BTT)_8$ (BTT = 1,3,5-tris(tetrazolate-5-yl)benzene). (c) Chain of pentacoordinated Zn atoms as found in Zn-MOF-74. (d) One unit cell of Zn-MOF-74. Color scheme: gray = carbon, red = oxygen, blue = nitrogen, brown = indium, purple = manganese, and light green = zinc.

hexagonal channels; these are the inorganic building units of the MOF-74 analogues (also known as CPO-27-M or M_2(dhbdc)) [20–24].

Another possibility to generate coordinatively unsaturated metal sites, first exploited by Chen and coworkers, is the use of Salen-type molecules as linkers [25]: In the mixed-metal MOF $Zn_3(bdc)_3Cu$(Pyen), square-planar copper centers are incorporated in the linker molecules. Finally, it should be mentioned that some MOFs possess a negatively charged framework, the charge being balanced by extra-framework cations (e.g., [26]). While the cations in these systems certainly play an important role for the gas adsorption characteristics, they do not correspond to coordinatively unsaturated metal sites in the sense defined above. Therefore, these systems are not discussed in the context of this chapter. Similarly, MOFs that are postsynthetically functionalized with groups containing lithium or other metal atoms will not be considered, although it should be mentioned that various theoretical studies have addressed these systems [27, 28].

In addition to their intriguing impact on the structural and electronic properties of an MOF, coordinatively unsaturated metal sites can be beneficial for various applications for which MOFs have been proposed: For example, the interaction of adsorbed molecules with the metal sites can help to enhance the gas uptake, which is important for the storage of gases such as hydrogen and methane. In gas separation, the preferential adsorption of one species at the metal sites may have a favorable impact on the adsorption selectivity. For applications in catalysis, the metal centers can act as catalytically active sites. Given the variety of potential applications, it is not surprising that these systems have been, and continue to be, widely studied experimentally and theoretically.

Prior to discussing the literature that addresses systems with accessible metal sites, previous review articles of modeling studies of MOFs should be mentioned: an early review that includes both electronic structure methods and molecular mechanics methods was published by Keskin et al. [29]. Similarly general, more recent overviews were compiled by Schmid and coworkers [30] and Jiang and coworkers [31]. Very recently, a review of theoretical studies of methane, hydrogen, and acetylene storage was published by Snurr and coworkers [32]. More specific surveys, explicitly focusing on simulations of hydrogen adsorption, were prepared by Goddard and coworkers [27] and by Sastre [33]. Hedin et al. concentrated on the modeling of carbon dioxide adsorption in different porous materials [34]. Finally, molecular mechanics studies of gas separation have been reviewed by Liu and Zhong [35].

The following section will give a brief overview of experimental evidence for the localized interaction of adsorbed molecules with coordinatively unsaturated metal sites. The next sections deal with theoretical approaches to model this interaction, starting with electronic structure methods, especially density functional theory (DFT) calculations, before moving on to more empirical approaches that use analytical expressions to model interatomic interactions, that is, molecular mechanics methods. Here, the discussion will concentrate on force field–based grand canonical Monte Carlo (GCMC) simulations, which are widely used to predict adsorption isotherms and adsorption energetics, among other properties. In the final section, some recent approaches to improve the description of localized metal–adsorbate interactions in Monte Carlo simulations are described.

3.2 A Short Review of Experimental Methods

While it is often possible to hypothesize that metal sites in an MOF may become accessible upon desolvation when inspecting the crystal structure (which is, in practically all cases, determined from fully solvated samples), direct experimental evidence to prove the presence of coordinatively unsaturated metal sites is more difficult to obtain. Some qualitative information may be gained from thermogravimetric methods. However, since the observable weight loss corresponding to coordinated solvent molecules typically overlaps with the loss of noncoordinated solvent, it is hardly possible to draw quantitative conclusions without further information from other experimental techniques.

Clearly, diffraction methods are in principle capable of directly determining whether a metal center has a free coordination site and in analyzing the interaction with adsorbed light molecules through in situ experiments. Since desolvated single-crystal samples are rarely available, the structure refinement needs to rely on powder data. In principle, both x-ray and neutron radiation can be used in these experiments. Due to the weak interaction of lights adsorbed molecules with x-ray radiation, it is often necessary to use neutron radiation to locate these molecules. For example, the first in situ diffraction studies of desolvated MOFs with unsaturated metal sites ($Cu_3(BTC)_2$ and $(Mn_4Cl)_3(BTT)_8$) used neutron radiation because the main aim of these investigations was the refinement of the

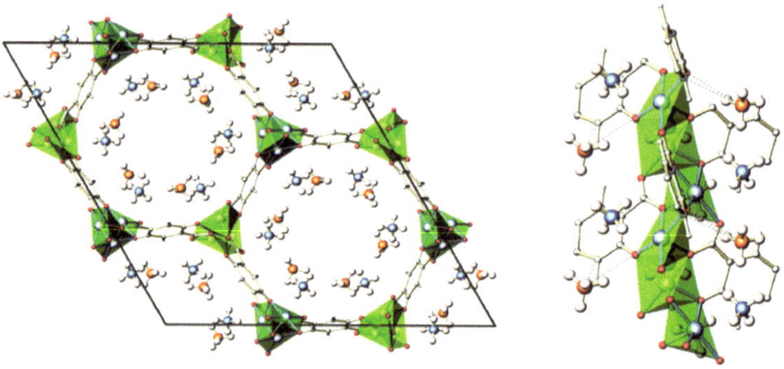

Figure 3.3 Visualization of the preferred methane adsorption sites in Mg-MOF-74, obtained using neutron diffraction [39]. (Left) Full unit cell with primary adsorption sites above the unsaturated metal sites (blue) and secondary adsorption sites above the organic linker (orange). (Right) Detailed view of adsorbed molecules to show their location and orientation with respect to the metal centers and the linker. Reprinted with permission from Ref. [39]. Copyright 2009 American Chemical Society.

positions of adsorbed deuterium molecules (D_2 was used instead of H_2 to avoid the large incoherent scattering of the ^1H isotope) [17, 36]. Later studies included Zn-MOF-74 [37], Fe-MOF-74 [38], and NOTT-101 as a representative member of the NOTT series [6], to name a few examples. In addition to structural information on the desolvated MOF structures, these studies also revealed the preferential adsorption of D_2 at the metal sites. It was found that the metal–D_2 distances, which typically range from 2.2 Å to 2.6 Å, are significantly shorter than the sum of the van der Waals radii, thereby providing experimental evidence for the specific interaction between deuterium/hydrogen and the metal sites. Subsequently, neutron diffraction experiments were also employed to investigate the adsorption of methane in MOF-74 analogues (Fig. 3.3) [39] and in $Cu_3(BTC)_2$ and other MOFs with Cu_2 paddlewheel building units [40, 41], the adsorption of carbon dioxide in $Cu_3(BTC)_2$ and Mg-MOF-74 [42], as well as the adsorption of oxygen in $Cr_3(BTC)_2$ [13]. Since the CO_2 molecule has a (comparatively) large number of electrons, the preferred carbon dioxide adsorption sites can also be obtained using x-ray diffraction experiments, as shown for the case of Ni-MOF-74 [43].

While diffraction methods provide direct information on the adsorption positions, spectroscopic methods give more indirect evidence for the interaction between adsorbed molecules and the host structure. However, a careful analysis of the observed spectra in comparison with the spectrum of the free molecule can also permit the identification of different adsorption "sites" that are distinguished by the interaction energy. The first spectroscopic evidence for a specific adsorption of hydrogen at the accessible metal sites of $Cu_3(BTC)_2$ was reported by Bordiga and coworkers and actually predates the neutron diffraction results mentioned above [4]. In later studies, the same group used the variable-temperature infrared (VTIR) method to determine the enthalpy of hydrogen adsorption in MOFs with unsaturated metal sites [44, 45]. In agreement with the heats of adsorption calculated from adsorption isotherms, it was found that MOFs like $Cu_3(BTC)_2$ and Ni-MOF-74 exhibit low-coverage adsorption enthalpies ranging from -10 kJ mol^{-1} to -13.5 kJ mol^{-1}, whereas the adsorption enthalpies of MOFs without unsaturated metal sites are much lower (e.g., -5 kJ mol^{-1} for the isoreticular metal–organic framework [IRMOF]-1). Therefore, these spectroscopic studies also point to the important role of coordinatively unsaturated metal sites in enhancing the interaction of the host lattice with adsorbed molecules. Further detailed IR spectroscopic investigations were directed at the adsorption of hydrogen in a number of MOFs with unsaturated metal sites [46–48] and various gases, such as carbon dioxide, carbon monoxide, and acetylene, in MOF-74 analogues [43, 49–51].

Though less frequently, ultraviolet/visible (UV/Vis) spectroscopy [4], x-ray absorption spectroscopy [52], nuclear magnetic resonance (NMR) spectroscopy [53], and inelastic neutron scattering [54] have also been employed to study the interaction of adsorbed molecules with accessible metal sites.

3.3 Electronic Structure Methods

While only very few electronic structure investigations related to the adsorption of guest molecules in MOFs were published until 2006, the surging interest in these systems for gas storage and separation, in conjunction with the ongoing development and improvement of computational chemistry techniques, has inspired a large number of computational studies using electronic structure methods in more recent years. The overwhelming majority of these studies relied on DFT, which permits the treatment of relatively large systems at a reasonable computational cost when compared to correlated wavefunction-based methods. A serious drawback of DFT is the inability to correctly model van der Waals interactions using standard exchange correlation (XC) functionals, because long-range correlation effects are not adequately accounted for (for a recent review, see Ref. [55]). Different approaches to introduce a dispersion correction in DFT have been proposed (reviewed in Refs. [56, 57], among others), and they have occasionally been employed for MOFs, as will be discussed below. However, standard DFT can also be used quite successfully in cases where mainly localized interactions are of interest, specifically in the case of adsorption at unsaturated metal sites.

3.3.1 *Hydrogen*

As mentioned previously, experimental evidence for the preferential adsorption of H_2 at coordinatively unsaturated metal sites has been obtained from various experimental methods. Due to their increased isosteric heats of hydrogen adsorption, these materials might be particularly promising for the storage of hydrogen under noncryogenic conditions, where other adsorbents like carbons and MOFs with no unsaturated metal sites adsorb only very limited amounts of hydrogen due to the weak solid–fluid interactions.

Early experimental results, as well as some theoretical considerations, were summarized in a review by Dinca and Long in 2008 [58]. An interesting point that was raised in this review (and various other publications on the topic) is the nature of the interaction. On the one hand, a side-on coordination of the H_2 molecule at the metal site causes an attractive electrostatic interaction, and the electron cloud of the H_2 molecule can to some extent be polarized by the metal center. On the other hand, there is the possibility of actual orbital interactions, which are well known as "Kubas type" bonding from the coordination chemistry of metal–H_2 complexes [59]. This bonding is based on an interaction of the occupied σ-orbital of the H_2 molecule with empty *d*-orbitals of the metal, as well as a possible back-donation from occupied metal *d*-orbitals to the H_2 σ^*-orbital. While it has been proposed to exploit this interaction for the development of materials with a higher heat of H_2 adsorption [60], the majority of experimental results do not indicate a significant amount of Kubas-type bonding for existing MOFs: For example, the metal–H_2 distances are longer than in actual metal–H_2 coordination complexes, and there is no pronounced elongation of the H–H bond. As it will be shown in the following discussion of DFT results, different groups have proposed different interpretations, either based on (rather weak) orbital interactions or electrostatic interactions. Clearly, the elucidation of the nature of the metal–H_2 interactions in MOFs remains a topic of ongoing investigation.

Due to the ubiquity of Cu_2 paddlewheels as inorganic MOF building units, it comes as no surprise that some of the earliest DFT studies of metal–hydrogen interaction treated these systems. As early as 2006, Yang and Zhong studied the adsorption of hydrogen in MOF-505 (also termed NOTT-100, an MOF with a 3,3',5,5'-biphenyltetracarboxylate linker) [61]. The calculations, which employed the Perdew–Burke–Ernzerhof (PBE) XC functional without dispersion corrections, correctly predicted the metal sites as the energetically most favorable H_2 adsorption sites. Side-on coordination of the H_2 dumbbell to the metal center was found to be considerably favored over end-on coordination. Calculations for other positions around the paddlewheel provided some insights into other preferential adsorption regions. Interestingly, despite the complete neglect of dispersion interactions, the binding energies are higher than the VTIR value of -10.1 kJ mol^{-1} obtained by Bordiga and coworkers for $Cu_3(BTC)_2$ [45] and also higher than DFT energies reported by Sastre and coworkers more recently, which were obtained with a more sophisticated, dispersion-corrected XC functional [62]. This might point to a significant amount of spurious long-range binding in the DFT-PBE calculations, a phenomenon that has been frequently observed for various XC functionals [63, 64].

Similar periodic calculations were presented later by Brown and coworkers for $Cu_3(BTC)_2$ [54]. The Cu–H_2 distance for the side-on

Table 3.1 DFT interaction energy E_{int} and distance between the Cu center and the center of mass of the H_2 molecule, d_{Cu-H2}, obtained by different authors for Cu_2 paddlewheel systems with a H_2 molecule in side-on coordination. Experimental reference values from VTIR spectroscopy (energy) and neutron diffraction (distance) are given for comparison

	E_{int} (kJ mol^{-1})	d_{Cu-H2} (Å)	Model system	XC functional
Yang and Zhong, 2006 [61]	−13.4	~2.44	periodic model, MOF-505	PBE
Brown et al., 2009 [54]	~ −7.5	2.47	periodic model, $Cu_3(BTC)_2$	GGA[1]
Fischer et al., 2010 [65]	−5.9	2.40	molecular model, Cu_2(benzoate)$_4$	PBE
Kim et al., 2010 [69]	~ −6.8	2.37	molecular model, Cu_2(benzoate)$_4$	PBE
Gomez et al., 2012 [62]	−10.8	2.60	molecular model, Cu_2(benzoate)$_4$	SSB-D
Experiment	−10.1 [45]	2.39 [36]	–	–

[1] The XC functional was not specified in the publication.

coordination of H_2 agreed well with the value obtained from neutron diffraction (Table 3.1). Moreover, the calculations approximately reproduced the small, albeit detectable red shift of the H–H stretching vibration observed spectroscopically. No significant elongation of the H–H bond and no charge transfer were found, indicating that there is no chemical metal–H_2 bond (as in Kubas-type complexes) but that the interaction is dominantly electrostatic. Finally, the DFT results were interpreted in conjunction with inelastic neutron scattering data and helped to understand the rotational dynamics of hydrogen molecules adsorbed at the metal sites.

Fischer et al. used a nonperiodic $Cu_2(L)_4$ ($L =$ benzoate or substituted benzoate) model system and performed DFT calculations with the PBE functional [65]. While the Cu–H_2 distance is in excellent agreement with the experimental value, the interaction energy is underestimated (−5.9 kJ mol^{-1} as compared to −10.1 kJ mol^{-1} obtained from IR spectroscopy). This observation was mainly explained by the lack of long-range dispersion interactions. A particular emphasis of this study was the evaluation of the influence of different substituents attached to the phenyl rings on the Cu–H_2

interaction. For this purpose, a variety of ligands L with electron-withdrawing and electron-donating substituents was compared. The calculations predicted electron-withdrawing substituents such as nitro and cyano groups to enhance the Cu–H_2 interaction energy by up to 20% with respect to the unsubstituted system, whereas electron-donating substituents had the opposite effect. A later, more detailed analysis of the results showed a correlation between the Hirshfeld charge of the Cu centers, that is, the degree of electron withdrawal from the metal center induced by the substituents and the DFT interaction energy [66]. These findings support the assumption that the Cu–H_2 interaction is dominantly electrostatic in nature. There is an interesting correspondence between these results and the substituent influence on the catalytic activity of functionalized Zr MOFs with unsaturated metal sites, which follows a strikingly similar pattern, as reported by De Vos, Van Speybroeck, and coworkers [67].

Calculations for an identical paddlewheel model, but different positions of the H_2 molecule were reported recently by Sastre and coworkers [62]. The Swart–Solà–Bickelhaupt (SSB)-D functional was used in their computations, an empirical "all round" functional that was derived with the aim of improving the description of weak interactions [68]. For a hydrogen molecule adsorbed at the Cu center, an interaction energy of -10.8 kJ mol^{-1} was obtained, which is in excellent agreement with the VTIR value for $Cu_3(BTC)_2$ quoted above.

DFT-PBE calculations for molecular model systems were also reported by Kim and coworkers [69]. A special emphasis of their study was the investigation of the influence of metal substitution on the strength of the metal–hydrogen binding. Strong binding was predicted for both Mg_2 and Zn_2 paddlewheel systems. From an analysis of the electronic structure, it was concluded that the metal–hydrogen interaction is due to a coupling of the occupied H_2 σ-orbital with unoccupied metal *s*-orbitals. This type of binding is distinct from Kubas-type binding (which is based on interactions of the H_2 σ-orbital with empty metal *d*-orbitals and a possible back-donation from occupied metal *d*-orbitals to the H_2 σ^*-orbital). In a later study, this approach was extended to the first transition metal row [70]. It was observed that the transition metal paddlewheels, except for CO_2, provide for a distinctly weaker metal–hydrogen interaction when compared to Zn_2 and Mg_2.

Cubic MOFs of the simplified stoichiometry $(M_4Cl)_3(BTT)_8$ have been reported for different divalent metals M, namely, manganese [17], copper [18], and iron [19]. The negative charge of the framework is compensated by additional cations that are not included in the stoichiometry given above, leading to the possibility of cation exchange [71]. Due to their high isosteric heat of hydrogen adsorption, these MOFs have received considerable attention by computational chemists.

In 2007, Sun et al. reported a DFT-PBE study of a molecular $(M_4Cl)(\text{tetrazolate})_8$ model system [72]. These authors compared several metals (Mn, Cu, and hypothetical systems with Sc, Ti, V, and Cr) and investigated the influence of the spin state of the metal center on the interaction with hydrogen. The strongest binding was observed for systems with the early transition metals, with interaction energies ranging from -22 kJ mol^{-1} to -47 kJ mol^{-1}. Compared to these values, the predicted binding is considerably weaker for manganese (-8.4 kJ mol^{-1}) and copper (-4.4 kJ mol^{-1}). In contrast to most other theoretical studies, a significant role of orbital interactions between the $H_2\sigma$-orbital and metal d-orbitals, namely, the d_{z2}-orbital, was postulated on the grounds of an analysis of the density of states: The interaction is strongest when the metal d_{z2}-orbital is completely unoccupied (Sc, Ti, V) and decreases for systems with a singly (Cr, Mn) and doubly (Cu) occupied d_{z2}-orbital. The same group later extended their approach to $(Fe_4Cl)_3(BTT)_8$ [73]. Here, it was found that the adsorption of hydrogen could induce a "spin flip" of the Fe(II) centers from a high-spin state in the bare MOF structure to a low-spin state for the structure with coordinated H_2 molecules. This behavior was explained by a change in the energetic order of the metal d-orbitals upon H_2 adsorption.

While the investigations by Sun et al. delivered a Kubas-type orbital interaction as the main source of the metal–hydrogen binding, quite different conclusions were drawn by Zhou and Yildirim in a study that used a similar computational methodology [74]. The authors compared different magnetic configurations of the $(Mn_4Cl)(\text{tetrazolate})_8$ cluster and found the "dimer" magnetic configuration to be the lowest in energy (whereas only the ferromagnetic configuration was considered by Sun et al. [72]). While the magnetic configuration has little influence on the strength of the interaction with coordinated hydrogen molecules, the spin state of the metal center plays a more important role, the high-spin state affording a significantly higher binding energy. Because the spin state of the metal center depends on the length of the M–X bond, where X is the central halogen, the authors proposed to tune the metal–hydrogen interaction strength by replacing $X = \text{Cl}$ with F or Br. An analysis of the H–H bond length of the coordinated molecules, the red shift of the H–H stretching vibration, and the charge density delivered no indications of a significant orbital interaction, leading Zhou and Yildirim to conclude that the metal–H_2 interaction is dominantly electrostatic.

As a final word of caution, it must be pointed out that none of the theoretical studies took the negative charge of the framework into account (which is, in real systems, balanced by extra-framework cations). Instead, all model systems were artificially defined to be charge neutral. This rather significant simplification might have a quite pronounced impact on the results, both from a quantitative and a qualitative point of view.

Frameworks having the MOF-74 structure type have been synthesized with a variety of metals, including $M = \text{Mg}$, Mn, Co, Fe, Ni, or Zn [20–24]. Therefore, these systems are very attractive for a systematic study of the influence of the nature of the metal center on the interaction with adsorbed molecules.

For the case of adsorbed hydrogen, this approach was followed by Zhou et al. in a study that comprised the experimental measurement of adsorption isotherms as well as DFT-PBE calculations [23]. Of the metals mentioned above, all except Fe were included in the investigation. The results are visualized in Fig. 3.4: While there is no quantitative agreement between the DFT interaction energies and the experimental isosteric heats of hydrogen adsorption due to the neglect of dispersive interactions in the computations, the qualitative trend is reproduced extremely well. Moreover, an inverse correlation between the metal–H_2 distance and the interaction energy was observed. It was found that the evolution of the interaction strength as a function of the nature of the metal center can be explained with the (empirical) Irving–Williams sequence,

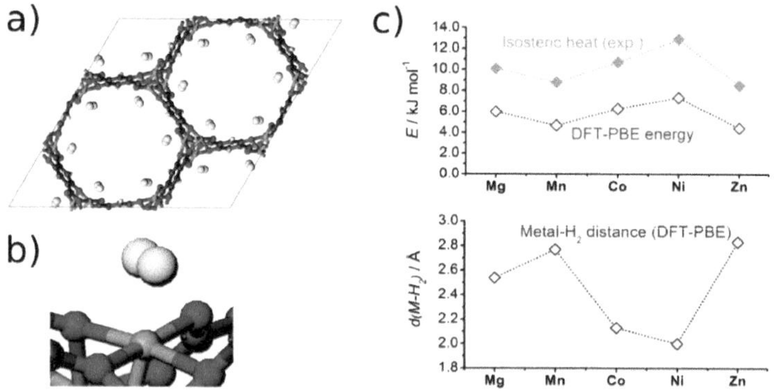

Figure 3.4 Visualization of key results from the DFT study of H_2 adsorption in MOF-74 analogues published by Zhou et al. [23]. (a) Structure of Ni-MOF-74 with one H_2 molecule per metal site (schematic representation). (b) Detailed view of one H_2 molecule coordinated to a Ni site. (c, top) Magnitude of DFT–PBE interaction energies compared to experimental heats of hydrogen adsorption. (c, bottom) Metal–H_2 equilibrium distances obtained from DFT-PBE calculations.

which relates the stability of octahedral complexes containing divalent cations to the ionic radius. From these observations, it was concluded that the metal–hydrogen interaction is dominantly electrostatic in nature and that it might be possible to predict the metal–hydrogen interaction strength as a function of the ionic radius, at least for the case of isostructural compounds.

Kong et al. reported a computational study of hydrogen adsorption in Zn-MOF-74 using a sophisticated van der Waals density functional (vdW-DF) that incorporates long-range dispersion [75]. The predicted binding energy for H_2 adsorbed at the metal center, corrected for the contribution of the zero-point vibrational energy (ZPVE), amounts to approximately -10 kJ mol^{-1}, a value that agrees much better with experimental data than the energy obtained using an uncorrected generalized gradient approximation (GGA) functional, which ranges below -5 kJ mol^{-1}. Interestingly, however, the metal–H_2 distance is overestimated quite significantly, with 2.9 Å compared to an experimental value of 2.6 Å [37]. In addition to the unsaturated metal site, three other H_2-binding sites were predicted from the computations. Furthermore, a detailed investigation of the orientational dependence of the potential permitted a prediction of the inelastic neutron scattering spectrum, which was found to be in very good agreement with experiment.

Before moving the focus to adsorbates other than hydrogen, a few other theoretical studies that employed somewhat more abstract model systems should be mentioned very briefly: For example, Head-Gordon and coworkers investigated the interaction of hydrogen with free metal cations and simple, positively charged metal–carbonyl and cyanide complexes [76], as well as metal–arene complexes that can be grafted to the MOF linker [77]. Kosa et al. used similar small complexes and compared two different metal centers (Mg^{2+} and Ni^{2+}) [78]. Finally, Kim et al. investigated the interaction of hydrogen with different porphyrin-type molecules, which could be incorporated in MOF linkers [79].

3.3.2 *Methane*

The methane molecule, which has a vanishing quadrupole moment due to its tetrahedral symmetry, typically interacts with other species virtually exclusively through van der Waals interactions. It could therefore be expected that methane will be preferentially adsorbed in narrow pores, where the molecule can interact with several framework atoms, rather than at unsaturated metal sites. Quite surprisingly, neutron diffraction measurements have shown that a preferential adsorption of methane occurs at the unsaturated metal sites of $Cu_3(BTC)_2$ and other MOFs with Cu_2 paddlewheel units [40, 41], as well as MOF-74 analogues [39], with typical metal–C(CH_4) distances ranging slightly above 3.0 Å, thus being significantly lower than the sum of the van der Waals radii.

DFT calculations for methane adsorbed in MOFs with unsaturated metal sites were reported by Zhou and coworkers for MOF-74 analogues [39], as well as different MOFs with accessible Cu sites [40]. In both cases, the computations were used to corroborate experimental observations obtained in adsorption measurements, as well as neutron diffraction experiments. In the first study, DFT calculations using the local density approximation (LDA) were performed for methane adsorbed at the metal sites of M-MOF-74, with M = Mg, Mn, Co, Ni, or Zn. It was found that the DFT calculations generally overestimate the interaction strength when compared to experimental heats of adsorption, a behavior that is in line with the well-known tendency of LDA to "overbind" weakly interacting systems [63]. While the calculations delivered Ni-MOF-74 as the system exhibiting the strongest interaction with methane, in line with experiment, the qualitative behavior for the other metals could not be captured. Moreover, the metal–C(CH_4) distance was underestimated by about 0.4 Å for Mg-MOF-74, the system for which diffraction experiments were carried out. In a subsequent study of methane adsorption in $Cu_3(BTC)_2$, PCN-11, and PCN-14 [40], it was found that, despite significant quantitative deviations, DFT-LDA delivers an approximately correct prediction of the energetic order of the different adsorption sites (i.e., positions at the metal centers compared to positions elsewhere in the structure). While DFT calculations using the LDA XC functional might therefore provide some insights on a qualitative level, these results indicate that it is not a particularly suitable method to study systems where relatively weak interactions are dominant. On the other hand, calculations performed by Chen and coworkers for a Cu MOF termed UTSA-20 showed that dispersion-corrected DFT (DFT-D) can deliver binding energies that are in reasonable agreement with experimental isosteric heats of CH_4 adsorption [80].

A more recent investigation by Düren, Nachtigall, and coworkers employed a sophisticated scheme to account for dispersion interactions in order to accurately model the adsorption of methane in $Cu_3(BTC)_2$ [81]. This contribution, which encompasses both DFT and GCMC calculations, will be discussed in more detail in section X.5.

3.3.3 *Carbon Dioxide*

Carbon dioxide molecules can coordinate to unsaturated metal sites end-on due to electrostatic interactions between one of the negatively polarized oxygen atoms and the metal center. This has been evidenced experimentally, for example, for $Cu_3(BTC)_2$ and different analogues of MOF-74 [42, 43]. In the neutron diffraction study of CO_2 in $Cu_3(BTC)_2$ and Mg-MOF-74 performed by Zhou, Yilidirim, and coworkers, DFT-LDA calculations were used to predict the vibrational modes of adsorbed CO_2 molecules. It was found that all vibrational modes of the free molecule are red-shifted ("softened") in the adsorbed state. In particular, the O–C–O bending mode softens, and a slightly bent configuration was found to be energetically most favorable. In total, the calculations reproduced the relatively large orientational freedom of the adsorbed CO_2 molecules, which was seen experimentally as a high degree of disorder.

A dispersion-corrected DFT (DFT-D) study of CO_2 (as well as CO and N_2) in Mg-MOF-74 was reported by Valenzano et al. [50], who used the hybrid B3LYP functional with a Grimme-type dispersion correction [82]. While CO and N_2 were found to coordinate to the metal center in a virtually linear fashion, the CO_2 molecule assumed a tilted orientation, with an angle Mg···O=C=O of 129° (however, no significant bending of the CO_2 molecule was reported). This deviation from end-on coordination was rationalized with a lateral electrostatic interaction between carboxylate oxygen atoms of the MOF linker and the (positively polarized) carbon atom of the CO_2 molecule (see Fig. 3.5). While a short Mg–O(CO_2) distance of 2.31 Å clearly evidences the importance of electrostatic interactions, it was found that about half of the total interaction energy stems from dispersion interactions. By correcting the DFT-D energies for the ZPVE and thermal contributions, enthalpy values that can be compared to the experimental adsorption enthalpy were obtained. Experimental adsorption enthalpies were determined using VTIR spectroscopy (it should be noted that the differential adsorption enthalpy corresponds to the negative equivalent of the more frequently used heat of adsorption [83]). The values of the calculated/experimental enthalpies amount to $-37.9/-47$ kJ mol^{-1} for CO_2, $-30.0/-29$ kJ mol^{-1} for CO, and $-25.2/-21$ kJ mol^{-1} for N_2. From the large differences in interaction strength, it was suggested that Mg-MOF-74 could be a suitable material for selective CO_2 adsorption, an application that has been explored experimentally by other groups [84, 85].

While this study highlights the successes of DFT-D in qualitative and semiquantitative predictions, the relatively large absolute deviations between calculated and experimental adsorption enthalpies for CO_2 suggest further space for methodological improvement. This

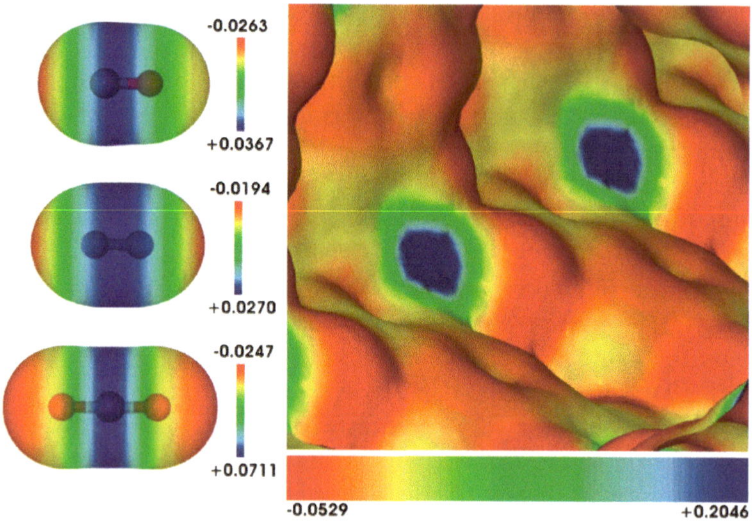

Figure 3.5 Visualization of the electrostatic potential from the DFT study published by Valenzano et al. [50]. Red, green, and blue represent negative, zero, and positive values (atomic units), respectively. (Left) Visualization of the electrostatic potential of CO, N_2, and CO_2. (Right) Visualization of the electrostatic potential of Mg-MOF-74, mapped on the charge density surface. The blue areas correspond to the accessible Mg sites. Reprinted with permission from Ref. [50]. Copyright 2010 American Chemical Society.

issue was addressed by Sauer and coworkers in a more recent study of CO_2 and CO adsorption in M-MOF-74 (M = Mg, Ni, or Zn) [86]. Here, dispersion-corrected DFT calculations (B3LYP-D functional) on periodic systems were compared to hybrid MP2:DFT calculations, which use MP2 perturbation theory calculations for a cluster to obtain a "high level" correction term that is then applied to the DFT-D results. It was found that the DFT-D calculations underestimate the heat of adsorption in most cases, with an average error of 5 kJ mol^{-1}. The hybrid scheme performs better, with an average error of 2 kJ mol^{-1}; however, the largest observed errors range near 10 kJ mol^{-1} for both methods. Both methods predict the same sequence of isosteric heats for the three cations, which is in agreement with the experimentally observed behavior.

Another correction scheme to improve the description of dispersion interactions was employed in a combined computational and experimental study of CO_2 adsorption in $Cu_3(BTC)_2$ by Nachtigall and coworkers [87]. Here, the DFT/CC method was used, which makes use of high-level coupled-cluster calculations for small model systems to calculate correction functions, which are then applied to correct the DFT energies. Having obtained these correction functions, the DFT calculations were carried out for a periodic unit cell of $Cu_3(BTC)_2$ using the PBE XC functional, assuming different adsorption sites. At low coverage, the unsaturated metal site was found to be the energetically preferred adsorption site, with a binding energy of -28.2 kJ mol^{-1}. Similar to the aforementioned investigation of Mg-MOF-74, a tilted orientation is preferred due to dispersive interactions with the linker, the angle Cu···O=C=O being 123°. Two other adsorption sites, at the center of the cage and at the cage window, are about 5 kJ mol^{-1} less favorable at low coverage. Due to lateral interactions with neighboring CO_2 molecules, the binding energies at these sites increase with increasing CO_2 coverage and reach approximately -29 kJ mol^{-1} when all metal sites are occupied. Therefore, there is no marked change in the heat of adsorption when the metal sites are filled, an observation that is in agreement with the relatively constant heat of adsorption measured by microcalorimetry. At high coverages, the isosteric heats even increase due to the large contribution of CO_2–CO_2 interactions, a feature that was also observed both experimentally and in the computations. These results show how detailed theoretical calculations can rationalize the seemingly peculiar evolution of the isosteric heat of CO_2 adsorption. Moreover, the quantitative agreement between calculated and experimental heats of adsorption was excellent.

Most recently, Poloni et al. studied the adsorption of CO_2 in different MOFs with unsaturated metal sites using dispersion-corrected DFT [88, 89]. In a systematic study of Mg-MOF-74 and Ca-BTT, an MOF with the $(M_4Cl)_3(BTT)_8$ structure and extra-framework Ca^{2+} cations, the performance of different dispersion-corrected DFT approaches was assessed, analyzing adsorption energies, geometries, and vibrational properties [88]. A strong dependence on the choice of functional was observed with the van der Waals density functional (vdW-DF) method by Langreth and coworkers [90], a slightly revised version of the vdW-DF method, and the DFT-D2 approach by Grimme [82] using the PBE functional exhibiting the best performance. These methods were then used in a study of CO_2 adsorption in a series of MOFs with the stoichiometry $M'_3[(M_4Cl)_3(BTT)_8]_2$, with M = Mg, Cu, or Ca and M' = Li, Na, or K [89]. A large electrostatic contribution of the ligand to the binding energy was predicted, leading to a preferred adsorption above the tetrazole moiety. Both changes of the metal and of the ligand (e.g., replacing tetrazole by triazole) were found to have a significant impact on the binding energy. Comparing the different metals considered, the smallest cations exhibit the highest binding energies due to the enhanced electrostatic interactions. Thus, $Na_3[(Mg_4Cl)_3(BTT)_8]_2$ emerged as the system with the highest affinity for CO_2, with a very high binding energy of -64.5 kJ mol^{-1}.

3.3.4 *Other Small Molecules*

Some of the results on carbon monoxide adsorption in MOF-74 analogues have already been mentioned above [50, 86]. A detailed DFT/CC study of CO adsorbed in $Cu_3(BTC)_2$ was published by Nachtigall and coworkers [91]. The performance of different methods was assessed for a nonperiodic model system. While dispersion-corrected DFT underestimates the interaction energy, MP2 tends to overestimate it with respect to the benchmark coupled-cluster (CCSD(T)) value. DFT/CC outperforms both methods. These findings were in essential agreement with an earlier study of water interacting with accessible copper sites published by the same group [92]. In both cases, the isosteric heats of adsorption calculated with the DFT/CC method are in excellent agreement with experiment, amounting to 28–30 kJ mol^{-1} for CO and 49 kJ mol^{-1} for H_2O. Moreover, it was found that the correct reproduction of the antiferromagnetic coupling of the two Cu sites, which is computationally demanding, is not necessary to obtain accurate energies for the interaction with adsorbed molecules and that the ferromagnetically coupled state can be used instead.

Liu and Zhong investigated the interaction of $Cu_3(BTC)_2$ and Cu(mip) with adsorbed CO molecules to assess the Lewis acidity of the metal sites employing DFT-B3LYP calculations. $Cu_3(BTC)_2$

exhibits higher Lewis acidity than Cu(mip) due to the stronger electron withdrawal from the metal center by the linker, which is also evidenced by a slightly higher partial charge on the Cu atom (calculated using the natural bond orbital method)

In their studies of acetylene adsorption in different MOFs with coordinatively unsaturated metal sites, Chen and coworkers used DFT calculations to elucidate the interaction between the C_2H_2 molecule and the metal center [93, 94]. In agreement with results for other guest molecules discussed previously, a DFT investigation of C_2H_2 adsorbed at the metal sites of $Cu_3(BTC)_2$ showed that the LDA functional overestimates the interaction strength, whereas GGA functionals underestimate it [93]. A basic analysis of the partial charges led the authors to conclude that the interaction is dominantly electrostatic in nature. However, it was not investigated in more detail whether orbital interactions between the Cu center and the π-system of the acetylene molecule contribute to the binding. In a later combined experimental and theoretical study of M-MOF-74 (M = Mg, Mn, Co, or Zn), the same computational approach was used [94]. The strongest interaction was observed for Co-MOF-74, which is in agreement with the high affinity observed experimentally. Again, the interpretation of the behavior was based exclusively on electrostatic effects.

Recently, MP2 calculations for propylene and propane adsorbed in different M-MOF-74 analogues (M = Mg, Mn, or Co) were reported by Snurr and coworkers in the framework of a gas separation study [95]. Both the binding energy for propylene and the difference in binding energy between propylene and propane are highest for Co-MOF-74, the system that exhibits the highest selectivity in adsorption experiments.

Finally, an exhaustive DFT study of a set of seven molecules (CO, NO, H_2O, C_2H_2, NH_3, H_2S, pyridine) adsorbed at the metal sites of $Cu_3(BTC)_2$ was published by Watanabe and Sholl [96]. A particular emphasis of this investigation was the behavior upon adsorption of pairs of unlike molecules at the two sites of a Cu_2 paddlewheel. Typically, the binding energies for mixed-species adsorption are not equal to the sum of the energies of single-component adsorption, indicating a complex interaction.

3.4 Molecular Mechanics Methods

The first force field-based GCMC simulations of gas adsorption in MOFs were reported between 2001 and 2005 [97–100]. Because GCMC simulations provide a computationally efficient means to predict adsorption isotherms, isosteric heats, and adsorption selectivities, among other quantities, this method has become extremely popular, and a vast number of GCMC studies of adsorption in MOFs has been published since then. It is beyond the scope of this chapter to give a complete account of all GCMC studies of MOFs with unsaturated metal sites, and more exhaustive reviews are available [29, 30, 32]. Instead, the focus will be put on those studies where the shortcomings of the method become apparent: As it is obvious from the preceding section, experimental studies and electronic structure calculations clearly show that the interaction between adsorbed molecules such as hydrogen, methane, or carbon dioxide and unsaturated metal sites is not exclusively dispersive and that the metal–adsorbate distances are often significantly shorter than the sum of the van der Waals radii. However, standard force field methods either exclusively consider dispersive interactions, or include long-range electrostatics by means of a point-charge model. In this approach, short-range electrostatic interactions at interatomic distances below the van der Waals contact are to a large extent cancelled out when a Lennard–Jones (LJ) potential is used, because the repulsion resulting from the LJ term will outweigh attractive electrostatic contributions. The missing—or, at least, highly inaccurate—representation of the localized interactions with the metal centers often leads to significant deviations between simulation and experiment, for example, in adsorption isotherms. These aspects will be discussed in the following, on the basis of a number of instructive examples, before some possibilities to account for these interactions in GCMC simulations are presented in the last section.

In addition to Monte Carlo simulations, molecular dynamics (MD) calculations are another very important force field-based method. MD calculations have been used extensively to study the diffusion of guest molecules in MOFs, as reviewed recently by Krishna [101]. However, the role of unsaturated metal sites for the diffusion properties has been investigated to a much lesser extent than their impact on the adsorption behavior, not least due to the relative scarcity of experimental data. Therefore, this section will virtually exclusively focus on GCMC studies.

3.4.1 *Hydrogen*

The first systematic study of H_2 adsorption in a variety of structurally different MOFs was published in 2005 by Johnson and coworkers. These authors performed computations using the two most widely used "generic" force fields for the framework atoms, namely, the universal force field (UFF) [102] and the DREIDING force field [103]. These parameters were combined with the hydrogen model developed by Buch [104], which approximates the H_2 dumbbell as a single sphere (united-atom model). Identical, or very similar parameter sets have been used in many subsequent studies. The issue of unsaturated metal sites was not discussed in detail in this publication, although two of the systems studied contain such sites. It is, however, a very interesting early example of how simulations can help to verify or falsify experimental data: In some cases where the simulations appeared to severely underestimate the H_2 uptake at room temperature, it was established later that the experimental data could not be reproduced [105, 106].

A GCMC study that explicitly addressed the impact of unsaturated metal sites on the hydrogen adsorption properties was published the following year by Yang and Zhong [61]. These authors modeled H_2 adsorption isotherms in MOF-505 at $T = 77$ K and room temperature, using a combination of the OPLS-AA force field [107] and the UFF, together with DFT-derived partial charges, to represent the framework atoms. Because the literature parameters were not able to reproduce experimental H_2 adsorption data, the parameters for the Cu and O framework atoms were adjusted empirically. While good agreement with available experimental low-pressure data was obtained, it was pointed out later by Keskin et al. that this procedure of empirically fitting selected parameters to isotherm data does not necessarily permit reliable predictions for other conditions, such as higher pressures [29].

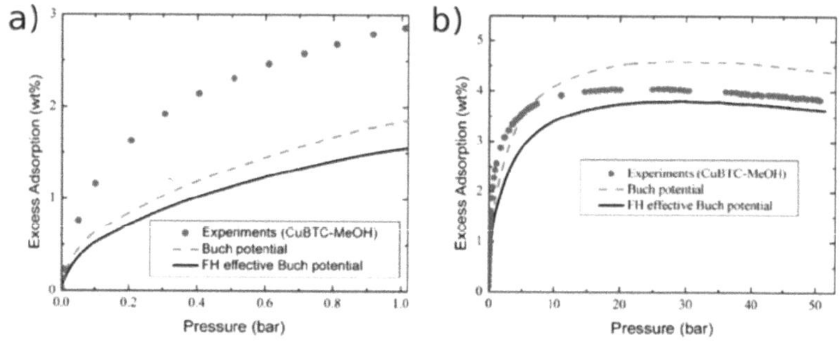

Figure 3.6 Results of hydrogen adsorption experiments and GCMC simulations reported by Johnson and coworkers for $Cu_3(BTC)_2$, $T = 77$ K [108]. MeOH designates the solvent used in the extraction procedure. (a) Isotherms for $p = 0–1$ bar. (b) Isotherms for $p = 0–50$ bar. Reprinted with permission from Ref. [108]. Copyright 2007 American Chemical Society.

A detailed combined experimental and theoretical investigation of H_2 (as well as N_2 and Ar) adsorbed in $Cu_3(BTC)_2$ was reported in 2007 by Johnson and coworkers [108]. Generic force field parameters (UFF, Buch parameters for H_2) were used. A detailed assessment highlighted that the inclusion of quantum effects is an important prerequisite to accurately model hydrogen adsorption isotherms at cryogenic temperatures. For $T = 77$ K, reasonable agreement of the simulations with experimental data was observed in the high-pressure regime (above 10 bar). On the other hand, a significant underestimation occurred at low pressures, the difference between simulation and experiment exceeding 1 wt% at $p = 1$ bar (Fig. 3.6).

While it was not attempted to improve the performance of the model, it was carefully discussed which factors could be responsible for the observed deviations, namely, (1) neglect of electrostatic contributions, (2) overcorrection of quantum effects, and (3) general errors in the solid-fluid potential parameters. Given the fact that even the uncorrected model drastically underestimates the loading, (2) can probably be ruled out, whereas the other two points warrant further study. Consequently, the same group investigated the impact of the inclusion of electrostatic interactions on the predicted adsorption and diffusion properties [109]. DFT-derived charges were used for the framework atoms, and a point-charge model was employed for the hydrogen molecule (for dispersive interactions, the same force field parameters as in the previous study were used). While the inclusion of electrostatic interactions increases the calculated amount of H_2 adsorbed by about 15% at low pressures (below 1 bar), it is by far not sufficient to close the gap between the calculated isotherm and experimental data. Another interesting aspect of this study is the comparison of different charge derivation schemes: Where charges based on the reproduction of the DFT electrostatic potential give realistic results at high pressures, charges based on the electron density may lead to severe overestimations of the saturation uptake by more than 50%. Finally, the effect of electrostatic interactions on the H_2 diffusivity was also investigated, and it was found that the model with charges predicts higher diffusivities at low loadings but lower diffusivities at high loadings than the dispersion-only model.

Two studies of hydrogen adsorption in several structurally different MOFs, some of them containing coordinatively unsaturated metal sites, were published by Fischer et al. [65, 110]. A slightly modified version of the UFF was used in both studies, and most calculations were carried out without accounting for electrostatic interactions. The first of these investigations was aimed at the prediction of hydrogen adsorption sites from GCMC simulations [110]. MOFs with accessible metal sites included $Cu_3(BTC)_2$, MOF-505, Zn-MOF-74, and $(Mn_4Cl)_3(BTT)_8$. For all systems except MOF-505, a significant underestimation of the low-pressure isotherms was observed, although the magnitude of the deviations between simulation and experiment varied significantly. The predicted adsorption sites were compared to available experimental neutron diffraction results. The simulations usually failed to predict the adsorption positions at the coordinatively unsaturated metal sites but were very successful in reproducing the large majority of other adsorption sites, often with very good positional accuracy. These observations clearly point to a direct relationship between the underestimation of the isotherms and the "missing" adsorption sites, and it was therefore concluded that simulations with standard force field parameters are not able to capture the localized interaction of adsorbed H_2 molecules with the metal centers. In a subsequent study, it was established that the magnitude of the underestimation of the isotherm is correlated to the number of accessible metal sites per weight unit [65]: For $Cu_3(BTC)_2$ and PCN-12, another system with Cu_2 paddlewheels, the deviations between simulation and experiment at $T = 77$ K and $p = 1$ bar amount to 1.07 wt% and 0.86 wt%, respectively. These values are in remarkable agreement with the amount of hydrogen that would correspond to the adsorption of one H_2 molecule per Cu site, which equals 0.99 wt% and 0.83 wt%.

Compared to the molecular mechanics studies summarized so far, a quite different methodological approach was employed by Belof et al. in a detailed study of H_2 adsorption in In-*soc*-MOF, a material with *soc* topology, trimeric In_3O building units and overall stoichiometry $(In_3O)(aBTC)_{1.5}(NO_3)$ (with aBTC = 3,3',5,5'-azobenzenetetracarboxylate) [111]. In addition to dispersive interactions and electrostatics, these authors included polarization effects by means of the Thole–Applequist model, which uses an iterative procedure to determine atom-centered dipole moments that represent induced polarization. Because the calculations become computationally very expensive, only one Monte Carlo simulation at fixed loading (*NVT* ensemble) was run. It was found that the inclusion of polarization effects has a significant impact on the calculated radial distribution function, which exhibits a much more pronounced peak close to the unsaturated indium sites (Fig. 3.7). Moreover, the majority of the adsorbed hydrogen molecules has a non-negligible induced dipole moment, leading the authors to conclude that hydrogen behaves as a dipolar fluid in MOFs that have a sufficiently polar framework (a subsequent study showed that the effect of including polarization is negligible for MOFs that are less polar, such as IRMOF-1 [112]). These results indicate that the inclusion of polarization effects may be one possible pathway to resolve the shortcomings of standard GCMC simulations in dealing with adsorption at unsaturated metal sites. However, due to the computational expense, this approach has not yet been widely followed.

3.4.2 *Methane*

Currently, there are a few potential applications that motivate the study of methane adsorption. Depending on the application in mind,

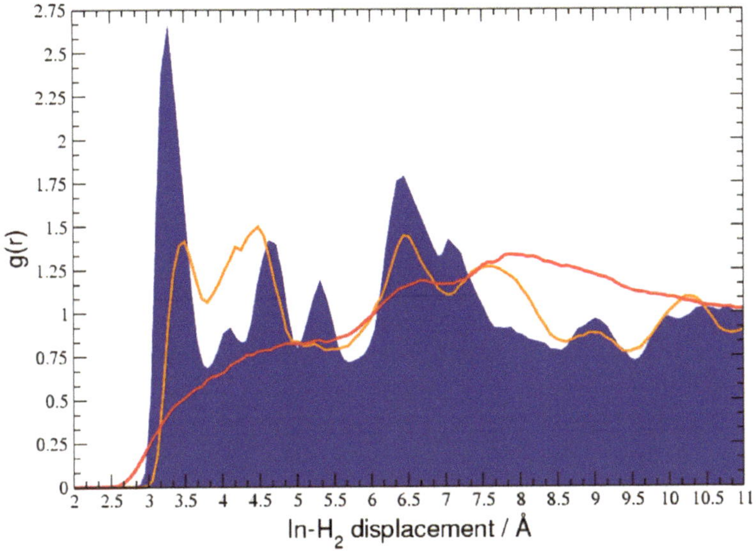

Figure 3.7 Radial distribution function of H_2 molecules around indium sites of In-*soc*-MOF, obtained in calculations including different types of interactions [111]. Dispersion interactions only (red line), dispersion and electrostatic interactions (yellow line), and the full model including polarization interactions (blue silhouette). Reprinted with permission from Ref. [111]. Copyright 2007 American Chemical Society.

different calculations are performed, and different quantities are studied. On the one hand, MOFs could be interesting materials for methane storage. On the other hand, the separation of methane-containing mixtures is a topic of relevance in different industrial processes, for example, in the separation of refinery off-gas (CH_4/H_2 mixtures) or the purification of natural gas (CO_2/CH_4 mixtures). In modeling investigations, the (ideal) adsorption selectivity α is defined as the molar ratio of the components in the adsorbed phase divided by the ratio of the partial pressures in the gas phase.

As early as 2004, Düren et al. assessed the applicability of MOFs for methane storage using GCMC simulations [99]. A comparative study of different MOFs including $Cu_3(BTC)_2$ was published by Wang [113]. From this work, $Cu_3(BTC)_2$ emerges as a very promising system compared to other MOFs. However, the results must be treated with some caution, as specifically derived parameters taken from earlier work by Yang and Zhong [114] were used for this MOF but not for the other systems considered. A more detailed investigation of methane adsorption in $Cu_3(BTC)_2$ using both simulations and experiments was carried out later by Kaskel and coworkers [41]. The computations used the generic UFF parameters for the framework atoms and the TraPPE (single-site) model for the CH_4 molecule [115]. While the simulated isotherms were found to agree reasonably well with experimental data at high pressures (up to 200 bar) and temperatures around room temperature, there are significant differences at cryogenic temperatures ($T = 77$ K, 87 K). These observations were taken as an indicator that the experimental adsorption mechanism is not correctly reproduced in the simulations. This was further substantiated by the fact that one adsorption site seen in neutron-diffraction experiments, namely, the position at the accessible copper centers, is not present among the GCMC-derived sites, much alike to the findings discussed above for hydrogen. Similar observations were made in a study of CH_4 adsorption in $Cu_3(BTC)_2$, PCN-11, and PCN-14 [40]. A very recent study of PCN-14 employed GCMC simulations with an empirically adjusted force field to develop a more detailed picture of the methane adsorption sites and the energy barriers between the sites [116]. Interestingly, an increased methane density at the metal sites was observed at $T = 150$ K but not at $T = 290$ K, leading the authors to conclude that the metal sites act as "weak" adsorption sites. However, this interpretation should possibly be treated with caution in the light of the experimental results mentioned above and the theoretical findings reported by Düren, Nachtigall, and coworkers discussed in section 3.5 [81].

GCMC studies of CH_4/H_2 mixture adsorption in *M*-MOF-74 analogues were reported by Gallo and Glossman-Mitnik (Zn-MOF-74) [117] and Keskin (Ni, Co-MOF-74) [118]. In both cases, the results for the systems with unsaturated metal sites were compared to other MOFs without such sites, especially IRMOFs. As pointed out by Keskin, no very accurate representation of the interaction with the metal sites can be expected from the generic force field used (UFF). Nevertheless, reasonable agreement with the available experimental data (single-component adsorption at room temperature) was reported in both studies, indicating that the parameters used may deliver sufficiently accurate results for the given systems and conditions. The MOF-74 analogues were found to exhibit a significantly higher CH_4/H_2 selectivity than noninterpenetrated IRMOFs [118] and a comparable selectivity to interpenetrated IRMOFs [117].

3.4.3 *Carbon Dioxide*

With growing environmental concerns, the removal of carbon dioxide from exhaust gases or other gas feeds has become a topic of high scientific interest, as it could be a crucial step toward a significant reduction of global CO_2 emissions [119]. MOFs are considered to be promising materials for adsorption-based CO_2 removal processes [120]. Therefore, the majority of force field–based modeling studies dealing with CO_2 adsorption have focused on mixture adsorption, especially the separation of CO_2/CH_4, CO_2/N_2, and CO_2/H_2 mixtures, typically at temperatures around room temperature.

Early contributions by Zhong and coworkers addressed the separation of CO_2/CH_4 and CO_2/H_2 mixtures [121], CO_2/N_2 mixtures [122], and mixtures of CO_2 with CO, CH_4, and C_2H_6 in $Cu_3(BTC)_2$ [114, 123]. Typically, DFT-derived charges were used in conjunction with parameters based on the OPLS-AA force field that were specifically adjusted to reproduce experimental data. As mentioned previously, this approach may have inherent problems with regard to the transferability of the parameters, and the extrapolation to conditions for which no experimental data is available. Moreover, the reliability of experimental data may vary considerably (an instructive comparison of adsorption isotherms obtained from different groups using different activation procedures has been carried out by Johnson and coworkers for the case of H_2 adsorption in $Cu_3(BTC)_2$ [108]). Despite this criticism, it must be emphasized that the studies by Zhong and coworkers delivered important insights: In particular, the importance of electrostatic interactions for the accurate modeling of CO_2 adsorption was firmly established. Taking the example of CO_2/CH_4 mixture adsorption, the adsorption selectivity rises from $\alpha(CO_2/CH_4) = 3–4$ for the dispersion-only case to $\alpha(CO_2/CH_4) = 6–9$ when electrostatics are included [114]. On the other hand, simulation snapshots reveal that the unsaturated metal sites are not the preferred adsorption sites at low pressures, an observation that is in contrast to experiment [42]. This indicates that, as observed for H_2, the inclusion of electrostatic interactions alone is not sufficient to obtain a correct representation of the interaction of CO_2 molecules with accessible metal sites.

A more recent combined experimental and theoretical study by Yazayidin et al. showed that the simulated CO_2 isotherms for $Cu_3(BTC)_2$ exhibit the typical underestimation when unmodified force field parameters are used [124], very similar to the findings discussed previously for hydrogen. Interestingly, a comparable degree of underestimation was observed for a dry system with accessible metal sites, and a hydrated system, where at least half of the metal sites were occupied by water molecules. Due to attractive electrostatic interactions between the carbon dioxide molecules and water, the affinity toward CO_2 is enhanced for the hydrated system. As this feature was correctly reproduced in the simulations, most of the observed differences between simulation and experiment are probably related to an inaccurate representation of the interaction of CO_2 with the copper centers. For some MOFs with accessible metal sites, quite significant deviations were observed in a subsequent comparative study of several MOFs by the same group [125]. The largest discrepancies between simulation and experiment occur for the MOF-74 analogues and low pressures (around 0.1 bar), whereas the agreement is more satisfactory for pressures around 1 bar.

The effect of the hydration of an MOF with unsaturated metal sites was also investigated by Jiang and coworkers for the case of MIL-101(Cr), an MOF with large pores and trimeric Cr_3O units [126]. A dehydrated model system, which contains one fluoride ion per trimer for charge compensation, was compared to a fully hydrated system (two water molecules and one fluoride ion per Cr_3O trimer). At low pressures, the calculated CO_2 uptake is higher for the hydrated system due to the interaction between carbon dioxide and the terminal water molecules, whereas the uptake at high pressures is slightly larger for the dehydrated system due to the larger free volume. An analysis of the radial distribution function showed the preferential adsorption of CO_2 at the chromium sites (or, in the case of the hydrated system, the water molecules), especially at pressures below 1 bar. Further GCMC simulations of CO_2/CH_4 mixture adsorption showed a modest enhancement of the selectivity upon hydration, highlighting that the hydration of MOFs could have a positive effect on their selectivity. This effect was more recently investigated experimentally by Llewellyn and coworkers for MIL-100(Fe) [127]. Here, a much more significant (up to five-fold) increase in CO_2 uptake with increasing relative humidity was observed.

A systematic comparison of the CO_2/CH_4 selectivity of various MOFs with structural features such as interpenetration, unsaturated metal sites, and a charged framework balanced by extra-framework anions was performed by Jiang and coworkers [128]. The MOFs with unsaturated metal sites comprised $Cu_3(BTC)_2$, the copper-containing PCN-6 and its noninterpenetrated counterpart PCN-6' [8], and the charged-framework In-*soc*-MOF. For the three Cu MOFs, the highest CO_2/CH_4 adsorption selectivity was observed for $Cu_3(BTC)_2$, which also has the smallest pores of the three systems, with $\alpha(CO_2/CH_4) = 4–7$ (note the difference to the value quoted above for the same system from a study by Yang and Zhong [114], highlighting the sensitivity to the choice of framework parameters). PCN-6', which is isostructural to $Cu_3(BTC)_2$, having a more extended linker molecule, exhibits a lower selectivity due to the weaker interaction in the large pores. In PCN-6, the unsaturated copper sites are not accessible due to the presence of the interpenetrating second framework, and the resulting selectivity falls between the two other systems. High selectivity of $\alpha(CO_2/CH_4) = 22–36$ was predicted for In-*soc*-MOF, which was rationalized with the strong electrostatic interaction with the framework and, especially, the extra-framework nitrate anions.

A general problem of the GCMC studies of CO_2 adsorption summarized here is the fact that they use literature force fields, or empirically adjusted parameters, to model the dispersive interaction with the framework. Similar to what has been outlined above for hydrogen and methane, these parameters are not necessarily able to capture the specific interaction with the metal sites, and this shortcoming often becomes evident when the simulated single-component isotherms are critically compared to experimental data (see, for example, the careful comparison with different sets of experimental isotherms provided by Yazayidin et al. [124]). Clearly, these shortcomings will affect the predictions of not only single-component data but also adsorption selectivities. While some approaches to represent these interactions more accurately will be discussed in the following section, most of the studies on this topic focus on the adsorption of hydrogen, methane, or small hydrocarbons. For the case of carbon dioxide, only relatively little effort has been made so far to develop improved parameter sets that allow for reliable predictions of adsorption in MOFs with accessible metal sites.

3.4.4 *Other Small Molecules*

The adsorption of carbon monoxide, and its separation from methane, nitrogen, and hydrogen, in $Cu_3(BTC)_2$ was investigated by Karra and Walton using GCMC simulations [129]. In this study, the empirically adjusted parameters from different publications by Zhong and coworkers were employed to model the interaction with the metal sites. As discussed above, this approach is quite problematic due to the limited parameter transferability and high dependence on the quality of the experimental data. Since carbon monoxide has a dipole moment, and a relatively large quadrupole moment, the inclusion of electrostatic interactions was found to have a very important impact on the resulting isotherms. An analysis of simulation snapshots showed that carbon monoxide molecules are preferentially adsorbed at the unsaturated metal sites, a behavior that can be attributed to the charge–dipole interaction. Calculations of mixture isotherms revealed high adsorption selectivity for CO over H_2, $\alpha(CO/H_2) = 28$–500; low selectivity over N_2; and virtually no preference over methane. In the last case, the stronger dispersion interactions with methane compensate the higher contribution of electrostatic interactions for CO. Due to the relatively large void volume, both species can coadsorb, resulting in a very low selectivity.

More recent computational investigations of CO adsorption (and, in the latter case, CO/H_2 mixture adsorption) in $Cu_3(BTC)_2$ were reported by Calero and coworkers [130], and Fischer et al.

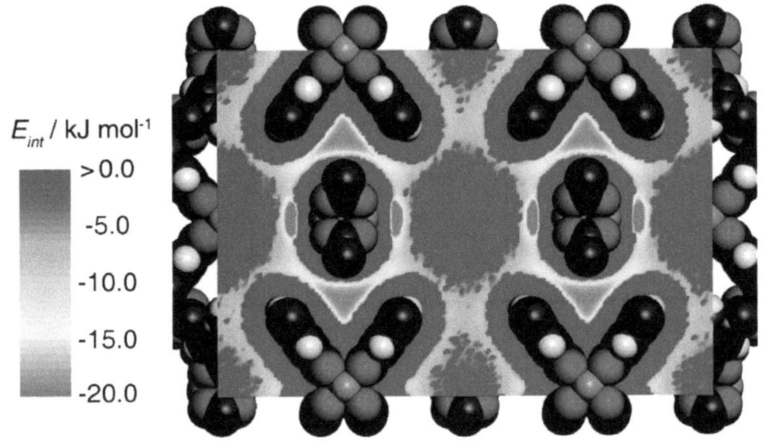

Figure 3.8 Potential energy map derived from GCMC simulations of carbon monoxide adsorption in $Cu_3(BTC)_2$ [131]. The section through the unit cell is oriented perpendicular to the crystallographic [110] direction. The strong energetic preference of CO to adsorb at the unsaturated copper sites is clearly visible. Reproduced by permission of The Royal Society of Chemistry 2012.

[131]. Calero and coworkers investigated the influence of the carbon monoxide model on the computed adsorption isotherm for $Cu_3(BTC)_2$ and IRMOF-1, and found remarkable differences among various models proposed in the literature. While different adsorption sites were analyzed, the possibility of adsorption at the unsaturated copper sites was not discussed in more detail. Fischer et al. reported their results for $Cu_3(BTC)_2$ in the framework of a comparative study of five microporous adsorbents. Of the systems studied, $Cu_3(BTC)_2$ has the largest free-pore volume and consequently the highest CO uptake. A visualization of the interaction energy as a function of the position in the pore revealed strongly preferred adsorption at the copper sites, although no specific adjustments of the parameters to account for the Cu–CO interaction were made for the simulations (Fig. 3.8). This behavior was explained by the large contribution of electrostatic (charge–dipole/charge–quadrupole) interactions to the total interaction. Despite the presence of localized CO adsorption sites, the predicted

CO/H_2 selectivity remains relatively low, with $\alpha(CO/H_2) = 10$–12 (note the discrepancy to the aforementioned results obtained by Karra and Walton for $Cu_3(BTC)_2$ with another parameter set). Adsorbents with smaller pores, such as Mg formate, were found to exhibit significantly higher selectivities because the dispersive interactions with CO are maximized in the narrow pores and because the coadsorption of H_2 is inhibited more efficiently.

The adsorption of water constitutes a difficult case for molecular modeling, especially due to the high dependence of the results on the charges employed to represent electrostatic solid–fluid interactions [132]. For the case of MOFs, this dependence was assessed, again using $Cu_3(BTC)_2$ as a prototypical system, by Castillo et al. [133]. Calculations with different sets of framework charges revealed that even small changes by 2% to 6% result in remarkable differences in the computed H_2O isotherm. A decomposition of the interaction energy into electrostatic and dispersive contributions showed the dominance of electrostatic effects, which make up for more than 80% of the total interaction energy. Having determined the best set of charges by reference to experiment, the authors also analyzed the adsorption sites. It was observed that the position at the unsaturated metal center is massively favored over all other sites (small cages, windows). A comparative study of several other guest molecules with no dipole moment showed no such preference for the metal site for nitrogen, oxygen, and small hydrocarbons [134]. Evidently, the qualitatively different behavior observed for water can be attributed to the strong charge–dipole interaction between the Cu centers and the water molecules.

Studies of more "exotic" guest molecules include the investigation of carbon tetrachloride adsorption, and its separation from air, in $Cu_3(BTC)_2$ by Calero and coworkers [135, 136], as well as the study of ammonia adsorption in the same system by Petit et al. [137]. While carbon tetrachloride is preferentially adsorbed in the larger cages of the structure (being unable to access the small cages), ammonia is strongly favored at the unsaturated metal sites, in a very similar fashion to water. Moreover, experimental results show that the reactive adsorption of ammonia leads to partial framework decomposition.

3.5 Bridging the Gap: Toward an Improved Modeling of Metal–Adsorbate Interactions

As it has become evident from the literature survey given in the previous section, force field-based simulations using literature parameters are not usually able to describe the specific adsorption of small molecules such as hydrogen, methane, or carbon dioxide at coordinatively unsaturated metal sites (although they sometimes seem to perform satisfactorily for dipolar molecules such as carbon monoxide or water). The GCMC simulation results are affected by this shortcoming, and evidence can be seen in the underestimation of adsorption isotherms, especially at low temperatures and low pressures, in the underestimation of heats of adsorption, and in the failure to predict preferred adsorption sites close to the metal centers. As discussed in the introduction to section 3.4, the inability of standard force fields to account for these interactions can be rationalized straightforwardly when it is considered how these force fields are constructed: They are designed to provide a reasonably accurate description of long-range interactions at interatomic distances larger than the sum of the van der Waals radii, whereas the repulsive term that dominates at lower distances is purely empirical. At these short interatomic distances, the repulsive part of the LJ potential will usually outweigh a possible attractive electrostatic contribution. Therefore, the construction of improved potentials to account for the metal-adsorbate interactions must deal with several problems: On the one hand, a reasonable way to quantify the interaction strength must be developed. On the other hand, it must also be ensured that the new potential is integrated into the existing force field in a sensible manner, taking into account the possibility of double counting or mutual cancellation of some interactions.

The main part of this section will deal with the implementation of improved metal-adsorbate potentials by integrating them in existing force fields, leaving the rest of the solid–fluid interaction parameters untouched. Primarily, two different ways to derive the potential parameters have been followed: They can either be obtained empirically, by fitting the parameters to match available experimental isotherms, or be obtained by making use of electronic structure calculations, for example, DFT methods. Most of this section will focus on the second possibility. Finally, a more sophisticated approach that eliminates the need to use empirical force field parameters by taking the whole potential energy surface from electronic structure calculations will be discussed.

As discussed previously, the empirical fitting of some force field parameters to match experimental isotherms is problematic for various reasons. Several studies using this approach by Zhong and coworkers [61, 114, 121–123], as well as the well-founded criticism by Keskin et al. [29], have already been mentioned. A critical evaluation of the merits and shortcomings of this approach was carried out by Jorge and coworkers, for the case of propylene adsorption in $Cu_3(BTC)_2$ [138, 139]. Here, initial GCMC simulations using standard parameters (DREIDING force field for framework atoms and TraPPE force field for guest molecules) delivered good agreement with experimental data for propane adsorption but a massive underestimation of the uptake of propylene. This was explained by a significant contribution of Cu–π interactions, which are not accounted for in the literature parameter set. The authors therefore modified the well depth of the potential representing the interaction between the copper sites and the sp^2-hybridized carbon atoms of propylene, leaving the Cu–CH_2 equilibrium distance as well as all other parameters untouched. The adjusted parameters afford good agreement of the simulated propylene isotherms with experimental data at all temperatures considered [138]. In a subsequent study, the authors extended their predictions to mixture adsorption [139]. Here, significant deviations of the simulated isotherms from the mixture adsorption isotherms calculated from experimental data using ideal adsorbed solution theory (IAST) were observed. Consequently, the propylene/propane adsorption selectivity obtained from the simulations ($\alpha \approx 1.5$) was found to be considerably lower than the IAST selectivity ($\alpha = 3$–4). This observation led the authors to conclude that a chemically more realistic description of the Cu–π interaction is necessary to obtain a sufficiently accurate prediction of both single-component and mixture adsorption. An attempt to develop such a potential model, based on DFT calculations, was reported recently and will be discussed in more detail below.

The strategy of deriving metal–adsorbate parameters from DFT calculations and combining them with literature parameters to model the interaction with all other frameworks was applied by Fischer, Kuchta and coworkers [140] as well as by Sastre and coworkers [62] to hydrogen adsorption in MOFs with Cu_2 paddlewheels, by Fischer, Hoffmann et al. to acetylene and carbon dioxide adsorption in $Cu_3(BTC)_2$ [141], and by Jorge, Fischer, and coworkers to propylene adsorption in $Cu_3(BTC)_2$ [142]. To explain the parameter derivation procedure, the approach presented in Ref. [140] will be discussed in most detail, and additional aspects following from the other publications will be addressed more briefly.

It has been discussed that the failure of standard force field parameters to reproduce experimental hydrogen adsorption data in MOFs with Cu_2 paddlewheels, such as $Cu_3(BTC)_2$ and PCN-12, can be attributed to localized Cu–H_2 interactions, which are not captured by the force field. These observations, presented in [108] and [65], among other publications, led Fischer, Kuchta and coworkers to the development of a new potential model to include this localized interaction. Their strategy, visualized in Fig. 3.9, can be summarized as follows:

(1) DFT calculations are carried out for a molecular model system to obtain a Cu–H_2 potential curve.
(2) A simple potential model (such as an LJ or Morse potential) is fitted to the DFT potential curve, possibly after correcting the curve for the double counting of some interactions.
(3) The potential obtained is used to replace the Cu–H_2 interaction term from the literature force field. All other force field parameters are retained.

Before discussing the performance of this approach, it is necessary to highlight the most important approximations involved: First, the energies obtained in the DFT calculations will depend on various factors, such as the basis set size, the choice of the XC functional, and the size of the model system. As long as no dispersion correction is employed, long-range dispersion interactions will be absent or represented in an inaccurate manner. Due to these uncertainties, it is not straightforward to establish a relationship between the DFT energies and the potential model, which is intended to represent exclusively the metal–adsorbate interaction. In their work, Fischer, Kuchta and coworkers used the simplest possible assumption, namely, taking the DFT energy (obtained for a Cu_2(benzoate)$_4$ model system, using the PBE functional) to directly correspond to the Cu–H_2 interaction. Thus, it was assumed that all long-range contributions to the DFT energy are negligible, and a Morse potential was directly fitted to the DFT results. However, one specific aspect that must be considered for hydrogen is the importance of different orientations: When using the united-atom model for H_2, it is inherently assumed that all orientations of the dumbbell are equally likely. However, the DFT energies used for the new potential model were obtained for the most favorable (side-on) coordination, whereas an end-on orientation results in negligible interaction. To account for this effect, the authors scaled their potential well depth by two-thirds and compared it with a model using the full DFT interaction strength.

Using the new potential parameters, hydrogen adsorption isotherms at cryogenic temperatures were predicted for $Cu_3(BTC)_2$ and PCN-12. An example of the simulation results for pressures up to 1 bar is included in Fig. 3.9. While the model using the full DFT energies overestimates the H_2 uptake, the orientation-averaged potential leads to excellent agreement of the low-pressure isotherms with experimental data. At high pressures, a moderate overestimation by up to 0.5 wt% occurs, which may be related to the "isotropic" nature of the Morse potential: The enhancement of the interaction energy is not confined to the region above the Cu sites, but a certain increase is also detectable in areas perpendicular to the Cu–Cu axis. However, the implementation of a nonisotropic potential model would be computationally considerably more demanding.

Having shown the good performance of their DFT-based potential model for two MOFs with Cu_2 paddlewheels, the authors performed additional calculations for $(Cu_4Cl)_3(BTT)_8$ as a Cu MOF with a different inorganic building unit. Again, a massive improvement of the predicted isotherms when compared to the original force field was observed, indicating a certain degree of

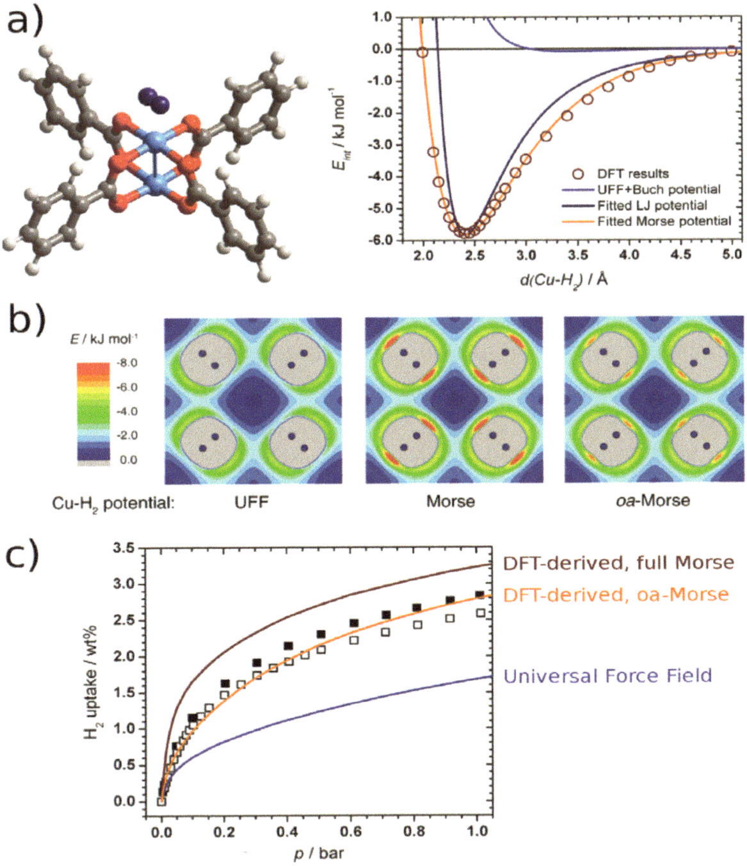

Figure 3.9 Visualization of the potential derivation strategy employed by Fischer, Kuchta, and coworkers [140]. (a, left) Model system used in DFT calculations. (a, right) Results of DFT calculations (open symbols), fit of different potential models to DFT results (orange and dark blue line), and Cu–H_2 interaction, as calculated with literature force field parameters (blue line). (b) Two-dimensional sections through the unit cell of $Cu_3(BTC)_2$, showing the potential energy for a H_2 molecule adsorbed in the structure, as calculated with different potential models: LJ potential with parameters from the UFF, full DFT-derived Morse potential, and orientation-averaged DFT-derived Morse potential. (c) Calculated H_2 adsorption isotherm for $Cu_3(BTC)_2$, $T = 77$ K, obtained from GCMC simulations using the three potential models. Experimental data from two different sources is included as full and open squares [108, 144]. Adapted and reprinted with permission from Ref. [140]. Copyright 2010 American Chemical Society.

transferability to other Cu MOFs with accessible metal sites. Further GCMC simulations of hydrogen adsorption at room temperature revealed only very little effect of the interaction with the unsaturated metal sites on the room-temperature H_2 isotherms, and calculations using an artificially enhanced potential model showed that a strong metal–H_2 interaction in the range of -15 kJ mol^{-1} would be necessary to significantly increase the room-temperature hydrogen uptake. However, even with this hypothetically increased interaction strength, the amount of hydrogen adsorbed remains drastically below technological benchmark values, as the number of available metal sites only accounts for an uptake of approximately 1 wt%, presuming that each site cannot be occupied by more than one H_2 molecule.

The potential model developed by Fischer, Kuchta and coworkers was also used in a subsequent study by Düren and coworkers [143]. In this work, Monte Carlo simulations were employed to predict thermal desorption spectra, and the inclusion of the metal–H_2 interaction in the simulations led to good agreement of the predicted spectra for $Cu_3(BTC)_2$ with experimental data. Interestingly, an analysis of the potential energy distribution revealed that the positions at the unsaturated metal sites are not the energetically most favorable sites but that the position inside the small cages is favored by about 1.0 kJ mol^{-1}. Consequently, H_2 molecules adsorbed at the metal sites desorb between 23 K and 28 K, whereas those adsorbed in the small cages desorb between 28 K and 35 K.

A similar attempt to derive metal–H_2 parameters from DFT calculations was reported by Sastre and coworkers, also for the case of MOFs with Cu_2 paddlewheels [62]. These authors used the SSB-D functional, which is intended to account for dispersion interactions. The parameters were derived on the basis of these calculations and validated by comparing the computed H_2 probability distribution with experimental results. Subsequently, the parameters were employed in MD calculations, from which the preferred adsorption sites were derived. In addition to the adsorption at the unsaturated metal sites, additional adsorption sites were found in narrow windows and small cavities, which were rationalized as being due to a confinement effect, that is, the overlap of attractive potentials stemming from several surrounding framework atoms.

DFT-derived parameters for the interaction of acetylene and carbon dioxide with unsaturated copper sites were derived by Fischer et al. in the framework of a study of C_2H_2 adsorption and C_2H_2/CO_2 separation in ordered microporous materials [141]. Having established the inadequacy of standard force field parameters to reproduce the experimental adsorption isotherms reported for $Cu_3(BTC)_2$, DFT-PBE calculations were carried out to quantify the interaction of both molecules with a $Cu_2(formate)_4$ cluster. The parameter derivation was carried out in a similar fashion as in the study of hydrogen discussed above, the main difference being the explicit inclusion of electrostatic interactions by means of a point-charge representation. To avoid double counting of interactions, the electrostatic interactions, as well as the repulsion stemming from the closest oxygen atoms, were subtracted from the DFT energy before the fit. GCMC simulations using the new parameters showed excellent agreement with the experimental isotherm for C_2H_2, whereas the CO_2 adsorption isotherm was still underestimated. Predictions of mixture adsorption revealed a low selectivity of $\alpha(C_2H_2/CO_2) = 2–4$, despite the preferential adsorption of acetylene at the metal sites, because the carbon dioxide molecules can coadsorb in the remaining pore space. However, a subsequent investigation of the selectivity as a function of the gas-phase composition showed a significant increase of the selectivity on decreasing C_2H_2 content, rendering $Cu_3(BTC)_2$ potentially interesting for the removal of trace amounts of acetylene [66].

While the DFT-based approach discussed in the preceding paragraphs certainly has a rather sound theoretical basis, especially when compared to empirical fitting procedures relying on experimental data, it nevertheless involves several assumptions and approximations. The most detailed account of the influence of different approximations was reported recently by Jorge, Fischer and coworkers, who investigated the adsorption of propylene in $Cu_3(BTC)_2$ [142]. In particular, the influence of the size of the model system used in the DFT calculations, and the inclusion of different structural degrees of freedom in these calculations, was assessed. DFT-PBE computations using different $Cu_2(L)_4$ model systems showed that the parameters obtained with the cluster that most resembled the actual MOF, namely, $Cu_2(BTC)_4$, performed better than parameters obtained with model systems containing formate or benzoate ligands. Moreover, it was found that the inclusion of flexibility of the Cu–Cu axis, as well as the C=C bond of the guest molecule, improved the results compared to a calculation of the potential energy curve with fixed subsystems. In the implementation of the Cu–π potential, a sophisticated functional form, using a Morse potential in conjunction with a power-law function, was used to cancel out repulsive interactions stemming from the framework atoms surrounding the copper site. The new potential model provides for excellent agreement of the simulated propylene adsorption isotherms with experimental data at several temperatures. However, the predicted propylene/propane mixture adsorption isotherm exhibits significant deviations when compared to the mixture isotherm derived from experiment using IAST, and the selectivity is underestimated, especially for propylene-rich mixtures. Assuming that the deviations are not due to inadequacies of IAST, these results indicate that even the "chemically realistic" potential models has difficulties in accurately predicting mixture adsorption and that further improvements are necessary to develop a Cu–π potential model that performs equally well for simulations of single-component and mixture adsorption.

All approaches discussed up to this point concentrate on the development of a "local" metal–adsorbate potential, while taking the parameters to describe the interactions of the guest molecule with all other framework atoms from a literature force field. A completely different strategy was pursued by Düren, Nachtigall, and coworkers, who studied the adsorption of methane in $Cu_3(BTC)_2$ [81]. Previous investigations had shown that standard force fields are not able to accurately predict the methane adsorption isotherm at low temperatures and do not deliver the correct sequence of occupation of the adsorption sites [41]. As an initial step, these authors performed a large number of periodic DFT/CC calculations for methane molecules located at different positions in the unit cell of $Cu_3(BTC)_2$, thereby obtaining the MOF–CH_4 interaction potential on a regular grid. Rather than fitting any type of empirical function to the results, this DFT/CC-based grid of interaction energy values was directly used as input for the GCMC simulations: For insertion of a molecule at a random position in the unit cell, the potential energy was calculated by interpolation between the pretabulated values stored on the grid.

GCMC simulations using the DFT-/CC-based potential map provide for excellent agreement with experimental adsorption isotherms obtained at cryogenic temperatures ($T = 77$ K) and exhibit a much superior performance to both classical force fields and alternative energy maps obtained with the (more empirical) DFT-D approach (Fig. 3.10). Moreover, the simulations correctly reproduce the relative occupancy of eight different adsorption sites that were refined experimentally at high coverage. In summary, this (almost) purely nonempirical approach leads to excellent agreement of predicted quantities with experiment and therefore holds much promise for future applications. There are, however, two main drawbacks that currently limit its applicability: First, the computational cost of the calculations that are necessary to build the potential energy map is too high for routine studies. Second, it is unclear how it can be extended from a spherical guest molecule (due to the small effect of the orientation on the interaction energy for methane, orientation effects were ignored) to other species, where it would be necessary to include several different orientations for each grid point.

3.6 Conclusions

The specific adsorption at the metal centers is an almost ubiquitous phenomenon in MOFs that contain coordinatively unsaturated metal sites. Experimentally, its importance has been established for a variety of guest molecules, ranging from methane, a molecule that has neither a dipole nor a quadrupole moment, over the weakly quadrupolar hydrogen molecule, to dipolar molecules such as water. In this chapter, the capabilities and limitations of different computational methods to capture this interaction have been summarized, drawing from the available literature up to early 2012.

Electronic structure methods, in particular DFT, are generally able to capture the specific interaction of adsorbed molecules with unsaturated metal sites, at least in a qualitative manner.

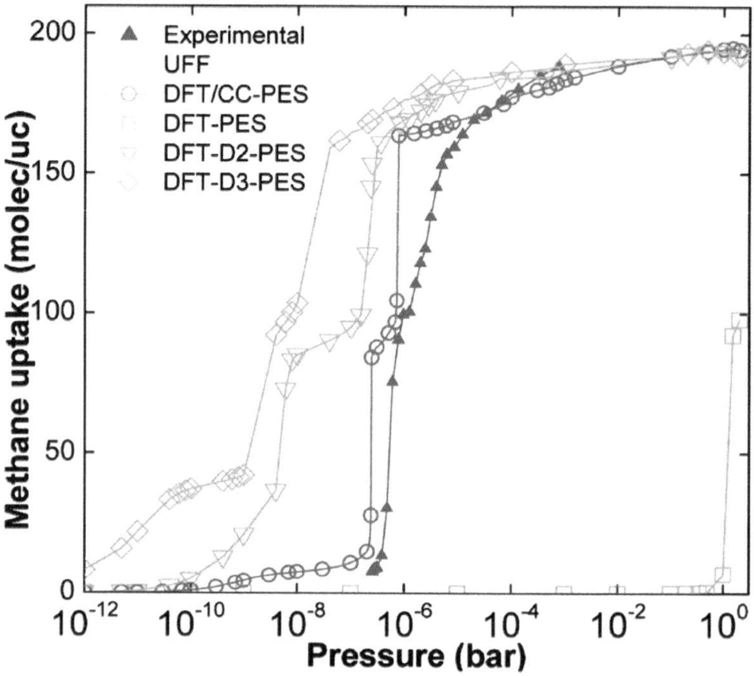

Figure 3.10 Calculated methane adsorption isotherms ($T = 77$ K) obtained with different theoretical approaches [81]. Experimental reference data from Ref. [41] is shown for comparison. PES designates calculations using the potential energy values obtained with different electronic structure methods: DFT/CC, uncorrected DFT, and dispersion-corrected DFT. UFF designates calculations using literature force field parameters. Reprinted with permission from Ref. [81]. Copyright 2011 American Chemical Society.

Using recently developed dispersion-correction approaches, or sophisticated schemes that make use of results obtained with correlated methods, a near quantitative agreement of the interaction energy with experimental values is attainable in the most favorable cases. Even if no quantitative agreement is reached, electronic structure calculations can give important qualitative insights into the nature of the interaction. However, the interpretation of the results is not always unambiguous, as it has become evident in the literature survey of computational studies of the metal–hydrogen interaction: While some authors interpreted the interaction as practically exclusively electrostatic in nature, others proposed a significant role of orbital interactions for the same (or very closely related) systems. Apparently, more detailed investigations are necessary to obtain a complete picture that reconciles these conflicting interpretations.

Force field-based simulations using literature parameters are usually not able to correctly represent specific metal–adsorbate interactions in a qualitative (let alone quantitative) manner. As these force fields are intended primarily to capture long-range dispersion effects, the stronger, more localized interactions with the metal sites are not accounted for. The inclusion of electrostatic effects is usually not sufficient to resolve this shortcoming, except for guest molecules having a dipole moment, such as water or carbon monoxide. While an approach that explicitly accounts for polarization interactions appears to be very promising, its applicability is currently limited by the additional computational expense. In general, simulations with standard force fields tend to underestimate the amount adsorbed when compared to experiment, the deviations being most pronounced at low temperatures and low pressures, conditions where the adsorption at the metal sites is most significant. In addition, the regions near the metal sites are not predicted to be preferential adsorption sites, in contrast to experimental observations. Nevertheless, GCMC simulations using literature force fields have been widely employed to predict single-component and mixture adsorption of various species in MOFs with unsaturated metal sites. To date, it remains largely unexplored to what extent the simulation results, for example, adsorption selectivities, may be affected by the inaccurate representation of metal–adsorbate interactions.

Different attempts have been made to improve upon the performance of literature force fields in molecular mechanics calculations. The simplest approach is the adjustment of the parameters representing the metal–adsorbate interaction to reproduce experimental adsorption isotherms. However, parameters derived with this strategy do not necessarily provide a physically reasonable description of the interatomic interactions, leading to an important potential source of error. An approach that has been followed for a number of sorbate species is the derivation of parameters for the metal–adsorbate interaction from DFT calculations for small, nonperiodic model systems. Typically, the remaining interaction parameters are taken from literature force fields. Although the parameter derivation has to rely on several, sometimes rather severe approximations, different examples have shown a drastically improved performance when compared to "generic" literature parameters. Future efforts in this direction should aim at a more precise description of all interactions in the electronic structure calculations, using advanced DFT methods that account for dispersion, or high-level wavefunction-based methods. In this instance, the parameter derivation also needs to be performed on a more sophisticated level, decomposing the total interaction into a "local" metal–adsorbate part and a part representing the interaction with all other atoms of the model systems. An even more elaborate possibility is the calculation of the potential energy of an adsorbed molecule on a grid covering the MOF unit cell using electronic structure methods. The energy map obtained from these calculations can then be used as input for GCMC simulations, without the need to resort to any empirical force field parameters. While this approach holds much promise for future applications that aim at high accuracy, it is currently too demanding for routine applications. The extension to nonspherical adsorbates is another challenge.

In total, the examples discussed in this chapter have shown that the interaction of adsorbed molecules with unsaturated metal sites is a current topic of high interest, which is widely studied with computational methods at different levels of theory. Important further insights into the nature of the interactions can be expected from the application of more advanced electronic structure methods. In the field of molecular mechanics methods, it is crucial to develop more accurate schemes to incorporate the localized metal–adsorbate interactions in the simulations in order to reliably predict macroscopically measurable adsorption properties. Computational methods are already widely employed to achieve a better understanding of experimentally observed phenomena, and future advances in method development will without a doubt help to further strengthen this trend.

Acknowledgments

We would like to thank Frank Hoffmann and Miguel Jorge for insightful discussions and for the critical reading of this manuscript. Thanks are also due to Daniela Frahm and Felix Brieler for providing the pictures of $Cu_3(BTC)_2$ samples used in Fig. 3.1. Michael Fischer gratefully acknowledges a postdoctoral fellowship by the German Research Foundation (DFG grant Fi 1800/1-1).

References

1. Li, H., Davis, C. E., Groy, T. L., Kelley, D. G., Yaghi, O. M. (1998). Coordinatively unsaturated metal centers in the extended porous framework of $Zn_3(BDC)_3{\cdot}6CH_3OH$ (BDC = 1,4-benzenedicarboxylate), *J. Am. Chem. Soc.*, **120**, 2186–2187.
2. Chui, S. S.-Y., Lo, S. M.-F., Charmant, J. P. H., Orpen, A. G., Williams, I. D. (1999). Chemically functionalizable nanoporous material $[Cu_3(TMA)_2(H2O)_3]_n$, *Science*, **283**, 1148–1150.
3. Zhang, X. X., Chui, S. S.-Y., Williams, I. D. (2000). Cooperative magnetic behavior in the coordination polymers $Cu_3(TMA)_2(L)_3$ ($L{=}H_2O$, pyridine), *J. Appl. Phys.*, **87**, 6007–6009.
4. Prestipino, C., Regli, L., Vitillo, J. G., Damin, A., Bonino, F., Zecchina, A., Lamberti, C., Solari, P. L., Kongshaug, K. O., Bordiga, S. (2006). Local structure of framework Cu(II) in HKUST-1 metallorganic framework: spectroscopic characterization upon activation and interaction with adsorbates, *Chem. Mater.*, **18**, 1337–1346.
5. Lin, X., Jia, J., Zhao, X., Thomas, K. M., Blake, A. J., Walker, G. S., Champness, N. R., Hubberstey, P., Schröder, M. (2006). High H_2 adsorption by coordination-framework materials, *Angew. Chem., Int. Ed.*, **45**, 7358–7364.
6. Lin, X., Telepeni, I., Blake, A. J., Dailly, A., Brown, C. M., Simmons, J. M., Zoppi, M., Walker, G. S., Thomas, K. M., Mays, T. J., Hubberstey, P., Champness, N. R., Schröder, M. (2009). High capacity hydrogen adsorption in Cu(II) tetracarboxylate framework materials: the role of pore size, ligand functionalization, and exposed metal sites, *J. Am. Chem. Soc.*, **131**, 2159–2171.
7. Chen, B., Ockwig, N. W., Millward, A. R., Contreras, D.S., Yaghi, O. M. (2005). High H_2 adsorption in a microporous metal-organic

framework with open metal sites, *Angew. Chem., Int. Ed.*, **44**, 4745–4749.

8. Ma, S., Sun, D., Ambrogio, M., Fillinger, J. A., Parkin, S., Zhou, H.-C. (2007). Framework-catenation isomerism in metal-organic frameworks and its impact on hydrogen uptake, *J. Am. Chem. Soc.*, **129**, 1858–1859.
9. Wang, X.-S., Ma, S., Forster, P. M., Yuan, D., Eckert, J., López, J. J., Murphy, B. J., Parise, J. B., Zhou, H.-C. (2008). Enhancing H_2 uptake by "close-packing" alignment of open copper sites in metal-organic frameworks, *Angew. Chem., Int. Ed.*, **47**, 7263–7266.
10. Yuan, D., Zhao, D., Sun, D., Zhou, H.-C. (2010). An isoreticular series of metal-organic frameworks with dendritic hexacarboxylate ligands and exceptionally high gas-uptake capacity, *Angew. Chem., Int. Ed.*, **49**, 5357–5361.
11. Zou, R.-Q., Sakurai, H., Han, S., Zhong, R.-Q., Xu, Q. (2007). Probing the Lewis acid sites and CO catalytic oxidation activity of the porous metal-organic polymer [Cu(5-methylisophthalate)], *J. Am. Chem. Soc.*, **129**, 8402–8403.
12. Lee, Y.-G., Moon, H. R., Cheon, Y. E., Suh, M. P. (2008). A comparison of the H_2 sorption capacities of isostructural metal-organic frameworks with and without accessible metal sites: $[\{Zn_2(abtc)(dmf)_2\}_3]$ and $[\{Cu_2(abtc)(dmf)_2\}_3]$ versus $[\{Cu_2(abtc)\}_3]$, *Angew. Chem., Int. Ed.*, **47**, 7741–7745.
13. Murray, L. J., Dincă, M., Yano, J., Chavan, S., Bordiga, S., Brown, C. M., Long, J. R. (2010). Highly-selective and reversible O_2 binding in Cr_3(1,3,5-benzenetricarboxylate)$_2$, *J. Am. Chem. Soc.*, **132**, 7856–7857.
14. Férey, G., Serre, C., Mellot-Draznieks, C., Millange, F., Surblé, S., Dutour, J., Margiolaki, I. (2004). A hybrid solid with giant pores prepared by a combination of targeted chemistry, simulation, and powder diffraction, *Angew. Chem., Int. Ed.*, **43**, 6296–6301.
15. Férey, G., Mellot-Draznieks, C., Serre, C., Millange, F., Dutour, J., Surblé, S., Margiolaki, I. (2005). A chromium terephthalate-based solid with unusually large pore volumes and surface area, *Science*, **309**, 2040–2042.
16. Liu, Y., Eubank, J. F., Cairns, A. J., Eckert, J., Kravtsov, V. C., Luebke, R., Eddaoudi, M. (2007). Assembly of metal-organic frameworks (MOFs) based on indium-trimer building blocks: a porous MOF with soc topology and high hydrogen storage, *Angew. Chem., Int. Ed.*, **46**, 3278–3283.

17. Dincă, M., Dailly, A., Liu, Y., Brown, C. M., Neumann, D. A., Long, J. R. (2006). Hydrogen storage in a microporous metal-organic framework with exposed Mn^{2+} coordination sites, *J. Am. Chem. Soc.*, **128**, 16876–16883.

18. M. Dincă, M., Han, W.S., Liu, Y., Dailly, A., Brown, C. M., Long, J. R. (2007). Observation of Cu^{2+}-H_2 interactions in a fully desolvated sodalite-type metal-organic framework, *Angew. Chem., Int. Ed.*, **46**, 1419–1422.

19. Sumida, K., Horike, S., Kaye, S. S., Herm, Z. R., Queen, W. L., Brown, C. M., Grandjean, F., Long, G. J., Dailly, A. Long, J. R. (2010). Hydrogen storage and carbon dioxide capture in an iron-based sodalite-type metal–organic framework (Fe-BTT) discovered via high-throughput methods, *Chem. Sci.,* **1**, 184–191.

20. Rosi, N. L., Kim, J., Eddaoudi, M., Chen, B., O'Keeffe, M., Yaghi, O. M. (2005). Rod packings and metal-organic frameworks constructed from rod-shaped secondary building units, *J. Am. Chem. Soc.*, **127**, 1504–1518.

21. Dietzel, P. D. C., Panella, B., Hirscher, M., Blom, R., Fjellvåg, H. (2006). Hydrogen adsorption in a nickel based coordination polymer with open metal sites in the cylindrical cavities of the desolvated framework, *Chem. Commun.*, 959–961.

22. Dietzel, P. D. C., Blom, R., Fjellvåg, H. (2008). Base-induced formation of two magnesium metal-organic framework compounds with a bifunctional tetratopic ligand, *Eur. J. Inorg. Chem.*, **2008**, 3624–3632.

23. Zhou, W., Wu, H., Yildirim, T. (2008). Enhanced H_2 adsorption in isostructural metal-organic frameworks with open metal sites: strong dependence of the binding strength on metal ions, *J. Am. Chem. Soc.*, **130**, 15268–15269.

24. Bloch, E. D., Murray, L. J., Queen, W. L., Chavan, S., Maximoff, S. N., Bigi, J. P., Krishna, R., Peterson, V. K., Grandjean, F., Long, G. J., Smit, B., Bordiga, S., Brown, C. M., Long, J. R. (2011). Selective binding of O_2 over N_2 in a redox-active metal-organic framework with open iron(II) coordination sites, *J. Am. Chem. Soc.*, **133**, 14814–14122.

25. Chen, B., Zhao, X., Putkham, A., Hong, K., Lobkovsky, E. B., Hurtado, E. J., Fletcher, A. J., Thomas, K. M. (2008). Surface interactions and quantum kinetic molecular sieving for H_2 and D_2 adsorption on a mixed metal-organic framework material, *J. Am. Chem. Soc.*, **130**, 6411–6423.

26. Nouar, F., Eckert, J., Eubank, J. F., Forster, P., Eddaoudi, M. (2009). Zeolite-like metal-organic frameworks (ZMOFs) as hydrogen storage platform: lithium and magnesium ion-exchange and H_2-(rho-ZMOF) interaction studies, *J. Am. Chem. Soc.*, **131**, 2864–2870.

27. Han, S. S., Mendoza-Cortés, J. L., Goddard, W. A. (2009). Recent advances on simulation and theory of hydrogen storage in metal-organic frameworks and covalent organic frameworks, *Chem. Soc. Rev.*, **38**, 1460–1476.
28. Getman, R. B., Miller, J. H., Wang, K., Snurr, R. Q. (2011). Metal alkoxide functionalization in metal–organic frameworks for enhanced ambient-temperature hydrogen storage, *J. Phys. Chem. C*, **115**, 2066–2075.
29. Keskin, S., Liu, J., Rankin, R. B., Johnson, J. K., Sholl, D. S. (2009). Progress, opportunities, and challenges for applying atomically detailed modeling to molecular adsorption and transport in metal-organic framework materials, *Ind. Eng. Chem. Res.*, **48**, 2355–2371.
30. Tafipolsky, M., Amirjalayer, S., Schmid, R. (2010). Atomistic theoretical models for nanoporous hybrid materials, *Microporous Mesoporous Mater.*, **129**, 304–318.
31. Jiang, J., Babarao, R., Hu, Z. (2011). Molecular simulations for energy, environmental and pharmaceutical applications of nanoporous materials: from zeolites, metal–organic frameworks to protein crystals, *Chem. Soc. Rev.*, **40**, 3599–3612.
32. Getman, R. B., Bae, Y.-S., Wilmer, C. E., Snurr, R. Q. (2012). Review and analysis of molecular simulations of methane, hydrogen, and acetylene storage in metal-organic frameworks, *Chem. Rev.*, **112**, 703–723.
33. Sastre, G. (2010). Hydrogen physisorption in metal–organic frameworks: concepts and quantum chemical calculations, *Theor. Chem. Acc.*, **127**, 259–270.
34. Hedin, N., Chen, L., Laaksonen, A. (2010). Sorbents for CO_2 capture from flue gas—aspects from materials and theoretical chemistry, *Nanoscale*, **2**, 1819–1841.
35. Liu, D., Zhong, C. (2010). Understanding gas separation in metal–organic frameworks using computer modeling, *J. Mater. Chem.*, **20**, 10308–10318.
36. Peterson, V. K., Liu, Y., Brown, C. M., Kepert, C. J. (2006). Neutron powder diffraction study of D_2 sorption in Cu_3(1,3,5-benzenetricarboxylate)$_2$, *J. Am. Chem. Soc.*, **128**, 15578–15579.
37. Liu, Y., Kabbour, H., Brown, C. M., Neumann, D. A., Ahn, C. C. (2008). Increasing the density of adsorbed hydrogen with coordinatively unsaturated metal centers in metal-organic frameworks, *Langmuir*, **24**, 4772–4777.
38. Queen, W. L., Bloch, E. D., Brown, C. M., Hudson, M. R., Mason, J. A., Murray, L. J., Ramirez-Cuesta, A. J., Peterson, V. K., Long, Jeffrey R.

(2012). Hydrogen adsorption in the metal–organic frameworks Fe_2(dobdc) and $Fe_2(O_2)$(dobdc), *Dalton Trans.*, **41**, 4180–4187.

39. Wu, H., Zhou, W., Yildirim, T. (2009). High-capacity methane storage in metal-organic frameworks M_2(dhtp): the important role of open metal sites, *J. Am. Chem. Soc.*, **131**, 4995–5000.
40. Wu, H., Simmons, J. M., Liu, Y., Brown, C. M., Wang, X.-S., Ma, S., Peterson, V. K., Southon, P. D., Kepert, C. J., Zhou, H.-C., Yildirim, T., Zhou, W. (2010). Metal-organic frameworks with exceptionally high methane uptake: where and how is methane stored?, *Chem. Eur. J.*, **16**, 5205–5214.
41. Getzschmann, J., Senkovska, I., Wallacher, D., Tovar, M., Fairen-Jimenez, D., Düren, T., van Baten, J. M., Krishna, R., Kaskel, S. (2010). Methane storage mechanism in the metal-organic framework $Cu_3(btc)_2$: an in situ neutron diffraction study, *Microporous Mesoporous Mater.*, **136**, 50–58.
42. Wu, H., Simmons, J. M., Srinivas, G., Zhou, W., Yildirim, T. (2010). Adsorption sites and binding nature of CO_2 in prototypical metal-organic frameworks: a combined neutron diffraction and first-principles study, *J. Phys. Chem. Lett.*, **1**, 1946–1951.
43. Dietzel, P. D. C., Johnsen, R. E., Fjellvåg, H., Bordiga, S., Groppo, E., Chavan, S., Blom, R. (2008). Adsorption properties and structure of CO_2 adsorbed on open coordination sites of metal-organic framework Ni_2(dhtp) from gas adsorption, IR spectroscopy and X-ray diffraction, *Chem. Commun.*, 5125–5127.
44. Bordiga, S., Regli, L., Bonino, F., Groppo, E., Lamberti, C., Xiao, B., Wheatley, P. S., Morris, R. E., Zecchina, A. (2007). Adsorption properties of HKUST-1 toward hydrogen and other small molecules monitored by IR, *Phys. Chem. Chem. Phys.*, **9**, 2676–2685.
45. Vitillo, J. G., Regli, L., Chavan, S., Ricchiardi, G., Spoto, G., Dietzel, P. D. C., Bordiga, S., Zecchina, A. (2008). Role of exposed metal sites in hydrogen storage in MOFs, *J. Am. Chem. Soc.*, **130**, 8386–8396.
46. Nijem, N., Veyan, J.-F., Kong, L., Wu, H., Zhao, Y., Li, J., Langreth, D. C., Chabal, Y. J. (2010). Molecular hydrogen "pairing" interaction in a metal organic framework system with unsaturated metal centers (MOF-74), *J. Am. Chem. Soc.*, **132**, 14834–14848.
47. Areán, C. O., Chavan, S., Cabello, C. P., Garrone, E., Palomino, G. T. (2010). Thermodynamics of hydrogen adsorption on metal-organic frameworks, *ChemPhysChem*, **11**, 3237–3242.
48. Fitzgerald, S. A., Burkholder, B., Friedman, M., Hopkins, J. B., Pierce, C. J., Schloss, J. M., Thompson, B., Rowsell, J. L. C. (2011). Metal-

specific interactions of H_2 adsorbed within isostructural metal-organic frameworks, *J. Am. Chem. Soc.*, **133**, 20310–20318.

49. Chavan, S., Bonino, F., Vitillo, J. G., Groppo, E., Lamberti, C., Dietzel, P. D. C., Zecchina, A., Bordiga, S. (2009). Response of CPO-27-Ni towards CO, N_2 and C_2H_4, *Phys. Chem. Chem. Phys.*, **11**, 9811–9822.
50. Valenzano, L., Civalleri, B., Chavan, S., Palomino, G. T., Areán, C. O., Bordiga, S. (2010). Computational and experimental studies on the adsorption of CO, N_2, and CO_2 on Mg-MOF-74, *J. Phys. Chem. C.* **114**, 11185–11191.
51. Chavan, S., Shearer, G. C., Bloch, E., Bordiga, S. (2012). Acetylene adsorption on CPO-27-M metal-organic frameworks (M = Fe, Co and Ni), *ChemPhysChem*, **13**, 445–448.
52. Bordiga, S., Bonino, F., Lillerud, K. P., Lamberti, C. (2010). X-ray absorption spectroscopies: useful tools to understand metallorganic frameworks structure and reactivity, *Chem. Soc. Rev.*, **39**, 4885–4927.
53. Gul-E-Noor, F., Jee, B., Pöppl, A., Hartmann, M., Himsl, D., Bertmer, M. (2011). Effects of varying water adsorption on a $Cu_3(BTC)_2$ metal-organic framework (MOF) as studied by ^1H and ^{13}C solid-state NMR spectroscopy, *Phys. Chem. Chem. Phys.*, **13**, 7783–7788.
54. Brown, C. M., Liu, Y., Yildirim, T., Peterson, V. K., Kepert, C. J. (2009). Hydrogen adsorption in HKUST-1: a combined inelastic neutron scattering and first-principles study, *Nanotechnology*, **20**, 204025.
55. Cohen, A. J., Mori-Sánchez, P., Yang, W. (2012). Challenges for density functional theory, *Chem. Rev.*, **112**, 289–320.
56. Johnson, E. R., Mackie, I. D., DiLabio, G. A. (2009). Dispersion interactions in density-functional theory, *J. Phys. Org. Chem.*, **22**, 1127–1135.
57. Foster, M. E., Sohlberg, K. (2010). Empirically corrected DFT and semi-empirical methods for non-bonding interactions, *Phys. Chem. Chem. Phys.*, **12**, 307–322.
58. Dincă, M., Long, J. R. (2008). Hydrogen storage in microporous metal-organic frameworks with exposed metal sites, *Angew. Chem., Int. Ed.*, **47**, 6766–6779.
59. Kubas, G. J. (2007). Fundamentals of H_2 binding and reactivity on transition metals underlying hydrogenase function and H_2 production and storage, *Chem. Rev.*, **107**, 4152–4205.
60. Hoang, T. K. A., Antonelli, D. M. (2009). Exploiting the Kubas interaction in the design of hydrogen storage materials, *Adv. Mater.*, **21**, 1787–1800.

61. Yang, Q., Zhong, C. (2006). Understanding hydrogen adsorption in metal-organic frameworks with open metal sites: a computational study, *J. Phys. Chem. B*, **110**, 655–658.
62. Gomez, D. A., Combariza, A. F., Sastre, G. (2012). Confinement effects in the hydrogen adsorption on paddle wheel containing metal-organic frameworks, *Phys. Chem. Chem. Phys.*, **14**, 2508–2517.
63. van Mourik, T., Gdanitz, R. J. (2002). A critical note on density functional theory studies on rare-gas dimers, *J. Chem. Phys.*, **116**, 9620–9623.
64. Johnson, E. R., Wolkow, R. A., DiLabio, G. A. (2004). Application of 25 density functionals to dispersion-bound homomolecular dimers, *Chem. Phys. Lett.*, **394**, 334–338.
65. Fischer, M., Hoffmann, F., Fröba, M. (2010). Molecular simulation of hydrogen adsorption in metal-organic frameworks, *Colloids Surf., A*, **357**, 35–42.
66. Fischer, M. (2011). *Molecular Simulations of Hydrogen Storage and Gas Separation in Metal-Organic Frameworks*, PhD thesis, University of Hamburg.
67. Vermoortele, F., Vandichel, M., Van de Voorde, B., Ameloot, R., Waroquier, M., Van Speybroeck, V., De Vos, D. E. (2012). Electronic effects of linker substitution on Lewis acid catalysis with metal-organic frameworks, *Angew. Chem., Int. Ed.*, **51**, 4887–4890.
68. Swart, M., Solà, M., Bickelhaupt, F. M. (2009). A new all-round density functional based on spin states and S(N)2 barriers, *J. Chem. Phys.*, **131**, 094103.
69. Kim, Y.-H-., Kang, J., Wei, S.-H. (2010). Origin of enhanced dihydrogen-metal interaction in carboxylate bridged Cu_2–paddle-wheel frameworks, *Phys. Rev. Lett.*, **105**, 236105.
70. Bak, J. H., Le, V.-D., Kang, J., Wei, S. H., Kim, Y.-H. (2012). First-principles study of electronic structure and hydrogen adsorption of 3d transition metal exposed paddle wheel frameworks, *J. Phys. Chem. C*, **116**, 7386–7392.
71. Dincă, M., Long, J. R. (2007). High-enthalpy hydrogen adsorption in cation-exchanged variants of the microporous metal-organic framework $Mn_3[(Mn_4Cl)_3(BTT)_8(CH_3OH)_{10}]_2$, *J. Am. Chem. Soc.*, **129**, 11172–11176.
72. Sun, Y. Y., Kim, Y.-H., Zhang, S. B. (2007). Effect of spin state on the dihydrogen binding strength to transition metal centers in metal-organic frameworks, *J. Am. Chem. Soc.*, **129**, 12606–12607.

73. Sun, Y. Y., Kim, Y.-H., Lee, K., West, D., Zhang, S. B. (2011). Altering the spin state of transition metal centers in metal-organic frameworks by molecular hydrogen adsorption: a first-principles study, *Phys. Chem. Chem. Phys.*, **13**, 5042–5046.
74. Zhou, W., Yildirim, T. (2008). Nature and tunability of enhanced hydrogen binding in metal–organic frameworks with exposed transition metal sites, *J. Phys. Chem. C.*, **112**, 8132–8135.
75. Kong, L., Román-Pérez, G., Soler, J. M., Langreth, D. C. (2009). Energetics and dynamics of H_2 adsorbed in a nanoporous material at low temperature, *Phys. Rev. Lett.*, **103**, 096103.
76. Lochan, R. C., Head-Gordon, M. (2006). Computational studies of molecular hydrogen binding affinities: the role of dispersion forces, electrostatics, and orbital interactions, *Phys. Chem. Chem. Phys.*, **8**, 1357–1370.
77. Lochan, R. C., Khaliullin, R. Z., Head-Gordon, M. (2008). Interaction of molecular hydrogen with open transition metal centers for enhanced binding in metal-organic frameworks: a computational study, *Inorg. Chem.*, **47**, 4032–4044.
78. Kosa, M., Krack, M., Cheetham, A. K., Parrinello, M. (2008). Modeling the hydrogen storage materials with exposed M^{2+} coordination sites, *J. Phys. Chem. C.*, **112**, 16171–16173.
79. Kim, Y.-H., Sun, Y. Y., Ih Choi, W., Kang, J., Zhang, S. B. (2009). Enhanced dihydrogen adsorption in symmetry-lowered metal–porphyrin-containing frameworks, *Phys. Chem. Chem. Phys.*, **11**, 11400–11403.
80. Guo, Z., Wu, H., Srinivas, G., Zhou, Y., Xiang, S., Chen, Z., Yang, Y., Zhou, W., O'Keeffe, M., Chen, B. (2011). A metal-organic framework with optimized open metal sites and pore spaces for high methane storage at room temperature, *Angew. Chem., Int. Ed.*, **50**, 3178–3181.
81. Chen, L., Grajciar, L., Nachtigall, P., Düren, T. (2011). Accurate prediction of methane adsorption in a metal–organic framework with unsaturated metal sites by direct implementation of an ab initio derived potential energy surface in GCMC simulation, *J. Phys. Chem. C*, **115**, 23074–23080.
82. Grimme, S. (2006). Semiempirical GGA-type density functional constructed with a long-range dispersion correction, *J. Comput. Chem.*, **27**, 1787–1799.
83. Myers, A. L., Monson, P. A. (2002). Adsorption in porous materials at high pressure: theory and experiment, *Langmuir*, **18**, 10261–10273.

84. Britt, D., Furukawa, H., Wang, B., Glover, T. G., Yaghi, O. M. (2009). Highly efficient separation of carbon dioxide by a metal-organic framework replete with open metal sites, *Proc. Natl. Acad. Sci., U. S. A.*, **106**, 20637–20640.

85. Mason, J. A., Sumida, K., Herm, Z. R., Krishna, R., Long, J. R. (2011). Evaluating metal-organic frameworks for post-combustion carbon dioxide capture via temperature swing adsorption, *Energy Environ. Sci.*, **4**, 3030–3040.

86. Valenzano, L., Civalleri, B., Sillar, K., Sauer, J. (2011). Heats of adsorption of CO and CO_2 in metal–organic frameworks: quantum mechanical study of CPO-27-M (M = Mg, Ni, Zn), *J. Phys. Chem. C*, **115**, 21777–21784.

87. Grajciar, L., Wiersum, A. D., Llewellyn, P. L., Chang, J., Nachtigall, P. (2011). Understanding CO_2 adsorption in CuBTC MOF: comparing combined DFT–ab initio calculations with microcalorimetry experiments, *J. Phys. Chem. C*, **115**, 17925–17933.

88. Poloni, R., Smit, B., Neaton, J. B. (2012). CO_2 capture by metal-organic frameworks with van der Waals density functionals, *J. Phys. Chem. A*, **116**, 4957–4964.

89. Poloni, R., Smit, B., Neaton, J. B. (2012). Ligand-assisted enhancement of CO_2 capture in metal-organic frameworks, *J. Am. Chem. Soc.*, **134**, 6714–6719.

90. Dion, M., Rydberg, H., Schröder, E., Langreth, D.C., Lundqvist, B. I. (2004). van der Waals density functional for general geometries, *Phys. Rev. Lett.*, **92**, 246401.

91. Rubeš, M., Grajciar, L., Bludský, O., Wiersum, A. D., Llewellyn, P. L., Nachtigall, P. (2011). Combined theoretical and experimental investigation of CO adsorption on coordinatively unsaturated sites in CuBTC MOF, *ChemPhysChem*, **20**, 488–495.

92. Grajciar, L., Bludský, O., Nachtigall, P. (2010). Water adsorption on coordinatively unsaturated sites in CuBTC MOF, *J. Phys. Chem. Lett.*, **1**, 3354–3359.

93. Xiang, S., Zhou, W., Gallegos, J. M., Liu, Y., Chen, B. (2009). Exceptionally high acetylene uptake in a microporous metal-organic framework with open metal sites, *J. Am. Chem. Soc.*, **131**, 12415–12419.

94. Xiang, S., Zhou, W., Zhang, Z., Green, M. A., Liu, Y., Chen, B. (2010). Open metal sites within isostructural metal-organic frameworks for differential recognition of acetylene and extraordinarily high acetylene

storage capacity at room temperature, *Angew. Chem., Int. Ed.*, **49**, 4615–4618.

95. Bae, Y.-S., Lee, C. Y., Kim, K. C., Farha, O. K., Nickias, P., Hupp, J. T., Nguyen, S. T., Snurr, R. Q. (2010). High propene/propane selectivity in isostructural metal-organic frameworks with high densities of open metal sites, *Angew. Chem., Int. Ed.*, **51**, 1857–1860.
96. Watanabe, T., Sholl, D. S. (2010). Molecular chemisorption on open metal sites in Cu_3(benzenetricarboxylate)$_2$: a spatially periodic density functional theory study, *J. Chem. Phys.*, **133**, 094509.
97. Kawakami, T., Takamizawa, S., Kitagawa, Y., Maruta, T., Mori, W., Yamaguchi, K. (2001). Theoretical studies of spin arrangement of adsorbed organic radicals in metal-organic nanoporous cavity, *Polyhedron*, **20**, 1197–1206.
98. Vishnyakov, A., Ravikovitch, P. I., Neimark, A. V., Bülow, M. Wang, Q. M. (2003). Nanopore structure and sorption properties of Cu–BTC metal–organic framework, *Nano Lett.*, **3**, 713–718.
99. Düren, T., Sarkisov, L., Yaghi, O. M., Snurr, R. Q. (2004). Design of new materials for methane storage, *Langmuir*, **20**, 2683–2689.
100. Garberoglio, G., Skoulidas, A. I., Johnson, J. K. (2005). Adsorption of gases in metal organic materials: comparison of simulations and experiments, *J. Phys. Chem. B*, **109**, 13094–13103.
101. Krishna, R. (2012). Diffusion in porous crystalline materials, *Chem. Soc. Rev.*, **41**, 3099–3118.
102. Rappe, A. K., Casewit, C. J., Colwell, K. S., Goddard, W. A., Skiff, W. M. (1992). UFF, a full periodic table force field for molecular mechanics and molecular dynamics simulations, *J. Am. Chem. Soc.*, **114**, 10024–10035.
103. Mayo, S. L., Olafson, B. D., Goddard, W. A. (1990). DREIDING: a generic force field for molecular simulations, *J. Phys. Chem.*, **94**, 8897–8909.
104. Buch, V. (1994). Path integral simulations of mixed para-D_2 and ortho-D_2 clusters: the orientational effects, *J. Chem. Phys.*, **100**, 7610–7629.
105. Rowsell, J. L. C., Millward, A. R., Park, K. S., Yaghi, O. M. (2004). Hydrogen sorption in functionalized metal-organic frameworks, *J. Am. Chem. Soc.*, **126**, 5666–5667.
106. Tedds, S., Walton, A., Broom, D. P., Book, D. (2011). Characterisation of porous hydrogen storage materials: carbons, zeolites, MOFs and PIMs, *Faraday Discuss.*, **151**, 75–94.
107. Jorgensen, W. L., Maxwell, D. S., Tirado-Rives, J. (1996). Development and testing of the OPLS all-atom force field on conformational

energetics and properties of organic liquids, *J. Am. Chem. Soc.*, **118**, 11225–11236.

108. Liu, J., Culp, J. T., Natesakhawat, S., Bockrath, B. C., Zande, B., Sankar, S. G., Garberoglio, G., Johnson, J. K. (2007). Experimental and theoretical studies of gas adsorption in $Cu_3(BTC)_2$: an effective activation procedure, *J. Phys. Chem. C*, **111**, 9305–9313.
109. Liu, J., Rankin, R., Johnson, J. K. (2009). The importance of charge-quadrupole interactions for H_2 adsorption and diffusion in CuBTC, *Mol. Simul.*, **35**, 60–69.
110. Fischer, M., Hoffmann, F., Fröba, M. (2009). Preferred hydrogen adsorption sites in various MOFs—a comparative computational study, *ChemPhysChem*, **10**, 2647–2657.
111. Belof, J. L., Stern, A. C., Eddaoudi, M., Space, B. (2007). On the mechanism of hydrogen storage in a metal-organic framework material, *J. Am. Chem. Soc.*, **129**, 15202–15210.
112. Belof, J. L., Stern, A. C., Space, B. (2009). A predictive model of hydrogen sorption for metal-organic materials, *J. Phys. Chem. C*, **113**, 9316–9320.
113. Wang, S. (2007). Comparative molecular simulation study of methane adsorption in metal-organic frameworks, *Energy Fuels*, **21**, 953–956.
114. Yang, Q., Zhong, C. (2006). Electrostatic-field-induced enhancement of gas mixture separation in metal-organic frameworks: a computational study, *ChemPhysChem*, **7**, 1417–1421.
115. Martin, M. G., Siepmann, J. I. (1998). Transferable potentials for phase equilibria. 1. United-atom description of *n*-alkanes, *J. Phys. Chem. B*, **102**, 2569–2577.
116. Lucena, S. M. P., Mileo, P. G. M., Silvino, P. F. G., Cavalcante, C. L. (2011). Unusual adsorption site behavior in PCN-14 metal-organic framework predicted from Monte Carlo simulation, *J. Am. Chem. Soc.*, **133**, 19282–19285.
117. Gallo, M., Glossman-Mitnik, D. (2009). Fuel gas storage and separations by metal–organic frameworks: simulated adsorption isotherms for H_2 and CH_4 and their equimolar mixture, *J. Phys. Chem. C*, **113**, 6634–6642.
118. Keskin, S. (2010). Comparing performance of CPO and IRMOF membranes for gas separations using atomistic models, *Ind. Eng. Chem. Res.*, **49**, 11689–11696.
119. Figueroa, J., Fout, T., Plasynski, S., McIlvried, H., Srivastava, R. (2008). Advances in CO_2 capture technology: the U.S. department of energy's carbon sequestration program, *Int. J. Greenhouse Gas Control*, **2**, 9–20.

120. Li, J.-R., Ma, Y., McCarthy, M. C., Sculley, J., Yu, J., Jeong, H.-K., Balbuena, P. B., Zhou, H.-C. (2011). Carbon dioxide capture-related gas adsorption and separation in metal-organic frameworks, *Coord. Chem. Rev.*, **255**, 1791–1823.

121. Yang, Q., Zhong, C. (2006). Molecular simulation of carbon dioxide/methane/hydrogen mixture adsorption in metal-organic frameworks, *J. Phys. Chem. B*, **110**, 17776–17783.

122. Yang, Q., Xue, C., Zhong, C., Chen, J.-F. (2007). Molecular simulation of separation of CO_2 from flue gases in CU-BTC metal-organic framework, *AIChE J.*, **53**, 2832–2840.

123. Wang, S., Yang, Q., Zhong, C. (2008). Adsorption and separation of binary mixtures in a metal-organic framework Cu-BTC: A computational study, *Sep. Purif. Technol.*, **60**, 30–35.

124. Yazaydın, A. Ö., Benin, A. I., Faheem, S. A., Jakubczak, P., Low, J. J., Willis, R. R., Snurr, R. Q. (2009). Enhanced CO_2 adsorption in metal-organic frameworks via occupation of open-metal sites by coordinated water molecules, *Chem. Mater.* **21**, 1425–1430.

125. Yazaydın, A. Ö., Snurr, R. Q., Park, T.-H., Koh, K., Liu, J., Levan, M. D., Benin, A. I., Jakubczak, P., Lanuza, M., Galloway, D. B., Low, J. J., Willis, R. R. (2009). Screening of metal-organic frameworks for carbon dioxide capture from flue gas using a combined experimental and modeling approach, *J. Am. Chem. Soc.*, **131**, 18198–18199.

126. Chen, Y.-F., Babarao, R., Sandler, S. I., Jiang, J. W. (2010). Metal-organic framework MIL-101 for adsorption and effect of terminal water molecules: from quantum mechanics to molecular simulation, *Langmuir*, **26**, 8743–8750.

127. Soubeyrand-Lenoir, E., Vagner, C., Yoon, J. W., Bazin, P., Ragon, F., Hwang, Y. K., Serre, C., Chang, J.-S., Llewellyn, P. L. (2012). How water fosters a remarkable 5-fold increase in low pressure CO_2 uptake within the mesoporous MIL-100(Fe), *J. Am. Chem. Soc.*, **134**, 10174–10181.

128. Babarao, R., Jiang, J., Sandler, S. I. (2009). Molecular simulations for adsorptive separation of CO_2/CH_4 mixture in metal-exposed, catenated, and charged metal-organic frameworks, *Langmuir*, **25**, 5239–5247.

129. Karra, J. R., Walton, K. S. (2008). Effect of open metal sites on adsorption of polar and nonpolar molecules in metal-organic framework Cu-BTC, *Langmuir*, **24**, 8620–8626.

130. Martín-Calvo, A., Lahoz-Martín, F. D., Calero, S. (2012). Understanding carbon monoxide capture using metal–organic frameworks, *J. Phys. Chem. C*, **116**, 6655–6663.

131. Fischer, M., Hoffmann, F., Fröba, M. (2012). Metal–organic frameworks and related materials for hydrogen purification: Interplay of pore size and pore wall polarity, *RSC Adv.*, **2**, 4382–4396.

132. Desbiens, N., Boutin, A., Demachy, I. (2005). Water condensation in hydrophobic silicalite-1 zeolite: a molecular simulation study, *J. Phys. Chem. B*, **109**, 24071–24076.

133. Castillo, J. M., Vlugt, T. J. H., Calero, S. (2008). Understanding water adsorption in Cu–BTC metal–organic frameworks, *J. Phys. Chem. C*, **112**, 15934–15939.

134. García-Pérez, E., Gascón, J., Morales-Flórez, V., Castillo, J. M., Kapteijn, F., Calero, S. (2009). Identification of adsorption sites in Cu-BTC by experimentation and molecular simulation, *Langmuir*, **25**, 1725–1731.

135. Calero, S., Martín-Calvo, A., Hamad, S., García-Pérez, E. (2011). On the performance of Cu-BTC metal organic framework for carbon tetrachloride gas removal, *Chem. Commun.*, **47**, 508–510.

136. Martín-Calvo, A., García-Pérez, E., García-Sánchez, A., Bueno-Pérez, R., Hamad, S., Calero, S. (2011). Effect of air humidity on the removal of carbon tetrachloride from air using Cu-BTC metal-organic framework, *Phys. Chem. Chem. Phys.*, **13**, 11165–11174.

137. Petit, C., Huang, L., Jagiello, J., Kenvin, J., Gubbins, K. E., Bandosz, T. J. (2011). Toward understanding reactive adsorption of ammonia on Cu-MOF/graphite oxide nanocomposites, *Langmuir*, **27**, 13043–13051.

138. Lamia, N., Jorge, M., Granato, M. A., Almeida Paz, F. A., Chevreau, H., Rodrigues, A. E. (2009). Adsorption of propane, propylene and isobutane on a metal–organic framework: Molecular simulation and experiment, *Chem. Eng. Sci.*, **64**, 3246–3259.

139. Jorge, M., Lamia, N., Rodrigues, A. E. (2010). Molecular simulation of propane/propylene separation on the metal–organic framework CuBTC, *Colloids Surf., A*, **357**, 27–34.

140. Fischer, M., Kuchta, B., Firlej, L., Hoffmann, F., Fröba, M. (2010). Accurate prediction of hydrogen adsorption in metal–organic frameworks with unsaturated metal sites via a combined density-functional theory and molecular mechanics approach, *J. Phys. Chem. C*, **114**, 19116–19126.

141. Fischer, M., Hoffmann, F., Fröba, M. (2010). New microporous materials for acetylene storage and C_2H_2/CO_2 separation: insights from molecular simulations, *ChemPhysChem*, **11**, 2220–2229.

142. Fischer, M., Gomes, J. R. B., Fröba, M., Jorge, M. (2012). Modeling adsorption in metal-organic frameworks with open metal sites: propane/propylene separations, *Langmuir*, **28**, 8537–8549.

143. Prosenjak, C., Banu, A.-M., Gellan, A. D., Düren, T. (2012). Hydrogen thermal desorption spectra: insights from molecular simulation, *Dalton Trans.*, **41**, 3974–3984.

144. Rowsell, J. L. C., Yaghi, O. M. (2006). Effects of functionalization, catenation, and variation of the metal oxide and organic linking units on the low-pressure hydrogen adsorption properties of metal-organic frameworks, *J. Am. Chem. Soc.*, **128**, 1304–1315.

Chapter 4

Accurate ab initio Description of Adsorption on Coordinatively Unsaturated Sites in Metal–Organic Frameworks

Lukáš Grajciar, Miroslav Rubeš, Ota Bludský, and Petr Nachtigall

Department of Physical and Macromolecular Chemistry, Faculty of Science, Charles University in Prague, Hlavova 8, 128 43 Prague, Czech Republic

petr.nachtigall@natur.cuni.cz

Accurate ab initio description of the adsorbate-adsorbent interactions in MOFs containing coordinatively unsaturated transition metal sites represents a significant challenge for computational chemists. The following complications are often faced: (i) spin coupling between the unpaired electrons on individual transition metal ions, (ii) complicated electronic structure of individual transition metal ions or metal-oxo-clusters resulting in multireference character of corresponding wavefunction, and (iii) the importance of dispersion interaction between adsorbate and adsorbent. As a consequence, a reliability of computational methods based on popular density functional theory (DFT) is often insufficient.

Metal–Organic Frameworks: Materials Modeling towards Potential Engineering Applications
Edited by Jianwen Jiang

ISBN 978-981-4613-45-3 (Hardcover), 978-981-4613-46-0 (eBook)
www.panstanford.com

The suitability, reliability, and accuracy of DFT methods for the description of adsorption in MOFs are discussed in this chapter on the basis of the comparison with highly accurate wavefunction-based methods and with accurate experimental data.

4.1 Introduction

There is a large group of MOFs where transition metal cations are not coordinatively saturated by organic linkers. Consequently, these metal–organic frameworks (MOFs) contain coordinatively unsaturated sites (*cus*) that exhibit unique adsorption properties. Cu–benzene-1,3,5-tricarboxylate (Cu-BTC) [1], CPO-27 [2], Matériaux de l'Institut Lavoisier (MIL)-100 [3], and MIL-101 [4] are only a few examples of MOFs with *cus* sites. Computational investigations of adsorption properties of MOFs mostly rely on the empirical interatomic potential functions (IPFs, often referred to as force fields) used in grand canonical Monte Carlo (GCMC) simulations [5]. While there are many successful applications of IPFs giving adsorption isotherms in good agreement with experiments [6, 7], the suitability of IPF-based approaches for the description of MOFs with *cus* sites is, at least, questionable [8, 9]. It is therefore understandable that interaction of various adsorbates with *cus* sites was investigated at the density functional theory (DFT) level. Not surprisingly it has been found that interaction of small molecules with MOFs was typically overestimated when local exchange correlation functionals (local density approximation [LDA]) were employed and underestimated when semilocal generalized gradient approximation (GGA)-type functionals were used [10–12]. There are two reasons for the failure of traditional exchange correlation functionals (LDA and GGA) to describe adsorbate interactions with *cus*-containing MOFs—first, the inability of these functionals to describe the van der Waals (vdW) interactions that are critical for correct description of adsorption in porous materials, and second, the inaccuracy of DFT when describing the electronic structure of transition metal cations, in particular those with open-shell electronic structure. While the former deficiency can be reasonably well overcome using either semiempirical dispersion corrections [13, 14] or nonempirical vdW density functionals [15, 16], the latter deficiency is more serious for MOFs containing transition metal *cus* sites.

Accurate ab initio calculations employing the reliable post-Hartree–Fock methods such as coupled cluster (CC) have been used for the description of adsorbate–*cus* site interactions (e.g., Refs. [17, 18]) employing relatively small cluster models representing the *cus* site. While the results obtained are highly accurate for the model used, they cannot be directly compared with experimental results since they cannot account for the long-range interactions. Instead, the results of CC (or similar) highly accurate calculations for cluster models can serve as important benchmark for evaluation of the accuracy of more approximate methods, such as DFT or empirical force fields.

The extent of this chapter does not allow for a broad overview of large number of applications of ab initio methods in the investigation of adsorption phenomenon in many different MOFs. Instead, this chapter is focused on one particular MOF, HKUST-1 (often denoted as Cu-BTC) [1], for which a large amount of experimental data is available and which is likely the most often computationally investigated MOF. The computational methods discussed within the chapter are described first. The computational difficulties related to the spin coupling issues are described next on the example of the Cu^{2+}–Cu^{2+} interaction in the Cu-BTC paddlewheel (PDWL). The reliability of computational methods is thoroughly discussed in Section 4.3.2, comparing cluster model results obtained at the CC level of theory with those obtained with less accurate electronic structure methods. The results obtained for water, CO, CO_2, CH_4, and C_3 hydrocarbons are reviewed in Sections 4.4.1, 4.4.2, 4.4.3, 4.4.4, and 4.4.5, respectively.

4.2 Methods and Models

4.2.1 *Cluster and Periodic Models of Cu-BTC*

The *cus* site in Cu-BTC consists of a pair of Cu^{2+} ions in close proximity (about 2.5 Å) connected with four carboxylic groups (a PDWL model; see Fig. 4.1b). The PDWL unit represents the smallest model capturing the electronic properties of the *cus* site.

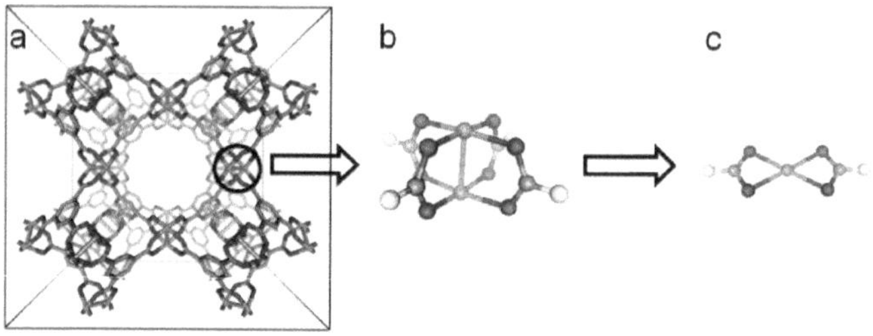

Figure 4.1 (a) Cu-BTC unit cell, (b) PDWL unit, and (c) $Cu(HCOO)_2$ model.

In situations where the coupling of unpaired electrons on Cu^{2+} can be neglected (such as physisorption of small molecules on *cus* site, see Section 4.3.1), a much smaller model, $Cu(HCOO)_2$ constrained at the geometry of PDWL, can be employed (Fig. 4.1c). The size of this model enables the benchmarking of density functionals (see Section 4.2.2.1) used in periodic calculations with respect to the reference level of theory (typically CCSD(T) level, Section 4.2.2.2).

While cluster models described above are extremely useful for explaining a rather complex electronic structure and properties of *cus* sites, an accurate description of adsorbate–MOF interactions accounting for long range dispersion and electrostatic contributions can only be obtained within the periodic model. The Cu-BTC structure was modeled either using the experimental rhombohedral primitive cell (RPC) [1] containing 156 framework atoms ($a = b = c = 18.627$ Å; $\alpha = \beta = \gamma = 60°$) or by the optimized RPC cell obtained from fitting the polynomial to the E(V) curve calculated at different volumes relaxing the cell shape and fractional coordinates ($a = b = c = 18.774$ Å; $\alpha = \beta = \gamma = 60°$) [17, 18].

4.2.2 *Methods*

4.2.2.1 Density functional theory

The cluster DFT calculations were carried out with several commonly used density functionals (LDA, GGA, hybrid, and nonlocal exchange correlation functionals, Fig. 4.2). The interaction energies

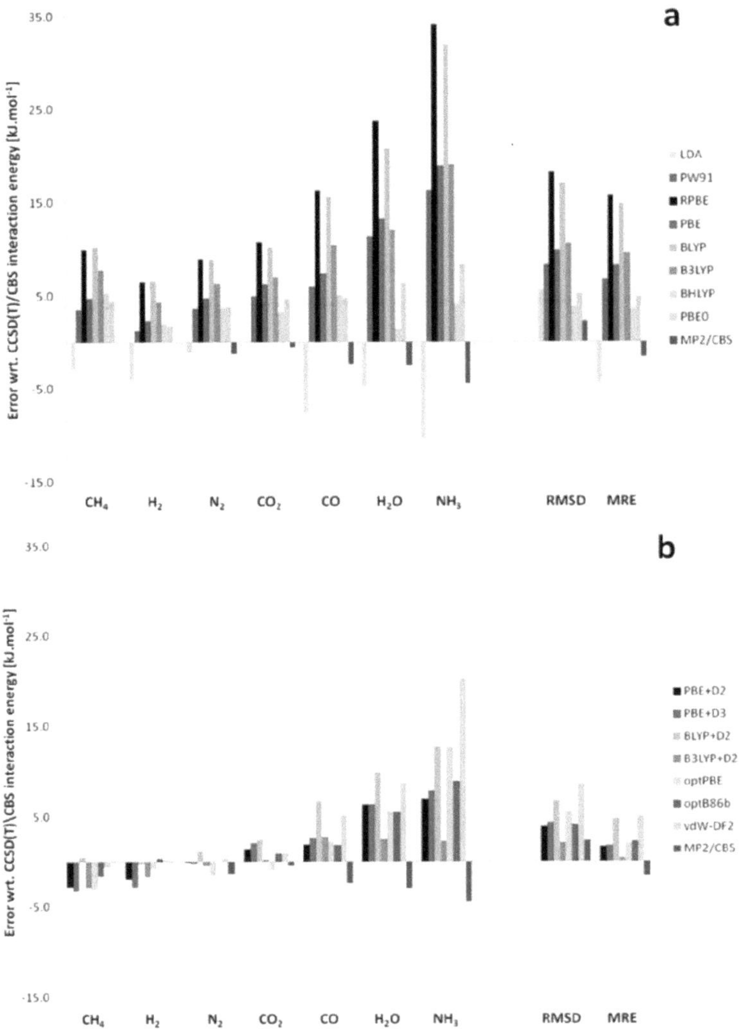

Figure 4.2 Error in interaction energies (in kJ mol^{-1}) with respect to CCSD(T)/CBS results. Interaction energies of various molecules with the $Cu(HCOO)_2$ model were calculated with the (a) DFT and (b) dispersion-corrected DFT functionals. MP2/CBS results are included for comparison.

were calculated using the supermolecular approach:

$$E_{\text{int}} = E_{\text{AB}} - E_{\text{A}} - E_{\text{B}} \tag{4.1}$$

corrected for the basis set superposition error (BSSE) by the counterpoise correction of Boys and Bernardi [19] within the frozen monomer approximation. The pseudopotential-based correlation-consistent valence-X-ζ basis set with polarization functions by Peterson et al. [20], cc-pVXZ-PP, were employed for Cu atoms, while the standard Dunning's correlation-consistent valence-X-ζ basis sets with polarization functions [21], cc-pVXZ, were used for other atoms. The corresponding augmented correlation-consistent basis sets were also employed, aug-cc-pVXZ-PP and aug-cc-pVXZ for Cu and other atoms, respectively.

The above-mentioned mixed basis sets are denoted throughout the text as VXZ and AVXZ for the standard and augmented basis sets, respectively. The periodic DFT calculations were performed with the Perdew–Burke–Ernzerhof (PBE) functional [22] for the ferromagnetic case (all unpaired spins on 12 copper atoms in the unit call were aligned in the same direction). The projector-augmented wave (PAW) [23] and the kinetic energy cutoff of 600 eV were used. The Γ-point sampling of the first Brillouin zone was found to be sufficient to yield converged results.

4.2.2.2 Post-Hartree-Fock methods

The CC method [24] with explicit single and double and perturbative triple excitations, CCSD(T), is currently the most reliable quantum-chemical method for the accurate description of the adsorbate-adsorbent interactions. Thanks to the recent progress in parallelizing CC equations, the CCSD(T) method is applicable even to moderately large systems. The only sizable errors in CCSD(T) calculations come from the basis set incompleteness (BSI). Two commonly employed approaches for dealing with the BSI error are the explicitly correlated CC method [25] and the extrapolation to the CBS limit [26]. Since the extrapolation technique requires the use of truly large basis sets, the actual CCSD(T)/CBS calculations for large cluster models (see Section 4.2.1) are usually performed in combination with more approximative (and less reliable) methods such as MP2. For models of the MOF adsorption complexes the estimation of the BSI error is performed at the MP2 level typically using the AVXZ (X = T, Q) basis sets.

An accurate theoretical description of the *cus* site in Cu-BTC by single reference methods (such as CCSD(T)) is complicated by the presence of Cu^{2+} ions with open-shell electronic structure. The multireference character of the singlet wavefunction and the stability in the corresponding singlet (S) and triplet (T) electronic states for the $Cu_2(HCOO)_4$cluster model (see Fig. 4.1b) were investigated by the complete active space second-order perturbation theory (CASPT2).

4.2.2.3 DFT/CC method

Reliable CC and multireference methods described in Section 4.2.2.2 cannot be applied directly, neither to large cluster models, nor to a periodic model of MOFs. Thus, there is a need for a computationally tractable yet reliable method for description of MOFs with transition metal *cus*s. DFT has been used extensively for description of molecular sieves. The GGA-type functionals, however, do not account for dispersion interactions (always important in extended systems). Various types of dispersion-corrected density functionals were employed for the periodic calculations of MOFs ranging from the nonlocal functionals to GGA-type functionals with damped r^{-6} dispersion corrections. However, the problem of DFT for the description of *cus* sites involves often unrealistically large spin and charge delocalization and cannot be fixed simply by adding a dispersion component.

Recently, a parameter-free DFT/CC correction scheme has been proposed for precise calculations (close to CCSD(T) accuracy) of weakly bound molecular systems [27]. The correction scheme has been successfully applied to the physical adsorption of molecules on graphene and various molecular sieves. The DFT/CC method is based on the pairwise representability of DFT error, ΔE, defined as

$$\Delta E = E_{\mathrm{CCSD(T)}} - E_{\mathrm{DFT}} \qquad (4.2)$$

where $E_{\mathrm{CCSD(T)}}$ and E_{DFT} are the interaction energy calculated at the CCSD(T)/CBS and DFT/AVQZ levels of theory, respectively. Therefore, the DFT/CC method can be used also for systems where the nature of the DFT error is more general than just a lack of nonlocal correlation. Within the DFT/CC method, the DFT error ΔE is expressed as a sum of atom–atom correction functions $\varepsilon_{ij}(R_{ij})$

$$\Delta E = \sum_{ij} \varepsilon_{ij}(R_{ij}) \tag{4.3}$$

where R_{ij} is the distance between atoms i and j. No explicit functional form for ε_{ij} correction functions is assumed; instead a reproducible kernel Hilbert space interpolation [28] is used. In addition to the assumption about pairwise representability of DFT error, Eq. 4.3, the transferability of the correction functions from the reference system to the system of interest is assumed. Correction functions ε_{ij} are evaluated from one-dimensional potential energy curves of suitably chosen reference complexes including complexes of the $Cu(HCOO)_2$ model described in Section 4.2.1 with small adsorbate molecules (CH_4, H_2, N_2, CO_2, CO, H_2O, and NH_3). The choice of the reference system is just critical for the reliability of the DFT/CC method. For example, the water–Cu-BTC correction functions ε_{OH} (ε_{HH}), ε_{OC} (ε_{HC}), ε_{OO} (ε_{HO}), and ε_{OCu} (ε_{HCu}) were evaluated from H_2O–H_2, H_2O–C_6H_6, H_2O–CO_2, and H_2O–$Cu(HCOO)_2$ complexes, respectively [17]. Details about the construction of other correction functions can be found elsewhere [18, 29, 30]. The assumption of the correction function transferability was tested by CCSD(T)/CBS and DFT/CC calculations carried out for the $Cu_2(HCOO)_4$ cluster model, and they are discussed below.

4.3 Electronic Structure of Coordinatively Unsaturated Sites in Cu-BTC

4.3.1 *Spin Coupling in Cu-BTC*

The PDWL units in Cu-BTC closely resemble some well-known molecular complexes of dimeric Cu^{2+} carboxylates, which show paramagnetic behavior at room temperature. The temperature lowering leads to a steady transition from paramagnetic to antiferromagnetic couplings of the two unapaired electrons on each Cu^{2+} pair [31–33]. Thus the ground state of the PDWL unit is a singlet state (diamagnetic), and the excited triplet state (paramagnetic) starts to be populated with increasing temperature. The experimental singlet–triplet splitting for such complexes is usually in the range of $-150\ cm^{-1}$ to $-200\ cm^{-1}$.

To reliably calculate the singlet–triplet splitting ($\Delta E_{ST} = E_S - E_T$) in a PDWL unit, the singlet and triplet states must be described with high accuracy. While the triplet state can be described with just a single Slater determinant, and thus "standard" methods of quantum chemistry such as CCSD(T), MP2, and DFT can be used, a multireference wavefunction (i.e., at least two Slater determinants) is required for the description of the singlet state [17, 30]. Within the DFT approach the energy of the singlet state is often approximated with the broken-symmetry solution. The calculated singlet–triplet splittings for PDWL and its complexes with H_2O and CO molecules are given in Table 4.1.

The ΔE_{ST} energy gap calculated at the CASPT2 level is consistent with experimental results and it does not change significantly upon the interaction with H_2O and CO. It follows that the calculated interaction energies for singlet and triplet states are very similar. Thus, accurate interaction energies can be obtained for the triplet electronic state for which an accurate approach such as CCSD(T) at the complete basis set (CBS) limit can be used.

The B3LYP and PBE functionals give correctly the singlet state (broken-symmetry solution) below the triplet state, and the ΔE_{ST} energy gaps are also not influenced significantly upon the interaction with H_2O or CO. The B3LYP results are clearly superior to the PBE ones, considering the strong overestimation of the ΔE_{ST} energy gap at the PBE level. Moreover in the case of CO interaction with *cus* sites there is a large change in Cu–Cu distance at PBE level with respect to B3LYP and CASPT2 results. This artificially increased Cu–Cu distance results in the overestimated σ-bonding interaction [30]. It is apparent that mixing of the exact exchange into the functional leads to better results.

The B3LYP and PBE closed-shell singlet-state solutions are not reported in Table 4.1 since corresponding energies are above the triplet state and, in addition, they strongly depend on the adsorbed molecule and geometry used. The strong dependence of the ΔE_{ST} energy gap upon adsorbed species and geometry can result in a largely overestimated difference (as much as 17 kJ mol^{-1}) between

Table 4.1 Calculated singlet–triplet (S,T) splitting (ΔE_{ST}) for bare PDWL, CO\PDWL, and H_2O\PDWL complexes along with optimized Cu–Cu equilibrium distances. The DFT (B3LYP, PBE) singlet corresponds to the broken-symmetry solution

		r(Cu–Cu) (Å)			ΔE_{ST} (cm^{-1})		
Complex[a,b]	State	CASPT2	B3LYP	PBE	CASPT2	B3LYP	PBE
PDWL	S	2.511	2.523	2.473	−155	−365	−953
	T	2.511	2.518	2.455			
H_2O\PDWL	S	2.600	2.618	2.548	−119	−300	−838
	T	2.600	2.614	2.532			
CO\PDWL	S	2.593	2.618	2.644	−135	−343	−1055
	T	2.593	2.614	2.604			

[a] The complexes have been described with a standard cc-pVDZ basis set; for Cu the cc-pVDZ-PP basis set with pseudopotential has been used. The size of the basis set should not significantly influence the singlet–triplet splitting [17].
[b] All complexes have D_{2h} symmetry, which implies that one H_2O (CO) molecule interacts with each of the *cus* sites of the PDWL unit.

interaction energies for closed-shell singlet and open-shell triplet (and properly described singlet) electronic states.

The experimental measurements on Cu-BTC indicated antiferromagnetic coupling on each PDWL unit at low temperatures [34]. The magnetic susceptibility measurements further suggested weak ferromagnetic coupling between PDWL units below 70 K, probably modulated by organic linkers. However, it was pointed out by the subsequent experimental study that presence of paramagnetic extra-framework Cu^{2+} cations can influence magnetic susceptibility measurements at very low temperatures [35].

It can be concluded that the antiferromagnetic solution in periodic DFT treatment is desirable. However, the presence of six PDWL units in the unit cell of Cu-BTC makes the convergence into the correct antiferromagnetic solution rather problematic. Corresponding convergence issues can be overcome by using a high-spin (ferromagnetic) solution, which can be taken as an analogy of the triplet state on each PDWL unit in the Cu-BTC unit cell. Considering weak coupling between PDWL units the resulting adsorption enthalpies obtained for a well-defined high-spin solution do not differ significantly from the correct antiferromagnetic solution. Furthermore, spin un-polarized solutions yield interaction energies also in a good agreement (within 2 kJ mol^{-1}) with the high-spin solution [17]. This apparent contradiction with the results obtained for the PDWL cluster (see above) is most likely due to different geometries of the PDWL cluster and the PDWL unit in Cu-BTC; some favorable error compensation takes place in the Cu-BTC structure, while such error cancellation it is not operative for the PDWL cluster. It follows that the calculations can be performed for adsorption of some adsorbates in Cu-BTC, even with the spin-unpolarized wavefunction; however, a spin-polarized solution should be preferred.

4.3.2 *Evaluation of Method Accuracy: The $Cu(COOH)_2$ Model*

The reliability of various DFT-based methods, including dispersion-corrected DFT, for the adsorbate–*cus* site interaction was investigated using the $Cu(HCOO)_2$ cluster model (Fig. 4.1c). The small size of this model enables the benchmarking of DFT functionals with respect to the highly accurate CCSD(T) results for a number of adsorbents (CH_4, H_2, N_2, CO_2, CO, H_2O, and NH_3); different types of adsorbate–adsorbent interactions, including dispersion, electrostatic, and partially covalent bonding, are represented in this set.

Interaction energies of the above-mentioned small molecules with the $Cu(HCOO)_2$ cluster model in PBE/AVTZ equilibrium geometry were first evaluated for several commonly used LDA, GGA, and hybrid functionals (Fig. 4.2a). For all the molecules considered herein the interaction energy is underestimated with tested exchange correlation functionals except for the LDA functional, which, on the contrary, systematically overestimates interaction energies. The root-mean-square deviation (RMSD) values averaged for hybrid exchange correlation functionals (B3LYP, BHLYP, and PBE0) are lower than those for the GGA functionals (PBE, PW91, revPBE, and BLYP), 7.1 and 14.0 kJ mol^{-1}, respectively. Note, however, that B3LYP has a larger RMSD than PBE, PW91, and LDA. Since none of these functionals accounts for dispersion interaction, it is not surprising that even for BHLYP (having the lowest RMSD) the relative errors are larger than 40% for adsorbates where the stabilization by the dispersion interaction is critical (CH_4, H_2, N_2). A notable trend, which has been also reported for Cu^{2+} cation interaction with other ligands [36–39], could be observed for the DFT functionals formed by the B88 exchange and LYP correlation functionals with an addition of a different amount of the exact HF exchange: the increasing amount of the exact HF exchange (BLYP < B3LYP < BHLYP) leads to improved accuracy.

The failure of the standard GGA or even hybrid functionals to provide reasonable agreement with the benchmarking data is not surprising as it is well-known that these DFT functionals do not account for dispersion interactions that should lead to a systematic underestimation of the interaction energies (Fig. 4.2a) Therefore, two different types of dispersion correction schemes were adopted: the semiempirical pairwise C_6 approach by Grimme (DFT+D2 or DFT+D3) [14, 40] and new nonlocal correlation functionals that capture long-range dispersion forces (vdW-DF2, optPBE, optB86b) [15, 16, 41]. The inclusion of dispersion interactions led to substantial improvement of the interaction energies (Fig. 4.2b) with an RMSD value decreasing as much as 7.4 kJ mol^{-1} upon adding the D2 correction; note that B3LYP+D2 has a smaller RMSD than MP2. However the improvement is dependent on the type of the adsorption complex. The following observations can be drawn: (i) the stability of dispersion-dominated complexes (H_2 and CH_4 in particular) is overestimated, (ii) interaction energies of N_2 and CO_2 are accurately described at the DFT-D level, and (iii) the interaction between the *cus* site and CO, H_2O, and NH_3 molecules is severely underestimated (except for the B3LYP-D2 level). Thus, adding the dispersion correction leads to improved accuracy; however, missing dispersion is not the only deficiency of exchange correlation functionals discussed herein and other effects (errors) play a role as well.

The accuracy of nonlocal functionals is somewhat disappointing; on the basis of the RMSD values it appears to be even worse than the accuracy of some hybrid functionals without the dispersion correction (PBE0 and BHLYP in particular). A closer inspection of the vdW-DF2 results reveals that this nonlocal functional performs excellently for systems where covalent bonding (and charge transfer; see Table 4.2) between adsorbate and adsorbent is negligible (CH_4,

Table 4.2 Mulliken charges computed employing various exchange correlation functionals and the reference averaged quadratic coupled-cluster (AQCC) method

				DFT functional			
				GGA	Hybrid		
System	Fragment	Molecule	LDA	BLYP	B3LYP	BHLYP	AQCC
$Cu(COOH)_2$ molecule	Cu atom	CH_4	0.49	0.57	0.70	0.88	0.87
		H_2	0.49	0.57	0.70	0.88	0.87
		N_2	0.47	0.55	0.68	0.86	0.86
		CO_2	0.50	0.58	0.71	0.88	0.88
		CO	0.47	0.56	0.69	0.87	0.87
		H_2O	0.49	0.59	0.72	0.90	0.89
		NH_3	0.48	0.58	0.72	0.90	0.88
	$Cu(COOH)_2$ monomer	CH_4	−0.01	0.00	0.00	0.00	0.00
		H_2	−0.01	0.00	0.00	0.00	0.00
		N_2	−0.04	−0.04	−0.03	−0.02	−0.02
		CO_2	−0.02	−0.02	−0.02	−0.01	−0.01
		CO	−0.06	−0.04	−0.04	−0.04	−0.04
		H_2O	−0.09	−0.08	−0.07	−0.06	−0.06
		NH_3	−0.13	−0.11	−0.10	−0.09	−0.09
$Cu(COOH)_2$			0.49	0.57	0.69	0.87	0.86

H_2, N_2 and CO_2). The error of vdW-DF2 increases with increasing covalency of adsorbate–adsorbent interactions (5 kJ mol^{-1}, 9 kJ mol^{-1}, and 20 kJ mol^{-1} for CO, H_2O, and NH_3, respectively).

The performance of dispersion-corrected functionals discussed above indicates that DFT cannot accurately describe interactions of adsorbates with *cus* sites. A problematic description of the Cu^{2+} cation interactions with small ligands at the DFT level has been already reported [38, 39]. An unrealistically large spin and charge delocalization, which is often connected with the incomplete cancellation of the self-interaction, was observed for GGA functionals in particular [38, 39]. This artificial delocalization was found to decrease with the increasing amount of the exact exchange mixed in hybrid exchange correlation functionals.

Similar dependence of the spin and charge delocalization on the amount of the HF exchange is reported in Table 4.2 for the adsorption complexes of small gas molecules with the $Cu(HCOO)_2$ cluster model. The description of the charge density improves from LDA to GGA and to hybrid functionals with respect to the reference high-level ab initio averaged quadratic coupled-cluster (AQCC) densities. All the DFT functionals considered, with the exception of the BHLYP, overestimate the spin and charge delocalization that leads to a too large charge density (and incorrect electronic structure description) on Cu^{2+} that is responsible for part of the DFT error. To quantify the DFT error due to an incorrect electronic density we have evaluated the density functionals on the precalculated AQCC densities for all the adsorption complexes considered. The use of the AQCC densities resulted in only minor (less than 1 kJ mol^{-1}) changes in the interaction energies for all the adsorption complexes with the exception of the water and ammonia complexes where a larger effect in the range of 1–8 kJ mol^{-1} (around 5%–15% of the CCSD(T)/CBS interaction energy) was observed. Furthermore, in the case of water and ammonia the AQCC densities led to an improvement of the interaction energies with the more pronounced effect on the GGA than hybrid functionals that is in line with the fact that the densities obtained using the hybrid functional are closer to the reference AQCC density. Although the discrepancy between DFT and reference CCSD(T)/CBS interaction energies can be to some extent attributed to either a missing part of the dispersion interaction at the DFT level or unrealistic charge distribution on the $Cu(HCOO)_2$ cluster model, significant uncertainties remain that could be related to the quality of the underlying exchange correlation functional, the transferability of the semiempirical dispersion corrections (DFT+D) and finally to the robustness of the combination of the exchange correlation functional with the scheme accounting for the dispersion (long-range correlation).

4.3.3 *Paddlewheel Model*

The assessment of various DFT methods presented in the previous section shows that none of the commonly used density functionals (local, semilocal, and nonlocal) provides satisfactory accuracy for the adsorbate–*cus* site interaction. Adding dispersion correction is

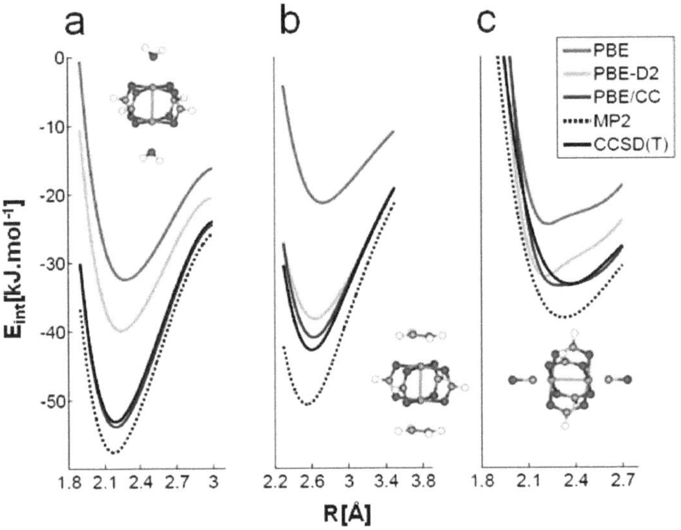

Figure 4.3 (a) PDWL/H2O, (b) CO, and (c) ethylene 1D dissociation curves at various levels of theory.

not sufficient, since the DFT deficiency stems not only from the lack of dispersion in the exchange correlation functionals but also from an incorrect electronic structure description of the *cus* site. The DFT/CC method described in Section 4.2.2.3 represents a viable alternative to the "standard" DFT approaches. While formally very similar to well established DFT-D methods, the DFT/CC approach can be used also for the description of adsorbate–*cus* site interaction since it does not assume any explicit functional form (such as r^{-6}).

The transferability of correction functions defined in Eq. 4.3 and obtained using the adsorbate-Cu(HCOO)$_2$ complexes was evaluated for the significantly more realistic PDWL model (see Fig. 4.1b). Full dissociation curves were considered for H_2O, CO, and ethylene molecules (Fig. 4.3), while only single-point energy calculations were carried out for CO_2, H_2, and CH_4. Interaction energies calculated at the PBE-optimized geometry of PDWL complexes with CO_2, H_2, and CH_4 molecules are -18.8 kJ mol^{-1}, -9.0 kJ mol^{-1}, and -13.5 kJ mol^{-1}, respectively, at the reference CCSD(T)/CBS level; they are -19.0 kJ mol^{-1}, -9.7 kJ mol^{-1}, and -13.5 kJ mol^{-1}, respectively, at the DFT/CC level.

Table 4.3 Interaction energies (in kJ mol^{-1}) of water with a PDWL model evaluated for various GGA, *meta*-GGA, and hybrid DFT functionals[a]

Functional type	Functional name	Dispersion correction None	D2	D3
GGA	PBE	−32.4	−39.5	−39.7
	RPBE	−19.2	−31.0	−33.1
	B-LYP	−23.7	−35.0	−38.1
meta-GGA	TPSS	−30.2	−39.6	−40.2
hybrid	B3LYP	−34.8	−44.7	−46.7
	BH-LYP	−47.6	–	−57.4

[a]The corresponding value obtained at the reference level of theory, CCSD(T)/CBS, is −51.2 kJ mol^{-1}.

The dissociation curves of the water/PDWL model (Fig. 4.3a) calculated at the DFT/CC and reference CCSD(T)/CBS levels are in excellent agreement for a whole range of intersystem distances. Results obtained with the semiempirical dispersion correction schemes show only partial improvement, being in a 10 kJ mol^{-1} error (Table 4.3). The dissociation curves for PDWL complexes with ethylene and carbon monoxide (Fig. 4.3b,c) also show very good agreement between DFT/CC and CCSD(T)/CBS levels (just slightly worse than in the case of water). The error of PBE-D2 for the ethylene/PDWL complex is about 4 kJ mol^{-1}; it is even smaller for the CO/PDWL complex. However, too short an equilibrium Cu–C distance (with respect to the CCSD(T)/CBS) observed for the PBE functional remains unchanged upon augmentation with the D2 correction.

In summary, on the basis of PDWL results none of the exchange correlation functionals (with or without dispersion corrections) can systematically describe the interaction between the adsorbate and the *cus* site in a Cu-BTC MOF. The dispersion correction accounted at the semiempirical level (D2 or D3) appears to be slightly overestimated (e.g., methane); however, the major source of error comes from the exchange part of the density functional. Increasing amount of the exact (Hartree–Fock) exchange mixed in the hybrid functional leads to improved accuracy. It follows that different functionals appear to be optimal for individual system; the

B3LYP-D2 gives the lowest RMSD among the functionals considered herein; however, accuracy results from partial error cancellation. It is concluded that the DFT/CC scheme outperforms other DFT methods and it can be used for the accurate description of the adsorption of application for the periodic model of Cu-BTC and compare its predictions with the available experimental data.

4.4 Adsorption in Cu-BTC: Electrostatic vs. Dispersion Interactions

4.4.1 H_2O

As synthesized Cu-BTC crystals contain H_2O molecules relatively firmly attached to the *cus* site [1]. The knowledge of the interaction energy of H_2O with Cu-BTC is thus important; however, experimental determination of adsorption heats is complicated due to the instability of the Cu-BTC at high humidity [42, 43]. Nevertheless, Henninger et al. [42] were able to obtain an adsorption enthalpy ($-\Delta H_{\text{ads}}$) of 50.7 $\pm$ 2.9 kJ mol^{-1} at 313 K using the thermogravimetry and differential scanning calorimetry (TG/DSC). The geometry of H_2O/Cu-BTC has been determined from single-crystal x-ray diffraction [1]; water is bound just on top of the Cu^{2+} *cus* site at a distance $R_{\text{Cu-water}} = 2.17$ Å. This experimental data can be used to evaluate the performance of various computational approaches. Structural characteristics of the water in Cu-BTC obtained at the DFT/CC level [17], $R_{\text{Cu-water}} = 2.19$ Å, are in very good agreement with experimental results. The adsorption heat of 49 kJ mol^{-1} for the zero-coverage limit has been obtained at the DFT/CC level. While this is in excellent agreement with experimental results (above) it must be noted that the TG/DSC measurements were carried out for the higher water coverage. The low-coverage adsorption behavior was investigated in the GCMC by Castillo et al. [44], who used interaction parameters fitted on the water adsorption isotherms in Cu-BTC from Wang et al. [45]; estimated adsorption enthalpy at the zero-coverage limit was 48.0 $\pm$ 6.6 kJ mol^{-1}.

An interaction energy of 52 kJ mol^{-1} was recently reported on the basis of the cluster model DFT calculations employing the

B3LYP functional [46]; this corresponds to $-\Delta H_{\text{ads}} = 46$ kJ mol^{-1} (considering the zero-point vibrational energy, ZPVE, correction of -6 kJ mol^{-1}). Using the periodic model and DFT calculations employing PW91 functionals interaction energies of 47.3 kJ mol^{-1} were reported [47]; this corresponds to $-\Delta H_{\text{ads}} = 41$ kJ mol^{-1}. These adsorption enthalpies based on DFT calculations are clearly underestimated (compared to experimental and DFT/CC values), which is understandable in the light of the results for the PDWL cluster model discussed in Section 4.3.2, although the magnitude of the underestimation is smaller than expected.

4.4.2 *CO*

The active catalytic sites in microporous materials are often investigated by means of IR spectroscopy [48]. The CO molecule is especially powerful probe due to the sensitive changes of CO-stretching frequency in different environments. Assignment of relatively rich spectra in the CO-stretching region in Cu-BTC has been discussed [49–51]; nevertheless, not all of the IR features were conclusively interpreted. There is an agreement on the interpretation of the main band located at about 2170–2178 cm^{-1} that is assigned to carbonyl complexes formed on Cu^{2+} sites in Cu-BTC [50, 52]. The small red shift of this band with increasing CO coverage can be explained by populating both *cus* sites of each PDWL unit with CO molecules [30]. This conclusion is also supported by theoretical estimates of adsorption enthalpy ($-\Delta H_{\text{ads}}$) 30 kJ mol^{-1} and 28 kJ mol^{-1} for the first and the second CO molecule on one PDWL unit of Cu-BTC. Theoretical calculations further ruled out the presence of dicarbonyl species on *cus* sites of Cu-BTC. The quite characteristic band at 2125 cm^{-1} is usually assigned to Cu^{+}–CO species due to the confirmed presence of such defects in Cu-BTC and its persistence upon evacuation [51]. The origin of two characteristic bands around 2192 cm^{-1} and 2148 cm^{-1} is still unclear, but they should not arise as a result of CO interacting with regular *cus* sites of the Cu-BTC framework.

Table 4.4 Adsorption heats (in kJ mol^{-1}) for a low-coverage regime

	Calculations/site type			
Adsorbate	*cus*	cage	window	Experiment
H_2O	49.0	–	–	50.7 ± 2.9[a]
CO	30.0	–	–	29.0[b]
CO_2	28.2	23.2	23.1	29.0[c]
CH_4	17.4	19.6	20.5	20.5[d]
C_3H_8	24.3	43.3	31.0	44.3[e]
C_3H_6	56.4	44.7	34.0	58.4[e]

[a] Ref. [42]
[b] Ref. [30]
[c] Ref. [29]
[d] Ref. [57]
[e] Ref. [58]

4.4.3 CO_2

The adsorption of CO_2 in Cu-BTC was addressed in a number of experimental studies [11, 29, 45, 53–56], including microcalorimetry [29] and the neutron diffraction study [11]. Thus the accuracy of computational approaches can be judged on the basis of the comparison with experimental data. The adsorption mechanism of the CO_2 molecule in Cu-BTC was systematically investigated at the DFT/CC level for a wide range of coverages (from zero-coverage limit to the CO_2:Cu ratio of 21:12). Calculations show that there are three types of adsorption sites in Cu-BTC: *cus*, the cage center, and the window sites (Fig. 4.4). Interaction energies calculated for the zero-coverage limit (Table 4.4) show that the most stable CO_2 adsorption complex is the one formed on the *cus* site (28.2 kJ mol^{-1}), whereas the adsorption complexes in the cage window and in the cage center are almost isoenergetic, being about 5 kJ mol^{-1} less stable than the complex on the *cus* site (see Fig. 4.4). Hence in the low-coverage regime (up to roughly one CO_2 per Cu), CO_2 molecules adsorb onto *cus* sites. Only one CO_2 molecule can interact with each *cus* site. CO_2 molecules on *cus* sites are tilted toward the adjacent *cus* site, and their arrangement is such that the lateral interactions are maximized (Fig. 4.5) [29].

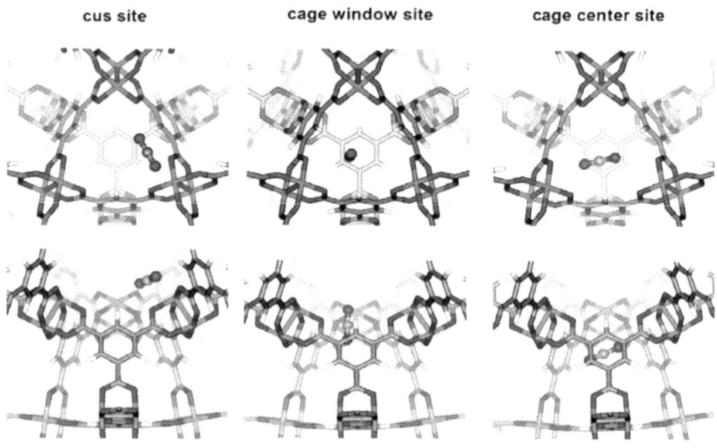

Figure 4.4 Site types in Cu-BTC, shown for CO_2 adsorption.

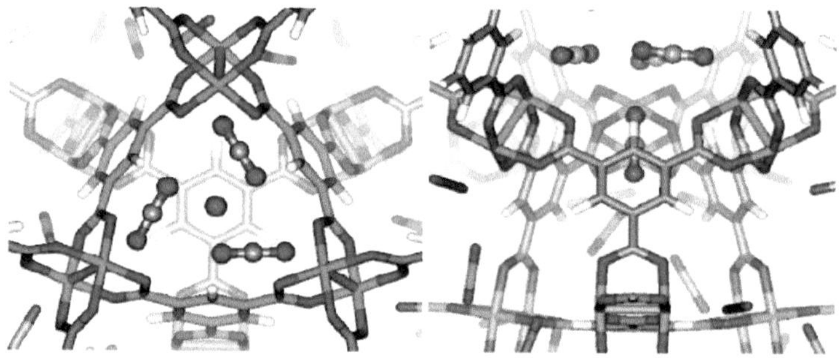

Figure 4.5 Structure of CO_2 adsorption complexes in Cu-BTC for higher loading.

At higher coverages, both secondary adsorption sites, the cage window and cage center sites, should be occupied, but it was observed that due to a proximity of the cage window sites to the *cus* sites the CO_2 complexes in the cage window sites are stabilized by lateral interaction with the already adsorbed CO_2 molecules on the *cus* sites to a larger extent than the ones in the cage center sites (Fig. 4.5). Up to $CO_2/CU = 5:3$ (the full occupation of *cus* and cage window sites), CO_2 molecules preferentially occupy sites in the cage windows of small cages. Lateral interactions between CO_2 molecules in cage window sites and those already adsorbed on *cus* sites (about 5 kJ mol^{-1}) result in increased stability of CO_2 in cage windows sites, and as a result, the adsorption enthalpy remains unchanged compared to the low-coverage regime. At even higher coverages, CO_2 molecules adsorb in the center of small cages and in large cages. The lateral interactions between these molecules and those already adsorbed in *cus* sites and cage windows sites (amounting to almost 10 kJ mol^{-1}) lead to a small increase (1–3 kJ mol^{-1}) in adsorption enthalpies in the high-coverage regime.

In line with theoretical predictions the adsorption enthalpies obtained by microcalorimetry measurements [29] were found to remain almost unchanged at about 29 kJ mol^{-1} for coverages up to almost two CO_2 molecules per *cus* site before increasing gradually by 2–3 kJ mol^{-1} for even higher coverages.

There is a number of experimental studies reporting both adsorption isotherms and heats of adsorption for CO_2 on Cu-BTC at various coverages [45, 53–56]; in many cases, the heats were obtained indirectly from adsorption isotherms measured at different temperatures using either the Clausius–Clapeyron equation [54, 56] or a temperature-dependent isotherm equation [53]. Nevertheless, with the exception of Ref. [54], all of the values reported in the literature are close to the enthalpies obtained by microcalorimetry and calculations discussed above being in the range of 25–35 kJ mol^{-1}. Furthermore the weak dependence of the adsorption heats on the coverage was observed in many studies, although some minor discrepancies were observed. Further confirmation of the accuracy of the theoretical description is due to an excellent agreement with the neutron diffraction data by Yildrim et al. [11], who reported the occupancies of *cus*, cage window and cage center adsorption sites for two different CO_2 loadings (1.07 and 1.47 CO_2/Cu) along with the structural characteristics of the CO_2 adsorption complexes. The calculated interaction energy values of −33 kJ mol^{-1} and −7 kJ mol^{-1} for the CO_2 interaction with the Cu-BTC at the zero-coverage limit employing LDA and PBE functionals, respectively, were reported [11]. These values could be well understood in light of the results obtained for the CO_2 interaction with the $Cu(HCOO)_2$ model in the Section 4.3.2. Lately a study devoted to the comparison of the newly implemented nonlocal density functionals for CO_2 adsorption in MOFs (including Cu-BTC) by Siegel et al. [10] reported an improvement of the energetics of the CO_2 interaction upon using nonlocal functional (with the vdW-DF1 and optB86b performing the best). This is again in the agreement with the data for the CO_2 interaction with the $Cu(HCOO)_2$ model, as shown in Fig. 4.2. But closer inspection of this figure indicates that the improvement is not fully transferable to other molecules like water, ammonia, or even methane.

In summary, very good agreement between DFT/CC and experimental results, along with the fact that the nature of the CO_2 adsorption is very different for *cus* (electrostatic) and cage window/center sites (vdW), implies that the DFT/CC method is able to provide a balanced description of complicated systems having various types of interactions.

4.4.4 CH_4

The adsorption of methane in Cu-BTC (besides other MOFs with the *cus* sites) has received considerable attention due to high methane storage capacities that were attributed, on the basis of the data obtained from the neutron diffraction studies of Yildirim [59] and Kaskel [60], to both the presence of the *cus* sites and the favorable structure of Cu-BTC containing small pockets (cages) significantly enhancing the vdW interaction. It has been observed [59, 60] that the grand-canonical Monte Carlo (GCMC) simulations employing standard empirical force fields fail to correctly describe the interaction of the methane with the *cus* site and leads thus to a prediction of an unrealistic adsorption mechanism [18, 60].

To address the problem of the interaction with the *cus* site and to take advantage of the detailed information about the adsorption sites available from the in situ neutron diffraction data and adsorption measurements by Kaskel [60], the potential energy surface (methane-framework interaction potential) calculated by the DFT/CC approach was directly implemented in the GCMC simulations (Fig. 4.6). Very good agreement was found with the experimental adsorption isotherms. In addition, the simulations based on the DFT/CC potential energy surface (PES) quantitatively capture the experimentally determined occupancies of the adsorp-

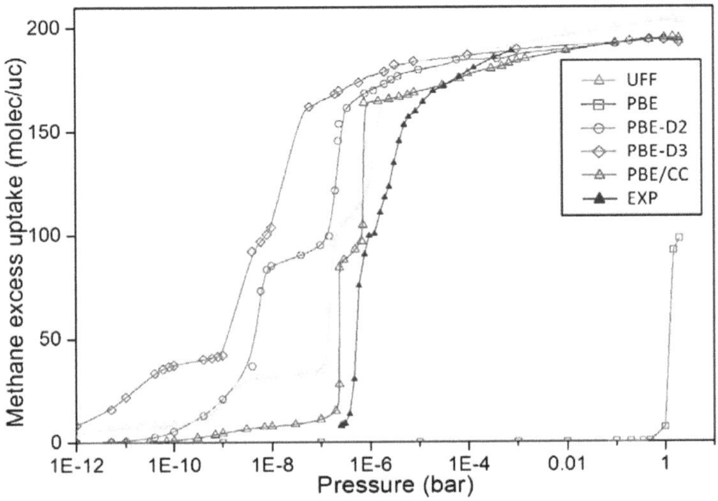

Figure 4.6 CH_4 adsorption isotherms in a semilog scale, calculated with various interaction potentials.

tion sites, including the open metal sites (Table 4.5) for which the simulated average methane–copper distance of 3.097 Å is well in line with the experimental value of 3.075 Å. It should be noted that neither DFT- or DFT-D2- nor DFT-D3-derived PES leads to a correct prediction of the adsorption isotherm (Fig. 4.6), with its under- or overestimation tendencies reflecting the trends observed for small cluster calculations (Fig. 4.2). The adsorption enthalpies calculated for the zero-coverage limit at the DFT/CC level (17.4 kJ mol^{-1}, 19.6 kJ mol^{-1}, and 20.5 kJ mol^{-1} for *cus*, cage, and window sites, respectively) are in good agreement with available experimental data (Table 4.4). Adsorption enthalpies calculated at the LDA level (25.5 kJ mol^{-1}, 18.1 kJ mol^{-1}, and 24.8 kJ mol^{-1} for *cus*, cage, and window sites, respectively) are clearly overestimated, and they are in incorrect order; overestimated interaction with the *cus* site is in line with the results reported for cluster models (Fig. 4.2).

4.4.5 *C_3H_8 and C_3H_6*

The separation of propane/propylene mixture in Cu-BTC has been extensively studied experimentally and theoretically [61–65].

Table 4.5 Loading of CH_4 molecules identified for each adsorption site at maximum loading

		176 molecules/u.c.		
Pore[a]	Location	DFT/CC	UFF	Exp.[b]
Small/large	Window	22.4	20.6	24.8
Medium	Corner	47.6	48.0	48.0
Large	*cus*	46.3	0.	44.6
Medium/large	Window	29.1	32.1	27.8
Small	Center	8.0	8.0	6.4
Medium/large	Window	4.9	19.9	4.0
Small/large	Window	10.1	12.2	11.8
Large	Center	3.5	31.3	4.0
Medium	Center	4.1	4.0	[c]

UFF: universal force field; u.c.: unit cell
[a]For some sites the same characterization is used as sites close to each other.
[b]Experimental results at 176 molecules/uc from Ref. [60], where a total of 171.21 molecules/uc were resolved.
[c]Not reported in Ref. [60].

Most of the theoretical studies used GCMC simulations to predict adsorption isotherms. However, the commonly used empirical potentials fail to describe the interaction with *cus* sites. The theoretical attempts to improve "standard" empirical force fields have been made. The main focus has been to improve description of the partially dative bond between Cu^{2+} and π-electron density on propylene. However, simple rescaling of Lennard–Jones parameters for Cu^{2+}–$C(sp^2)$ interaction or parameterizations based on the GGA functionals do not yield a desired accuracy (Fig. 4.3b). The interaction energy between ethylene and the *cus* site (using a PDWL cluster) calculated at the PBE level is only about half of the reference CCSD(T)-/CBS-level value of -41.3 kJ mol^{-1}.

Adsorption enthalpies ($-\Delta H_{\mathrm{ads}}$) of propane and propylene calculated for the zero-coverage limit are summarized in Table 4.4. The agreement between DFT/CC predictions and experimentally measured adsorption enthalpy is excellent. A small underestimation of the DFT/CC adsorption enthalpy on the *cus* sites is consistent with the cluster model results discussed above. Propylene preferably adsorbs on the *cus* sites ($-\Delta H_{\mathrm{ads}} \approx 58$ kJ mol^{-1}) due to the formation of partial dative bond (Fig. 4.7b). Calculations also

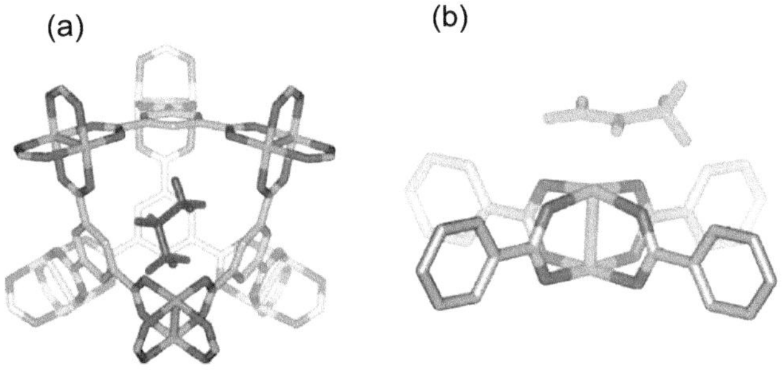

Figure 4.7 (a) Propane in the cage and (b) propylene on the *cus* site.

indicated that upon adsorption of second propylene molecule on the same PDWL unit the adsorption enthalpy decreases by 9 kJ mol^{-1}. This indicates that with increasing propylene loading the *cus* sites are no longer equivalent. For cage center and cage widow sites the propane and propylene adsorption enthalpies do not differ significantly. The adsorption of propane is driven by vdW interactions, thus, the topology of the Cu-BTC framework determines the stability of each adsorption site. The most stable propane adsorption complex is formed in the cage center site ($-\Delta H_{ads} \approx 44$ kJ mol^{-1}) due to the favorable dimension (maximal dispersion interactions) of a small Cu-BTC octahedral cage (Fig. 4.7a). Once the cages in Cu-BTC are occupied by propane molecules the adsorption enthalpy decreases due to a smaller stability of adsorption complexes formed on other adsorption sites. However, the importance of lateral interactions increases at large propane loadings.

4.5 Summary

The application of accurate post-Hartree–Fock and DFT-based methods has been reviewed above for the particular case of a Cu-BTC MOF. It is obvious that an accurate description of MOFs having *cus* sites with partially filled *d*-orbitals is rather involved. Accurate post-Hartree–Fock methods are computationally too demanding to be routinely used for MOFs, and computationally tractable DFT-based methods do not have sufficient accuracy. The dispersion interactions between adsorbate and MOFs are no longer a problem; they can be relatively well described either by nonlocal exchange correlation functionals or by one of the correction schemes (DFT-D or DFT/CC). The problematic part is the description of the transition metals with an open-shell electronic structure. It is therefore always very important to verify the accuracy of the particular DFT method, and it is clear that the accuracy decreases with increasing complexity of the electronic structure of the particular transition metal. All these conclusions are based on our systematic investigation of the Cu-BTC MOF's interaction with various adsorbates. They are in line with other calculations reported for Cu-BTC (e.g., [10–12, 47]) and for other MOFs (e.g., [6, 66–72]).

Acknowledgments

The research leading to these results has received funding from the Grant Agency of the Czech Republic (Center of Excellence - P106/12/G015) and from the European Community's Seventh Framework Programme (FP7/2007–2013) under grant agreement n° 228862. MACADEMIA is a large-scale integrating project under the Nanosciences, Nanotechnologies, Materials and New Production Technologies Theme (www.macademia-project.eu).

References

1. Chui, S. S. Y., Lo, S. M. F., Charmant, J. P. H., Orpen, A. G., Williams, I. D. (1999). A chemically functionalizable nanoporous material Cu-3(TMA)(2)(H_2O)(3) (n), *Science*, **283**, 1148–1150.
2. Rosi, N. L., Kim, J., Eddaoudi, M., Chen, B., O'Keeffe, M., Yaghi, O. M. (2005). Rod packings and metal-organic frameworks constructed from rod-shaped secondary building units., *J. Am. Chem. Soc.*, **127**, 1504–1518.
3. Férey, G., Serre, C., Mellot-Draznieks, C., Millange, F., Surblé, S., Dutour, J., Margiolaki, I. (2004). A hybrid solid with giant pores prepared by a combination of targeted chemistry, simulation, and powder diffraction, *Angew. Chem., Int. Ed.*, **43**, 6296–6301.

4. Férey, G., Mellot-Draznieks, C., Serre, C., Millange, F., Dutour, J., Surblé, S., Margiolaki, I. (2005). A chromium terephthalate-based solid with unusually large pore volumes and surface area, *Science*, **309**, 2040–2042.
5. Duren, T., Bae, Y.-S., Snurr, R. Q. (2009). Using molecular simulation to characterise metal-organic frameworks for adsorption applications, *Chem. Soc. Rev.*, **38**, 1237–1247.
6. Dzubak, A. L., Lin, L. C., Kim, J., Swisher, J. A., Poloni, R., Maximoff, S. N., Smit, B., Gagliardi, L. (2012). Ab initio carbon capture in open-site metal-organic frameworks, *Nat. Chem.*, **4**, 810–816.
7. Getman, R. B., Bae, Y. S., Wilmer, C. E., Snurr, R. Q. (2012). Review and analysis of molecular simulations of methane, hydrogen, and acetylene storage in metal-organic frameworks, *Chem. Rev.*, **112**, 703–723.
8. Garcia-Perez, E., Gascon, J., Morales-Florez, V., Castillo, J. M., Kapteijn, F., Calero, S. (2009). Identification of adsorption sites in Cu-BTC by experimentation and molecular simulation, *Langmuir*, **25**, 1725–1731.
9. Krishna, R., van Baten, J. M. (2011). In silico screening of metal-organic frameworks in separation applications, *Phys. Chem. Chem. Phys.*, **13**, 10593–10616.
10. Rana, M. K., Koh, H. S., Hwang, J., Siegel, D. J. (2012). Comparing van der Waals density functionals for CO_2 adsorption in metal organic frameworks, *J. Phys. Chem. C*, **116**, 16957–16968.
11. Wu, H., Simmons, J. M., Srinivas, G., Zhou, W., Yildirim, T. (2010). Adsorption sites and binding nature of CO_2 in prototypical metal-organic frameworks: a combined neutron diffraction and first-principles study, *J. Phys. Chem. Lett.*, **1**, 1946–1951.
12. Xiang, S., Zhou, W., Gallegos, J. M., Liu, Y., Chen, B. (2009). Exceptionally high acetylene uptake in a microporous metal-organic framework with open metal sites, *J. Am. Chem. Soc.*, **131**, 12415–12419.
13. Grimme, S. (2006). Semiempirical GGA-type density functional constructed with a long-range dispersion correction, *J. Comput. Chem.*, **27**, 1787–1799.
14. Grimme, S., Antony, J., Ehrlich, S., Krieg, H. (2010). A consistent and accurate ab initio parametrization of density functional dispersion correction (DFT-D) for the 94 elements H-Pu, *J. Chem. Phys.*, **132**, 154104.
15. Klimes, J., Bowler, D., Michaelides, A. (2010). Chemical accuracy for the van der Waals density functional., *J. Phys.: Condens. Matter*, **22**, 022201.

16. Lee, K., Murray, E., Kong, L., Lundqvist, B., Langreth, D. (2010). Higher-accuracy van der Waals density functional, *Phys. Rev. B*, **82**, 081101.
17. Grajciar, L., Bludsky, O., Nachtigall, P. (2010). Water adsorption on coordinatively unsaturated sites in CuBTC MOF, *J. Phys. Chem. Lett.*, **1**, 3354–3359.
18. Chen, L., Grajciar, L., Nachtigall, P., Düren, T. (2011). Accurate prediction of methane adsorption in a metal–organic framework with unsaturated metal sites by direct implementation of an ab initio derived potential energy surface in GCMC simulation, *J. Phys. Chem. C*, **115**, 23074–23080.
19. Boys, S. F., Bernardi, F. (1970). Calculation of small molecular of small molecular interactions by differences of separate total energies: some procedures with reduced errors, *Mol. Phys.*, **19**, 553–566.
20. Peterson, K. A., Puzzarini, C. (2005). Systematically convergent basis sets for transition metals. II. Pseudopotential-based correlation consistent basis sets for the group 11 (Cu, Ag, Au) and 12 (Zn, Cd, Hg) elements, *Theor. Chem. Acc.*, **114**, 283–296.
21. Dunning, T. H. (1989). Gaussian-basis sets for use in correlated molecular calculations 1. The atoms boron through neon and hydrogen, *J. Chem. Phys.*, **90**, 1007–1023.
22. Perdew, J. P., Burke, K., Ernzerhof, M. (1996). Generalized gradient approximation made simple, *Phys. Rev. Lett.*, **77**, 3865–3868.
23. Blochl, P. E. (1994). Projector augmented-wave method, *Phys. Rev. B*, **50**, 17953–17979.
24. Čížek, J. (1969). *On the Use of the Cluster Expansion and the Technique of Diagrams in Calculations of Correlation Effects in Atoms and Molecules*, Vol. 14 (John Wiley & Sons).
25. Adler, T. B., Werner, H. J. (2009). Local explicitly correlated coupled-cluster methods: efficient removal of the basis set incompleteness and domain errors, *J. Chem. Phys.*, **130**, 241101.
26. Halkier, A., Helgaker, T., Jorgensen, P., Klopper, W., Olsen, J. (1999). Basis-set convergence of the energy in molecular Hartree-Fock calculations, *Chem. Phys. Lett.*, **302**, 437–446.
27. Bludsky, O., Rubes, M., Soldan, P., Nachtigall, P. (2008). Investigation of the benzene-dimer potential energy surface: DFT/CCSD(T) correction scheme, *J. Chem. Phys.*, **128**, 114102.
28. Soldan, P., Hutson, J. M. (2000). On the long-range and short-range behavior of potentials from reproducing kernel Hilbert space interpolation, *J. Chem. Phys.*, **112**, 4415–4416.

29. Grajciar, L., Wiersum, A. D., Llewellyn, P. L., Chang, J.-S., Nachtigall, P. (2011). Understanding CO_2 adsorption in CuBTC MOF: comparing combined DFT–ab initio calculations with microcalorimetry experiments, *J. Phys. Chem. C*, **115**, 17925–17933.

30. Rubeš, M., Grajciar, L., Bludský, O., Wiersum, A. D., Llewellyn, P. L., Nachtigall, P. (2012). Combined theoretical and experimental investigation of CO adsorption on coordinatively unsaturated sites in CuBTC MOF, *Chemphyschem*, **13**, 488–495.

31. Jotham, R., Kettle, S., Marks, J. (1972). Antiferromagnetism in transition-metal complexes. Part IV. Lowlying excited states of binuclear copper (II) carboxylate complexes, *J. Chem. Soc., Dalton Trans.*, 428.

32. Kahn, O. (1993). *Molecular Magnetism* (John Wiley & Sons).

33. Melník, M. (1982). Structural isomerism of copper (II) compounds, *Coord. Chem. Rev.*, **47**, 239–261.

34. Zhang, X. X., Chui, S. S.-Y., Williams, I. D. (2000). Cooperative magnetic behavior in the coordination polymers [Cu3(TMA)2L3] (L=H2O, pyridine), *J. Appl. Phys.*, **87**, 6007–6009.

35. Poppl, A., Kunz, S., Himsl, D., Hartmann, M. (2008). CW and pulsed ESR spectroscopy of cupric ions in the metal-organic framework compound Cu3(BTC)2, *J. Phys. Chem. C*, **112**, 2678–2684.

36. Ducéré, J.-M., Goursot, A., Berthomieu, D. (2005). Comparative density functional theory study of the binding of ligands to Cu^+ and Cu^{2+}: influence of the coordination and oxidation state, *J. Phys. Chem. A*, **109**, 400–408.

37. El-Nahas, A. A. M., Tajima, N., Hirao, K. (2000). Do $Cu^{2+}NH_3$ and $Cu^{2+}OH_2$ exist? Theory confirms "yes"! *Chem. Phys. Lett.*, **318**, 333–339.

38. Jeanvoine, Y., Spezia, R. (2010). Cu^{2+} binding chalcogen–chalcogen bridges: a problematic case for DFT, *THEOCHEM*, **954**, 7–15.

39. Poater, J., Solà, M., Rimola, A., Rodríguez-Santiago, L., Sodupe, M. (2004). Ground and low-lying states of Cu^{2+}–H_2O. A difficult case for density functional methods, *J. Phys. Chem. A*, **108**, 6072–6078.

40. Antony, J., Grimme, S. (2006). Density functional theory including dispersion corrections for intermolecular interactions in a large benchmark set of biologically relevant molecules, *Phys. Chem. Chem. Phys.*, **8**, 5287–5293.

41. Klimes, J., Bowler, D., Michaelides, A. (2011). van der Waals density functionals applied to solids, *Phys. Rev. B*, **83**, 1–13.

42. Henninger, S. K., Schmidt, F. P., Henning, H.-M. (2010). Water adsorption characteristics of novel materials for heat transformation applications, *Appl. Therm. Eng.*, **30**, 1692–1702.
43. Küsgens, P., Rose, M., Senkovska, I., Fröde, H., Henschel, A., Siegle, S., Kaskel, S. (2009). Characterization of metal-organic frameworks by water adsorption, *Microporous Mesoporous Mater.*, **120**, 325–330.
44. Castillo, J. M., Vlugt, T. J. H., Calero, S. (2008). Understanding water adsorption in Cu-BTC metal-organic frameworks, *J. Phys. Chem. C*, **112**, 15934–15939.
45. Min Wang, Q., Shen, D., Bülow, M., Ling Lau, M., Deng, S., Fitch, F. R., Lemcoff, N. O., Semanscin, J. (2002). Metallo-organic molecular sieve for gas separation and purification, *Microporous Mesoporous Mater.*, **55**, 217–230.
46. Lukose, B., Supronowicz, B., St. Petkov, P., Frenzel, J., Kuc, A. B., Seifert, G., Vayssilov, G. N., Heine, T. (2012). Structural properties of metal-organic frameworks within the density-functional based tight-binding method, *Phys. Status Solidi B*, **249**, 335–342.
47. Watanabe, T., Sholl, D. S. (2010). Molecular chemisorption on open metal sites in Cu(3)(benzenetricarboxylate)(2): a spatially periodic density functional theory study, *J. Chem. Phys.*, **133**, 094509.
48. Hadjiivanov, K. I., Vayssilov, G. N. (2002). Characterization of oxide surfaces and zeolites by carbon monoxide as an IR probe molecule, *Adv. Catal.*, **47**, 307–511.
49. Alaerts, L., Seguin, E., Poelman, H., Thibault-Starzyk, F., Jacobs, P., De Vos, D. (2006). Probing the Lewis acidity and catalytic activity of the metal-organic framework [$Cu_3(btc)_2$] (BTC=benzene-1,3,5-tricarboxylate), *Chem. Eur. J.*, **12**, 7353–7363.
50. Bordiga, S., Regli, L., Bonino, F., Groppo, E., Lamberti, C., Xiao, B., Wheatley, P., Morris, R., Zecchina, A. (2007). Adsorption properties of HKUST-1 toward hydrogen and other small molecules monitored by IR, *Phys. Chem. Chem. Phys.*, **9**, 2676–2685.
51. Prestipino, C., Regli, L., Vitillo, J., Bonino, F., Damin, A., Lamberti, C., Zecchina, A., Solari, P., Kongshaug, K., Bordiga, S. (2006). Local structure of framework Cu(II) in HKUST-1 metallorganic framework: spectroscopic characterization upon activation and interaction with adsorbates, *Chem. Mat.*, **18**, 1337–1346.
52. Drenchev, N., Ivanova, E., Mihaylov, M. (2010). CO as an IR probe molecule for characterization of copper ions in a basolite C300 MOF sample, *Phys. Chem. Chem. Phys.*, **12**, 6423–6427.

53. Aprea, P., Caputo, D., Gargiulo, N., Iucolano, F., Pepe, F. (2010). Modeling carbon dioxide adsorption on microporous substrates: comparison between Cu-BTC metal-organic framework and 13X zeolitic molecular sieve, *J. Chem. Eng. Data*, **55**, 3655–3661.
54. Farrusseng, D., Daniel, C., Gaudillère, C., Ravon, U., Schuurman, Y., Mirodatos, C., Dubbeldam, D., Frost, H., Snurr, R. Q. (2009). Heats of adsorption for seven gases in three metal-organic frameworks: systematic comparison of experiment and simulation, *Langmuir*, **25**, 7383–7388.
55. Liang, Z., Marshall, M., Chaffee, A. L. (2009). CO_2 adsorption-based separation by metal organic framework (Cu-BTC) versus zeolite (13X), *Energy Fuels*, **23**, 2785–2789.
56. Moellmer, J., Moeller, a., Dreisbach, F., Glaeser, R., Staudt, R. (2011). High pressure adsorption of hydrogen, nitrogen, carbon dioxide and methane on the metal–organic framework HKUST-1, *Microporous Mesoporous Mater.*, **138**, 140–148.
57. Getzschmann, J., Senkovska, I., Fairen-Jimenez, D., Düren, T., Kaskel, S. (unpublished results).
58. Rubeš, M., Wiersum, A. D., Llewellyn, P. L., Grajciar, L. S., Bludský, O., Nachtigall, P. (2013). Adsorption of Propane and Propylene on CuBTC MOF: combined theoretical and experimental investigation, *J. Phys. Chem. C*, **117**, 11159–11167.
59. Wu, H., Simmons, J. M., Liu, Y., Brown, C. M., Wang, X.-S., Ma, S., Peterson, V. K., Southon, P. D., Kepert, C. J., Zhou, H.-C., Yildirim, T., Zhou, W. (2010). Metal-organic frameworks with exceptionally high methane uptake: where and how is methane stored? *Chem. Eur. J.*, **16**, 5205–5214.
60. Getzschmann, J., Senkovska, I., Wallacher, D., Tovar, M., Fairen-Jimenez, D., Düren, T., van Baten, J. M., Krishna, R., Kaskel, S. (2010). Methane storage mechanism in the metal-organic framework $Cu_3(btc)_2$: an in situ neutron diffraction study, *Microporous Mesoporous Mater.*, **136**, 50–58.
61. Ferreira, A., Santos, J., Plaza, M., Lamia, N., Loureiro, J., Rodrigues, A. (2011). Suitability of Cu-BTC extrudates for propane–propylene separation by adsorption processes, *Chem. Eng. J.*, **167**, 1–12.
62. Fischer, M., Gomes, J. R. B., Froba, M., Jorge, M. (2012). Modeling Adsorption in metal-organic frameworks with open metal sites: propane/propylene separations, *Langmuir*, **28**, 8537–8549.
63. Jorge, M., Lamia, N., Rodrigues, A. (2010). Molecular simulation of propane/propylene separation on the metal–organic framework CuBTC, *Colloids Surf., A*, **357**, 27–34.

64. Lamia, N., Jorge, M., Granato, M., Almeida Paz, F., Chevreau, H., Rodrigues, A. (2009). Adsorption of propane, propylene and isobutane on a metal–organic framework: molecular simulation and experiment, *Chem. Eng. J.*, **64**, 3246–3259.
65. Yoon, J., Jang, I., Lee, K., Hwang, Y., Chang, J. (2010). Adsorptive separation of propylene and propane on a porous metal-organic framework, copper trimesate, *Bull. Korean Chem. Soc.*, **31**, 220–223.
66. Bae, Y. S., Lee, C. Y., Kim, K. C., Farha, O. K., Nickias, P., Hupp, J. T., Nguyen, S. T., Snurr, R. Q. (2012). High propene/propane selectivity in isostructural metal-organic frameworks with high densities of open metal sites, *Angew. Chem., Int. Ed.*, **51**, 1857–1860.
67. Chen, L., Morrison, C. A., Düren, T. (2012). Improving predictions of gas adsorption in metal–organic frameworks with coordinatively unsaturated metal sites: model potentials, ab initio parameterization, and GCMC simulations, *J. Phys. Chem. C*, **116**, 18899–18909.
68. Peralta, D., Chaplais, G., Simon-Masseron, A., Barthelet, K., Chizallet, C., Quoineaud, A.-A., Pirngruber, G. D. (2012). Comparison of the behavior of metal-organic frameworks and zeolites for hydrocarbon separations, *J. Am. Chem. Soc.*, **134**, 8115–8126.
69. Poloni, R., Smit, B., Neaton, J. B. (2012). CO_2 capture by metal-organic frameworks with van der Waals density functionals, *J. Phys. Chem. A*, **116**, 4957–4964.
70. Valenzano, L., Civalleri, B., Sillar, K., Sauer, J. (2011). Heats of adsorption of CO and CO_2 in metal–organic frameworks: quantum mechanical study of CPO-27-M (M = Mg, Ni, Zn), *J. Phys. Chem. C*, **115**, 21777–21784.
71. Valenzano, L., Vitillo, J. G., Chavan, S., Civalleri, B., Bonino, F., Bordiga, S., Lamberti, C. (2012). Structure–activity relationships of simple molecules adsorbed on CPO-27-Ni metal–organic framework: in situ experiments vs. theory, *Catal. Today*, **182**, 67–79.
72. Watanabe, T., Sholl, D. S. (2012). Accelerating applications of metal-organic frameworks for gas adsorption and separation by computational screening of materials, *Langmuir*, **28**, 14114–14128.

Chapter 5

Modeling Sorbate Equilibria and Transport in Porous Coordination Polymers

George K. Papadopoulos
School of Chemical Engineering, National Technical University of Athens, Athens, GR 15780, Greece
gkpap@chemeng.ntua.gr

In this chapter, a statistical mechanics-based modeling of the sorption thermodynamics and dynamics of selected guest sorbate molecules in porous coordination polymers is presented. The proposed methodology is applied to two zeolitic imidazolate frameworks of different symmetries, growing into different pore networks, and aims at presenting a way of simulating phase equilibria and kinetics of guest molecules sorbed in the PCP networks in a manner that allows strict thermodynamic consistency between the quantities calculated via computer experiments and their measured homologs by means of conventional experiments.

Metal–Organic Frameworks: Materials Modeling towards Potential Engineering Applications
Edited by Jianwen Jiang

ISBN 978-981-4613-45-3 (Hardcover), 978-981-4613-46-0 (eBook)
www.panstanford.com

5.1 Introduction

The subject of computer-aided simulation processes involving equilibria and dynamics of guest sorbates within metal–organic frameworks (MOFs) is of growing importance because of the promising role of MOFs as molecular traps for carbon-neutral, clean-energy, or biological applications. Their synthesis procedure is rooted in a reticular strategy by means of combinatorial methodologies toward target structures [1] that comprise a large category of crystalline networks, coined as porous coordination polymers (PCPs). These materials, mainly because of their high uptake capacities to hydrogen, methane, and carbon dioxide, have become one of the most popular classes of sorbent materials. Another prominent series of MOFs recently synthesized, Matériaux de l'Institut Lavoisier (MIL) [2], may serve in addition as candidates for drug delivery in medical applications. A noteworthy computational approach aiming at the construction of a large number of conceivable MOFs from a given chemical library of building blocks based on the structures of known MOFs can be found in Ref. [3].

The great success of PCPs as molecular binders is based on the appropriate tailoring of selected molecular blocks together, which results in geometrically well-defined morphologies [1–7]. Moreover, the high availability of these compounds for further chemical modification of their linking units has led to the creation of a whole series of homologues possessing various pore widths. Such a group is the zeolitic imidazolate frameworks (ZIFs), formed by transition metals, which replace T-atoms, and imidazolate links, which replace oxygen bridges in the conventional zeolitic structures [8–10]. Further systematic chemical modification of their substituents has created a series of ZIF structures over the past few years, characterized by high chemical and thermal stability and based on framework types of either known zeolite topologies such as sodalite, RHO, MER, or even predicted structures. Although their thermal stability—a key characteristic for a technologically important sorbent—can be considered moderate compared to zeolites, the reticulated process of their synthesis makes them superior compared to the conventional aluminosilicate crystals, in the sense that the resulted materials are characterized by large pore sizes, high apparent surface areas, and chemically active interiors, which provide them with exceptional properties as sorbents being capable of hosting large amounts of sorbates of a broad span of sizes.

The remarkable performance of the PCP structures as molecular capturers have attracted the interest of scientists from the field of theory, simulation, and engineering over the last few years in several metal and covalent organic frameworks (COFs) [11–20]. Also, sources reporting experimentally measured high uptake capacities regarding either hydrogen or greenhouse gases in several ZIF structures can be found in Refs. [8, 9, 20–27] and references therein.

In this chapter indicative examples of statistical mechanics-based modeling procedures are presented toward predicting phase equilibria and kinetics of guest molecules in recently synthesized [8, 20, 28] imidazolate structures such as the two types of frameworks, ZIF-3 and ZIF-8, differing in the structural topology, the pore network, and the nature of the organic ligand [29]. The objective of the presented study is the investigation of the structure–property relationships of this class of sorbents with respect to storage and selective separation of technologically important guest molecules, firstly by examining the evolution of the sorption thermodynamics over a wide range of occupancies and secondly by elucidating the rate and mechanism of motion inside the MOF'spores.

A crucial input for any kind of atomistic modeling is a force field (FF) that in principle should be capable of capturing all the essential physicochemical properties of the system under study. However, utilization of common FFs rarely yields accurate results with respect to real experiments; this is especially true for the PCP materials because of the complexity and plurality of their atomic framework. In the presented study two generic FFs were employed, the DREIDING FF [30] and the universal force field (UFF) [31], which, in the absence of a case-specific FF, have been used in several previous simulations of various isoreticular MOF structures [13]. Full-scale ab initio quantum mechanical calculations, focusing on a certain material or, much better, on a class of materials, can serve as a reasonable basis toward developing more specialized parameters for the potential functions. Such an FF, which appeared in the literature after the end of the computational part of the work the current chapter is based on, can be found in Ref. [32], wherein a density functional theory–based symmetry-adapted perturbation theory (DFT-SAPT) was tailored so as to accurately model CO_2 and N_2 adsorption in several ZIFs.

The experience gained by the majority of atomistic simulations in zeolitic crystals where a rigid model was used is that framework flexibility mainly affects the thermodynamics and kinetics of long molecules when they are subject to tight fitting. Nonetheless, simulations performed lately by incorporating framework flexibility in the modeling of some MOFs have led to somewhat contradictory diffusivity trends regarding the effect of flexibility [33, 34]. However, a recent modeling work on collective and self-diffusion coefficients of methane and carbon dioxide within an imidazolate framework highlighted the effect of the atomic framework mobility and especially of the bridging ligands due to their susceptibility to conformation altering, acting like "saloon doors" on the moving guest sorbate molecules [27].

The present study aims at shedding light on the influence of the framework dynamics of the aforementioned ZIFs on the equilibrium and mobility properties of sorbates by means of molecular dynamics (MD) through properly adapted Widom averaging, thus avoiding dependence on stochastic methodologies for modeling sorbed-phase equilibria under the (practically necessary) assumption of a rigid ZIF model.

Furthermore, by virtue of MD, the structural effects on the materials' physicochemical properties are explored in a way that conventional experiments often are not able to capture. Such an example is provided by considering the degree of hindrance of the 2-methylimidazolate links on the sorbate mobility; the latter groups being located at the ZIF-8 hexagonal apertures (see Fig. 5.1) intervene in the cage-to-cage paths, hence constituting the only regions allowing mass transport within the ZIF-8 unit cell. For this reason various loadings of methane and carbon dioxide were hosted in the materials' unit cell in order to compute the self-diffusivity, as well as the collective diffusivities (Maxwell–Stefan [M-S] and transport) [35–37]. Moreover, modeling options such as the flexibility of the entire matrix versus a rigid framework version, the mobility of the imidazolate ligand, and the inclusion of quantum

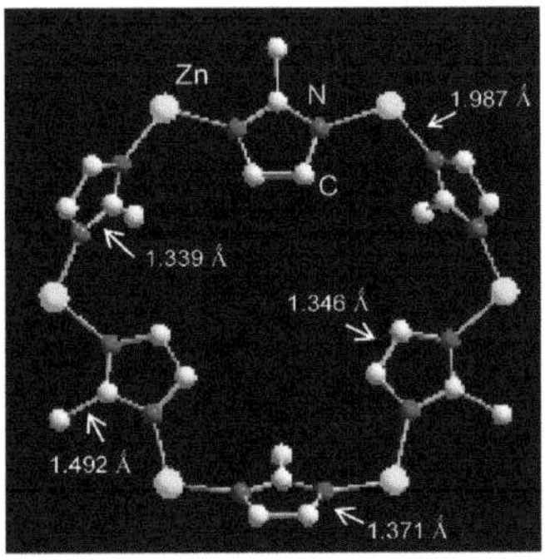

Figure 5.1 Detail of the atomic framework of the ZIF-8 crystal, showing the 2-methylimidazolate organic group link and the Zn and N atoms; H atoms are omitted in the figure.

mechanical effects [38] in the potential functions for the case of light (quantum) molecules, are examined in detail with no adjustment of any of the FF parameters with respect to measured quantities by means of real experiments. It must be stressed here that the presented in this chapter simulation methodology focuses primarily upon the thermodynamics consistency between the computer and physical experiments for all systems under consideration.

5.2 Sorbent Digitization

The proper modeling of a sorbent material is of substantial importance in computer simulation studies, since it is a prerequisite for the accurate representation and hence implementation of the guest framework FF. Depending on the medium, for example, a unique crystal or a cluster of crystals forming a supported membrane or a catalyst bed, the representation of the material can vary from an exact atomistic one up to a mesoscopic digitized rendition of the spatial distribution of solid and void phases on the basis of suitable experimental input. The reader can find details of such methodologies in Ref. [39]. In the following subsection the digitization routine of an MOF on an atomistic basis is described through the examples of the mentioned ZIFs.

5.2.1 *Atomistic Reconstruction*

The ZIF-3 and ZIF-8 crystals can be reconstructed in atomistic level using data from single-crystal x-ray diffraction (XRD) analysis [8, 20, 29]. In both structures, each zinc atom is tetrahedrally coordinated by four nitrogen atoms (ZnN_4); the bridging groups in the structures are the imidazolate (IM) and the 2-methyl-imidazolate (mIM) in ZIF-3 and ZIF-8, respectively (Figs. 5.1 and 5.2). ZIF-8 has a composition described by the formula $Zn(mIM)_2$. The ZIF-8 unit cell belongs to the sodalite topology; for digital representation it is started by means of (i) XRD analysis data to locate the atomic positions in the unit cell lattice and (ii) the $I\bar{4}3m$ space group for the final reproduction of the atomic coordinates, according to their symmetry operations of this space group and their occupancy probabilities in the framework. The obtained structure has each zinc ion tetrahedrally coordinated by four nitrogen atoms (ZnN_4) belonging to the imidazolate rings, as illustrated in Fig. 5.1. The shared area formed by the hexagonal windows (effective width: 0.34 nm) and the neighboring cages constitute the channels that run along the four body-diagonal directions of the ZIF-8 cubic lattice. The cubic unit cell (edge length: 1.6990 nm) has the formula $C_{96}H_{120}N_{48}Zn_{12}$.

The digitized crystal contains no water molecules. The reconstruction procedure results in a structure with the sod topology consisting of a hexagonal periodic building unit (PerBU) that contains an array of nonconnected six rings; neighboring PerBUs are connected through tilted four rings forming truncated octahedra denoted as sodalite units or β-cages. The shared area formed by the six-ring apertures and the neighboring cages constitutes the pore channels in the ZIF-8 cubic lattice.

XRD spectra [8] have identified six positions for the hydrogen atoms as a consequence of the two equiprobable orientations of the methyl group; thus, each position has an occupancy probability of 0.5; all the other framework atoms have occupancy probability equal to 1. These probabilities have been properly taken into account in the potential formulae employed for the calculations of the host–guest interactions.

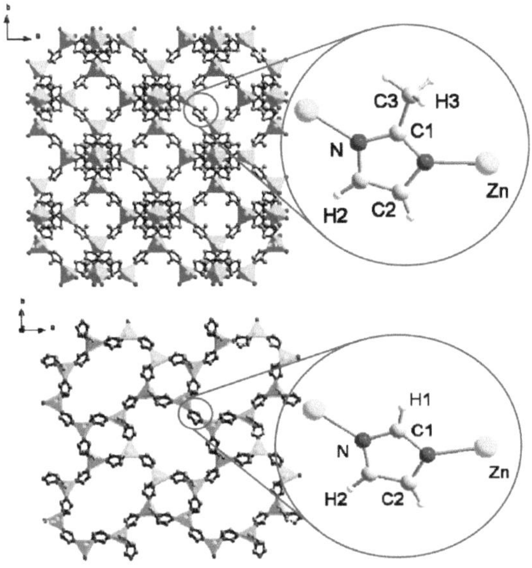

Figure 5.2 Face views of the simulation box (eight unit cells) of (top) ZIF-3 and (bottom) ZIF-8, illustrating the framework atoms and bridging organic groups.

The ZIF-3 unit cell ($C_{96}H_{96}N_{64}Zn_{16}$) belongs to the tetragonal symmetry with edge lengths [20], $a = b = 1.89701$ and $c = 1.6740$ nm. Utilizing the symmetry operations of the $P4_2/mnm$ (No. 136) space group [40], one results in a digitalized structure of the dft topology. The dft framework type can be constructed using double crankshaft chains of the narsarsukite type running along the c axis. This type of chain, also denoted as a bifurcated hexagonal square chain (bhs), consists of twofold (1, 3)-connected double four rings joined to form two six rings with four common tetrahedral atoms. These chains are arranged parallel to the [110] plane, forming two sets of eight-ring channels; each bhs chain is connected directly to the neighboring two bhs chains, generating a sheet.

These sheets are stacked along the [110] direction to form a three dimensional framework with eight-ring channels (Fig. 5.2). The 3D framework can also be described as constructed from the same sheets that are oriented parallel to the [−110] plane and are stacked along [−110] direction, giving rise to eight-ring channels in the

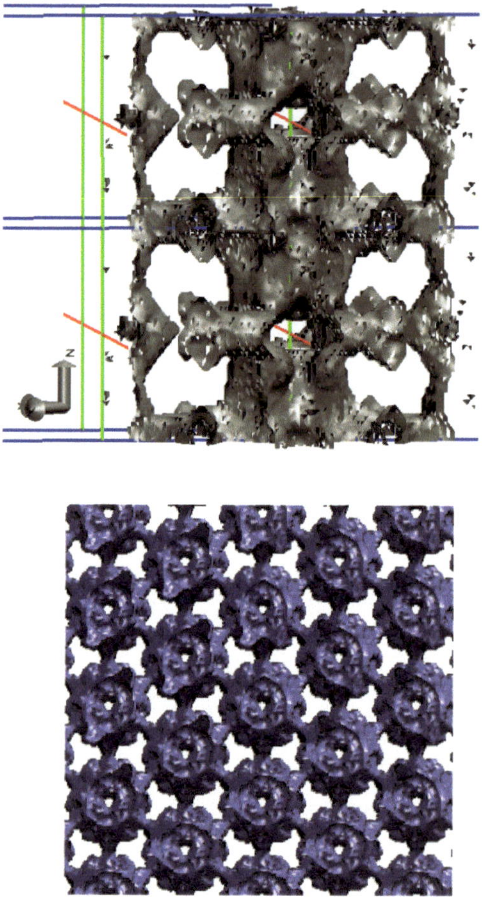

Figure 5.3 The ZIF-3 pore system wherein (top) the channels [0 0 1] (green), [1 1 0] (blue), [1 −1 0] (red), and their two intersections are depicted and (bottom) the ZIF-8 network of cages.

[−110] direction. The structure can also be described on the basis of the 4.8^2 net when viewed down the *c* axis, generating channels with elliptical eight-ring apertures along *c* axis. The channels in the [110] and [−110] directions are identical and intersect with those running in the [001] direction but not with each other [40, 41], as can be seen in Fig. 5.3.

Thus, ZIF-3 possesses two channel-like pore systems intersecting each other as can be seen in the same figure, depicted as blue-green and red-green crossing points, whereas ZIF-8 constitutes a 3D network of cages having a connectivity value eight.

It must be noted here that two options concerning the mobility of the bridging ligand of ZIF-8 were employed—one where a zero-value torsion potential acts on the dihedral angle Zn–N–C_1–C_3 (Fig. 5.2), thus allowing for a quasi-free motion of the ligand (coded as FLEX1), and another where the aforementioned potential does exist (coded as FLEX2), hence partly restricting the motion of the link. Obviously, as both options represent two somewhat extreme situations chosen so that to meet the modeling needs in the absence of a precise value for the aforementioned potential, neither of them can be fully realistic. Nonetheless, we believe that after considering the modeling predictions with respect to experimental results appeared recently [27], a less stiff version than the one represented by FLEX2 simulates more reasonably the motion of the link in this MOF material. However, both options do appear in this chapter for the sake of comparison.

5.3 Basics of Sorbate Equilibria and Transport

The aim of a sorption equilibrium simulation is to compute the concentration of sorbed species in the metal–organic interior, usually expressed as the number of sorbate molecules per unit cell, as a function of the chemical potential (or fugacity) or the partial pressure of the sorbate in the surrounding bulk phase of fluid at a given temperature. In terms of statistical thermodynamics-based computer experiments, this presupposes efficient sampling of the phase space of the system comprised by the guest sorbate molecules plus the sorbent host framework atoms in a particular statistical ensemble using importance sampling methods. The most commonly used scheme for sorption simulation in the guest sorbate–host sorbent system under equilibrium, is the Metropolis Monte Carlo in the grand canonical ensemble (grand canonical Monte Carlo [GCMC]). The method generates a stochastic chain of states fully compatible with the probability density function of the aforementioned statistical ensemble under any prescribed set of the macroscopically imposed parameters fugacity f, volume V, and temperature T for the guest sorbate–MOF system. Detailed descriptions of the GCMC method as well as of more advanced versions thereof, employing, for example, configurational-bias Monte Carlo, cavity, and/or energy bias moves to increase sampling efficiency, can be found in detail in Ref. [42].

In a classically described sorbate–sorbent model system, the primary tool for probing the molecular motion of the sorbed phase inside the host lattice at an atomistic scale is MD simulation. MD involves tracking the molecular positions and momenta as functions of time in the microcanonical statistical ensemble, namely, under a constant number of particles, N, volume, and total energy, E, (NVE) via numerical integration of the Newtonian equations of motion. It is possible to formulate the equations of motion so as to constrain other thermodynamic variables than the energy, for example, pressure, P, or temperature, thereby conducting MD in other ensembles, such as the canonical (NVT) or isobaric (NPT) ensembles. In modeling transport within MOFs, the later isochoric or isobaric MD schemes are often a convenient alternative to NVE-MD, provided care is taken in the choice of inertial parameters of the extended ensemble algorithm, so as not to perturb the intrinsic dynamics of the system.

A disadvantage of MD is that currently available computational means do not allow accessing times longer than μs; as a consequence, diffusivities smaller than 10^{-10} m^2/s cannot be easily computed via the conventional MD techniques. This problem can be overcome by invoking methods for the analysis and simulation of infrequent events. Typically, for bulky molecules one estimates the rates of elementary jumps from a site of minimum (free) energy to another, across an energy barrier that is high relative to $k_B T$. Transition-state theory (TST) approximates the rate constant, $k_{i \to j}$, of the rare dynamical event between two states i and j in configuration space as a ratio of two configuration integrals, one taken over the dividing surface between states i and j and the other taken over the origin state i, times a mean velocity across the dividing surface. Dynamically corrected TST provides a better estimate of $k_{i \to j}$ by accounting for dynamical recrossing events across the dividing surface via short MD simulations initiated on the dividing surface. Once all relevant rate constants have been calculated, they can be used in a kinetic Monte Carlo (KMC) simulation scheme to extract the diffusivity [42, 43].

5.3.1 *Sorption Thermodynamics*

Adaptation of Monte Carlo methodology to a mobile framework, usually in course of grand canonical ensemble, would be inefficient entailing a large number of stochastic moves over the total configuration degrees of freedom (generalized coordinates) for the large number of atoms that constitute the framework of coordination polymers. Hybrid schemes combining Monte Carlo stochastic and deterministic dynamics moves have been developed [44], however, without having been applied in the simulation of PCP matrices. Instead, we have proposed [26, 27, 29] a combination of the Gibbs' phase coexistence rules with properly adapting Widom averaging along the trajectory of the system, as produced by the MD computer experiment; the method leads to a rigorous computational scheme for predicting the isotherm of guests in a flexible host matrix.

5.3.1.1 Rigid sorbent model

In this subsection, assuming an immobile sorbent atomic framework, some Monte Carlo computations of sorption thermodynamics will be presented, focusing on comparisons with experimental data where available. In particular, sorption computer experiments of argon, methane and hydrogen in ZIF-8, carried out in the grand canonical ensemble, by means of the metropolis importance sampling technique for the guest configurational space, are presented. Also, computations of the enthalpies of adsorption, as well as guest probability densities within the metal–organic lattice are computed along the Monte Carlo stochastic chain.

The relative frequencies of displacement, insertion, and deletion moves during the Monte Carlo run are 1:1:1, respectively for an optimum convergence time. The sorption isotherm is obtained by computing averages of the number of sorbate molecules, N, over a large number of equilibrated configurations, as a function of the fugacity, f, converted to pressure in the graphs, via one of the procedures described in the next MD section. The relation of the fugacity of the bulk phase with the externally imposed chemical potential of the sorbed phase, μ_s, at the temperature T, is derived through the equality of the chemical potentials of the two phases leading to the relations

$$\mu_{\mathrm{s}} = \mu_{\mathrm{b}} = k_{\mathrm{B}}T \ln \left[\frac{f\Lambda^3}{k_{\mathrm{B}}T}\right] \qquad (5.1)$$

where Λ is the de Broglie thermal wavelength. The potential energy due to cross guest–host interactions can be pretabulated on a 3D grid of 0.02 nm spacing over the entire unit cell; the latter technique reduces the CPU time significantly thus allowing efficient configurational space sampling by including a larger number of unit cells in the simulation box.

Simulation results for argon in ZIF-8 at 87 K, using FF parameters from Ref. [27], are shown in the Fig. 5.4. It is seen that the simulated isotherms through both the generic FFs mentioned in the preceding section, although capturing the steep step of the experimental isotherm, eventually overpredict the measured points over the whole range of pressures.

An immediate conclusion of the graphs of Fig. 5.4 is that DREIDING is more suitable for modeling this metal–organic structure, although it fails to predict the first step at its experimental pressure. A comment on this—it can be the effect of the mobile methyl-imidazolate link on the sorption thermodynamics when inflexions are present [45]. The observed discrepancy between the measured occupancies and simulation results at saturation may be attributed to the susceptibility of the real framework toward more favorable conformations when higher sorbate amounts are to be hosted in the unit cell. As a consequence, the measured saturation uptake is higher than the predicted value of the simulations in the rigid matrix. The reversibility of the sorption–desorption curve shown in Fig. 5.4, gives no indication of hysteresis loop either as a consequence of structural variations (impossible under the issue of a rigid model) or sorbate condensation phenomena.

In the following lines a computational study on the energetic content of the adsorption sites of ZIF-8, with respect to the

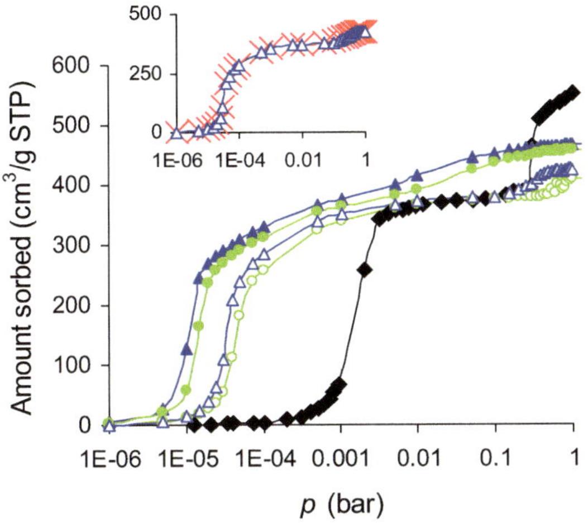

Figure 5.4 Sorption isotherms of Ar in ZIF-8 at 87 K obtained by experiment [8] (black symbols) and indicative GCMC simulations using various model parameters for guests and hosts [27]: DREIDING (open blue triangles and open green circles), UFF (filled blue triangles and filled green circles); inset shows simulated sorption (open blue triangles)–desorption (red crosses) branches using the DREIDING parameters set.

thermodynamic predictions shown in the above graph will be presented. A usual measure of the energetic inhomogeneity from experimental perspective is the partial molar heat of sorption (isosteric heat), q_{st}, defined as

$$-q_{\mathrm{st}} \equiv \bar{H}_{\mathrm{s}} - H_{\mathrm{b}} \tag{5.2}$$

In a single-component system, H_{s} is the molar enthalpy of the bulk phase; $\bar{H}_{\mathrm{s}}$ is the partial molar enthalpy of the sorbed phase (in the presence of host species) given by the relation, $\bar{H}_{\mathrm{s}} = \bar{U}_{\mathrm{s}} + P\bar{V}_{\mathrm{s}}$; and macron diacritics denote partial molar quantities for the internal energy and volume of the sorbate molecules under constant pressure and temperature. Whereas in conventional experiments the quantity q_{st} is measured, in the simulation, $\bar{H}_{\mathrm{s}}$ can be directly calculated by means of statistical mechanics from the covariance between number of molecules, N, and enthalpy, divided by the variance in the number of molecules, as follows:

$$\bar{H}_{\mathrm{s}} = \frac{\langle NU_{\mathrm{s}}\rangle - \langle N\rangle\langle U_{\mathrm{s}}\rangle}{\langle N^2\rangle - \langle N\rangle^2} + P\frac{\langle NV_{\mathrm{s}}\rangle - \langle N\rangle\langle V_{\mathrm{s}}\rangle}{\langle N^2\rangle - \langle N\rangle^2} \tag{5.3}$$

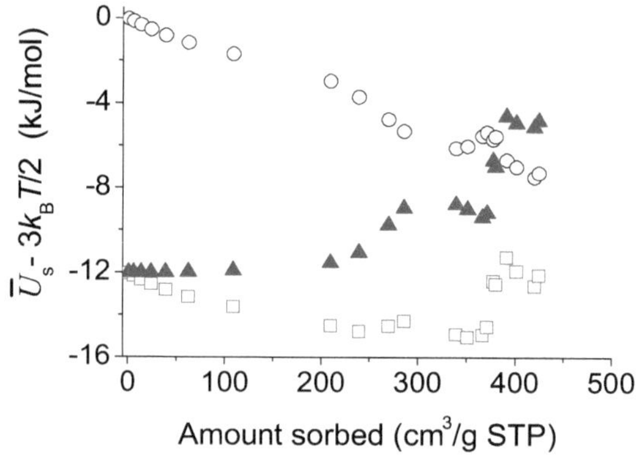

Figure 5.5 Partial molar configurational internal energy of Ar ($s = 3$) calculated at 87 K versus occupancy (squares); sorbate–sorbent (triangles) and sorbate–sorbate interactions (circles) are also shown.

The ratio in the second term of the right part of the above equation expresses the partial molar volume $\bar{V}_s$, which in the course of a GCMC simulation vanishes; therefore, by splitting internal energy to its kinetic and configurational, $\mathcal{V}$, (potential energy) components, Eq. 5.3 can be put in the form

$$\bar{U}_s - \frac{sk_B T}{2} = \frac{\langle N\mathcal{V}\rangle - \langle N\rangle\langle\mathcal{V}\rangle}{\langle N^2\rangle - \langle N\rangle^2} \tag{5.4}$$

where s is the degrees of freedom of the guest molecules and $sk_B T/2$ their partial molar kinetic energy.

The left part of Eq. 5.4 expresses the partial molar configurational internal energy of the sorbed phase. In Fig. 5.5 this quantity is plotted as a function of the argon occupancy in ZIF-8. In the same graph, the sorbate–sorbate and sorbate–sorbent contributions to the total energy appear; the latter amount is equivalent to the experimentally measured excess energy, reflecting the actual binding energy of the guest molecules with the framework atoms. Around the value of 100 cm^3/g STP the sorbate–sorbent part starts to increase, exhibiting a peak at the upper value of the isotherm step.

Subsequently, the energy reaches a plateau and increases steeply again; at the point where the second step of the isotherm appears. The observed trend in the calculated partial molar internal energy of argon within the ZIF-8 matrix conforms to the isotherm shape, indicating adsorption sites possessing different energy content.

One- and two-particle probability density distributions $\rho^{(1)}(\mathbf{r}_1)$ and $\rho^{(2)}(\mathbf{r}_1,\mathbf{r}_2)$, respectively, have been utilized in this study in order to map the location of Ar molecules with varying pressure in this imidazolate framework:

$$\rho^{(n)}(\mathbf{r}_1\ldots,\mathbf{r}_n) = \frac{1}{\Xi(\mu,V,T)}\sum_{N=n}^{\infty}\frac{\exp[\mu N/k_\mathrm{B}T]}{\Lambda^{3N}(N-n)!} \times \int d^3r_{n+1}\ldots d^3r_N \exp[-\mathcal{V}(\mathbf{r})/k_\mathrm{B}T] \quad (5.5)$$

The above relation expresses the probability for finding a state with one guest molecule located at a certain position $\mathbf{r}_1$ $(n = 1)$ or the probability for finding a state with one guest molecule located at $\mathbf{r}_1$ and a second one located at $\mathbf{r}_2$ $(n = 2)$, averaged over all the momentum and configuration space, $\mathbf{r} = (\mathbf{r}_1, \mathbf{r}_2, \ldots, \mathbf{r}_N)$, of the N molecules inside the ZIF-8 unit cell.

In the graphs of Fig. 5.6 the probability density $\rho^{(2)}(\mathbf{r}_1,\mathbf{r}_2)$ was normalized by $\langle N\rangle^2 V^{-2}$ to give the pair distribution function $g^{(2)}(\mathbf{r}_1,\mathbf{r}_2)$. Considering the magnitude, $r = |\mathbf{r}_1 - \mathbf{r}_2|$, the two-particle distribution function takes the form of the radial distribution function $g(r)$. Similarly, in the special case where the singlet probability density, $\rho^{(1)}(\mathbf{r}_1)$, of finding a molecule at positions radially distributed with respect to a certain point, $\mathbf{r}_\mathrm{ref}$, within the framework so that $r = |\mathbf{r}_\mathrm{ref} - \mathbf{r}_2|$, $\rho^{(1)}(\mathbf{r}_1)$ is replaced by $\rho(r)$; otherwise, $\rho^{(1)}(\mathbf{r}_1)$ is being calculated (see Figs. 5.6–5.8).

From the comparison of the effect of pressure before, within, and after the first step of the argon isotherm, it is seen that the fluid becomes slightly more structured with increasing pressure, resulting in a narrower nearest-neighbor peak at about 0.36 nm (see Fig. 5.6). In fact, this distance is close to the nearest-neighbor distance of 0.37 nm measured in solid argon [46]. Increase of occupancy up to the second isotherm step causes mainly a rise in the coordination number of the fluid. The observed behavior of the radial distribution function with increasing pressure indicates a rather structured sorbed phase inside the ZIF-8 pores, eliciting existence of adsorption sites with varying degrees of occupancy as a result of energetic inhomogeneity effects.

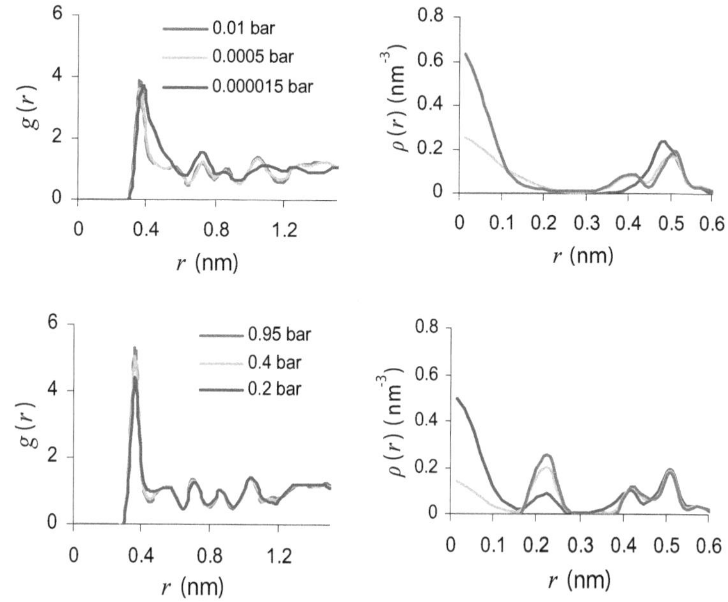

Figure 5.6 Density distribution functions for various occupancies of Ar in ZIF-8 at 87 K chosen around the first (top graphs) and second inflection (bottom graphs) of the isotherm shown in Fig. 5.4. (Left) Pair distribution functions $g(r)$; (right) density probability distributions $\rho(r)$, calculated with respect to the cage centers.

The probability density, $\rho(r)$, of finding a molecule at positions radially distributed around a certain point inside the zeolite framework, may provide information on the location of these sites inside ZIF-8. The evolution of probability density of argon around the center of cages for pressure values before, inside and after the formation of the first transition, shows a steep increase of the number of molecules in the center of cage as pressure increases, with a simultaneous appearance of two shorter double peaks at 0.4 nm and 0.5 nm that are close to the square and hexagonal window necks, respectively. Namely, as loading (sorbate chemical potential) increases, molecules start migrating to a new energetically less favored site at the cavity center. The evolution of the probability density at higher pressure exhibits a different trend from the sitting transition at the lower pressure (see Fig. 5.6). That

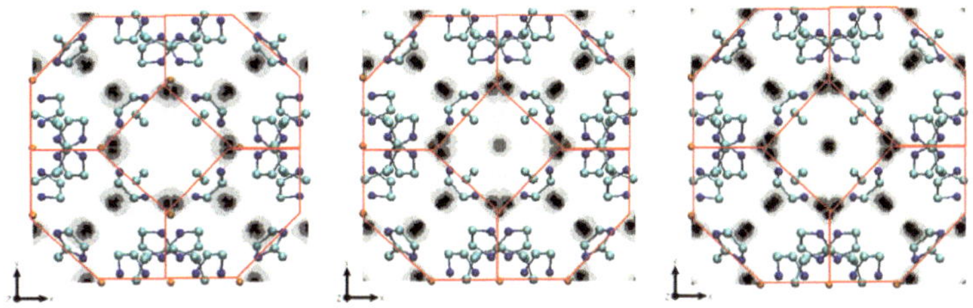

Figure 5.7 Evolution of isodensity surfaces of Ar at 87 K projected in the *xy* plane of the ZIF-8 unit cell, under p = 2.5 × 10^{-5} bar (left), $p = 5 \times 10^{-4}$ bar (middle), and $p = 1.0 \times 10^{-2}$ (right), corresponding, respectively, to pressure values before, within, and after the low-pressure transition step (cf. Figs. 5.4 and 5.6); the isovalues used are 1.0 nm^{-3} (light gray), 3.0 nm^{-3} (gray), 15.0 nm^{-3} (dark gray), and 32.0 nm^{-3} (black); the red frame indicates hexagonal and square faces.

is to say, as argon occupancy increases, the density probability at the cavity center decreases until it finally empties after 0.9 bar; at the same time, enhancement of the probability density at 0.2 nm indicates a new site, being located within a distance of about one argon sigma from the bidisperse distribution shown in Fig. 5.6, thus indicating layer formation.

The reader is reminded that to obtain probability distributions, particles have been binned over volumes between successive spherical cells surrounding a particular reference position. A complementary view to this aspect can be given through the actual singlet probability distribution, $\rho^{(1)}(\mathbf{r}_1)$.

Figure 5.7 presents singlet probability density distributions in the form of isodensity surfaces, for the three occupancies used in Fig. 5.6 in order to give a pictorial explanation of the low and higher pressure sorption thermodynamics of argon guest molecules. Thus, the population of argon is depicted for various loadings at positions close to hexagonal and square faces inside the ZIF-8 cavity.

In the Fig. 5.8 isodensities for carbon dioxide molecules sorbed in ZIF-8 are depicted; it is seen that they occupy primarily sites close to the double bonds of the imidazolate rings that form the hexagonal windows. Therefore, the singlet probability density is higher at those positions (smaller free-energy barrier, see Ref. [27] and Eq. 5.19) compared to methane molecules at the same area.

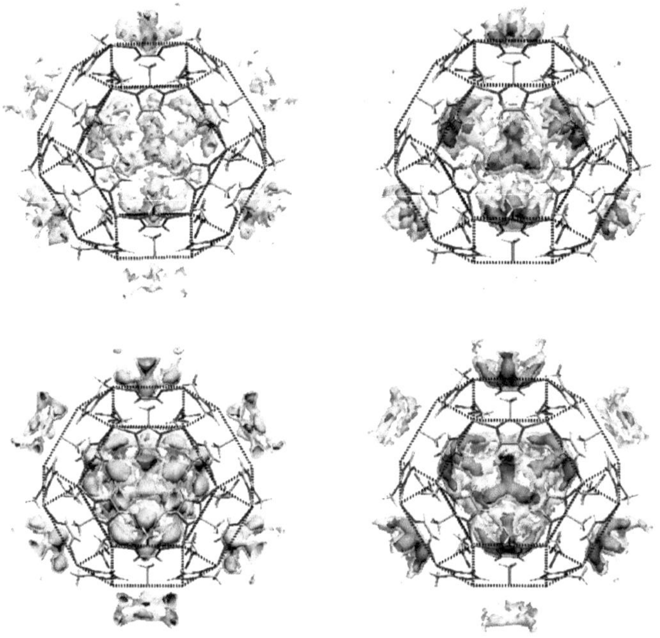

Figure 5.8 (Top) Isodensity surfaces calculated in the vicinity of the hexagonal windows of ZIF-8 for 1.4 CH_4 molecules/cage (left) and 1.7 CO_2 molecules/cage (right). (Bottom) The same as above for 10.1 CH_4 molecules/cage (left) and 10.2 CO_2 molecules/cage (right). Low to high density values follow the color code light to dark, respectively.

5.3.1.2 Flexible sorbent model

In the present subsection of sorption equilibria it is discussed the possibility of studying guest–host equilibria simultaneously with MD, which is discussed in the next section, without the need for assuming a rigid sorbent model due to GCMC "constraints" discussed previously. The statistical mechanics-based simulation methodology for modeling mobile frameworks of the PCP sorbents relies closely upon the modeling procedure of polymer chains [27]. Along this line, nonbonded interactions between skeletal and pendant atoms of the framework, as well as interactions depending on the deviations of bond lengths (stretching), bond angles (bending), and dihedral angles (torsion) from their equilibrium values, have been taken into account. For this reason we have tested several fully flexible ZIF-8 crystal versions under different charge distributions determined by either a coarse united atoms representation of the framework (UM) or an explicit atomistic digitization of the ZIF structure (EM). Then, integration of the equations of motion may be performed over the entire degrees of freedom for the host framework atoms, N_H, and guest sorbate molecules, N_G, in the isochoric–isothermal ($N_H + N_G, V, T$) and isobaric–isothermal ($N_H + N_G, P, T$) ensembles. Carbon dioxide was modeled as a flexible three-site molecule bearing partial charges to represent its quadrupole moment; methane was modeled as a neutral sphere by neglecting its octapole moment.

Then, we followed a twofold strategy in applying the values for the various potential functions:

(i) In the isochoric–isothermal case, by making use of them without any kind of reparameterization with respect to measurements of the present work or to relevant experiments found elsewhere and also with no adulteration with values from other known FFs.
(ii) In the isobaric–isothermal case, by setting the crystallographic equilibrium values [8] in all potentials mentioned in the preceding paragraph and, moreover, increasing the force constants from their original values [30] by an order of magnitude for the stretching potential along the bond Zn–N and the angle bending in the Zn–N–C_1 and Zn–N–C_2 (see Fig. 5.2).

This type of adjustment proved to be the optimum selection in the isobaric–isothermal ensemble to preserve the unit cell volume [8], causing a maximum reduction up to 4%, however. The latter parameters set was selected among a long series of (N_H, P, T)-MD runs applied to the bare ZIF-8. On the contrary, employment of the DREIDING values results in unphysical contraction of the ZIF cages.

The first option was adopted for the isochoric–isothermal MD computer experiments, with the fixed value V being the crystallographic volume of the ZIF-8 unit cell [8], whereas the latter option was chosen so that to enable us a more realistic modeling of the presented physical experiments by allowing volume fluctuations in the guest–framework system under constant pressure, temperature, and chemical potential as imposed on by the Gibbs thermodynamic phase equilibrium rules.

The excess chemical potential, μ^{ex}, of the guest phase (actual chemical potential minus the value of ideal gas of the same density, $\rho = N_G/V$, at T), namely, $\mu^{ex}(\rho, T) \equiv \mu(\rho, T) - \mu^{ig}(\rho, T)$, can be calculated through the Widom method by taking the ensemble average of the Boltzmann factor of the change in potential energy brought about by the insertion of a "ghost" sorbate molecule (interacting with the framework atoms and the other sorbate molecules without perturbing the current host-guest system configuration), at the given temperature, $\beta = 1/k_B T$. This routine in the isobaric–isothermal ensemble is prescribed by the following formula:

$$\frac{\int d\mathbf{r}_t \int dV \exp(-\beta PV) \int d\mathbf{r} \exp[-\beta V(\mathbf{r})] \exp[-\beta \Delta V(\mathbf{r}_t; \mathbf{r}) - \beta V(\mathbf{r}_t)]}{\int d\mathbf{r}_t \int dV \exp(-\beta PV) \int d\mathbf{r} \exp[-\beta V(\mathbf{r})]}$$
$$= \frac{\int d\mathbf{r}_t \langle \exp[-\beta \Delta V(\mathbf{r}_t; \mathbf{r}) - \beta V(\mathbf{r}_t)] \rangle}{V}$$
$$= \exp[-\beta \mu^{ex}(\rho, T)] \tag{5.6}$$

where the vector $\mathbf{r}$ denotes the configuration space of the guest-host phase; $\mathbf{r}_t$ is the configuration space of the internal degrees of freedom of the "ghost" (test sorbate); $\Delta V(\mathbf{r}_t; \mathbf{r})$ stands for the potential energy change of the system upon the insertion of the ghost (ghost–sorbate and ghost–framework interparticle energy); and $V(\mathbf{r}_t)$ stands for the intramolecular energy of the ghost. More specifically, the computational procedure of Eq. 5.6 entails evaluation of the following integrals: (a) over the total states of the host–guest system as they are created during the dynamical trajectory, weighted according to the probability density of the ensemble under consideration; (b) over the internal configurations of the "ghost" weighted according to the Boltzmann factor of its intramolecular energy (e.g., the CO_2 case); and (c) a spatial integral over a sequence of random insertions of the "ghost" within the current ZIF volume created each time through the dynamics evolution. It must be noticed that the latter procedure may create a ghost molecule physically inaccessible by the sorbate molecule region within the imidazolate framework; therefore, such values are not taken into account in the above sampling.

In the isochoric–isothermal ensemble for spherical molecules Eq. 5.6 takes the form

$$\frac{\int d\mathbf{r}_t \int d\mathbf{r} \exp\left[-\beta V(\mathbf{r})\right] \exp\left[-\beta \Delta V(\mathbf{r}_t; \mathbf{r}) - \beta V(\mathbf{r}_t)\right]}{\int d\mathbf{r}_t \int \exp\left[-\beta V(\mathbf{r})\right] d\mathbf{r}}$$
$$= \exp[-\beta\mu^{\mathrm{ex}}(\rho, T)] \quad (5.7)$$

Subsequently, by invoking the phase coexistence rule for any sorbate–PCP system (host–guest–bulk phases) under thermodynamic equilibrium, one requires that pressure, P, temperature and chemical potential of the guest molecules in the sorbed and bulk phases remain equal. After invoking the more convenient quantity, the fugacity f, defined by the relation

$$f(\rho, T) = \frac{\rho}{\beta \exp\left[-\beta\mu^{\mathrm{ex}}(\rho, T)\right]} \quad (5.8)$$

the corresponding pressure P at the prevailing T and ρ remains to be determined.

For MD under the $(N_{\mathrm{H}} + N_{\mathrm{G}}, P, T)$ statistical ensemble, making the logical assumption that the vapor pressure of the sorbent material (host species) is negligible at the employed temperatures (nonvolatile), the above requirement of phase equilibrium is attained by executing the computer experiment in an iterative manner so that the evaluated conjugate pressure of the fugacity, is introduced to the barostat algorithm of a subsequent MD run, whereby integration of the equations of motion is carried out (see Section 5.3.2), until the currently evaluated pressure converges to the value of pressure imposed to the algorithm of the previous computer run; the latter value represents the said pressure, P.

In the $(N_{\mathrm{H}} + N_{\mathrm{G}}, V, T)$-MD, P is evaluated after calculating the specific volume of the bulk phase,v, at the given T, from experimental charts of the form $v = v(P)$, found elsewhere (NIST Chemistry WebBook [47]), through numerical evaluation of the following integral, assuming standard states, $f^{\mathrm{o}} = P^{\mathrm{o}}$, that is,

$$RT \int_{f^{\mathrm{o}}}^{f} d\ln f = \int_{P^{\mathrm{o}}}^{P} v dP \quad (5.9)$$

Alternatively, working out the fugacity difference isothermally between the bulk state with molar volume, υ, at T and P, and a

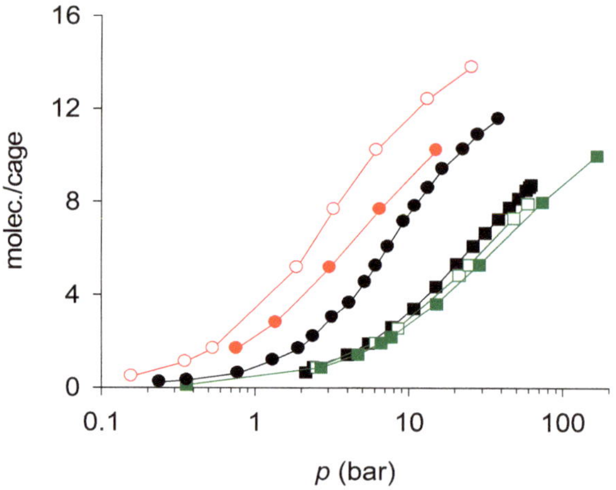

Figure 5.9 Isotherms computed via molecular dynamics for CH_4 at 300 K (green squares) and CO_2 at 303 K (red circles) in ZIF-8 versus measurements of CH_4 at 300 K [21] and CO_2 at 303 K [48] (similar symbols in black); simulations were conducted under isochoric–isothermal (open symbols) and isobaric–isothermal ensembles (filled symbols).

state comprising the same number of molecules with infinite molar volume, the following relationship can be proved:

$$\ln \frac{f\upsilon}{RT} - \frac{P\upsilon}{RT} - \frac{1}{RT}\int_{\infty}^{V}\left(\frac{RT}{\upsilon} - P\right)d\upsilon + 1 = 0 \qquad (5.10)$$

By virtue of the pressure-explicit Peng–Robinson (PR) equation of state, the integrals of Eq. 5.10 are calculated analytically and eventually it is solved numerically to find the molar volume, υ. Subsequently, substituting the value of υ back in the PR equation of state, the pressure, P, is obtained.

An example study of the CO_2 and CH_4 phase equilibria in ZIF-8 via $(N_H + N_G, V, T)$ and $(N_H + N_G, P, T)$-MD for various types of flexible structures, as a function of sorbate fugacity is presented in Fig. 5.9.

The thermodynamics of the sorbed phase in a ZIF-8 rigid model as well as in a variety of flexible versions, and also for two cases of charge distribution over the framework atoms, is found to be less influenced by the sorbent model compared to the effect caused by the kind of guest–guest interactions [27]. An analogous result has also been found in a previous study of other MOFs [14]. This finding becomes clearer in view of the fact that CO_2 sorption isotherms remain almost unchanged when assuming a neutral atomic framework; whereas, swapping between different FFs for the CO_2 molecule, significant variations in the sorption capacity are observed [27]. It is also shown that the above-described averaging procedure through isobaric–isothermal MD results in somewhat higher fugacity values, and hence pressure, compared to the averaging under the isochoric–isothermal ensemble at the same occupancies for carbon dioxide, whereas the spherical methane seems to be less affected by the ensemble-weighting factor. A reason for this could be associated with the ZIF cell conformations yielded along the system's dynamic trajectory, thereby creating the sequence of volumes wherein the averaging is performed. That is to say, conformations of the flexible linear carbon dioxide quadrupoles not achievable by the constant volume imposition may now be favored.

Another example are the sorption isotherms of hydrogen and deuterium in ZIF-3 and ZIF-8 at 77 K, computed via the above described procedure in the two frameworks, modeled either as rigid or flexible, are presented in Fig. 5.10. The H_2 and D_2 molecules were modeled as spheres; the sorbate–sorbate and sorbate framework–atom dispersion interactions were calculated via the classical Lennard–Jones (LJ) effective potential, U_{LJ}, by adding to this the first two terms of the series

$$\frac{\hbar^2}{24k_{\mathrm{B}}T\,m_{\mathrm{r}}}\left[\frac{\partial^2 U_{\mathrm{LJ}}\left(r_{ij}\right)}{\partial r_{ij}^2}+\frac{2}{r_{ij}}\frac{\partial U_{\mathrm{LJ}}\left(r_{ij}\right)}{\partial r_{ij}}+\cdots\right]$$

due to Feynman and Hibbs (F-H) [38] to account for the quantum mechanical nature of all interactions where protons are involved.

The above correction to U_{LJ} is introduced in view of the formulation of the evolution of a quantum mechanical system as a path integral between two states; this integral encompasses all the possible paths by which the system can travel between two different configurations in the time interval $\hbar/k_{\mathrm{B}}T$ at temperature T, where $\hbar = h/2\pi$ is the reduced Planck's constant, h, and m_r is the reduced mass of the interacting pair of atoms l and k, defined as m_{r}^{-1} =

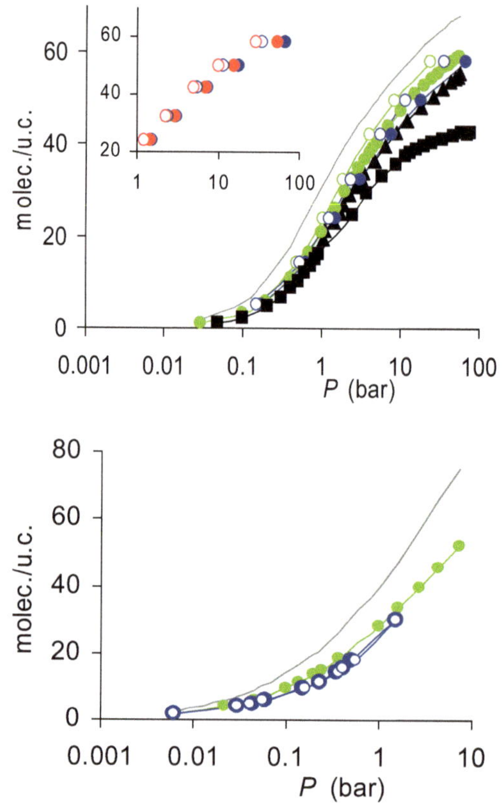

Figure 5.10 Experimental (squares[8] and triangles [21]) and predicted via molecular dynamics sorption isotherms at 77 K for H_2 (filled circles) and D_2 (open circles) in ZIF-3 (top) and ZIF-8 (bottom), with rigid (green), FLEX1 (blue), and FLEX2 (red) models, with all interactions described by the F-H potential; H_2-simulated isotherms within rigid ZIF models, by means of the conventional LJ potential (gray line).

$m_l^{-1} + m_k^{-1}$. Then, by considering a pairwise type potential, like the LJ, the mean path can be replaced by the position vector distance, $\mathbf{r}_{ij}$, between the centers of mass of molecules i and j; the correction elicits that paths located at $\mathbf{r}_j$ are enveloped inside a Gaussian width. Therefore, for low temperatures and atomic masses, the distribution around $\mathbf{r}$ becomes widespread, hence a classical description is not adequate anymore [29, 35, 38, 49].

Both model options using the F-H potential predict satisfactorily the experimental isotherm showing that the framework mobility does not influence significantly the thermodynamics of the system.

In the above figures the effect of the quantum mechanical correction is presented; thus, employing the F-H potential in place of the classical LJ one observes that fewer guest molecules for the same fugacity values are computed; this is explainable on the basis of the lower localization of the hydrogen molecules as predicted via the F-H theory, thereby a lower potential well and a higher effective sigma value cause a lower number density than the one predicted by the LJ potential function.

In Fig. 5.11, the evolution of the probability density distribution as the hydrogen loading increases within ZIF-3 shows that the areas close to the ligands in the interior of the three channel systems constitute primary adsorption sites over the whole occupancy range. It is also observed that the vicinity of the intersections become progressively populated until these regions of the framework practically tend to minimize the communication between the two pore systems; thus, it is verified the finding of the trajectory analysis presented in the form of MSD plots of the next section, which shows the overall mass flow at high loadings to be undertaken mostly through the z axis pore channel [0 0 1].

5.3.2 *Molecular Dynamics*

The diffusivity studies of fluids sorbed in PCPs mainly deal with the computation of the intracrystalline orientationally averaged ($d_o = 3$) self-diffusivity via equilibrium MD computations by tracking the displacements of the guest molecules for a long period of time, t, as imposed by the Einstein equation:

$$D_\mathrm{s} = \frac{1}{2d_\mathrm{o}} \lim_{t\to\infty} \frac{\mathrm{d}}{\mathrm{d}t} \left\langle \frac{1}{N} \sum_{i=1}^{N} [\mathbf{r}_i(t) - \mathbf{r}_i(0)]^2 \right\rangle \tag{5.11}$$

where N is the number of guest sorbate molecules and $\mathbf{r}_i$ the position vector of the center of mass of molecule i. It must be noted that in anisotropic media, such as the metal–organic frameworks in general, the self-diffusivity is a tensor, $\mathbf{D}_s$. The components of the tensor can be computed from an equation

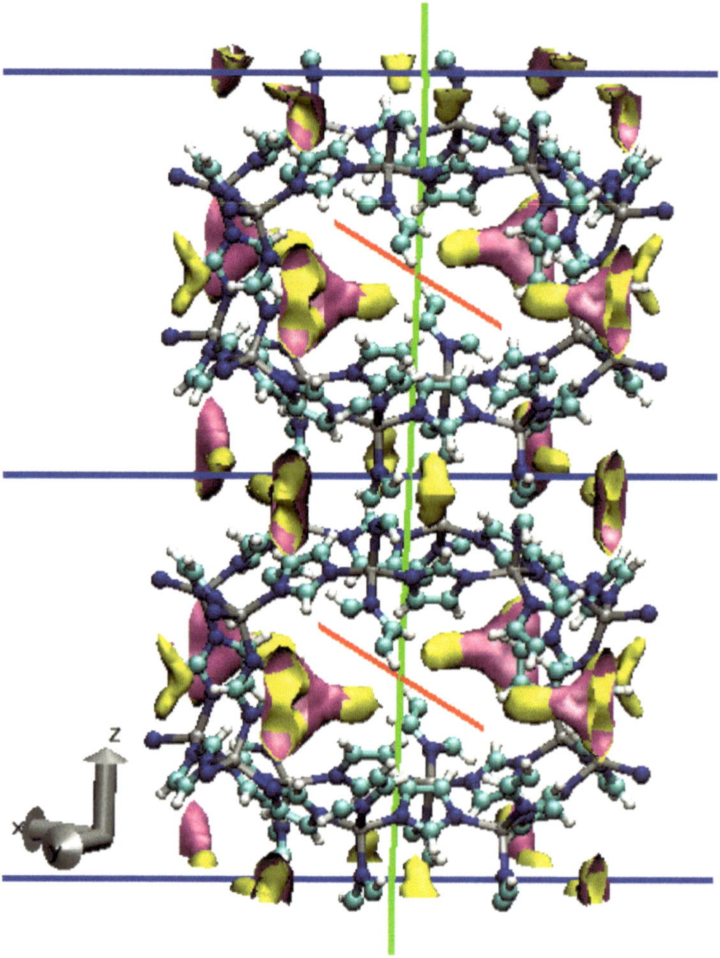

Figure 5.11 Singlet probability density distribution of H_2 within ZIF-3 in the form of isodensity contours for two indicative loadings of 5.75 molecules/unit cell (magenta) and 15.5 molecules/unit cell (yellow); colored lines denote the three channel systems, as depicted in Fig. 5.3.

analogous to Eq. 5.11. with the mean squared displacement replaced by the corresponding product of components of the molecular displacement vector along the three Cartesian directions and $d_o = 1$. Experimentally measured values of self-diffusivity from pulsed-field gradient nuclear magnetic resonance (PFG-NMR) and incoherent quasi-elastic neutron scattering (QENS) are directly compared to self-diffusivities computed from MD. This comparison is more legitimate with QENS, since this experimental method operates over MD length and time scales.

In most engineering applications, it is the so-called transport diffusivity, D_t, that is of main interest, that is,

$$D_\mathrm{t} = D_0 \frac{\partial \ln f}{\partial \ln c} \tag{5.12}$$

where f and c are the fugacity and concentration of the guest molecules, respectively; the corrected or M-S diffusivity, D_0, is a collective property of the swarm of diffusing particles, subject to linear response theory, and is computed through the relation

$$D_0 = \frac{V^2}{d_\mathrm{o} N} \int_0^\infty \mathrm{d}t \, \langle \mathbf{j}(t) \cdot \mathbf{j}(0) \rangle \tag{5.13}$$

where $\mathbf{j}(t)$ is the instantaneous microscopic flux vector defined as

$$\mathbf{j}(t) = \frac{1}{V} \sum_{i=1}^{N} \mathbf{v}_i(t) \tag{5.14}$$

with $\mathbf{v}_i(t)$ being the center-of-mass velocity of sorbate molecule i. From Eqs. 5.13 and 5.14 it is obvious that transport diffusivity is also a collective property and therefore suffers from poor statistics compared to self-diffusivity (cf. Eq. 5.11).

An interesting nonequilibrium MD scheme that computes directly the transport diffusivity is the dual-control volume grand canonical molecular dynamics (DCV-GCMD) [50]. This technique utilizes a simulation box divided into three sections in the flux direction; constant chemical potential is maintained in the two end sections by Metropolis Monte Carlo creation and destruction trials; MD is performed on all molecules in the system, in such a way that the ratio of stochastic trials to dynamics steps is fairly large, resulting in the development of a concentration gradient between the two end sections of constant density at steady state [51]. By counting the total numbers of particles lost, N_-, and gained, N_+, in each end section and after applying a mass balance condition, the molecular flux is obtained by the formula

$$\mathbf{J} = \frac{1}{A_\mathrm{c}} \frac{\mathrm{d}(N_+ - N_-)}{\mathrm{d}t} \tag{5.15}$$

where A_c is a cross sectional area. Then, by plotting the quantity $N_+ - N_-$ against time, an effective transport diffusion coefficient at steady state is calculated by integrating the Fick equation, that is,

$$\int_{x=0}^{x=\ell} J dx = J\,\ell = \int_{x=0}^{x=\ell} D_t dc = D_t(c_{x=\ell} - c_{x=0}) \qquad (5.16)$$

Further references on nonequilibrium MD inside or beyond the linear response regime in the form of transient time correlation functions can be found in Ref. [52].

The availability of powerful computer resources has enabled the use of equilibrium MD in conjunction with Eq. 5.13 for the calculation of M-S and hence transport diffusivities. Thus, equilibrium runs can be conducted for sufficiently long times, ensuring acceptable statistics. Indeed, comparisons of DCV-GCMD and equilibrium MD utilizing Eq. 5.13 have given results within experimental error [51, 52].

An attempt toward a quantitative explanation of the concentration dependence of the collective diffusivities described above, on the basis of Quasichemical (QC) Mean Field Theory and a simple model for surface diffusivity [51, 53], leads to the following equation

$$D_0(\theta) = D_0(0)\left[\frac{\zeta+1}{2-2\theta}\right]^{-z}\left[1+\frac{\zeta-1+2\theta}{2-2\theta}\exp(2w/k_B T z)\right]^{z-1} \qquad (5.17)$$

where θ is the fractional occupancy of sorbate molecules in the zeolite, being considered as possessing a uniform lattice of sorption sites with coordination number z, and ζ is defined by the expression

$$\zeta = \{1 - 4\theta(1-\theta)[1-\exp(-2w/k_B T z)]\}^{1/2} \qquad (5.18)$$

with w expressing the pair interactions between neighboring molecules in the sorbed phase.

Equation 5.17 provides a satisfactory representation of the occupancy dependence of the corrected diffusivity, however, the mass transport in PCPs follows a much complicated route; this is mainly due to their significantly inhomogeneous energetic interior as a results of the charge distribution of the organic ligands as well as of the steric phenomena exerted by these links on the diffusing guest molecules.

The results of the elaboration of the trajectories of the center of mass of the individual guest molecules via the Einstein equation

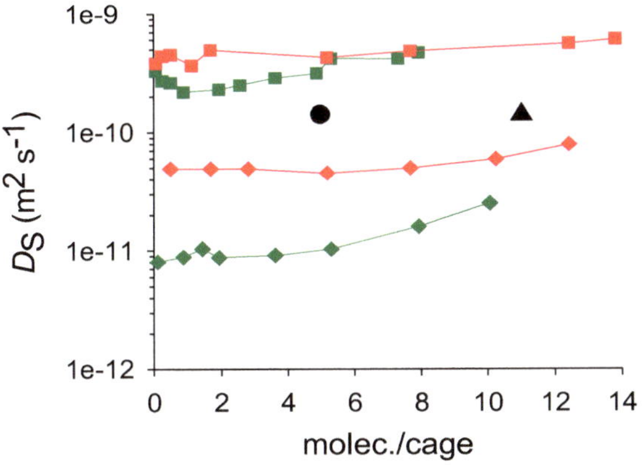

Figure 5.12 Self-diffusivity of CH_4 at 300 K (green symbols) and CO_2 at 303 K (red symbols) as a function of their loading in ZIF-8 via isochoric–isothermal (squares) and isobaric–isothermal (rhombi) MD and via PFG-NMR for CH_4 (black circle) and CO_2 (black triangle) at 298 K.

in order to obtain self-diffusivities for the two sorbates within the host lattice are presented in Fig. 5.12. In the same figure two values of the intracrystalline self-diffusivity measured by PFG NMR under equilibrium conditions in a bed of large ZIF-8 crystals (mean edge length: 20 μm), are also included [27].

For the occupancies covered by both sorbates it is shown that no significant variation in the self-diffusivity is observed. It is noticeable that for CH_4 the discrepancy between the PFG NMR value and the isobaric MD data is much higher compared to the isochoric MD. Nonetheless, the kind of ensemble applied for the generation of the dynamics evolution causes no alteration of the self-diffusivity trend as can be seen from the curves of Fig. 5.12. The observed quantitative differences are explainable on the basis of the effect of the framework contraction on the windows' aperture (crystallographic width [8]: 0.34 nm). That is to say, they might be considered as a simulation artifact due to the "compliant" DREIDING FF that causes the mobility of the bulkier methane molecule (kinetic diameter: 0.38 nm) to be much affected, compared to carbon dioxide (kinetic diameter: 0.33 nm). Investigations on the influence of lattice flexibility on self-diffusivity can be found in Refs 37, 54 and 55, as well. Furthermore, zeroing of the torsion potential of the dihedral angle Zn–N–C_1–C_3 (see Fig. 5.2) that controls the stiffness of the 2-methylimidazolate gate-like links, located at the cage-to-cage paths of ZIF-8, causes a considerable shift of the diffusivity curves shown in Fig. 5.12, to higher values for both sorbates. This finding highlights the effect of the organic group mobility on the mass transport within imidazolate frameworks possessing pore sizes commensurate with the guest molecular dimensions.

In the plots of Fig. 5.13 is shown the mean squared displacement within the rigid framework model appears methane to be entrapped in the ZIF-8 cages at all times, whereas the less bulky carbon dioxide, being able to pass the cage-to-cage windows, exhibits a tiny diffusivity value even in the rigid structure. Computer experiments conducted for similar flexible structures, differing only in the inclusion of the torsion potential due to dihedral angle Zn–N–C_1–C_3, indicate that omission of this interaction that is responsible for the torsion of the ligand (quasi-free motion of this gate-like group), located at the said critical for transport hexagonal apertures, leads to diffusivity values up to three times higher than the ones obtained after adopting a nonzero value for the mentioned potential. Additional case studies show that the framework charge distribution as results from the models examined, namely, a united atom representation and an explicit modeling of all atoms in the framework, affects primarily the kinetic behavior of sorbate molecules bearing charged centers like the carbon dioxide.

In view of the above tested model features (also being guided by the order of magnitudes of the measured values), it transpires that a logical choice for modeling the flexible ZIF framework is through an explicit presence of its constituting atoms, thus taking also into account partial charges to all atoms, taken from Ref. [48].

To examine further the host structural effects on the guest dynamics we compare the simulation predictions with a transient diffusion experiment for measuring the transport diffusion coefficient, D_t, the infrared microscopy (IRM); the IRM measurements [27] were carried out in a gigantic single crystal (average length scale to be covered by guest molecules of about 100 μm). For the sake of such a comparison, we computed the nonequilibrium collective M-S diffusion coefficient, D_0, by elaborating the trajectory of the center of mass of the entire swarm of guest particles and

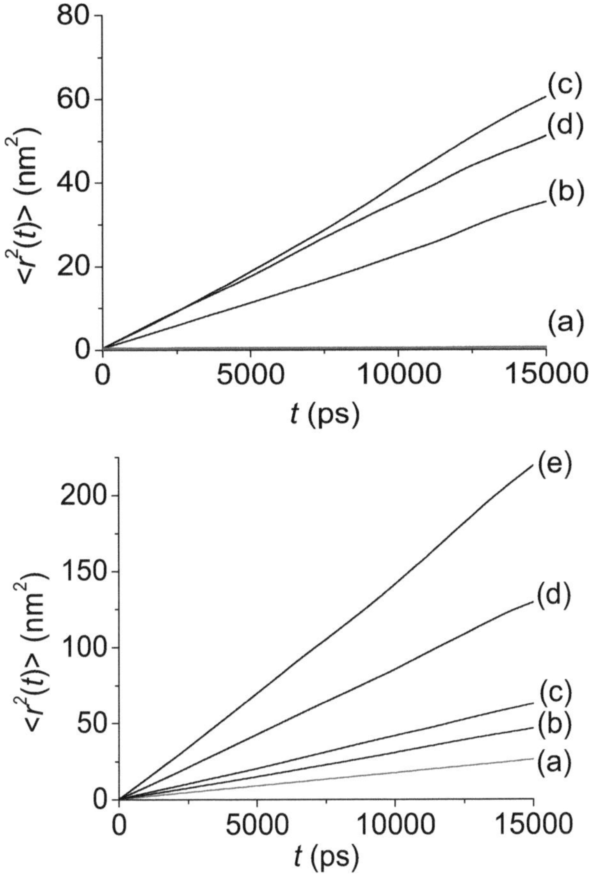

Figure 5.13 (Top) Indicative MSD plots of CH_4 (5.3 molecules/cage) at 300 K within ZIF-8 modeled as rigid (a); explicitly represented flexible framework with organic ligands controlled either via a torsion potential (see text) (b) or in the absence of such a potential (quasi-freely moving link) (c); the previous case in a united atoms framework (d). (Bottom) MSD plots of CO_2 (13.8 molecules/cage) within ZIF-8 modeled as previously (a) to (d); the case (d) in a neutral framework (e).

then the transport coefficient, D_t, through the thermodynamic factor. Conversion of the experimentally measured D_t to D_0 was done via the above mentioned inversed thermodynamic factor extracted from the isotherm in the flexible explicitly modeled ZIF-8 model

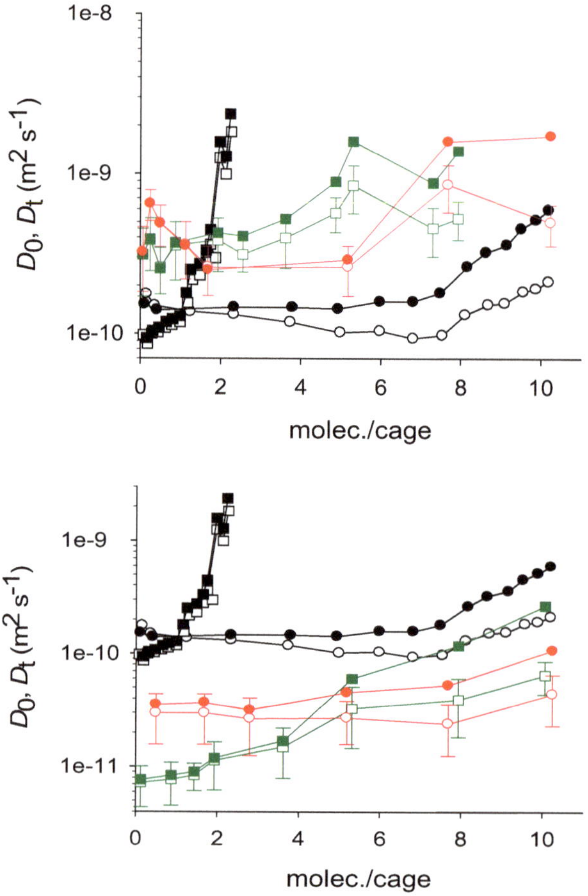

Figure 5.14 (Top) Collective diffusion coefficients from isochoric–isothermal MD (open symbols: D_0; filled symbols: D_t) as a function of CH_4 (green squares) and CO_2 (red circles) occupancies. (Bottom) Same as top graphs from isobaric–isothermal MD. IRM measurements in a single ZIF-8 crystal at 298 K (black symbols), following the same symbol notation as in MD plots, appear in both graphs for the sake of comparison.

(see Eq. 5.12). The results of this simultaneous study are presented in Fig. 5.14. Both collective coefficients predicted by MD are subject to inherent poor statistics as indicated from the error bars in the M-S diffusivity, D_0, (the propagated error on D_t is omitted for clarity reasons). The evolution of these quantities as the guest concentration increases exhibits a different quantitative trend depending on the statistical ensemble used; in agreement with self-diffusivity, the smaller transport rates computed via the isobaric–isothermal MD reflect volume fluctuations about a slightly smaller mean as mentioned in the preceding sections.

However, performing the computer experiments by means of the previously discussed averaging procedure, namely, combining properly adjustment of the fugacity of species and the pressure of the system through isobaric–isothermal MD, less noisy diffusivity curves are obtained being in satisfactory qualitative agreement with the IRM measurements. Also, being in line with the thermodynamics predictions, up to about 3 and 4 molecules per cage of CO_2 and CH_4 respectively (Henry's low regime), equal values for both the self and collective diffusivities as expected from theory are observed, thus showing self-consistency of the simulation results.

In Fig. 5.15, MD results in ZIF-8 explicitly modeled and bearing a link group released from the torsion potential (FLEX1-EM, see Sections 1.2.1 and 1.3.1.2) are presented to simulate the effect of the mobility of the organic group on the collective transport coefficients. The MD runs were carried out under isochoric conditions in an attempt to reveal solely the role of ligand excluding possible competitive effects on the hexagonal rings due to volume contraction. The results show considerable augmentation of the diffusivity values for both sorbates with respect to the cases where the mentioned above dihedral angle is hindered (see plots in Fig. 5.13), by means of FLEX2-EM and FLEX1 models. Moreover, for methane it is demonstrated that quasi-free motion of the organic link gives rise to a steeper transition that occurs at earlier occupancies, in agreement with the IRM measurements (cf. Figs. 5.14 and 5.15). That is to say, application of a nonzero torsion potential to the ligand makes methane molecules remain behind the windows until the chemical potential increases at some higher value of pressure, so that eventually to overcome the free-energy barrier and land to the next cage. Whereas, the limiting case of a quasi-freely moving link (weaker hindrance, hence, lower free energy at the window, which means lower cavity-to-window free-energy barrier), causes transition at an earlier stage, namely, at lower chemical potentials.

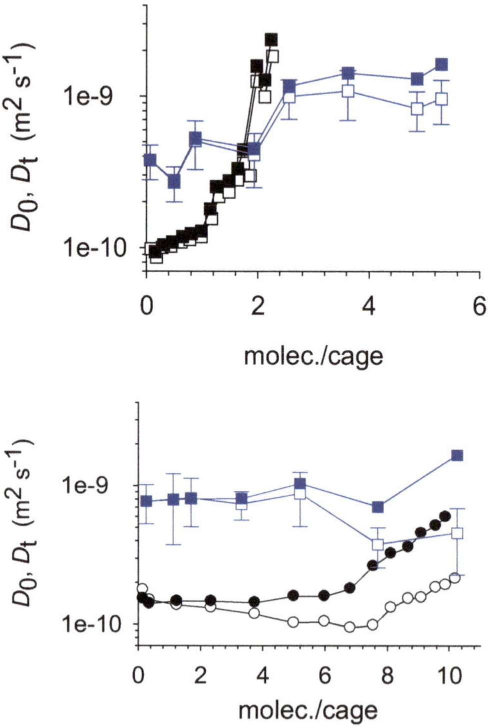

Figure 5.15 Predicted collective diffusivities (blue squares) of CH_4 (top) and CO_2 (bottom) as a function of loading in a flexible ZIF-8 model with its bridging ligand group (2-methylimidazolate) capable of quasi-free motion (cf. Fig. 5.14) where the ligand is subject to nonzero torsion potential), versus IRM measurements.

In essence, it is seen that the jumping rate of CH_4 molecules is lower in the beginning; as the loading of CO_2 increases further, sites at the cage interior start filling without significant change of its probability density at the windows, thus, diffusivity remains almost constant over a loading range up to about 7 molecules per cage, until it finally starts rising, probably because molecular congestion around the openings lowers further the aforementioned free-energy barriers (see Figs. 5.14 and 5.15).

It must be noted here that the probability density, $p(z)$, can be expressed as a function of the free-energy profile, $A(z)$, along, say, coordinate z, through the relation

$$A(z) = -k_{\mathrm{B}}T \ln[p(z)] \tag{5.19}$$

On the other hand, the density of the nonpolar CH_4 develops gradually retaining its relative ratio between the sites in the cage interior and the window areas throughout the employed loading as found in a previous work [26]. Therefore, CH_4 molecules experience higher cage-to-cage free-energy barriers in the beginning, which gradually decrease as the particular sites start being populated, thus making methane diffusivity rise monotonically with loading (see Figs. 5.14 and 5.15). This fall of the free-energy barrier due to methane probability density evolution around the windows, obviously incorporates the synergic influence of the steric hindrance caused by the bridging organic ligands. In other words, the lowering of the "stiffness" of the gate-like organic link, as seen in Fig. 5.15, tends to accelerate substantially the diffusivity transition to higher values when molecular dimensions approach or exceed the aperture width, like methane. On the other hand, the effect on the transport rate of carbon dioxide exerted by said link brings about a much smoother transition.

The conclusions discussed above are in line with the pictorial result for the singlet probability density, $\rho^{(1)}(\mathbf{r}_1)$, presented in Fig. 5.8, which depicts the CO_2 molecules to occupy primarily sites close to the double bonds of the imidazolate rings forming the hexagonal windows; therefore, $\rho^{(1)}(\mathbf{r}_1)$ is higher at those positions (smaller free-energy barrier), compared to CH_4 molecules at the same area.

In Fig. 5.16, the computed hydrogen self-diffusivity as a function of concentration within the two ZIFs, being modeled as rigid and flexible crystals is presented, along with a few QENS experimental measurements for the sake of comparison. The appeared trend of insignificant variation of diffusivity values as concentration increases has also been observed in the bulkier and weakly adsorbed ethane at 298 K in ZIF-8, showing an analogous to the hydrogen-D_s curve at the same temperature, remaining almost independent of the concentration up to high loadings, whereas the D_s of the polar methanol strongly bound to the host material exhibited a more pronounced loading dependence [56].

From the modeling perspective, contrary to the sorption equilibria, the hydrogen MD computations clearly indicate that the mobility

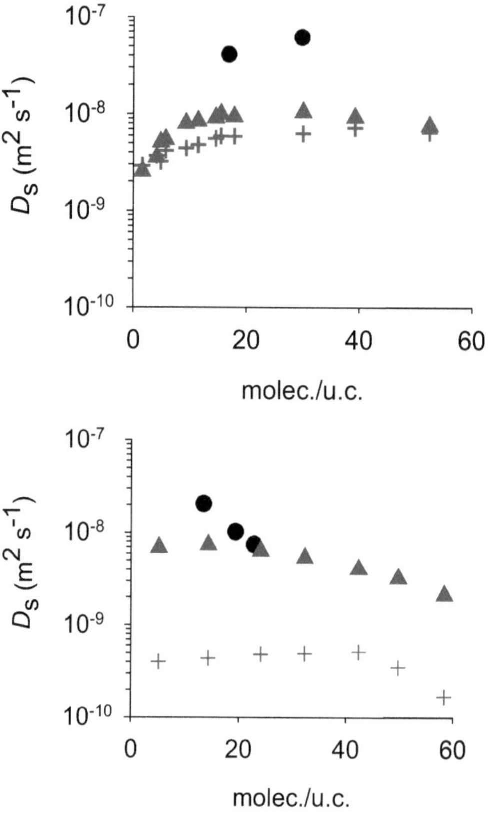

Figure 5.16 Computed hydrogen self-diffusivity values compared with QENS measurements (circles) at 77 K for various loadings in ZIF-3 (top) and ZIF-8 (bottom) in a rigid (crosses) and a FLEX1 (triangles) framework model.

of the framework influences substantially the predicted diffusivity results; thus, the calculated self-diffusivity in the mobile host matrix becomes larger, approaching the measured values, as shown in the graphs of Fig. 5.16. Moreover, the choice of the framework modeling type appears particularly important in ZIF-8 compared to ZIF-3, probably as a consequence of the tighter fitting that the hydrogen molecule experiences while passing through the ZIF-8 hexagonal windows, which, along with the hindrance due to the surrounding methyl groups, constitute the main free-energy barrier for the cavity-to-cavity molecular jumps.

Concerning the macroscopic aspect of diffusivity it should be reminded here that a concentration gradient applied to an anisotropic crystalline material in any direction (coordinate system) relative to the unit cell axes, must elicit a flux vector that remains invariant to any coordinate system transformation. Therefore, application of the Fick's phenomenological equation to the ZIF-3 cell, respecting its tetragonal symmetry arguments, namely, the existence of two equivalent unit vectors lying on a plane with the third different unit vector at a right angle to that plane, can easily prove that the diffusion tensor, $\mathbf{D} = [D_{ij}]$, contains two independent elements with its off-diagonal elements vanishing; that is to say, $D_{11} = D_{22} \neq D_{33}$ and $D_{ij} = 0$ for $i \neq j$, where the indices $i, j = 1, 2, 3$ denote the x, y, and z axes, respectively. The diffusion tensor of ZIF-8, on the other hand, is a symmetric tensor with $D_{11} = D_{22} = D_{33}$, with its off-diagonal elements vanishing, as stated by the cubic symmetry group operations.

The MSD plots presented in Fig. 5.17 for three selected loadings obtained from Fig. 5.16 reflect the different influence of the hydrogen concentration, ρ, in ZIF-3, on the computed elements D_{11}, D_{22}, and D_{33} of the tensor, $\mathbf{D}(\rho)$; they were calculated from the slopes of the presented lines produced by postprocessing the MD trajectories after a sufficiently long time as imposed by the Einstein equation. In these plots is shown the D_{11} and D_{22} elements to be equally evolving over the whole loading range used in the simulation, as imposed by the crystal symmetry. In particular, an increase in their values from low to moderate hydrogen densities is observed, and as the density increases further their values become lower. Elaboration of the z axis component of the mean squared displacement of the hydrogen molecules for correlation times over 10 ns, as was done for the other axes, provided the D_{33}, which slightly diverges from the two other diagonal elements within the high dilution regime and then rises steeply as the hydrogen density increases.

Comparison of the two upper plots of Fig. 5.17 may explain the appeared maximum of self-diffusivity curve up to about 30

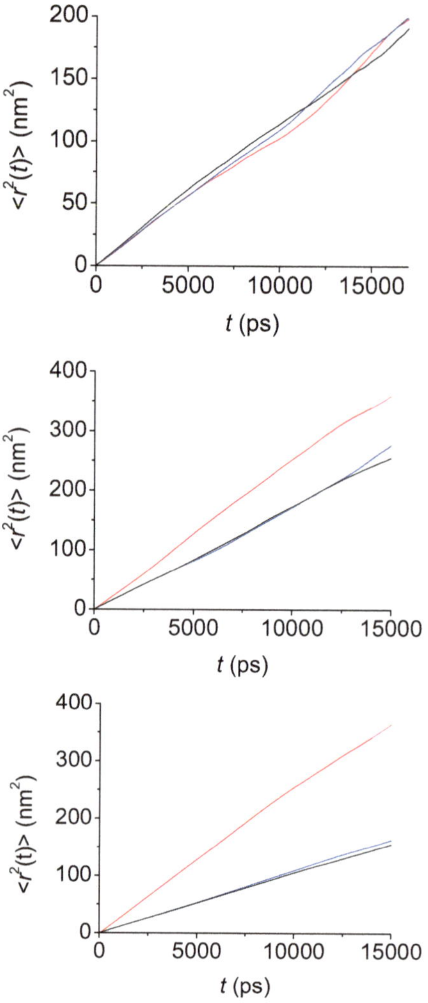

Figure 5.17 MSD plots of H_2 at 77 K within ZIF-3 for (a) 5.75, (b) 15.5, and (c) 52.5 molecules/unit cell along the *x* (black line), *y* (blue line), and *z* axes (red line).

molecules H_2 per unit cell observed from QENS and simulation, as can be seen in Fig. 5.16. Therefore, the initial increase of the hydrogen diffusivity in ZIF-3, may be attributed to the drastic rise of D_{33}, as a result of the gradually increasing mobility along the

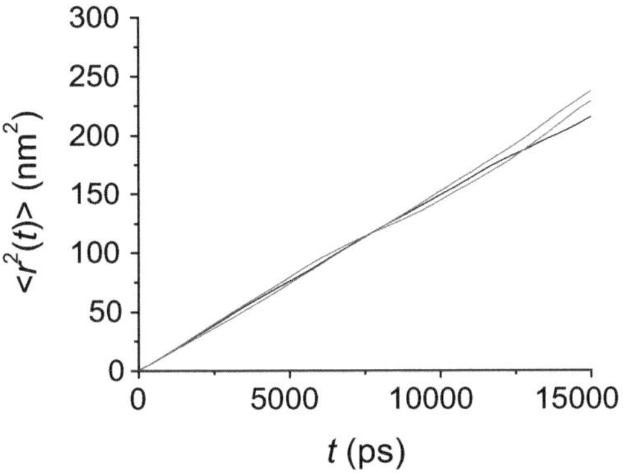

Figure 5.18 Same as Fig. 5.14 for ZIF-8 for a H_2 loading of 14.38 molecules/unit cell at 77 K.

z axis pore channel, whereas as shown in the same plots, a tiny contribution to the overall flux vector from the other two pore channels at this loading range is observed. Then, as the hydrogen population grows further the interparticle collisions cause a gradual fall of the molecular mobility.

Also, these conclusions are in accord with the pictorial result for the singlet probability density, $\rho^{(1)}(\mathbf{r}_1)$, calculations presented in Fig. 5.11, where the evolution of the probability density distribution as the hydrogen loading rises within ZIF-3, shows that the areas close to the ligands in the interior of the three channel systems constitute primary adsorption sites over the whole occupancy range. It is also observed that the vicinity of the intersections become progressively populated until these regions of the framework practically tend to reduce the communication between the two pore systems, thus verifying the findings of the trajectory analysis via the MSD plots that show the overall mass flow at high loadings to be undertaken mostly through the z axis pore channel, [1–10], (see Fig. 5.11).

In ZIF-8, the rise in the sorbed phase concentration causes a uniform evolution of the three elements of the diffusion tensor presented in Fig. 5.18, in accord with the theory, making eventually their orientation average lower smoothly as concentration increases further (see Fig. 5.16).

5.4 Conclusions

In this chapter a statistical mechanics-based modeling of the sorption thermodynamics and dynamics of selected guest sorbate molecules in PCPs is presented; the methodology has been constructed in a manner that allows strict thermodynamic consistency between the computer experiments and the experimentally measured homologous quantities.

As a paradigm of this methodology, the sorption and kinetics properties of Ar, CH_4, and H_2 in ZIF-3 and ZIF-8 over a wide range of occupancies are presented. Secondly, we investigated the mechanism of guest molecular siting and motion under the complicated metal–organic structures; then, the effect of hindrance phenomena exerted by the motion of organic ligands on the guest mobility in the two digitized unit cells was examined on the basis of various modeling options.

The partial molar configurational internal energy was calculated for Ar, CH_4, and H_2 as a function of their loading in the above frameworks via GCMC simulations. Because the computed quantity is related to the covariance of the potential energy and number of particles at a certain loading and temperature, it can map the strength of sorbate–sorbate and sorbate–sorbent interactions over a wide guest loading range. The results showed a higher perception of energetic inhomogeneity of argon in comparison with methane and hydrogen.

One- and two-point probability density functions calculated by postprocessing the equilibrated configurations produced during the Monte Carlo experiments provide further insight into the distribution of adsorption centers inside the ZIF-8 network. Moreover, interpretation of these results on the basis of free-energy distribution offers a plausible explanation of the mobility of the sorbed phase as a function of the chemical potential inside these PCPs. Although for hydrogen and argon the agreement between simulated and measured isotherms is satisfactory, the freezing of the degrees of freedom of the organic links in one of the tested flexible options of the ZIF-8 model may cause the siting transition step of the predicted isotherm of argon to appear earlier than the one observed experimentally. It is also shown that such inflections observed in sorption isotherms can be related to sorbate surpassing over free-energy barriers located at positions near the six-ring window faces; therefore, the bonding flexibility at these areas influences these barriers and hence the dynamics of guest molecules even at low temperatures. The molecular mobility of Ar at 87 K and H_2 at 77 K inside these metal–organic sorbents was estimated by virtue of equilibrium MD experiments in the canonical ensemble. Also, we observed the displacement of argon at 87 K over a broad range of occupancies in conjunction with the preceding thermodynamics findings, in order to understand the kinetics behavior of the sorbed phase with respect to the calculated spatial partitioning of the sorbate probability density.

In view of the quantum nature of hydrogen at those temperatures, the path-integral formulation was employed for the quantum mechanical description of the intermolecular interactions, which leads to the quadratic Taylor expanded pairwise potential function as follows from the temperature-dependent potential energy function of Feynman and Hibbs; their work states that paths located at a certain state point in the configurational space are enveloped inside a Gaussian width. For low temperatures and atomic masses, the distribution around this position becomes widespread; hence a classical description is not adequate any more. The self-diffusivity of hydrogen in ZIF-8 at 77 K, as revealed by the MD simulations, showed a marked difference between the values obtained from the classical and quantum mechanical descriptions of the energetics. The predicted self-diffusivity presents a shallow maximum as a function of occupancy due to the loading-dependent free-energy barrier that separates the adsorption sites located close to the hexagonal windows and around the cavity center of the ZIF-8 unit cell.

At this point we should stress that the rigidity of the frameworks should be seriously considered when simulation is carried out at high temperatures or/and the diffusing guests possess a molecular size commensurate with the effective size of critical paths for diffusion within the PCP network. In the presented study these issues are mainly associated with the mobility of the pendant organic group of the imidazolate-type link, oftentimes acting as "saloon doors" controlling the guest transport. The analysis about the diffusivity tensor concentration dependence relies on the density profiles and hence the free-energy evolution of the guest sorbed phase in the anisotropic interiors of ZIFs. On the absence of experimental data from specialized techniques able to probe the spatial variation of mass transport, such as interference microscopy, only simulation can provide such information. However, phenomena that may be related to the structural topology of these anisotropic metal–organic materials that are likely to affect the resistance to mass transport at various directions within the host matrix, as can be found, for instance, in some zeolite crystals, do deserve further investigation.

References

1. Yaghi, O. M., Li, G., Li, H. (1995). *Nature*, **378**, 703.
2. Horcajada, P., Serre, C., Vallet-Regi, M., Sebban, M., Taulelle, F., Férey, G. (2006). *Angew. Chem., Int. Ed.*, **45**, 5974.
3. Wilmer, Ch. H., Leaf, M., Lee, C., Farha, O., Hauser, B., Hupp, J. T., Snurr, R. Q. (2012). *Nat. Chem.*, **4**, 83.
4. Li, H., Eddaoudi, M., Groy, T. L., Yaghi, O. M. (1998). *J. Am. Chem. Soc.*, **120**, 8571.
5. Yaghi, O. M., O'Keeffe, M., Ockwig, N. W., Chae, H. K., Eddaoudi, M., Kim, J. (2003). *Nature*, **423**, 705.
6. Chae, H. K., Siberio-Pérez, D. Y., Kim, J., Go, Y., Eddaoudi, M., Matzger, A. J., O'Keeffe, M., Yaghi, O. M. (2004). *Nature*, **427**, 523.
7. Côté, A. P., Benin, A. I., Ockwig, N. W., Matzger, A. J., O'Keeffe, M., Yaghi, O. M. (2005). *Science*, **310**, 1166.
8. Park, K. S., Côté A. P., Choi J. Y., Huang R., Uribe-Romo F. J., Chae H. K, O'Keeffe M., Yaghi O. M. (2006). *PNAS*, **103**, 10186.
9. Hayashi, H., Côté, A. P., Furukawa, H., O'Keeffe, M., Yaghi, O. M, (2007). *Nat. Mater.*, **6**, 501.
10. Banerjee, R., Phan, A., Wang, B., Knobler, C., Furukawa, H., O'Keeffe, M., Yaghi, O. M. (2008). *Science*, **319**, 939.
11. Düren, T., Sarkisov, L., Yaghi, O. M., Snurr, R. Q. (2004). *Langmuir*, **20**, 2683.
12. Dubbeldam, D., Frost, H., Walton, K. S., Snurr, R. Q. (2007). *Fluid Phase Equil.*, **261**, 152.

13. Frost, H., Snurr, R. Q. (2007). *J. Phys. Chem. C*, **111**, 18794.
14. Walton, K. S., Millward, A. R., Dubbeldam, D., Frost, H., Low, J. J., Yaghi, O. M., Snurr, R. Q. (2008). *J. Am. Chem. Soc.*, **130**, 406.
15. Garberoglio, G. (2007). *Langmuir*, **23**, 12154.
16. Han, S. S., Furukawa, H., Yaghi, O. M., Goddard, W. A. (2008). *J. Am. Chem. Soc.*, **130**, 11580.
17. Furukawa, H., Yaghi, O. M. (2009). *J. Am. Chem. Soc.*, **25**, 8876.
18. Greathouse, J. A., Kinnibrugh, T. L., Allendorf M. D. (2009). *Ind. Eng. Chem. Res.*, **48**, 3425.
19. Babarao, R., Jiang, J. W., Sandler, S. I. (2009). *Langmuir*, **25**, 5239.
20. Huang, X. C., Lin, Y. Y., Zhang, J. P., Chen, X. M. (2006). *Angew. Chem., Int. Ed.*, **45**, 1557.
21. Zhou, W., Wu, H., Hartman, M. R., Yildirim, T. (2007). *J. Phys. Chem. C*, **111**, 16131.
22. Wang, B., Côté, A. P., Furukawa, H., O'Keeffe, M., Yaghi, O. M. (2008). *Nature*, **453**, 207.
23. Morris, W., Doonan, Ch. J., Furukawa, H., Banerjee, R., Yaghi, O. M, (2008). *J. Am. Chem. Soc.*, **130**, 12626.
24. Liu, D., Zheng, Ch., Yang, Q., Zhong, Ch. (2009). *J. Phys. Chem. C*, **113**, 5004.
25. Zhou, M., Wang, Q., Zhang, L., Liu, Y., Kang, Y. (2009). *J. Phys. Chem. B*, **113**, 11049.
26. Pantatosaki, E., Pazzona, F. G., Megariotis, G., Papadopoulos, G. K. (2010). *J. Phys. Chem. B*, **114**, 2493.
27. Pantatosaki, E., Megariotis, G., Pusch, A.-K., Chmelik, C., Stallmach, F., Papadopoulos, G. K. (2012). *J. Phys. Chem. C*, **116**, 201.
28. Yaghi's group resynthesized (see Ref. [8]) the ZIF-8 using a more efficient synthesis route than the one originally reported in Ref. [20].
29. Pantatosaki, E., Kolokolov, D. I., Karmakar, S., Biniwale, R., Jobic, H., Papadopoulos, G. K. (2013). *J. Chem. Phys.*, **138**, 034706.
30. Mayo, S. L., Olafson, B. D., Goddard III, W. A. (1990). *J. Phys. Chem.*, **94**, 8897.
31. Rappé, A. K., Casewit, C. J., Colwell, K. S., Goddard III, W. A., Skiff, W. M. (1992). *J. Am. Chem. Soc.*, **114**, 10024.
32. McDaniel, J. G., Schmidt, J. R. (2012). *J. Phys. Chem. C*, **116**, 14031.
33. Amirjalayer, S., Tafipolsky, M., Schmid, R. (2007). *Angew. Chem., Int. Ed.*, **46**, 463.
34. Greathouse, J. A., Allendorf, M. D. (2008). *J. Phys. Chem. C*, **112**, 5795.

35. Pantatosaki, E., Papadopoulos, G. K. (2007). *J. Chem. Phys.*, **127**, 164723.
36. Papadopoulos, G. K., Theodorou, D. N. (2009). *Mol. Simul.*, **35**, 79.
37. Seehamart, K., Nanok, T., Krishna, R., van Baten, J.-M., Remsungnen, T., Fritzsche, S. (2009). *Microporous Mesoporous Mater.*, **125**, 97.
38. Feynman, R. P., Hibbs, A. R. (1965). *Quantum Mechanics and Path Integrals* (McGraw-Hill, New York).
39. Papadopoulos, G. K., Theodorou, D. N. (2008). Computer simulation of sorption and tramsport in zeolites, in *Handbook of Heterogeneous Catalysis*, eds. Ertl, G. Knözinger, Schuth, F., Weitkamp, J. (Wiley-VCH, Verlag GmbH & Co. KGaA), 1662–1676.
40. (a) International Tables for Crystallography (2005). *Volume A: Space-Group Symmetry*, 5th Ed., ed. Hahn, T. (Springer, New York); (b) http://www.iza-structure.org/databases/.
41. Zhao, Y., Jin, J., Kwon, Y.-U. (2005). *Bull. Korean Chem. Soc.*, **26**, 1277.
42. Kärger, J., Ruthven, D., Theodorou, D. N. (2012). *Diffusion in Nanoporous Materials*, 1st Ed. (Wiley-VCH, Weinheim).
43. Kolokathis, P., Pantatosaki, E., Gatsiou, C.-A., Jobic, H., Theodorou, D., Papadopoulos, G. K. (2014). *Mol. Simul.*, **40**, 80.
44. Chempath, S., Clark, L. A., Snurr, R. Q. (2003). *J. Chem. Phys.*, **118**, 7635.
45. Vlugt, T. J. H., Schenk M. (2002). *J. Phys. Chem. B*, **106**, 12757.
46. Guse, M. P., Kunz, A. B. (1975). *Phys. Status Solidi*, **71**, 631.
47. http://webbook.nist.gov/chemistry/.
48. Pérez-Pellitero J., Amrouche H., Siperstein F. R., Pirngruber G., Nieto-Draghi C., Chaplais G., Simon-Masseron A., Bazer-Bachi D., Peralta D., Bats N. (2010). *Chem. Eur. J.*, **16**, 1560.
49. Kowalczyk, P., Gauden, P. A., Terzyk, A. P., Pantatosaki, E., Papadopoulos, G. K., (2013). *J. Chem. Theory Comput.*, **9**, 2922.
50. Heffelfinger, G. S., van Swol, F. J. (1994). *Chem. Phys.*, **100**, 7548.
51. Papadopoulos, G. K. (2005). *Mol. Simul.*, **31**, 57.
52. Arya, G., Chang, H. C., Maginn, E. J. (2001). *J. Chem. Phys.*, **115**, 8112.
53. Reed, D. A., Ehrlich, G. (1981). *Surf. Sci.*, **102**, 588.
54. Seehamart, K., Nanok, T., Kärger, J., Chmelik, C., Krishna, R., Fritzsche, S. (2010). *Microporous Mesoporous Mater.*, **130**, 92.
55. Hertag, L., Bux, H., Caro, J., Chmelik, C., Remsungnen, T., Knauth, M.,Fritzsche, S. (2011). *J. Membr. Sci.*, **377**, 36.
56. Chmelik, C., Bux, H., Caro, J., Heinke, L., Hibbe, F., Titze, T., Kärger, J. (2010). *Phys. Rev. Lett.*, **104**, 085902.

Chapter 6

Modeling Quantum Effects on Adsorption and Diffusion of Hydrogen in Metal–Organic Frameworks

Giovanni Garberoglio
Interdisciplinary Laboratory for Computational Science (LISC), Fondazione Bruno Kessler and University of Trento, via Sommarive 18, I-38123 Povo (TN), Italy
garberoglio@fbk.eu

6.1 The Hydrogen Molecule and Its Isotopologues

H_2 is the simplest molecule, yet a proper understanding of its properties requires the application of all the principles of quantum mechanics. The chemical bond between two hydrogen atoms is determined by the electronic wavefunction, which, in its ground state, is characterized by zero electronic spin. The total energy of the H_2 molecule as a function of nuclear separation is shown in Fig. 6.1. This potential energy curve has a minimum for a separation $\ell_0 = 0.741$ Å and a well depth $U_0/k_B = 55,090$ K. The nuclear degrees of freedom of the H_2 molecule are conveniently separated into the center-of-mass coordinates and the relative coordinate along of the axis joining the two atoms.

Metal–Organic Frameworks: Materials Modeling towards Potential Engineering Applications
Edited by Jianwen Jiang

ISBN 978-981-4613-45-3 (Hardcover), 978-981-4613-46-0 (eBook)
www.panstanford.com

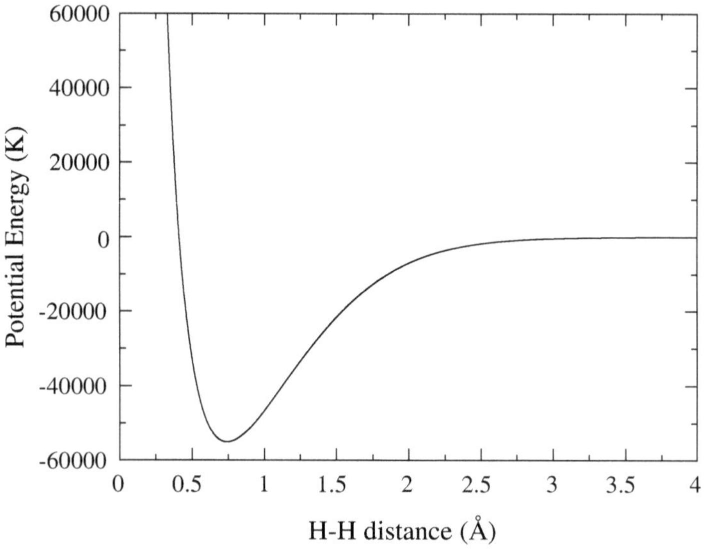

Figure 6.1 Bond-stretching potential of H_2 as a function of the bond length in Born–Oppenheimer approximation [Mielke et al. (2002)].

H_2 has six different isotopologues obtained by replacing the H (protium) atoms with the other isotopes, deuterium (D) and tritium (T). In the case of homonuclear isotopologues, the overall wavefunction must have well-defined properties upon the exchange of the two identical particles. H and T are fermions with nuclear spin $I = \frac{1}{2}$, whereas D is a boson with nuclear spin $I = 1$.

In the case of H_2 and T_2 the overall wavefunction must be odd under the exchange of the two nuclei, and therefore its spatial symmetry is correlated with the nuclear state, which can be either odd ($I_{tot} = 0$) or even ($I_{tot} = 1$) under exchange. The parity under exchange of the eigenstates of rotational angular momentum is $(-1)^L$ [Landau and Lifshitz (1997)], and therefore molecules with $I_{tot} = 0$ can have only even rotational angular momenta, while molecules with $I_{tot} = 1$ can have only odd rotational angular momentum. The former case is known as *para*-H_2, whereas the second case is known as *ortho*-H_2, with analogous definitions in the case of T_2. Due to the very slow rate of ortho–para transitions (in the absence of catalyzers), a gas of H_2 molecules behaves as if it were a mixture of these two species in proportions given by the ratio of the nuclear degeneracies [Hill (1960)].

In the case of D_2 the situation is reversed due to the bosonic nature of deuterium: molecules with even nuclear spin ($I_{tot} = 0$ or $I_{tot} = 2$) are associated with even rotational angular momenta L, whereas molecules with odd nuclear spin ($I_{tot} = 1$) are associated with odd values of L. The former case is known as *ortho*-D_2 the latter as *para*-D_2.

Assuming instantaneous equilibration of the electronic wavefunction (i.e., the usual Born–Oppenheimer approximation) the potential energy shown in Fig. 6.1 acts as a bond stretching potential among the two hydrogen atoms, or isotopes thereof. Due to the light mass of hydrogen isotopes, the motion along the molecular axis is quantized. As a consequence, the energy and bond length of the molecular ground state, as well as the energy of the excited vibrational states, depend on the nature of the hydrogen isotopes making up the molecule. The principal characteristics of the H_2 isotopologues are summarized in Table 6.1. There are two principal features worth pointing out: first of all, different isotopologues have different equilibrium bond lengths, although the difference is $\sim$1% only. Second, the energy required to excite higher vibrational states is of the order of several thousand degrees Kelvin.

As a consequence of these facts, in modeling of H_2 (or its isotopologues) for phenomena occurring at room temperature or below, it can be safely assumed that the hydrogen molecule is a rigid linear rotor, with bond length $\ell \approx 0.76$ Å.

Table 6.1 Principal properties of the H_2 isotopologues, calculated using the potential developed by [Mielke et al. (2002)].

	Ground-state bond length (Å)	Zero-point Energy (K)	First excited energy (K)
H_2	0.766	3136	5989
D_2	0.759	2205	4308
T_2	0.756	1821	3546
HD	0.763	2720	5227
HT	0.762	2566	4943
DT	0.757	2033	3947

6.1.1 *Models for the Hydrogen Molecule*

Although very precise potential-energy surfaces have been developed using ab initio quantum chemical calculations to describe the interaction between two hydrogen molecules, [Patkowski et al. (2008)] a similar approach cannot in general be adopted to calculate the interaction of H_2 with metal–organic frameworks (MOFs), although in a few cases some form of first-principles calculations have been used to shed light on the details of the interaction of hydrogen molecules with MOFs, particularly those having open metal sites (for an updated review, see [Getman et al. (2012)]).

Usually, this latter approach involves the identification of a relevant moiety within the MOF, which is then extracted and used to perform quantum chemical calculations at various level of sophistication (such as dispersion-corrected density functional theory or post-Hartree–Fock methods). The results of these investigation provide useful insights on the H_2–MOF interaction, but unfortunately ab initio calculations cannot be used directly to perform simulations aimed at characterizing the full thermodynamics of adsorption, due to their prohibitively large computational cost. This limitation is particularly severe when the quantum nature of the hydrogen molecule must be taken into account.

By far, the most effective simulations are those performed using a classical model to describe the interactions between hydrogen and MOF and between the hydrogen molecules themselves. The solid–fluid interactions are usually taken from some general-purpose classical force field, which can be refined using the results of the quantum chemical calculations described above. The DREIDING force field [Mayo et al. (1990)] and the universal force field (UFF) [Rappé et al. (1992)] have been used with considerable success in the modeling efforts so far [Garberoglio et al. (2005)]. These force fields are based on atomic-centered potentials of the Lennard–Jones (LJ) form:

$$U(r) = 4\varepsilon\left[\left(\frac{\sigma}{r}\right)^{12} - \left(\frac{\sigma}{r}\right)^{6}\right], \tag{6.1}$$

and are optionally augmented with partial charges, obtained with either charge-equilibration procedures or from ab initio calculations.

Similar models, based on LJ centers and point charges, have also been developed for the hydrogen molecule and are usually parameterized so to reproduce with high accuracy the thermodynamic properties of the pure fluid.

In general, H_2 models can be classified according to the number of interaction centers used. The simplest model has only one interaction center of the LJ form. The underlying assumption is that the rotational and translational degrees of freedom of the hydrogen molecule can be considered decoupled and H_2 approximated as a spherically symmetric object. This is usually a good assumption even for hydrogen in the liquid or solid state. However, as discussed below, this assumption fails in situations where the hydrogen molecule is strongly confined. The most used spherical LJ approximation is the one proposed by Buch and having $\varepsilon_{\mathrm{H}_2}/k_{\mathrm{B}} = 34.2$ K and $\sigma_{\mathrm{H}_2} = 2.96$ Å [Buch (1994)].

Two-center LJ models have also been developed. The most used potential is probably the one developed by Murad and Gubbins [Murad and Gubbins (1978)]: H_2 is modeled as a rigid rotor, with two atoms placed at the classical equilibrium distance ℓ_0. An LJ center with $\varepsilon_{\mathrm{H}}/k_{\mathrm{B}} = 8.6$ K and $\sigma_{\mathrm{H}} = 2.81$ Å is associated with each atom.

Recently, a five-center model for H_2 has been developed [Belof et al. (2008)]. This model uses point charges placed in suitable points along the molecular axis to reproduce the electric quadrupole moment of the hydrogen molecule.

When these models for the hydrogen molecule are used in computer simulations, the interaction terms between the solid (S) and the fluid centers (F) are usually fixed using the Lorentz–Berthelot mixing rules:

$$\varepsilon_{\mathrm{SF}} = \sqrt{\varepsilon_{\mathrm{S}}\varepsilon_{\mathrm{F}}} \tag{6.2}$$

$$\sigma_{\mathrm{SF}} = \frac{1}{2}\left(\sigma_{\mathrm{S}} + \sigma_{\mathrm{F}}\right), \tag{6.3}$$

where typical values of σ_{S} and $\varepsilon_{\mathrm{S}}/k_{\mathrm{B}}$ lie in the range of 3–4 Å and 30–200 K, respectively.

6.2 Computer Simulations of Quantum Statistical Mechanics

The quantum statistical mechanics of a system of N particles at temperature T in a fixed volume V is determined by the canonical partition function

$$Z(N, V, T) = \frac{1}{N!} \sum_{n} \sum_{\pi} \langle n | e^{-\beta H} \mathcal{P}_{\pi} | n \rangle, \tag{6.4}$$

where H is the Hamiltonian of the system, $\beta = (k_{\mathrm{B}} T)^{-1}$, and the states $|n\rangle$ are a complete basis set in the N-particle Hilbert space. In Eq. (6.4) π runs over all the possible $N!$ permutations of N objects, and $\mathcal{P}_{\pi}$ is the corresponding quantum mechanical operator, weighted with the sign of the permutation in the case of fermions. In what follows, we will not be concerned with exchange effects in hydrogen molecules, and therefore the sum over π in Eq. (6.4) will be restricted to the identity permutation only.

In the case of free particles, when the Hamiltonian H reduces to the kinetic energy K, Eq. (6.4) can be evaluated analytically. In particular, using the eigenstates $|\mathbf{r}\rangle$ of the position operator as a complete basis set, one can show that

$$\langle \mathbf{r}' | e^{-\beta K} | \mathbf{r} \rangle = \frac{1}{\Lambda^3} \exp\left[-\frac{\pi}{\Lambda^2} |\mathbf{r} - \mathbf{r}'|^2 \right], \tag{6.5}$$

where

$$\Lambda = h / \sqrt{2\pi M k_{\mathrm{B}} T} \tag{6.6}$$

is known as the thermal de Broglie wavelength. Consequently, the partition function becomes $Z_1 = V/\Lambda^3$, which gives the physical interpretation of Λ as the average size of quantum fluctuations for a particle of mass M at temperature T. In the case of H_2, one has that $\Lambda(300\ \mathrm{K}) = 0.7$ Å, which is a sizable fraction of the LJ diameter $\sigma_{\mathrm{H_2}} = 2.98$ Å. At $T = 77$ K one obtains $\Lambda = 1.41$ Å, which is almost one half of the hard-core diameter. These estimates show that quantum diffraction effects cannot in general be neglected in the case of hydrogen molecules, even at room temperature.

Finally, we notice that $\Lambda(15\ \mathrm{K})/\sigma_{\mathrm{H_2}} \sim 1$. For temperatures $T \lessapprox 15$ K one might therefore expect a sizable superposition of the single-particle wavefunctions. In this regime, exchange effects cannot be neglected and the bosonic or fermionic nature of the molecules becomes apparent. In this case all the permutations appearing in Eq. (6.4) should be taken into account.

6.2.1 *Diagonalization of the Hamiltonian*

In principle, all the thermodynamic observables of a system of quantum particles can be obtained from the knowledge of the partition function Eq. (6.4). For example, the average energy is given by

$$E(N, V, T) = -\frac{\partial \ln Z}{\partial \beta} = \frac{1}{Z}\sum_n \langle n|He^{-\beta H}|n\rangle = \frac{1}{Z}\sum_\alpha E_\alpha e^{-\beta E_\alpha} \tag{6.7}$$

where the last equality has been obtained by taking as $|n\rangle$ the eigenstates $|\alpha\rangle$ of the Hamiltonian, given by $H|\alpha\rangle = E_\alpha|\alpha\rangle$.

In practice, the diagonalization of the Hamiltonian can be performed only in a few cases. For example, this technique has been used to calculate the states of an H_2 molecule, approximated as a rigid rotor, confined in carbon nanotubes [Lu et al. (2003, 2006)] and in graphitic slit pores [Garberoglio (2009b)]. In these calculations, the confining region has been assumed as possessing cylindrical or translational symmetry, allowing a considerable reduction of the size of the numerical representation of the Hamiltonian.

In the case of adsorption within a MOF, the memory requirements for the study of adsorption of just a single rigid rotor are quite demanding. Let us consider for example hydrogen adsorption in the prototypical MOF-5 (also known as isoreticular metal–organic framework [IRMOF-1]). The unit cell of this substance is cubic, with a side length of 25.7 Å. Setting as target a spatial resolution of $l = 0.1$ Å, and considering only the $L = 0$ and $L = 2$ rotational states of the molecule, one would have to evaluate matrices with a typical size of $\sim 10^8 \times 10^8$. Although these requirements could probably be made much smaller by a judicious use of symmetry and by introducing a suitably chosen basis set, they remain quite substantial. In any case, this reduction of the computational effort would require a thorough analysis to be performed in any specific case. To the best of my knowledge, no one has ever attempted this kind of calculation for MOFs.[a]

However, even assuming that a single-particle calculation could be carried out, one is generally interested in having many hydrogen molecules (tens or even hundreds) within the unit cell of the system. The memory requirements of this approach grow exponentially with the number of molecules, thus making it quite unfeasible to study many-body phenomena.

6.2.2 *The Path-Integral Approach*

The path-integral method has been developed to overcome the difficulties arising from the direct application of quantum mechanical methods detailed above. This approach, combined with the use of Monte Carlo–based sampling procedures, concentrates the computational effort into considering only those states that are more likely to contribute to averages such as Eq. (6.7).

The path-integral approach to quantum statistical mechanics [Feynman and Hibbs (2010)] is a formal rewriting of Eq. (6.4) where the sum over a complete set of quantum states ($\sum_n$) is replaced by an integral over all the possible closed paths covered by particles in three-dimensional space and the expectation value of the Boltzmann factor ($\langle n|e^{-\beta H}|n\rangle$) is replaced by the Boltzmann factor of a suitably defined action integral to be evaluated on each of the paths.

This equivalence is best seen by using the Trotter formula, which states that for any pair of operators A and B one has

$$e^{A+B} = \lim_{P\to\infty} \left(e^{A/P} e^{B/P}\right)^P . \tag{6.8}$$

The path-integral formulation of quantum statistical mechanics is then obtained by inserting Eq. (6.8) in Eq. (6.4). For the sake of simplicity we will illustrate the derivation in the case of a single particle in an external potential, using the eigenfunctions of the position operator as the complete basis set. In this case the operators A and B is Eq. (6.8) correspond to $-\beta K = \hbar^2\nabla^2/(2Mk_\text{B}T)$

[a] *Note added in proof.* A clever way to perform the diagonalization of the Hamiltonian describing a rigid H_2 molecule in an MOF has been described in [Matanovic et al. (2012)].

and $-\beta U(\mathbf{r})$, respectively. Substituting Eq. (6.8) in Eq. (6.4) and inserting a series of completeness relations

$$1 = \int d^3\mathbf{r}_i \, |\mathbf{r}_i\rangle\langle\mathbf{r}_i|$$

between each of the factors defined on the right-hand side of Eq. (6.8), one obtains

$$Z = \lim_{P\to\infty} \int \prod_{i=1}^{P} d^3\mathbf{r}_i \, \langle\mathbf{r}_i|e^{-\beta K/P}|\mathbf{r}_{i+1}\rangle \, \exp\left(-\frac{\beta}{P}\sum_{k=1}^{P} U(\mathbf{r}_k)\right), \quad (6.9)$$

where we have defined $\mathbf{r}_{P+1} = \mathbf{r}_1$. The matrix elements of the Boltzmann factors of the kinetic energy can be evaluated using Eq. (6.5), giving

$$\langle\mathbf{r}_i|e^{-\beta K/P}|\mathbf{r}_{i+1}\rangle = \frac{P^{3/2}}{\Lambda^3} \exp\left[-\frac{\pi P}{\Lambda^2}|\mathbf{r}_i - \mathbf{r}_{i+1}|^2\right],$$

and, therefore,

$$Z = \lim_{P\to\infty} \int \frac{P^{3P/2}}{\Lambda^{3P}} \prod_{i=1}^{P} d^3\mathbf{r}_i \, \exp\left[-\beta\sum_{k=1}^{P}\frac{MP\,|\mathbf{r}_{k+1} - \mathbf{r}_k|^2}{2\hbar^2\beta^2} + \frac{U(\mathbf{r}_k)}{P}\right]. \quad (6.10)$$

Equation (6.10) is the principal result of this chapter, as well as the starting point to all the numerical implementations of the path-integral approach. It can be interpreted as the partition function of a classical system: the original quantum particle, described by the coordinate $\mathbf{r}$, has been mapped into a classical polymer made by P beads, the coordinates of which are the new $\mathbf{r}_k$ variables. The kth bead interacts only with the nearest neighbors along the polymer backbone (i.e., those denoted by the indices $k-1$ and $k+1$) with a harmonic potential having a spring constant $K = MP/(\hbar\beta)^2$. The polymer is closed onto itself (because of the condition $\mathbf{r}_{P+1} = \mathbf{r}_1$), and the interaction of each of the P beads with the external potential is rescaled by a factor $1/P$.

The generalization to the case of N interacting particles is straightforward [Landau and Binder (2005)]. One ring polymer is associated to each of the original quantum particles and each of the beads of any ring polymer—the kth, say—interacts only with the kth bead of all the other polymers with the original interparticle interaction rescaled by a factor of P.

Finally, let us notice that, defining $d\tau = \beta\hbar/P$, the expression in the exponential on the right-hand side of Eq. (6.10) is actually the discretized form of the integral

$$\mathcal{A}[\mathbf{r}(\tau)] = -\frac{1}{\hbar}\int_0^{\beta\hbar} \frac{M}{2}\left|\frac{d\mathbf{r}(\tau)}{d\tau}\right|^2 + U(\mathbf{r}(\tau))\, d\tau \tag{6.11}$$

where $\mathbf{r}(0) = \mathbf{r}(\beta\tau)$ due to the ring-polymer condition. In the $P \to \infty$ limit, the ring polymers are represented by continuous paths $\mathbf{r}(\tau)$, whence the name of the method. One can define a measure in the space of paths

$$\mathcal{D}\mathbf{r}(t) = \lim_{P\to\infty} \frac{P^{3P/2}}{\Lambda^{3P}} \prod_{i=1}^{P} d^3\mathbf{r}_i, \tag{6.12}$$

so that the partition function becomes

$$Z = \int \mathcal{D}\mathbf{r}(\tau) \exp\left[-\frac{1}{\hbar}\int_0^{\beta\hbar} \frac{M}{2}\left|\frac{d\mathbf{r}(\tau)}{d\tau}\right|^2 + U(\mathbf{r}(\tau))\, d\tau\right], \tag{6.13}$$

which has a straightforward interpretation as an integral over paths, each of which is weighted by the exponential of the action $\mathcal{A}$ defined in Eq. (6.11).

6.2.3 *Probability Distribution for the Ring-Polymer Configurations*

The free-particle case of Eq. (6.10) (i.e., when $U(\mathbf{r}) = 0$) is particularly instructive, and we will now discuss it in some detail. This analysis will provide an expression for the probability of a ring-polymer configuration, P_{RP}, that will be widely used in the following applications.

In the case of a free particle, one of the coordinates, $\mathbf{r}_1$ say, can be used to identify the location of one of the beads within the integration volume. The other $P-1$ coordinates can be chosen as the relative position of one bead with respect to the previous one, that is $\Delta\mathbf{r}_k = \mathbf{r}_{k+1} - \mathbf{r}_k$. Notice that these coordinates describe the internal configuration of the ring polymer and that the condition of closeness implies

$$\Delta\mathbf{r}_P = -\sum_{k=1}^{P-1} \Delta\mathbf{r}_k. \tag{6.14}$$

We note that Eq. (6.10) in the free-particle case implies that P_{RP} is proportional to

$$P_{\mathrm{RP}} \propto \exp\left[-\beta \sum_{k=1}^{P} \frac{MP}{2\hbar^2\beta^2}|\Delta\mathbf{r}_k|^2\right], \tag{6.15}$$

which is a standard Gaussian probability distribution. Each of the Cartesian coordinates of $\Delta\mathbf{r}_k$ is distributed according to

$$\exp\left(-\frac{1}{2}\sum_{k,j=0}^{P-1} \Delta x_k \mathsf{A}_{kj}\Delta x_j\right) \tag{6.16}$$

where the elements of the matrix A are given by

$$\mathsf{A}_{kj} = \frac{MP}{\beta\hbar^2}\left(\delta_{kj}+1\right). \tag{6.17}$$

Using the Gaussian integral

$$\int \prod_{i=1}^{N} dx_i \exp\left(-\frac{1}{2}\sum_{k,j=0}^{N} x_i \mathsf{M}_{kj} x_j\right) = \frac{(2\pi)^{N/2}}{(\det \mathsf{M})^{1/2}} \tag{6.18}$$

and the fact that

$$\det\left[\frac{MP}{\hbar^2\beta}\left(\delta_{kj}+1\right)\right] = \left(\frac{MP}{\hbar^2\beta}\right)^{P-1} P, \tag{6.19}$$

then the properly normalized ring-polymer distribution probability becomes

$$P_{\mathrm{RP}}(\Delta\mathbf{r}_1,\ldots,\Delta\mathbf{r}_{P-1}) = \Lambda^3\left(\frac{P^{3/2}}{\Lambda^3}\right)^P \exp\left[-\beta\sum_{k=1}^{P}\frac{MP}{2\hbar^2\beta^2}|\Delta\mathbf{r}_k|^2\right]. \tag{6.20}$$

This equation is very important from a numerical point of view, because ring polymers distributed according to P_{RP} can be generated using a formula due to Levy [Levy (1954)]. This approach has indeed been used in the pioneering path-integral calculations by Fosdick and Jordan in the 1960s [Fosdick and Jordan (1966); Jordan and Fosdick (1968)]. Levy's formula states that

$$\mathbf{r}_{k+1} = \frac{P-k}{P-k-1}\mathbf{r}_k + \sqrt{\frac{P-k}{P(P-k-1)}}\,\xi \tag{6.21}$$

where $\mathbf{r}_1 = \mathbf{0}$ and ξ is a vector made by three independent Gaussian random variables with average zero and standard deviation $\Lambda/\sqrt{2\pi}$.

Notice that the quantum partition function of a single particle in an external potential, Eq. (6.10) can be written using Eq. (6.20) as

$$Z = \int \frac{d^3\mathbf{r}_1}{\Lambda^3} \prod_{i=1}^{P-1} d^3\Delta\mathbf{r}_i \, P_{\text{RP}}(\Delta\mathbf{r}_1, \ldots, \Delta\mathbf{r}_{P-1}) \exp\left[-\frac{\beta}{P}\sum_{k=1}^{P} U(\mathbf{r}_k)\right] \tag{6.22}$$

which says that the probability density of finding an ideal-gas ($U(\mathbf{r}) = 0$) ring polymer in a certain configuration within the volume under consideration is $\Pi = P_{\text{RP}}\,\Lambda^3/V$.

6.2.4 *Using the Path-Integral Method: Radial Distribution Functions of Adsorbed Hydrogen*

As a first application of the fundamental Eq. (6.10), we will now calculate the radial distribution functions of hydrogen adsorbed in MOFs. For the time being we will not be concerned with large occupancies, that is we will neglect the H_2–H_2 interaction terms: our results are therefore valid in the limit of zero pressure.

The radial distribution functions express the ratio between the local density of adsorbed hydrogen at a distance r from a fixed reference point (one of the atoms of the framework, say) and the average density of hydrogen within the unit cell of the system. Equivalently, they can be defined as the quantities $g_A(r)$ given by

$$\frac{N}{V} g_A(r)\, 4\pi r^2 dr = \frac{1}{N_A}\left\langle \sum_{i=1}^{N}\sum_{j=1}^{N_A} \delta\left(r - \left|\mathbf{r}_i - \mathbf{R}_j\right|\right) dr \right\rangle, \tag{6.23}$$

where N and N_A are the number of hydrogen molecules and reference atoms present in the system, respectively. The right-hand side of Eq. (6.23) simply counts the average number of H_2 molecules present in a spherical shell of thickness dr centered in the N_A reference sites. The positions of the H_2 molecules are denoted by $\mathbf{r}_i$ ($i = 1, \ldots, N$), whereas the position of the reference sites on the framework are denoted by $\mathbf{R}_j$ ($j = 1, \ldots, N_A$).

The function $g_A(r)$ is defined as the ratio between the average number of H_2 molecules in a shell of radius r and the number of molecules expected in the same shell assuming an average density N/V. Since molecules do not overlap, $g_A(r)$ is equal to zero for values of r smaller than a suitably defined interatomic diameter (see

Eq. (6.3)). At large distances from the reference center, one expects to find molecules arranged in a disordered manner and having the average density, so that $\lim_{r\to\infty} g_A(r) = 1$.

Equation (6.23) is valid in both the classical and quantum case, by a suitable definition of the average value $\langle \ldots \rangle$. The zero-pressure limit is obtained by setting $N = 1$ and $V \to \infty$ in Eq. (6.23).

In the classical case, the expression for $g_A(r)$ at low coverage then becomes

$$g_A(r) = \frac{1}{N_A} \frac{\sum_{j=1}^{N_A} \int \frac{d\Omega}{4\pi} \exp\left[-\beta U(\mathbf{r} + \mathbf{R}_j)\right]}{\int \frac{d^3\mathbf{r}}{V} \exp[-\beta U(\mathbf{r})]}, \tag{6.24}$$

where $d\Omega = d\cos\theta d\phi$ denotes the solid angle pointed by the vector $\mathbf{r}$ that has a fixed modulus $|\mathbf{r}| = r$. The denominator of Eq. (6.24) is the average value of the Boltzmann factor of the solid–fluid potential energy in a suitably large supercell and, as such, can be straightforwardly calculated. The numerator is the average value of the Boltzmann factor within a spherical shell of radius r centered on a reference atomic sites, and averaged over all the position of the chosen atomic sites. Also this quantity can be calculated straightforwardly.

We show in Fig. 6.2 the classical radial distribution functions for H_2 adsorption in MOF-5 at $T = 77$ K. These results have been obtained using the Buch potential to represent the hydrogen molecules, and by calculating the H_2–MOF-5 interaction using UFF. Close inspection of these results show that in the limit of zero pressure, the hydrogen density is higher close to the zinc oxide clusters present in the material, although a significant adsorption is also observed close to the carbon atoms of the organic linker. These results are consistent with the outcome of computer simulations performed using ab initio quantum chemical methods, [Mueller and Ceder (2005)] as well as with the experimental results obtained using the neutron scattering technique [Yildirim and Hartman (2005)].

However, at low temperature quantum diffraction is expected to modify appreciably the classical results. To investigate how the classical results compare with the quantum ones, we have to use the

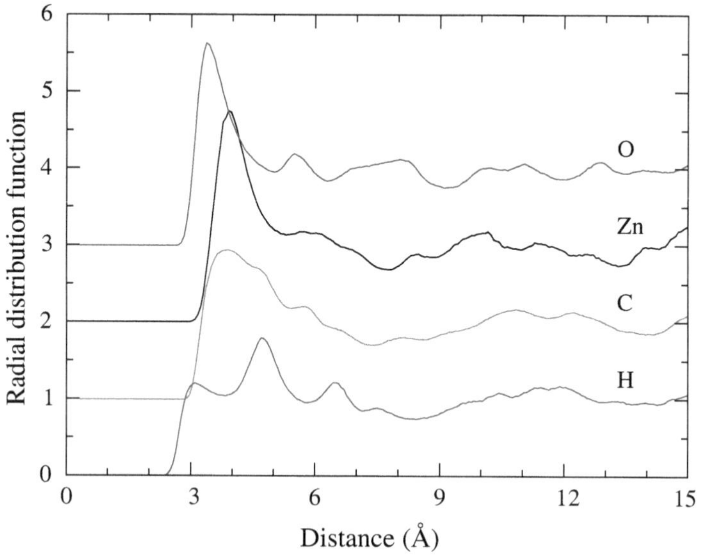

Figure 6.2 Classical radial distribution functions for H_2 adsorbed in MOF-5 at $T = 77$ K, calculated using UFF. The various curves correspond to different reference sites, chosen as the atomic species of the MOF, and have been translated along the y axis for clarity. From bottom to top: H, C, Zn, and O.

quantum mechanical expression for $g_A(r)$. Starting from Eq. (6.23), where the average is to be interpreted in the quantum mechanical sense, and using the Trotter theorem, together with Eq. (6.20), one can write $g_A(r)$ as:

$$g_A(r) = \frac{1}{N_A\, Z_Q} \sum_{j=1}^{N_A} \int d^3\mathbf{r}_1 \, \frac{\delta\left(r - |\mathbf{r}_1 - \mathbf{R}_j|\right)}{4\pi r^2} \, \Lambda^3 \langle \mathbf{r}_1 | e^{-\beta H} | \mathbf{r}_1 \rangle, \tag{6.25}$$

where we have defined

$$Z_Q = \int \frac{d^3\mathbf{r}_1}{V} \, \Lambda^3 \langle \mathbf{r}_1 | e^{-\beta H} | \mathbf{r}_1 \rangle. \tag{6.26}$$

Notice that by using Trotter theorem and Eq. (6.20) one can write

$$Z_Q = \int \frac{d^3\mathbf{r}_1}{V} \prod_{i=1}^{P-1} d^3\Delta\mathbf{r}_i \, P_{RP}(\ldots) \exp\left[-\frac{\beta}{P} \sum_{i=1}^{P} U\left(\mathbf{r}_1 + \rho_i\right)\right], \tag{6.27}$$

where we have defined $\rho_i = \mathbf{r}_i - \mathbf{r}_1$ as the relative position of the ith bead of the ring polymer with respect to the first bead. Eq. (6.27) can be interpreted as the average over the unit cell of the Boltzmann factor of the ring polymers, which is in turn averaged over all the possible ring-polymer configurations. Recalling that the latter can be efficiently generated using Eq. (6.21), the numerical evaluation of Eq. (6.27) is straightforward.

With analogous manipulations on the integral in Eq. (6.25), the radial distribution function can be rewritten as

$$g_A(r) = \frac{1}{N_A Z_Q} \sum_{j=1}^{N_A} \int \frac{d\Omega}{4\pi} \prod_{i=1}^{P-1} d^3 \Delta\mathbf{r}_i P_{RP}(\ldots) \exp \times \left[-\frac{\beta}{P} \sum_{i=1}^{P} U(\mathbf{r} + \mathbf{R}_j + \rho_i) \right], \tag{6.28}$$

where one has $|\mathbf{r}| = r$ and the integration is performed on the orientations of the vector $\mathbf{r}$. The translation of this formula into a computational algorithm is straightforward. One generates a series of directions uniformly distributed over a sphere (corresponding to the integration over $d\Omega$). For each of these directions, a fixed number of ring polymers is generated using Eq. (6.21), and they are used to perform the integration over $P_{RP}(\ldots)$. For each of these polymers, the solid–fluid interaction potential is averaged over the position of the beads, and the Boltzmann factor of the resulting value is summed in the integration.

In performing a path-integral calculation, one has to always check that the Trotter index P has been chosen large enough that convergence in the limit procedure indicated by Eq. (6.8) has been attained.

In general, one should perform a series of calculations by progressively increasing P until the results obtained are converged with sufficient accuracy. Although the actual value of P depends on the fine detail of the system under consideration, and possibly also on the nature of the observable being investigated, it is possible to make a rough order-of-magnitude a priori estimate on how the optimal value of P scales with the main parameters of the system under consideration.

We start by noticing that Eq. (6.8) states that, for sufficiently large values of P, one has

$$\exp\left[-\frac{\beta}{P}(K+U)\right] \simeq \exp\left(-\frac{\beta}{P}K\right)\exp\left(-\frac{\beta}{P}U\right), \qquad (6.29)$$

which is tantamount to assume that the operators K and V commute. Indeed, the left-hand side of Eq. (6.29) can be interpreted as the Boltzmann factor of a system at the temperature PT. Since commuting operators are equivalent to classical operators, one can see that P is sufficiently large when the system at temperature TP can be considered classical. As a consequence, one expects that the optimal value of P is—for a given system—roughly inversely proportional to the absolute temperature.

Moreover, we notice that the quantum mechanical Planck's constant $\hbar$ appears in ring-polymer distribution probability of Eq. (6.20) only through the combination $MP\beta/\hbar^2$. Therefore one could expect the optimal value of P to be inversely proportional to the mass M of the isotope under consideration, for a given system and at a fixed temperature.

The results of the numerical evaluation of Eq. (6.28) are reported in Fig. 6.3 The quantum radial distribution functions have been calculated by setting the Trotter index $P = 16$, which is tantamount to saying that H_2 can be considered classical for $T > 1200$ K. We notice that, in general, the peak values of $g_A(r)$ are reduced by quantum effects. This is due to the zero-point motion, which tends to delocalize the particles on a region of typical size Λ. Consequently, the quantum radial distribution functions present secondary structures that are less pronounced than their classical counterparts.

As a final remark, let us reexamine the classical approximation Eq. (6.29) used to factorize the Boltzmann factor for large values of the Trotter index, which is generally referred to as the *primitive* approximation. A more evolved approach consists in using a semiclassical (i.e., $\mathcal{O}(\hbar^2)$) expression in factorizing the left-hand side of Eq. (6.29), and the corresponding expressions have been worked out by Li and Broughton, [Li and Broughton (1987)] whose main

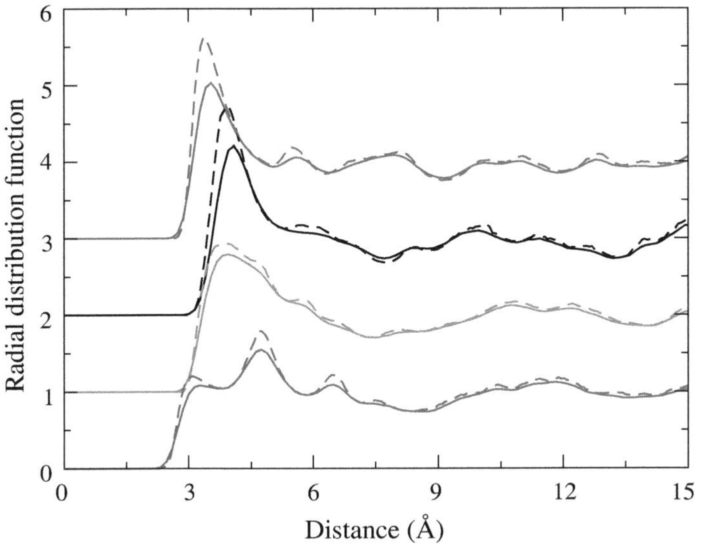

Figure 6.3 Quantum radial distribution functions for H_2 adsorbed in MOF-5 at $T = 77$ K, calculated using UFF (solid lines). The classical results, reported in Fig. 6.2, are shown as dashed lines. The various curves correspond to different reference sites, chosen as the atomic species of the MOF, and have been translated along the y axis for clarity. From bottom to top: H, C, Zn, and O.

result is:

$$\exp\left[-\frac{\beta}{P}(K+U)\right] \simeq \exp\left(-\frac{\beta}{P}K\right)\exp\left[-\frac{\beta}{P}\left(U+\frac{\beta^2\hbar^2}{24mP^2}|\nabla U|^2\right)\right]. \tag{6.30}$$

The use of Eq. (6.30) in the path-integral expressions, usually results in convergence being achieved for smaller values of P. It also enables one to simulate particles interacting with the Coulomb potential, which are notably very difficult to study using the primitive approximation.

6.2.5 *Monte Carlo Sampling of the Path-Integral Expressions*

In the case of finite loadings, where the H_2–H_2 interaction cannot be neglected, the direct evaluation of multidimensional integrals leading to $g_A(r)$ and other observables rapidly becomes impractical. The principal reason is that the largest contribution to the integrals comes from a relatively small subspace where the integrand is large.

In this case, the Monte Carlo integration techniques is most useful. Instead of uniformly sampling the (rapidly increasing) integration hypervolume, the Monte Carlo technique concentrates the sampling in those regions giving the largest contribution to the observable of interest.

We will illustrate the method by considering the calculation of the radial distribution functions $g_A(r)$ between adsorbed hydrogen and the framework centers A at finite loadings. The definition of these quantities is still given by Eq. (6.23), but the average is now performed using an Hamiltonian that includes H_2–H_2 interaction. For the sake of simplicity, we will assume that the fluid–fluid interaction can be approximated by a two-body potential only and that this potential depends only on the relative distance between (the center of mass of) two molecules.

In this case, denoting the pair potential by $u(r)$, the Hamiltonian of the system is:

$$H = \sum_{i=1}^{N} \frac{\mathbf{p}^{(i)} \cdot \mathbf{p}^{(i)}}{2m} + \sum_{i=1}^{N} U(\mathbf{r}_1^{(i)}) + \sum_{i<j}^{N} u(|\mathbf{r}_1^{(i)} - \mathbf{r}_1^{(j)}|). \tag{6.31}$$

Average values such as those appearing in Eq. (6.23) are given by

$$\langle f(\mathbf{R}_1) \rangle = \frac{1}{Z} \int d^{3N}\mathbf{R}_1\, f(\mathbf{R}_1) \langle \mathbf{R}_1 | e^{-\beta H} | \mathbf{R}_1 \rangle \tag{6.32}$$

where f is a general observable (dependent on positions only), $Z = \langle 1 \rangle$ is the normalization, and we denoted with $\mathbf{R}_1$ a vector containing the $3N$ coordinates $\mathbf{r}^{(1)}, \ldots, \mathbf{r}^{(N)}$ of the particles. The matrix element $\langle \mathbf{R}_1 | e^{-\beta H} | \mathbf{R}_1 \rangle$ can be evaluated using again the Trotter expansion Eq. (6.8) and a sufficiently large discretization index P: in this way one introduces one ring polymer having P beads for each particle. Denoting by $\mathbf{R}_k^{(i)}$ the position of the kth bead of the ith polymer and defining the average potentials

$$\overline{U}_i = \frac{1}{P} \sum_{k=1}^{P} U(\mathbf{R}_k^{(i)}) \tag{6.33}$$

$$\overline{u}_{ij} = \frac{1}{P} \sum_{k=1}^{P} u(|\mathbf{R}_k^{(i)} - \mathbf{R}_k^{(j)}|), \tag{6.34}$$

one has

$$\langle f(\mathbf{R}_1)\rangle = \frac{1}{Z}\int \prod_{i=1}^{N}\prod_{k=1}^{P} d^3\mathbf{R}_k^{(i)}\, f(\mathbf{R}_1)\mathcal{P}(\mathbf{R}_1,\ldots,\mathbf{R}_P), \tag{6.35}$$

where

$$\mathcal{P}(\mathbf{R}_1,\ldots,\mathbf{R}_P) = \prod_{i=1}^{N} P_{RP}(\Delta\mathbf{R}_k^{(i)}) \exp\left[-\beta\left(\sum_a \overline{U}_a + \sum_{a<b}\overline{u}_{ab}\right)\right]. \tag{6.36}$$

Notice that $\mathcal{P}(\mathbf{R}_1,\ldots,\mathbf{R}_P)$ is invariant for permutations of its argument. Using the dummy nature of the integration variables in Eq. (6.35), one can see that the average value of f can be written as

$$\langle f\rangle = \frac{1}{P}\sum_{i=1}^{P}\langle f(\mathbf{R}_i)\rangle. \tag{6.37}$$

The Monte Carlo method is an efficient way to evaluate the multidimensional integral Eq. (6.35). The idea is to generate a Markov chain (random walk) in the space of the $3NP$ coordinates $\mathbf{X} = (\mathbf{R}_1,\ldots,\mathbf{R}_P)$ whose equilibrium distribution is just the function $\mathcal{P}(\mathbf{X})/Z$. In this way, the average value $\langle f\rangle$ can be evaluated as the arithmetic average

$$\langle f\rangle = \frac{1}{n}\sum_{l=1}^{n} f(\mathbf{X}_l), \tag{6.38}$$

where $\mathbf{X}_l$ is the lth sample generated by the Markov chain, after a suitable equilibration procedure. A Markov chain is characterized by the transition probability $W(\mathbf{Y}|\mathbf{X})$ of reaching the configuration $\mathbf{Y}$ from a given configuration $\mathbf{X}$. The form of W can be obtained by requiring that it satisfies the *detailed balance* condition:

$$W(\mathbf{Y}|\mathbf{X})\mathcal{P}(\mathbf{X}) = W(\mathbf{X}|\mathbf{Y})\mathcal{P}(\mathbf{Y}). \tag{6.39}$$

In actual applications, the transition probability is written in the form $W = TA$, where T is a trial probability and A is an acceptance probability. The flexibility of the Monte Carlo method comes essentially on the freedom available in performing this decomposition. In particular, once $T(\mathbf{Y}|\mathbf{X})$ has been chosen, the acceptance probability is given by Eq. (6.39)

$$\frac{A(\mathbf{Y}|\mathbf{X})}{A(\mathbf{X}|\mathbf{Y})} = \frac{T(\mathbf{X}|\mathbf{Y})\,\mathcal{P}(\mathbf{Y})}{T(\mathbf{Y}|\mathbf{X})\,\mathcal{P}(\mathbf{X})} = R(\mathbf{X},\mathbf{Y}), \tag{6.40}$$

a condition that can be satisfied, for example, by the standard Metropolis choice [Landau and Binder (2005)]:

$$A(\mathbf{Y}|\mathbf{X}) = \min\left[1,\ R(\mathbf{X},\mathbf{Y})\right] \tag{6.41}$$

$$A(\mathbf{X}|\mathbf{Y}) = \min\left[1,\ \frac{1}{R(\mathbf{X},\mathbf{Y})}\right]. \tag{6.42}$$

Notice that in this way it is not necessary to calculate the normalization factor Z appearing in Eq. (6.32), but the probabilistic interpretation of the method assumes that $\mathcal{P}(\mathbf{X})$ (which depends on the translational density matrix Eq. (6.5)) is a positive quantity.

Although in principle any choice of the transition probability $T(\mathbf{X}|\mathbf{Y})$ subject to the above mentioned would do, in practical applications one is limited by the available computational resources. The efficiency of the sampling can be gauged by the value of the acceptance probability: too low a value means that the (usually very large) configuration space is not sampled efficiently, resulting in poor equilibration and biased results. On the other hand, an acceptance probability always very close to 1 might indicate that the distance (in the $3NP$-dimensional space) between two consecutive steps of the Markov chain is too small, resulting again in poor sampling of the configuration space. As a rule of thumb, any choice of algorithms—or parameters thereof—resulting in $A \approx 1/2$ is considered acceptable.

Let us consider, for example, the following possible choice: let us take for $T(\mathbf{X}|\mathbf{Y})$ a constant value when a single bead of a ring polymer (i.e., one of the coordinates $\mathbf{R}_k^{(i)}$ introduced above) is displaced within a box of side ℓ from its initial position. A program implementing this choice would generate two random integers $1 \leq i \leq N$ and $1 \leq k \leq P$ to identify the bead to be moved and a random vector $\mathbf{v}$ with coordinates $\mathbf{v}_i$ uniformly distributed within $-\ell/2$ and $\ell/2$.

In this way the quantity $R(\mathbf{X},\mathbf{Y})$ in Eq. (6.40) would be very small for displacements ℓ exceeding the average bead-to-bead distance, which, from Eq. (6.20), can be seen to be of the order $\hbar\sqrt{\beta/(mP)}$. One is thus forced to choose $\ell \approx \hbar\sqrt{\beta/(mP)}$, but it is easy to see that in the deep quantum limit (i.e., when a large value of P is needed to reach convergence in the path-integral results), this would lead to a very inefficient sampling, because only displacements much smaller than the average interpolymer distance would be performed at each iteration.

One way to overcome this difficulty would be to devise a scheme with two different kinds of moves. In practice one could choose to perform a trial moves made by either a rigid translation of a whole ring polymer, or the replacement of a ring polymer with another one, generated using Eq. (6.21). This is indeed a much better choice than the naive one described above.

However, in doing so, one is still assuming that the distribution of confined ring polymers is quite similar to the distribution in the ideal gas. This assumption breaks down in the case of strong confinement, and specific approaches have been developed in this case [Garberoglio (2008)].

One approach that has found widespread adoption is the so-called hybrid Monte Carlo method, [Duane et al. (1987)] where molecular dynamics (MD) with a large time step is used to generate trial configurations. In this scheme, fictitious masses m^* are assigned to the ring polymers' beads, and a classical Hamiltonian H_{HMC} is built using the kinetic energies $\mathbf{p}_k^{(i)} \cdot \mathbf{p}_k^{(i)}/(2m^*)$ and the ring-polymer classical potential energies of Eq. (6.33) and Eq. (6.34). The whole Hamiltonian H_{HMC} is used in the detailed balance condition Eq. (6.39). This approach can be efficiently coupled to multiple-time-scale integration methods for the equation of motions, [Tuckerman et al. (1992)] resulting in a very efficient sampling of path-integral configurations [Tuckerman et al. (1993)].

As an illustration of the results that can be obtained with this approach, we show in Fig. 6.4 how the radial distribution functions between H_2 and selected atoms in MOF-5 depend on the cell loading. In general we notice that for small loadings, the path-integral Monte Carlo (PIMC) result is very similar to the calculation performed above in the zero-pressure limit. It is only when the number of molecules adsorbed reaches saturation (which corresponds to $\sim$250 molecules per unit cell, in the case of MOF-5) that the solid–fluid radial distribution functions change appreciably. The principal effect observed at finite loading is a reduction of the value of the principal peak, together with the corresponding appearance of a larger probability of finding the molecules far from the framework.

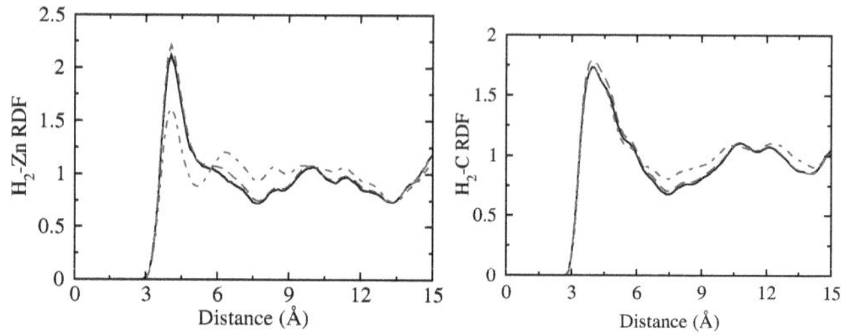

Figure 6.4 Radial distribution functions between H_2 adsorbed in MOF-5 at $T = 77$ K and the zinc (left) or carbon (right) atoms of the frameworks. Solid line: zero-pressure value. Dashed line: loading of 10 H_2 molecules per unit cell. Dot-dashed line: loading of 100 H_2 molecules per unit cell.

The Monte Carlo approach allows also to calculate the H_2–H_2 radial distribution function in the finite-loading conditions. From the results shown in Fig. 6.5, one can see that there is a well-defined first shell of neighbors and that after that the atoms tend to a uniform distribution characterized by $g(r) = 1$. The sharpness of the first neighbor shell, measured by the difference between the values of

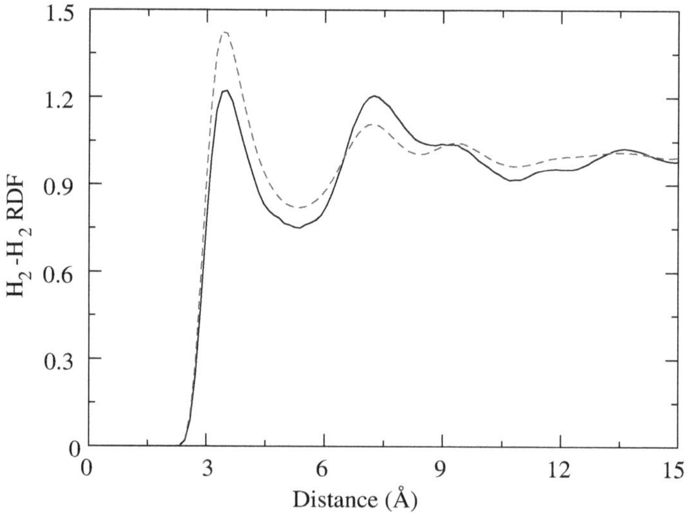

Figure 6.5 H_2–H_2 radial distribution functions for hydrogen adsorbed in MOF-5 at $T = 77$ K for different loadings. Solid line: 10 molecules per unit cell. Dashed line: 100 molecules per unit cell.

$g(r)$ at the first maximum and subsequent minimum, increases with increasing density, which is the usual behavior observed in dense fluids.

6.2.6 *Approximating the Path Integral: Centroid and Semiclassical Approaches*

Although Monte Carlo methods are an efficient way to evaluate quantum statistical mechanical averages, effective approximations to the calculation of the partition function Eq. (6.10) have been developed. These approaches are usually valid in the high-temperature regime (although the meaning of high temperature usually depends on the desired accuracy) and are based on the development of an effective classical potential $U_{\text{eff}}(\mathbf{R})$. When $U_{\text{eff}}(\mathbf{R})$ is used in classical statistical simulations, the results reproduce as close as possible those obtained using the actual quantum mechanical partition function.

One of the most successful approximations can be derived by rewriting the partition function Eq. (6.13) as

$$Z = \int \frac{d^3\mathbf{R}}{\Lambda^3} \left\langle \exp\left[-\frac{1}{\hbar}\int_0^{\beta\hbar} U\left(\mathbf{R}+\mathbf{r}(\tau)\right) d\tau\right]\right\rangle_{\text{RP}} \tag{6.43}$$

where $\langle \cdots \rangle_{\text{RP}}$ indicates the average over closed-ring polymers with the first bead at the origin—whose discretized form is given by Eq. (6.20)—and $\mathbf{R}$ is the coordinate of the first bead.

Next, we notice that for any given closed path $\mathbf{r}(\tau)$ there exists a value $\overline{\tau}$ so that

$$\int_0^{\beta\hbar} U\left(\mathbf{R}+\mathbf{r}(\tau)\right) d\tau = U\left(\mathbf{R}+\mathbf{r}(\overline{\tau})\right) \tag{6.44}$$

and assume, with Feynman, [Feynman and Hibbs (2010); Garberoglio (2012)] that $\mathbf{r}(\overline{\tau})$ is given for each path by the path centroid $\overline{\mathbf{r}}$

$$\mathbf{r}(\overline{\tau}) \approx \overline{\mathbf{r}} = \frac{1}{\beta\hbar}\int_0^{\beta\hbar} \mathbf{r}(\tau)\, d\tau = \frac{1}{P}\sum_{i=1}^{P-1}(P-i)\Delta\mathbf{r}_i, \tag{6.45}$$

where the last equality is its discretized form.

Finally, by performing the approximation $\langle e^{-U}\rangle_{\text{RP}} \approx e^{-\langle U\rangle_{\text{C}}}$, the centroid potential $U_{\text{C}}(\mathbf{R})$ can be written as $U_{\text{C}}(\mathbf{R}) = \langle U\left(\mathbf{R}+\overline{\mathbf{r}}\right)\rangle_{\text{C}}$, where the last average is performed on the distribution function of the centroid position.

Since the distribution of the distances $\Delta\mathbf{r}_i$ between adjacent beads in the ring polymers is Gaussian—cf. Eq. (6.20)—and the centroid position is a linear combination thereof, it follows that the distribution of the centroid position is also Gaussian. The average $\langle\bar{\mathbf{r}}\rangle$ is zero, and the variance of each of its Cartesian components is given by the standard Gaussian integral:

$$\frac{1}{3}\langle\bar{\mathbf{r}}\cdot\bar{\mathbf{r}}\rangle_{\mathrm{RP}} = \sum_{i,j=1}^{P-1}\left(\frac{P-i}{P}\right)\left(\frac{P-j}{P}\right)A_{ij}^{-1} = \frac{\beta\hbar^2}{12M}, \tag{6.46}$$

where $A_{ij}^{-1} = \beta\hbar^2(\delta_{ij} - 1/P)/(MP)$ is the inverse of the matrix A defined in Eq. (6.17).

Finally, the expression of the centroid potential can be written as

$$U_{\mathrm{C}}(\mathbf{R}) = \left(\frac{6M}{\pi\beta\hbar^2}\right)^{3/2}\int U(\mathbf{R}+\bar{\mathbf{r}})\exp\left(-\frac{6M}{\beta\hbar^2}\bar{\mathbf{r}}^2\right)d^3\bar{\mathbf{r}}. \tag{6.47}$$

Notice that we have assumed that U is an external potential, that is, the interaction potential between the H_2 molecule and the framework. In the case of the H_2–H_2 pair potential, the mass M of the hydrogen molecule appearing in Eq. (6.47) must be replaced with the reduced mass $m = M/2$ of the interacting pair.

Equation (6.47) states that the effective potential in the centroid approximation is obtained as the average of the original pair potential with a Gaussian distribution of size $\lambda^2 \approx \beta\hbar^2/(12M)$. If the potential $U(\mathbf{R})$ is such that $U''(\mathbf{R})\lambda^2 \ll 1$ in the region where the Boltzmann factor $e^{-\beta U(\mathbf{r})}$ is significantly different from zero, then Eq. (6.47) can be further simplified by expanding the integrand around the position $\mathbf{R}$. Assuming that the potential is spherically symmetric, one has

$$U(\mathbf{R}+\bar{\mathbf{r}}) = U(\mathbf{R}) + \sum_{i=1}^{3}\bar{\mathbf{r}}_i\nabla_i U(\mathbf{R}) + \frac{1}{2}\sum_{i,j=1}^{3}\bar{\mathbf{r}}_i\bar{\mathbf{r}}_j\nabla_{ij}^2 U(\mathbf{R}) + \ldots \tag{6.48}$$

Substituting the previous expression in Eq. (6.47) and performing the integrations, one obtains the semiclassical—that is, $O(\hbar^2)$—approximation [Feynman and Hibbs (2010)]:

$$U_{\mathrm{SC}}(R) = U(R) + \frac{\beta\hbar^2}{24M}\mathrm{Tr}\left[\nabla_{ij}^2 U(\mathbf{R})\right]. \tag{6.49}$$

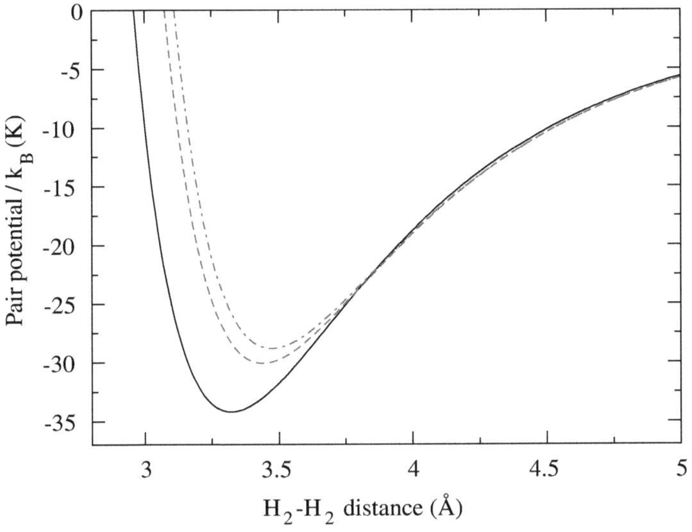

Figure 6.6 Classical and effective quantum H_2–H_2 pair potentials at $T = 77$ K. Solid line: LJ classical potential using Buch parameters. Dashed line: Feynman–Hibbs $O(\hbar^2)$ potential (see Eq. (6.49)). Dot-dashed line: centroid potential (see Eq. (6.47)).

Notice that in the case of a spherically symmetric potential one has

$$\mathrm{Tr}\left[\nabla_{ij}^2 U(\mathbf{R})\right] = U''(R) + \frac{2}{R}U'(R), \tag{6.50}$$

where we have denoted $R = |\mathbf{R}|$.

We show in Fig. 6.6 the classical, Feynman–Hibbs, and centroid potentials for a pair of interacting H_2 molecules at a temperature $T = 77$ K. There are two features worth noticing. First, quantum mechanical diffraction effects (due to the Heisenberg principle) result in particles with a larger effective size. This can be measured, for example, by the value of the distance where the potential crosses zero. Second, the absolute value of the minimum of the potential energy is smaller for quantum particles. This is due to the zero-point motion, which is again a manifestation of the Heisenberg principle.

Generally speaking the effective quantum potentials have a more repulsive character than the corresponding classical ones. This effective repulsion is larger for lighter isotopes.

At the temperature $T = 77$ K, the effective size of an hydrogen molecule grows from 2.96 to 3.1 Å (which is a difference of the order of $\beta\hbar^2/(24m) = 0.15$ Å) and the minimum of the potential raises by about 5.5 K. The differences between the quantum effective potential and the classical one become larger and larger as the temperature is decreased.

6.2.7 *Path-Integral Approach to Rotational Degrees of Freedom*

In the previous sections we have always assumed that H_2 can be treated as a spherical (structureless) object. This approximation is indeed quite good even in the condensed phase, but it is known to fail under very confining conditions.

In these cases, the linear structure of the H_2 molecule has to be taken into account in any effective modeling. In the zero-pressure limit and when the confining potential has a symmetric form, the direct approach of diagonalizing the Hamiltonian can be carried out. Rotational effects on H_2 have been studied in this way when adsorbed within carbonaceous materials, such as fullerene, [Xu et al. (2008)] carbon nanotubes [Brown et al. (2000); Hathorn et al. (2001); Trasca et al. (2003); Lu et al. (2006); Garberoglio et al. (2006)] or slit pores made by graphene sheets [Garberoglio (2009b)].

These calculations point out that under strong confinement there is a significant coupling between rotational and translational degrees of freedom. As a result, the ground state of the system is characterized by a nonzero value of the average angular momentum, a condition that leads to a very strong isotopic effect, termed extreme confinement [Lu et al. (2006); Garberoglio (2009b)].

Although this peculiar condition has not been reported to be present when hydrogen isotopologues are confined in an MOF, we briefly present in this section, for the sake of completeness, the general theory of path-integral simulation of rigid rotors, limiting ourselves to the case of a single molecule. The interested reader will find more detailed discussions in the literature [Müser (1996); Marx and Müser (1999); Garberoglio and Johnson (2010)].

Denoting by $R = \mathbf{L}^2/(2I)$ the rotational kinetic energy, where $\mathbf{L}$ is the angular momentum and I the moment of inertia of the H_2 molecule, the Hamiltonian has the general form: $H = K + R + V$, where the potential energy V depends also on the polar angles θ and ϕ describing the orientation of the rotor with respect to the quantization axis. In this case, Trotter's theorem Eq. (6.8) can be generalized to

$$\exp\left[-\beta\left(K + R + V\right)\right] = \lim_{P\to\infty}\left[e^{-\beta K/P}e^{-\beta R/P}e^{-\beta V/P}\right]^P, \qquad (6.51)$$

and the completeness relations to be inserted between each of the P products on the right-hand side of the previous equation have the form

$$1 = \int d^3\mathbf{r}d\Omega\,|\mathbf{r},\,\Omega\rangle\langle\mathbf{r},\,\Omega| \qquad (6.52)$$

where Ω denotes the angular direction and $d\Omega = d\cos\theta d\phi$. The matrix element of the rotational kinetic energy can be written by

$$\langle\Omega'|e^{-\beta R/P}|\Omega\rangle = \sum_{l=0}^{\infty}{}'\sum_{m=-l}^{l} e^{-Bl(l+1)/(TP)}Y_{lm}^{*}(\Omega')Y_{lm}(\Omega) \qquad (6.53)$$

where $B = \hbar^2/(2k_{\mathrm{B}}I)$ is the rotational temperature and $Y_{lm}(\Omega)$ is a spherical harmonics. Using the addition theorem of the spherical harmonics, and denoting by Θ the angle between the direction Ω and Ω', one can rewrite

$$\langle\Omega'|e^{-\beta R/P}|\Omega\rangle \equiv \varrho(\Theta) = \sum_{l=0}^{\infty}{}'\frac{2l+1}{4\pi}e^{-Bl(l+1)/(TP)}P_l(\cos\Theta). \qquad (6.54)$$

Equation (6.54) is the rotational analogue of Eq. (6.5) in the translational case. No further analytical simplification is possible, but in actual numerical calculations Eq. (6.54) can be precalculated at the beginning of the run and interpolated using, for example, cubic splines. The prime over the summation of the angular momentum signifies that in the case of homonuclear diatomics not all the angular momenta must be taken into account.

As discussed in Section 6.1, only even or odd angular momenta must be taken into account, according to the fermionic or bosonic nature of the nuclei and the nuclear spin state. When only the odd angular momenta are added in Eq. (6.54), one finds that

$\varrho(\Theta) = -\varrho(\pi - \Theta)$. The single-rotor density matrix is not positive definite and particular care must be taken in performing the Monte Carlo sampling [Müser (1996)]. In the other cases, $\varrho(\Theta)$ is positively definite, and the Monte Carlo sampling can be carried out straightforwardly, using the methods described in Section 6.2.5 [Garberoglio and Johnson (2010)].

6.3 Modeling Adsorption of Quantized Molecules

Up to this point, we have been considering the computer simulation of quantum systems made by a finite and constant number of particles, that is, those described by the canonical ensemble.

However, in many cases one is interested in calculating how the number of molecules adsorbed in a material depends on externally imposed conditions, such as temperature or pressure of the adsorbate, as well as its composition. In these cases, the relevant statistical ensemble is the grand canonical one, which describes systems with a finite volume that can exchange energy and particles with an external reservoir.

As is well known, the statistical properties of these systems are described by the grand canonical partition function

$$\Xi(\mu, V, T) = \sum_{N=0}^{\infty} Z(N, V, T) e^{\beta \mu N}, \tag{6.55}$$

where we denoted by μ the chemical potential. This quantity can be related to the actual pressure p of the adsorbate via an equation of state. In the case of an ideal-gas system, one has [Hill (1960)]

$$\mu = k_B T \ln \left(\frac{p\Lambda^3}{k_B T} \right). \tag{6.56}$$

Notice that, in general, one can always associate a pressure to a given chemical potential using Eq. (6.56), even when nonideal gases are being considered. In this case the quantity p appearing in Eq. (6.56) is referred to as the *fugacity* of the adsorbate.

The vast majority of grand canonical calculations on quantized systems are based on Monte Carlo methods quite similar to those described in Section 6.2.5, [Landau and Binder (2005)] properly extended to the path-integral formulation of quantum mechanics

[Wang et al. (1997); Garberoglio (2008)]. The procedure that can be used to sample quantum grand canonical states comes directly from considering the fact that Eq. (6.55) is the normalization condition for the probability $\Pi(N, \mathbf{X}_N)$ of finding N molecules (actually, N ring polymers in the path-integral formulation) having a configuration described by the $3NP$-dimensional vector $\mathbf{X}_N$. This quantity is given by

$$\Pi(N, \mathbf{X}_N) = \mathcal{P}_{\mathrm{RP}}(N, \mathbf{X}_N) e^{-\beta U(N, \mathbf{X}_N)} e^{\beta \mu N} \, d\mathbf{X}_N, \tag{6.57}$$

where $\mathcal{P}_{\mathrm{RP}}(N, \mathbf{X}_N)$ is the product of N ideal-gas ring-polymer probability distributions, defined in Eq. (6.20) and $U(N, \mathbf{X}_N)$ is the potential energy of the N ring-polymer system.

In a grand canonical PIMC simulation one performs two kinds of moves. The first kind is the canonical Monte Carlo sampling typical of a systems with a fixed number of particles and discussed in Section 6.2.5. In the second kind of move, one tries to modify the number of molecules, by inserting or deleting one of them. The simplest kind of insertion move corresponds to placing an ideal-gas ring polymer with uniform probability within the simulation box. According to the discussion following Eq. (6.22), the probability density associated with this attempted move is $T(N+1|N) = P_{\mathrm{RP}} \Lambda^3 / V$. The corresponding deletion move has $T(N|N+1) = 1/(N+1)$, since any particle that can be removed is chosen with the same probability. Deletion moves have to be attempted with the same frequency of insertion moves in order to respect the detailed balance condition.

In the case of insertion and deletion move, the ratio between the acceptance probabilities is

$$\frac{A(N+1|N)}{A(N|N+1)} = \frac{T(N|N+1)\, \Pi(N+1, \mathbf{X}_{N+1})}{T(N+1|N)\, \Pi(N, \mathbf{X}_N)} = \frac{V e^{\beta\mu}}{(N+1)\Lambda^3} e^{-\beta \Delta U} \tag{6.58}$$

where $\Delta U = U(N+1, \mathbf{X}_{N+1}) - U(N, \mathbf{X}_N)$ is the difference in the potential energy of the system between the configurations with $N+1$ and N ring polymers. Also in this case, the Metropolis prescription Eq. (6.42) can be used to define the acceptance probability for the attempted moves [Wang et al. (1997); Garberoglio (2008)].

The same approach can of course be used to either perform classical grand canonical simulations or approximate quantum grand canonical simulations, using either the Feynman–Hibbs or centroid potentials discussed in Section 6.2.6.

Finally, we would like to point out the fact that we have written Eq. (6.57) without the correct-Boltzmann-counting (cBc) factor of $N!$ at the denominator, that is we have been considering distinguishable particles. We could have obtained the same results assuming indistinguishability, but in this case one should have assumed the trial probability for a deletion move as being *independent* on the number of particles in the system, that is $T(N|N+1) = 1$, because only the overall quantum state is the physical entity from this point of view. Nevertheless, Eq. (6.58) would have turned out to be the same, the factor $N + 1$ at the denominator originating in this case from the ratio $(N+1)!/N!$ of the cBc factors. The considerations based on distinguishable particles are closer to the actual simulation practice, where particles are distinguished (if only by the location of their coordinates in the computer's memory) and can be selected individually.

6.3.1 *Quantum Effects on Hydrogen Adsorption in Metal–Organic Frameworks*

As a first application of these methods, we show in Fig. 6.7 the adsorption isotherms of H_2 adsorbed in MOF-5 at the temperature $T = 77$ K, calculated in the classical approximation, using the quantum effective potentials and using the path-integral grand canonical Monte Carlo method [Garberoglio et al. (2005)].

We notice that the quantum nature of hydrogen results in an adsorbed amount that is smaller than what would be predicted using classical statistical mechanics. In the case under consideration, hydrogen adsorption is reduced by $\sim$20% by quantum mechanical effects. Moreover, these results show that at these temperatures, the effective potentials, both in the centroid and the Feynman–Hibbs approximation, are very effective in taking into account the quantum nature of H_2. The results obtained with the effective classical potentials are within the statistical uncertainty of the path-integral calculation. At room temperature, the difference between the classical and quantum simulation of adsorption is very small. In

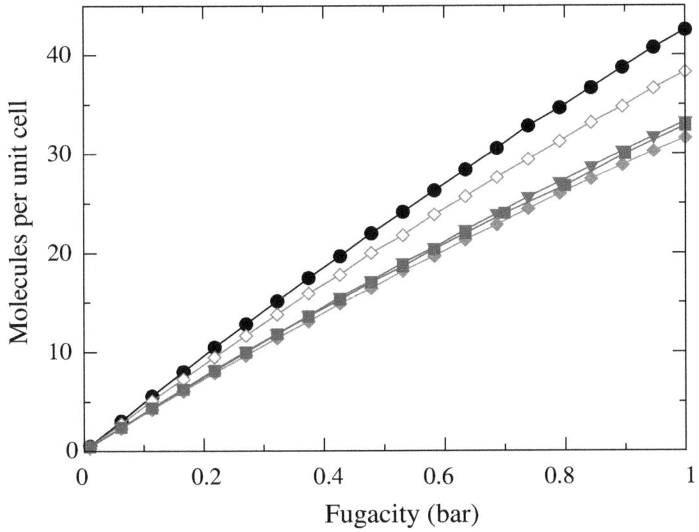

Figure 6.7 Simulated adsorption isotherms of H_2 in MOF-5 at $T = 77$ K using UFF and Buch potentials. Circles: classical approximation. Triangles: Feynman–Hibbs potential. Full diamonds: centroid potential. Squares: path-integral grand canonical Monte Carlo simulations. The empty diamonds show the D_2 isotherm obtained using the centroid-based quantum effective potential.

the case of MOF-5 the observed decrease of H_2 uptake in MOF-5 is $\sim$1%.

The reduction of the amount of adsorbed hydrogen due to quantum effects can be easily understood using the properties of the quantum effective potentials. First of all, the effective size of the adsorbate is slightly larger, resulting in a reduction of the available space within the framework, and secondly, the zero-point energy reduces the strength of the attraction between H_2 and the framework.

We also show, in Fig. 6.7, the isotherm of D_2 under the same conditions, calculated using the centroid approximation. In the classical case H_2 and D_2 would have the same isotherm (assuming a spherically symmetric potential), but the quantum nature of these isotopes results in mass-dependent properties. Recalling that the magnitude of quantum effects is roughly proportional—at least in the semiclassical case—to the quantity $\beta\hbar^2/(24M)$, the heavier isotopes are less affected. In fact, the isotherm of D_2 is closer to the classical one than the H_2 one, but the quantum mechanical nature of the heavier isotopologue is still evident, resulting in a $\sim$10% reduction in the adsorption with respect to the classical prediction.

6.3.2 *Isotopic Effects on Adsorption and Quantum Sieving*

The results of the preceding section suggest that mass-dependent quantum effects upon adsorption could be used to perform isotopic separation and/or isotopic enrichment of a given mixture.

In this respect, the effective classical potential of Eq. (6.47) suggests that the main consequence of quantum diffraction is that molecules have an effective size that is larger than the classical one by a quantity roughly proportional to the thermal de Broglie wavelength Λ of Eq. (6.6).

Therefore, one could expect that the difference in adsorption between different isotopes on the same material be 1) enhanced at low temperatures, because of a larger difference in the thermal de Broglie wavelength, and 2) enhanced in materials presenting pores not much larger than the classical molecular size (e.g., the LJ σ), because under these circumstances the lighter isotopes can be severely hindered in entering the adsorbent.

This possibility was first suggested by Beenakker et al. as a way to effectively separate helium isotopes using selective adsorption in small-pore materials, [Beenakker et al. (1995)] and this phenomenon was called quantum sieving. In Beenakker's paper the nature of the small pore was left undetermined, but the subsequent discovery of carbon nanotubes—which could, in principle, be created with diameters as small as 3 to 4 Å–revived the interest in a more detailed characterization of the conditions needed to enter the quantum sieving regime.

Many studies have been devoted to investigate the difference in the adsorption of hydrogen isotopologues in carbon nanotubes. All of the computational techniques outlined in the previous sections have been used, starting from semiclassical calculations, [Hathorn et al. (2001); Garberoglio et al. (2006)] to diagonalization of the Hamiltonian, [Wang et al. (1999); Lu et al. (2003); Trasca et al. (2003); Lu et al. (2006); Garberoglio (2009b)] and—finally—path-integral simulations [Challa et al. (2001); Kowalczyk et al. (2008); Garberoglio (2008); Kowalczyk et al. (2009); Garberoglio and Johnson (2010)].

The quantity used to measure the preferential adsorption of one isotopologue with respect to the other is the selectivity of species B with respect to specie A, defined as

$$S(B|A) = \frac{x_B/x_A}{y_B/y_A} \tag{6.59}$$

where x_i and y_i are the mole fractions of specie i in the adsorbed and bulk phases, respectively, once thermodynamical equilibrium has been attained.

In the zero-pressure limit, where the interaction among the adsorbates can be neglected, the selectivity is independent of the composition of the bulk mixture and is given by the expression [Challa et al. (2001)]

$$S_0(B|A) = \frac{Z_A^{\rm id}}{Z_B^{\rm id}} \frac{Z_B}{Z_A} \tag{6.60}$$

where $Z^{\rm id}$ is the ideal-gas partition function and Z is the partition function of a single molecule in the adsorbed phase. In the quantum mechanical case, Eq. (6.60) depend on the energy levels of the adsorbed species.

In general, the selectivity can be efficiently calculated using path-integral grand canonical Monte Carlo methods extended to mixtures [Challa et al. (2002); Garberoglio (2008); Garberoglio and Johnson (2010)]. Multicomponent grand canonical Monte Carlo methods are a straightforward generalization of the algorithms presented in Section 6.3. In this case one has to consider a chemical potential for each of the adsorbed species. In the case of binary mixtures, these chemical potentials are related to the molar composition and the total pressure of the bulk gas and are usually fixed using an equation of state. In this case the grand canonical partition function is:

$$\Xi(\mu_A, \mu_B, V, T) = \sum_{N_A=0}^{\infty} \sum_{N_B=0}^{\infty} Z(N_A, N_B, V, T) e^{\beta(\mu_A N_A + \mu_B N_B)}. \tag{6.61}$$

The calculation of selectivity in carbon nanotubes shows that it depends strongly on the diameter D of the pores. When $D \leq$ 3.5 Å, the selectivity for the T_2/H_2 mixture can become quite large, exceeding the value 100 at $T = 20$ K. In even smaller pores, the nonspherical nature of the H_2 isotopologues cannot be neglected, and this results in an exceptionally high selectivity, which is estimated to be in the order of 10^8 [Lu et al. (2006); Garberoglio and Johnson (2010)].

Although these exceptional values of the selectivity have not been measured yet, mostly due to the difficulty of creating adsorption structures resembling the ideal cases studied with computer simulations, there are many experimental observation of quantum sieving in carbon-based microporous materials [Hattori et al. (2006); Bondarenko and Alekseev (2007); Noguchi et al. (2010)].

The possibility of quantum sieving in MOFs has also been investigated, both experimentally, [Noguchi et al. (2008)] and using the quantum effective potentials discussed in Section 6.2.6 [Xu et al. (2009); Liu et al. (2012)]. In all of these cases, the selectivity was nowhere close to the exceptional values predicted for small-diameter carbon nanotubes, and selectivities of the order of $\sim$10 have been predicted. More rigorous path-integral calculations confirmed these findings [Garberoglio (2009a)].

The size of isotopic effects in MOFs are illustrated in Fig. 6.8, where the isotherms of adsorption of H_2 and D_2 adsorbed in the Zn-TBIP MOF [Pan et al. (2006)] at $T = 20$ K are shown. Due to the smallness of its pores, Zn-TBIP has been indicated as one of the materials where a significant quantum sieving effect can be expected [Garberoglio (2009a)].

The analysis of the isotherms reported in Fig. 6.8 shows that at the small temperature investigated, any approximation based on classical effective potentials breaks down. In particular, the results based on the centroid potential result in isotherms shifted to significantly larger pressures. Moreover, the centroid-based effective potential predicts a shift in the isotherm due to isotopic effects by almost 2 orders of magnitude in pressure, whereas the more rigorous PIMC calculations show that the adsorption isotherm of H_2 is displaced from the D_2 one by just a factor of 10 in pressure. Path-integral calculations show that $S_0(D_2/H_2) \simeq 10$ in this case.

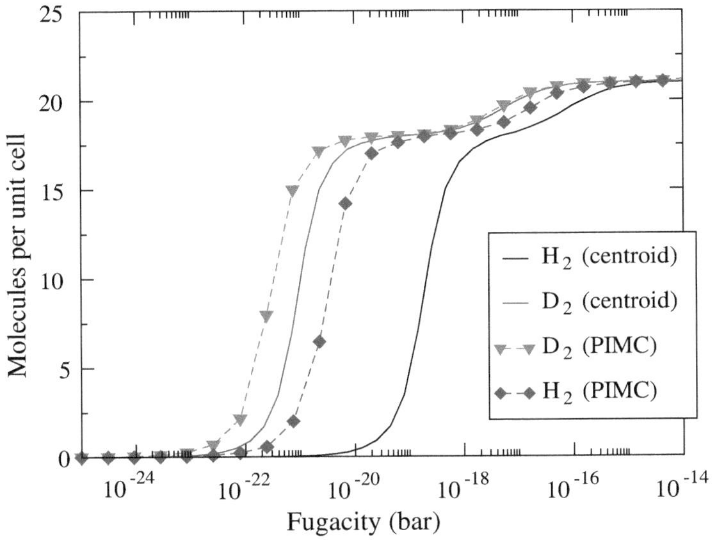

Figure 6.8 Simulated adsorption isotherms of H_2 and D_2 in Zn-TBIP at $T = 20$ K using UFF and Buch potentials. Solid lines: centroid potential. Diamonds: H_2 from PIMC simulations. Triangles: D_2 from PIMC simulations. The dashed lines connecting the symbols are a guide to the eye.

6.4 Modeling the Dynamics of Quantized Molecules

The dynamical properties of quantum particles are described by the time-dependent statistical operator $\hat{\rho}$, which evolves according to the von Neumann–Liouville equation

$$i\hbar\frac{\partial\hat{\rho}}{\partial t} = [\hat{H}, \hat{\rho}], \tag{6.62}$$

whose solution are completely determined by the knowledge of the eigenstates of the Hamiltonian operator, as well as the value of $\hat{\rho}$ at a given time. Although this approach is feasible for one-dimensional systems, the computational resources needed to carry out the diagonalization of $\hat{H}$ in the case of MOFs are quite demanding, as discussed in Section 6.2.1. Moreover, even if this procedure could be carried out, the number of states to be considered would grow exponentially with the number of particles N present in the system.

To the best of my knowledge, no efficient procedure has been devised yet to reduce the computational efforts needed obtain solutions of Eq. (6.62) in the large N case. The Monte Carlo methods that have been so successful in quantum statistical mechanics cannot be applied to study quantum dynamics, because the appearance of imaginary factors makes the probabilistic approach of Monte Carlo methods unfeasible.

Consequently, many approximate ways of solving Eq. (6.62) have been proposed. These efforts have mostly concentrated on the calculation of Kubo-transformed time-dependent correlation functions, from which many observable quantities (e.g., transport coefficients) can be extracted [Toda et al. (1992)].

One of the first approaches used the fact that the dynamical propagator $e^{-it\hat{H}/\hbar}$ becomes the statistical Boltzmann factor $e^{-\beta\hat{H}}$ when the real time is transformed into a fictitious imaginary time $it \rightarrow \tau = \beta\hbar$, and therefore time-dependent correlation functions can be obtained by analytical continuation of temperature-dependent ones. The Wick rotation from imaginary to real time is usually performed by using maximum entropy methods to overcome the ill-poseness of the inverse Laplace transform needed to perform the analytic continuation [Rabani et al. (2002)].

A lot of effort has been devoted to extend the path-integral Monte Carlo approach to time-dependent problems, and many approximate schemes have been devised. In particular, we mention the ring-polymer MD [Craig and Manolopoulos (2004)] and the centroid MD [Cao and Voth (1994a,b); Jang and Voth (1999)] methods.

Although all of these approaches try to include the full coherent dynamics embodied in Eq. (6.62), one expects that in some cases a simple classical MD approach using the quantum effective potentials introduced in Section 6.2.6 could be a viable solution. In particular, this approximation is expected to be valid in those cases when the incoherent thermal fluctuations can induce jumps among the quantum mechanical levels, therefore disrupting the coherent dynamics. This regime can be quantitatively expressed by the condition

$$\hbar\Delta\omega\,\beta < 1, \tag{6.63}$$

where we denoted by $\Delta\omega$ a typical separation among quantum levels in the system under consideration. In the case of hydrogen isotopologues moving within a MOF, an order-of-magnitude estimate of $\Delta\omega$

is provided by $\sqrt{K/M}$ where $K \approx \varepsilon/\sigma^2$ is the second derivative of the solid–fluid potential at the bottom of the potential well. Using this result, the inequality of Eq. (6.63) can be expressed as

$$\frac{\Lambda}{\sigma}\sqrt{\beta\varepsilon} < 1, \tag{6.64}$$

which implies that the temperature has to be high enough so that the de Broglie thermal wavelength is smaller than the molecular size *and* the thermal energy larger than the solid–fluid attraction. Using typical values of σ and ε (see Eq. (6.2) and Eq. (6.3)), one therefore expects this semiclassical approach to be valid for $T >$ 50 K. As we will see in the next section, this level of approximation is enough to reveal unexpected features in the quantum dynamics of H_2 isotopologues adsorbed in nanoporous materials.

6.4.1 *Reverse Kinetic Isotope Effect in Diffusion*

Dimensional analysis considerations about the properties of classical unstructured particles of mass M, interacting with potentials having an energy scale ε and a length scale σ, show that dynamical processes are expected to occur on a typical timescale t_0 of the order of

$$t_0 \approx \sqrt{\frac{M\sigma^2}{\varepsilon}}, \tag{6.65}$$

and, consequently, the self-diffusion coefficient is expected to scale as

$$D \propto \frac{\sigma^2}{t_0} = \sqrt{\frac{\varepsilon}{M}}\sigma. \tag{6.66}$$

so that, in the case of two isotopologues (H_2 and D_2 say) one expects

$$\frac{D_{H_2}}{D_{D_2}} = \sqrt{\frac{M_{D_2}}{M_{H_2}}}, \tag{6.67}$$

that is, lighter isotopes diffuse faster that heavier ones.

However, quantum mechanics alters this finding. In fact, it has been shown theoretically [Anil Kumar and Bhatia (2005); Anil Kumar et al. (2006, 2007)] that under certain conditions the heavier isotope can diffuse faster than the lighter one. This effect has been termed reverse kinetic isotope effect, and subsequent

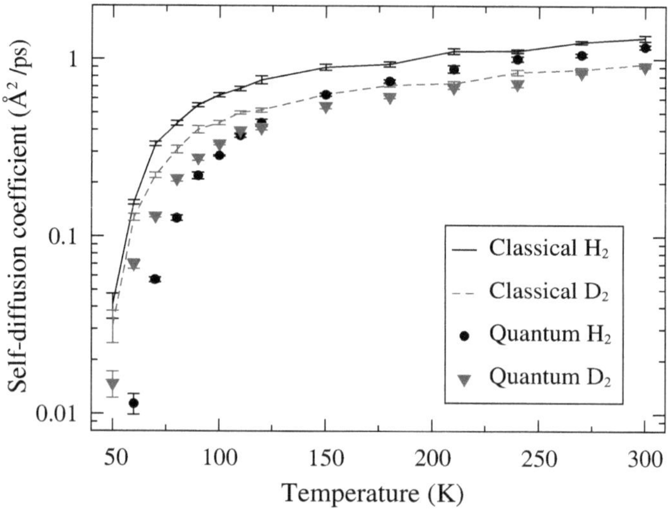

Figure 6.9 Self-diffusion coefficient of H_2 and D_2 diffusing in ZIF-8 as a function of temperature in the zero-pressure limit. The solid and dashed lines are the classical values for H_2 and D_2 respectively. The circles (triangles) denote the quantum results for H_2 (D_2), obtained using the centroid effective potential.

experimental investigations supported its existence [Zhao et al. (2006); Chen et al. (2008); Nguyen et al. (2010)].

This result can be qualitatively understood using the semiclassical approach to quantum dynamics discussed in the previous section. To this end, we show in Fig. 6.9 the self-diffusion coefficients of H_2 and D_2 confined in the microporous material zeolitic imidazolate framework ZIF-8. The results are valid at zero pressure and have been obtained by neglecting the fluid–fluid interaction.

There are many features worth noting. First of all, the classical self-diffusion coefficients of H_2 and D_2 obey Eq. (6.67) in the whole temperature range shown. Secondly, the introduction of quantum effects in the description of the system generally *decreases* the value of the self-diffusion coefficient.

In the case of diffusion in ZIF-8, one observes that for temperatures lower than about 120 K, the self-diffusion coefficient of the heavier species becomes larger than that of the lighter one.

This is the reverse kinetic isotope effect introduced above, which is of a purely dynamical origin. This phenomenon goes in the same direction of the adsorptive quantum sieving effect described above and therefore would contribute to increase the selectivity toward the heavier isotope of a nanoporous membrane.

Finally, we would like to notice that even at the larger temperature studied here, the diffusion coefficient of H_2 does not reach the classical value, whereas the quantum self-diffusion coefficient of D_2 under the same conditions is compatible with the classical one (within the uncertainties of this calculation). This is yet another indication that quantum mechanical effects are crucial in determining the statistical and dynamical properties of H_2 even at room temperature.

References

Anil Kumar, A. Bhatia, S. (2005). Quantum effect induced reverse kinetic molecular sieving in microporous materials, *Phys. Rev. Lett.*, **95**(24), 245901.

Anil Kumar, A., Jobic, H. Bhatia, S. (2006). Quantum effects on adsorption and diffusion of hydrogen and deuterium in microporous materials, *J. Phys. Chem. B*, **110**(33), 16666–16671.

Anil Kumar, A., Jobic, H. Bhatia, S. (2007). Quantum effect induced kinetic molecular sieving of hydrogen and deuterium in microporous materials, *Adsorption*, **13**(5), 501–508.

Beenakker, J., Borman, V. Krylov, S. (1995). Molecular transport in subnanometer pores: zero-point energy, reduced dimensionality and quantum sieving, *Chem. Phys. Lett.*, **232**(4), 379–382.

Belof, J., Stern, A., Space, B. (2008). An accurate and transferable intermolecular diatomic hydrogen potential for condensed phase simulation, *J. Chem. Theo. Comput.*, **4**(8), 1332–1337.

Bondarenko, S. D., Alekseev, I. A. (2007). *Hydrogen Materials Science and Chemistry of Carbon Nanomaterials*, chap. Study of Isotopic Effect for Hydrogen and Deuterium Adsorption on Nanoporous Carbon (NPC) at 67–78 K (Springer), 493–497.

Brown, C., Yildirim, T., Neumann, D., Heben, M., Gennett, T., Dillon, A., Alleman, J., Fischer, J. (2000). Quantum rotation of hydrogen in single-wall carbon nanotubes, *Chem. Phys. Lett.*, **329**(3), 311–316.

Buch, V. (1994). Path integral simulations of mixed para-D2 and ortho-D2 clusters: the orientational effects, *J. Chem. Phys.*, **100**, 7610–7629.

Cao, J., Voth, G. (1994a). The formulation of quantum statistical mechanics based on the feynman path centroid density. II. Dynamical properties, *J. Chem. Phys.*, **100**, 5106.

Cao, J., Voth, G. (1994b). The formulation of quantum statistical mechanics based on the feynman path centroid density. IV. Algorithms for centroid molecular dynamics, *J. Chem. Phys.*, **101**, 6168.

Challa, S. R., Sholl, D. S., Johnson, J. K. (2001). Light isotope separation in carbon nanotubes through quantum molecular sieving, *Phys. Rev. B*, **63**(24), 245419.

Challa, S. R., Sholl, D. S., Johnson, J. K. (2002). Adsorption and separation of hydrogen isotopes in carbon nanotubes: Multicomponent grand canonical Monte Carlo simulations, *J. Chem. Phys.*, **116**, 814.

Chen, B., Zhao, X., Putkham, A., Hong, K., Lobkovsky, E., Hurtado, E., Fletcher, A., Thomas, K. (2008). Surface interactions and quantum kinetic molecular sieving for H2 and D2 adsorption on a mixed metal-organic framework material, *J. Am. Chem. Soc.*, **130**(20), 6411–6423.

Craig, I., Manolopoulos, D. (2004). Quantum statistics and classical mechanics: real time correlation functions from ring polymer molecular dynamics, *J. Chem. Phys.*, **121**, 3368.

Duane, S., Kennedy, A., Pendleton, B., Roweth, D. (1987). Hybrid Monte Carlo, *Phys. Lett. B*, **195**(2), 216–222.

Feynman, R. P., Hibbs, A. (2010). *Quantum Mechanics and Path Integrals* (Dover, New York), emended by D. F. Styer.

Fosdick, L. D., Jordan, H. F. (1966). Path-integral calculation of the two-particle Slater sum for He4, *Phys. Rev.*, **143**, 58.

Garberoglio, G. (2008). Boltzmann bias grand canonical Monte Carlo, *J. Chem. Phys.*, **128**(13), 134109.

Garberoglio, G. (2009a). Quantum sieving in organic frameworks, *Chem. Phys. Lett.*, **467**(4–6), 270–275.

Garberoglio, G. (2009b). Quantum states of rigid linear rotors confined in a slit pore: quantum sieving of hydrogen isotopes and extreme one dimensional confinement, *Eur. Phys. J. D*, **51**(2), 185–191.

Garberoglio, G. (2012). Quantum effects on virial coefficients: a numerical approach using centroids, *Chem. Phys. Lett.*, **525–526**, 19.

Garberoglio, G., DeKlavon, M. M., Johnson, J. K. (2006). Quantum sieving in single-walled carbon nanotubes: effect of interaction potential and rotational-translational coupling. *J. Phys. Chem. B*, **110**(4), 1733–1741.

Garberoglio, G., Johnson, J. K. (2010). Hydrogen isotope separation in carbon nanotubes: calculation of coupled rotational and translational states at high densities, *ACS Nano*, **4**, 1703.

Garberoglio, G., Skoulidas, A. I., Johnson, J. K. (2005). Adsorption of gases in metal organic materials: comparison of simulations and experiments. *J. Phys. Chem. B*, **109**(27), 13094–13103.

Getman, R. B., Bae, Y.-S.,Wilmer, C. E., Snurr, R. Q. (2012). Review and analysis of molecular simulations of methane, hydrogen, and acetylene storage in metalorganic frameworks, *Chem Rev.*, **112**(2), 703–723.

Hathorn, B., Sumpter, B., Noid, D. (2001). Contribution of restricted rotors to quantum sieving of hydrogen isotopes, *Phys. Rev. A*, **64**(2), 022903.

Hattori, Y., Tanaka, H., Okino, F., Touhara, H., Nakahigashi, Y., Utsumi, S., Kanoh, H., Kaneko, K. (2006). Quantum sieving effect of modified activated carbon fibers on H2 and D2 adsorption at 20 K, *J. Phys. Chem. B*, **110**(20), 9764–9767.

Hill, T. (1960). *An Introduction to Statistical Thermodynamics* (Dover Publications).

Jang, S., Voth, G. (1999). Path integral centroid variables and the formulation of their exact real time dynamics, *J. Chem. Phys.*, **111**, 2357.

Jordan, H. F., Fosdick, L. D. (1968). Three-particle effects in the pair distribution function for He4 gas, *Phys. Rev.*, **171**(1), 128.

Kowalczyk, P., Gauden, P., Terzyk, A. (2008). Cryogenic separation of hydrogen isotopes in single-walled carbon and boron-nitride nanotubes: insight into the mechanism of equilibrium quantum sieving in quasi-one-dimensional pores, *J. Phys. Chem. B*, **112**(28), 8275–8284.

Kowalczyk, P., Gauden, P., Terzyk, A., Furmaniak, S. (2009). Impact of the carbon pore size and topology on the equilibrium quantum sieving of hydrogen isotopes at zero coverage and finite pressures, *J. Phys.: Condens. Matter*, **21**, 144210.

Landau, D., Binder, K. (2005). *A Guide to Monte Carlo Simulations in Statistical Physics* (Cambridge University Press).

Landau, L., Lifshitz, E. (1997). *Quantum Mechanics, Theoretical Physics Course*, Vol. 3 (Butterworth–Heinemann).

Levy, P. (1954). *Memorial des Sciences Mathematiques* (Gauthier Villars, Paris), fascicule 126.

Li, X., Broughton, J. (1987). High-order correction to the Trotter expansion for use in computer simulation, *J. Chem. Phys.*, **86**, 5094.

Liu, D., Wang, W., Mi, J., Zhong, C., Yang, Q., Wu, D. (2012). Quantum sieving in metal-organic frameworks: A computational study, *Ind. Eng. Chem. Res.* **51**(1), 434–442.

Lu, T., Goldfield, E., Gray, S. (2003). Quantum states of molecular hydrogen and its isotopes in single-walled carbon nanotubes, *J. Phys. Chem. B,* **107**(47), 12989–12995.

Lu, T., Goldfield, E., Gray, S. (2006). Quantum states of hydrogen and its isotopes confined in single-walled carbon nanotubes: dependence on interaction potential and extreme two-dimensional confinement, *J. Phys. Chem. B,* **110**(4), 1742–1751.

Marx, D. Müser, M. (1999). Path integral simulations of rotors: theory and applications, *J. Phys.: Condens. Matter*, **11**, R117.

Matanovic, I., et al. (2012). Hydrogen adsorbed in a metal organic framework-5: coupled translation-rotation eigenstates from quantum five-dimensional calculations, *J. Chem. Phys.*, **137**, 014701.

Mayo, S., Olafson, B., Goddard III, W. (1990). DREIDING: a generic force field for molecular simulations, *J. Phys. Chem.,* **94**, 8897.

Mielke, S., Garrett, B., Peterson, K. (2002). A hierarchical family of global analytic Born–Oppenheimer potential energy surfaces for the H+H reaction ranging in quality from double-zeta to the complete basis set limit, *J. Chem. Phys.,* **116**, 4142.

Mueller, T., Ceder, G. (2005). A density functional theory study of hydrogen adsorption in MOF-5, *J. Phys. Chem. B,* **109**(38), 17974–17983.

Murad, S., Gubbins, K. (1978). Molecular dynamics simulation of methane using a singularity-free algorithm, *Computer Modelling of Matter*, **86**, 62–71.

Müser, M. (1996). The path-integral Monte Carlo of rigid linear molecules in three dimensions, *Mol. Sim.*, **17**(3), 131–141.

Nguyen, T., Jobic, H., Bhatia, S. (2010). Microscopic observation of kinetic molecular sieving of hydrogen isotopes in a nanoporous material, *Phys. Rev. Lett.,* **105**(8), 85901.

Noguchi, D., Tanaka, H., Fujimori, T., Kagita, H., Hattori, Y., Honda, H., Urita, K., Utsumi, S., Wang, Z., Ohba, T., Kanoh, H., Hata, K., Kaneko, K. (2010). Selective D2 adsorption enhanced by the quantum sieving effect on entangled single-wall carbon nanotubes, *J. Phys.: Condens. Matter*, **22**, 334207.

Noguchi, D., Tanaka, H., Kondo, A., Kajiro, H., Noguchi, H., Ohba, T., Kanoh, H., Kaneko, K. (2008). Quantum sieving effect of three-dimensional

Cubased organic framework for H2 and D2, *J. Am. Chem. Soc.*, **130**(20), 6367–6372.

Pan, L., Parker, B., Huang, X., Olson, D., Lee, J., Li, J. (2006). Zn(tbip) (H2tbip = 5-tert-butyl isophtalic acid): a highly stable guest-free microporous metal organic framework with unique gas separation capability, *J. Am. Chem. Soc.*, **128**, 4180.

Patkowski, K., Cencek, W., Jankowski, P., Szalewicz, K., Mehl, J., Garberoglio, G., Harvey, A. (2008). Potential energy surface for interactions between two hydrogen molecules. *J. Chem. Phys.*, **129**(9), 094304.

Rabani, E., Reichman, D., Krilov, G., Berne, B. (2002). The calculation of transport properties in quantum liquids using the maximum entropy numerical analytic continuation method: Application to liquid para-hydrogen, *Proc. Natl. Acad. Sci.* **99**(3), 1129.

Rappé, A., Casewit, C., Colwell, K., Goddard III, W., Skiff, W. (1992). UFF, a full periodic table force field for molecular mechanics and molecular dynamics simulations, *J. Am. Chem. Soc.*, **114**, 10024.

Toda, M., Kubo, R., Hashitsume, N. (1992). *Statistical Physics: Nonequilibrium Statistical Mechanics*, Vol. 2 (Springer).

Trasca, R., Kostov, M., Cole, M. (2003). Isotopic and spin selectivity of H2 adsorbed in bundles of carbon nanotubes, *Phys. Rev. B*, **67**(3), 035410.

Tuckerman, M., Berne, B., Martyna, G. (1992). Reversible multiple time scale molecular dynamics, *J. Chem. Phys.*, **97**, 1990.

Tuckerman, M., Berne, B., Martyna, G., Klein, M. (1993). Efficient molecular dynamics and hybrid Monte Carlo algorithms for path integrals, *J. Chem. Phys.*, **99**, 2796.

Wang, Q., Challa, S. R., Sholl, D. S., Johnson, J. K. (1999). Quantum sieving in carbon nanotubes and zeolites, *Phys. Rev. Lett.*, **82**(5), 956.

Wang, Q., Johnson, J. K., Broughton, J. Q. (1997). Path integral grand canonical Monte Carlo, *J. Chem. Phys.*, **107**, 5108.

Xu, M., Sebastianelli, F., Gibbons B., Bacic, Z., Lawler, R., Turro, N. (2008). Quantum dynamics of coupled translational and rotational motions of H2 inside C60, *J. Chem. Phys.*, **128**, 011101.

Xu, Q., Liu, D., Yang, Q., Zhong, C. (2009). Molecular simulation study of the quantum effects of hydrogen adsorption in metal-organic frameworks: influences of pore size and temperature, *Mol. Sim.*, **35**(9), 748–754.

Yildirim, T., Hartman, M. R. (2005). Direct observation of hydrogen adsorption sites and nanocage formation in metal-organic frameworks, *Phys. Rev. Lett.*, **95**, 215504.

Zhao, X., Villar-Rodil, S., Fletcher, A., Thomas, K. (2006). Kinetic isotope effect for H2 and D2 quantum molecular sieving in adsorption/desorption on porous carbon materials, *J. Phys. Chem. B*, **110**(20), 9947–9955.

Chapter 7

Molecular Modeling of Gas Separation in Metal–Organic Frameworks

Qingyuan Yang, Dahuan Liu, and Chongli Zhong
Department of Chemical Engineering, Beijing University of Chemical Technology, Beijing 100029, People's Republic of China
qyyang@mail.buct.edu.cn, zhongcl@mail.buct.edu.cn

7.1 Introduction

Over the past decades, the associated research on metal–organic frameworks (MOFs) has been rapidly developed into one of the most prolific areas in chemistry, chemical engineering, and material science [1–5]. The intrinsic nature of MOFs allows the ability of systematically modulating the pore dimensions, surface areas, topologies and surface chemistry within them in an extremely broad range. This remarkable feature of MOFs is largely absent in conventional porous materials such as zeolites [6], which permits MOFs to serve as an ideal platform for various specific targets and thus puts them in a unique position. For this reason, one of significant efforts has been devoted in recent years to explore and improve the performance of MOFs for gas separation [5]. From

Metal–Organic Frameworks: Materials Modeling towards Potential Engineering Applications
Edited by Jianwen Jiang

ISBN 978-981-4613-45-3 (Hardcover), 978-981-4613-46-0 (eBook)
www.panstanford.com

an experimental point of view, it is not only very time consuming to explore the thousands of MOFs reported in the literature, but also the direct measurements on the related behaviors of mixtures confined in MOFs remain challenging. In addition, highly detailed information that leads to the macroscopic properties observed in MOFs is as well not easy to be addressed using experimental methods. In contrast, molecular modeling provides a very valuable complement to experimental studies of MOFs, which can give deep insights into the mechanisms that control the gas separation capability of MOFs at a molecular level. Compared with experiments, this method can be utilized to isolate the key influencing factors and quantify their separate contributions to the separation behaviors of a large population of MOFs [7], as well as the cooperative effects among a set of selected factors [8]. Furthermore, due to a variety of inorganic and organic moieties, it offers a theoretically unlimited number of possible structures of MOFs. Molecular modeling is then a particularly attractive tool for large-scale virtual screening of the existing MOFs and new hypothetical MOFs designed by computer [7, 9–11], allowing experimental endeavors to only concentrate on the candidates with the most promising separation performance.

In this chapter, we concentrate our attention on molecular modeling of gas separation in MOFs. The main methods adopted in modeling studies include quantum mechanics (QM) calculations, Monte Carlo (MC) and molecular dynamics (MD) simulations, etc. Indeed, before the naissance of MOFs, these methodologies have already been quite well established and have been widely used to probe the separation behaviors of gas mixtures in other conventional porous solids. For details of these modeling tools, we suggest to the reader to refer the related textbooks and therein [12–14]. Here we just intend to introduce some advances in the development of research methods with respect to MOFs. In addition, from a survey of the literature, the separation of gas mixtures in MOFs could be classified into two fundamentally different concepts, adsorption-based and membrane-based processes. Thus, some representative simulation studies based on these two concepts are highlighted in this chapter, followed by a brief review of the strategies used for enhancing the separation performance of MOFs. Finally, this chapter is ended with a summary of current challenges of molecular modeling and some suggestions for prospective studies on MOFs.

7.2 Development of Research Methods

7.2.1 *Charge Calculations*

For molecular modeling with classical force fields, the potential energy of a quadrupolar or polar molecule adsorbed in MOFs is commonly expressed as a sum of van der Waals (vdW) and electrostatic interactions. As a consequence, multiple strategies have been used in MOF studies to take into account the electrostatic interactions. Generally speaking, they can be grouped into two categories: one is to assign partial charges for the framework atoms of MOFs [15], while the other is to directly incorporate the tabulated three-dimensional electrostatic potential energy surface (EPES) into calculations [16]. Although the second strategy may be a rigorous way to evaluate the electrostatic interactions in nanoporous crystalline materials, it is not suitable for MOFs with flexible framework. Further, it also becomes very computation-expensive to tabulate fine electrostatic potential (ESP) grids for MOFs with very large unit cell sizes. Thus, only the first strategy is briefly described here.

Currently, several types of atomic partial charges obtained from QM calculations have been used in the computational studies of MOFs. The first type is the ESP-derived charges. To compute ESP charges, a density functional theory (DFT) calculation with the ChelpG [15] or the Merz–Singh–Kollman (MSK) [17] method is usually performed on a representative cluster cleaved from the matrix of one certain MOF. Then, a curve-fitting procedure is used to determine the set of partial charges on the nuclei that would reproduce the resultant electrostatic potential surrounding this cluster. The second type is the Mulliken charges, which are extracted from the Mulliken population analysis of the DFT calculation results conducted on the periodic structures of MOFs [18]. The fundamental assumption used by this scheme for partitioning the wave function is that the overlap between two orbitals is shared equally [19].

Another type is a so-called density-derived electrostatic and chemical (DDEC) charges. DDEC charges can be obtained using the periodic structures of MOFs, which optimize the net atomic charges (NACs) to be chemically meaningful and to reproduce the electrostatic potential outside the electron distribution by simultaneously optimizing atomic electron density distributions to be close to spherically symmetric and to resemble reference state atoms [20]. Although none of these methods give an unambiguous definition of the resulting atomic partial charges, ESP charges obtained with ChelpG method are the most popular ones adopted in the simulation studies of MOFs [21–23].

Aiming at computational screening of MOFs, there is considerable value to have a general protocol that can be used to quickly estimate the atomic partial charges with reasonable accuracy for any MOFs, both existing and hypothetical ones. This can stimulate significantly the progress in computational modeling of MOFs and thus guide experiments. To solve this obstacle, our group proposed a strategy named "connectivity-based atom contribution" (CBAC) [24, 25]. It was built on the assumption that the atomic partial charge of an atom in the framework of an MOF is determined by its bonding environment and the atoms with same connectivity have identical charges in different MOFs. This approximate method is similar to the well-known group contribution method that has been successfully used for calculating thermodynamic properties of fluids. Excellent agreements were achieved between the simulated adsorption isotherms using CBAC-based and QM-based charges [24–26]. As commented by others [27], the CBAC method can greatly save both time and computational cost of quantum calculations, further enabling a theoretically rapid screening of MOFs for the targeted property. Moreover, this simple method can be readily extended by including new atom types, making it possible to estimate framework charges in any MOF solely on the basis of its structure.

Another semiempirical technique that can benefit screening efforts is the charge equilibration (Qeq) method proposed by Rappé and Goddard in 1991 [28], which is based on electronegativity equalization of the individual atoms of a compound with the ionization potentials and electron affinities of the atoms taken into account for the electrostatic contributions to the cohesive energy. To obtain better agreement with the partial charges of heavy atoms ($Z > 20$) or atoms with partial charges typically greater in magnitude than 1, Willmer and Snurr [29] suggested a revised version of the Qeq method by simply expanding in a Taylor series around the charged atom and using the ionization potentials and electron affinities corresponding to the metal atoms. Very recently, Sholl et al. [11] proposed a periodic charge equilibration (PQeq) method, which is applicable to the full periodic structure of MOFs, while it often leads to an underestimation of the charges on the metal atoms and the atoms connected to them [11].

7.2.2 *Evaluation of Electrostatic Characteristic of MOFs*

Due to the structural nature of MOFs that contain inorganic clusters and organic moieties, there exist two distinct structural microdomains in MOFs. These can induce a strong electrostatic potential gradient inside the pores of MOFs, and such an electrostatic characteristic may become an important factor that influences the separation behavior of gas mixtures that have components with large difference in dipole and/or quadrupole moments. To that purpose, we conducted a systematic study of the separation of equimolar CO_2/CH_4 and C_2H_6/CH_4 mixtures in a representative material Cu–benzene-1,3,5-tricarboxylate (Cu-BTC) [30]. Due to the unique structure of this MOF, two ordered microdomains with different electrostatic field strengths were identified: strong regions of the tetrahedron-shaped pockets and weak ones of the large square-shaped pores. Grand canonical Monte Carlo (GCMC) simulation results showed that the selectivity for C_2H_6 over CH_4 decreases with increasing pressure, while the selectivity for CO_2 over CH_4 shows the opposite trend. To find out the underlying separation mechanisms, we compared the adsorption behaviors of a CO_2/CH_4 mixture in three different circumstances: case 1, switching off all the electrostatic interactions; case 2, only switching off the electrostatic interactions between CO_2 molecules and Cu-BTC; and case 3, considering all the electrostatic interactions. The corresponding simulation results are presented in Fig. 7.1, as a function of pressure. Obviously, in cases 1 and 2, the selectivities in the low-pressure range show a decreasing trend as found for

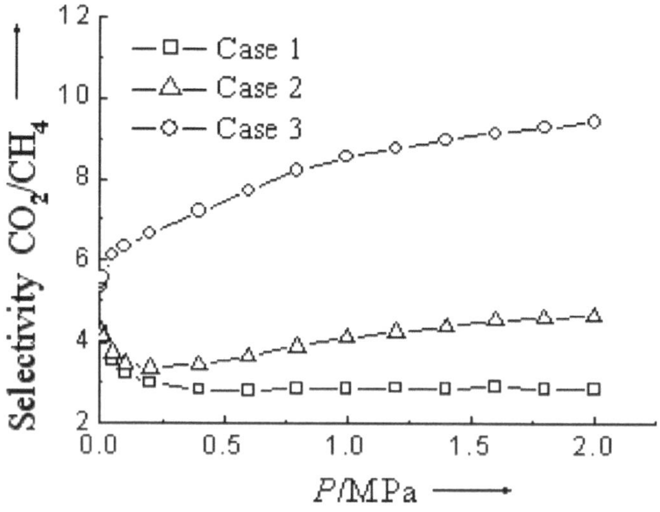

Figure 7.1 Effect of electrostatic interactions on the selectivity of carbon dioxide from the equimolar mixture of methane and carbon dioxide in Cu-BTC at 298 K. Reprinted with permission from Ref. [30]. Copyright 2006, Wiley-VCH.

C_2H_6 over CH_4, whereas this trend is reversed in case 3. In addition, at low pressures, the amount of molecules adsorbed in the material is very small and the effects of the electrostatic interactions between adsorbed molecules on the selectivity are negligible, which explains the almost identical selectivities in cases 1 and 2. Compared with the monotonically decreasing trend observed in case 1 (only the dispersive interactions exist in this system), in case 2, the electrostatic interactions between CO_2 molecules can enhance the separation slightly at high pressures and the enhancement becomes larger with increasing pressure. In contrast, the selectivity in case 3 increases moderately with increasing pressure in the whole pressure range by further including the CO_2–Cu-BTC electrostatic interactions. Similar effects have also been reported on the separation of a CO_2/N_2 mixture in Cu-BTC [31] and a CO_2/H_2 mixture in isoreticular metal–organic frameworks (IRMOFs) [32]. Therefore, it is clear that the electrostatic characteristic of MOFs should be taken into account when MOFs are used for the separation of gases with different quadrupole moments or polarities. This work also provides a useful methodology to study how the electrostatic properties of MOFs influence their separation performances for gas mixtures.

7.2.3 *Methodology for Large-Scale Screening of MOFs*

Given the large population of the existing MOFs as well as their theoretically infinite number of structures, it would be very useful to find an efficient approach that could rationalize the performance of a large series of MOFs for a targeted application. Deriving a quantitative structure–property relationships (QSPR) model satisfying this requirement would be invaluable for not only predicting the performances of a given MOF without the need to perform experiments or molecular simulations but also for further guiding the tuning/design of advanced materials with outstanding performance. To this target, we recently proposed a new concept "adsorbility" (*AD*) [7] for nanoporous materials defined as

$$AD = \frac{Q_{\mathrm{st}}^0}{\varphi} \tag{7.1}$$

where Q_{st}^0 is the isosteric heat of adsorption at infinite dilution in unit of kJ/mol for an adsorbate and φ is the free volume per unit volume in units of cm^3/cm^3, which is also known as the porosity of one material. The physical meaning of parameter *AD* is a measure of adsorption energy density for a gas in a given nanoporous material. As the concept of *AD* is enlightened from the solubility parameter in solution theory, it is expected this new parameter can also serve as a measure of the "compatibility" of the two gases adsorbed in a nanoporous material, and thus the difference in *AD* (ΔAD) of the components can be used to characterize the selectivity.

To illustrate the applicability of this concept, the limiting selectivities of CH_4 from an equimolar CH_4/H_2 mixture in 105 selected MOFs as a function of $1/\Delta AD$ of the two components are shown in Fig. 7.2, which reflects the intrinsic separation capability of the material. A comparison with the correlation in this figure indicates that the selectivities in the limiting condition correlate well with ΔAD, demonstrating that *AD* can grasp the main feature of a gas/MOF system and approximately represent the interplay of

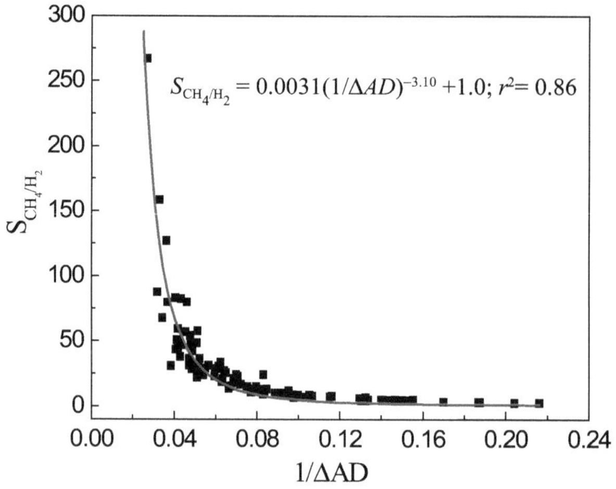

Figure 7.2 Limiting selectivity vs. $1/\Delta AD$ from the equimolar mixture CH_4/H_2. Reprinted with permission from Ref. [7]. Copyright 2011, American Institute of Chemical Engineers (Wiley-VCH).

various influencing factors. This figure also points out that to obtain a large ΔAD, the differences in the interactions between the two adsorbates and the MOF should be as large as possible and the MOF should have small porosity. As a result, it can be concluded that synthesizing MOFs with a large ΔAD can serve as a general design criterion for the development of MOFs as well as for preliminary large-scale screening of MOFs for targeted separations.

7.3 Adsorption-Based Separation

Adsorption-based separation is a physisorptive operation governed by thermodynamic equilibrium process, which relies on the fact that gases reversibly adsorb in nanopores at densities that far exceed the bulk densities of the gas sources equilibrated with the porous solids [33]. The separation behavior is mainly controlled by the competitive adsorption affinity of the pore surface of the adsorbents for various species in the mixtures. For adsorption-based mixture separation, the performance of a porous material is usually characterized by adsorption selectivity. The selectivity for species A relative to B is calculated with the following expression, where x_i and y_i (i = A, B) are the mole fractions of species i in the adsorbed and bulk phases, respectively.

$$S_{\text{ads}} = \frac{x_{\text{A}}/x_{\text{B}}}{y_{\text{A}}/y_{\text{B}}} \tag{7.2}$$

7.3.1 CO_2-Related Systems

As a subject of widespread public concerns, the reduction of the anthropogenic CO_2 emission in the atmosphere has become one of the most urgent climate issues. On the basis of the large capacity for gas adsorption and easy structural regulation of MOFs, significant efforts have been made in recent years to examine the performance of such type materials for carbon capture and sequestration (CCS) from various gas mixtures. One scenario that received the heaviest studies using molecular modeling is to assess MOFs for precombustion removal of CO_2 from its gas mixtures with CH_4 or H_2. The separations of these two mixtures are also of great economical and technological importance for the upgrading of low-quality natural gas (such as biogas and landfill gases) or the purification of H_2-rich syngas obtained from steam re-forming of natural gas. Another important scenario is to probe the applicability of MOFs for postcombustion CO_2 capture from flue gas emitted from power plants, which is mainly composed of CO_2 and N_2.

Our group has carried out a series of molecular modeling studies to evaluate the performance of different MOFs [30–32, 34–38] for the separation of various gas mixtures involving CO_2. In 2006, we conducted the first computational study on the subject of CO_2 capture, in which the separation behaviors of $CO_2/CH_4/C_2H_6$ mixtures were investigated in a representative MOF called Cu-BTC [30]. In the whole pressure range examined in this study, the selectivity for CO_2 over CH_4 increases moderately with increasing pressure, as already shown in Fig. 7.1. The selectivity for CO_2 over C_2H_6 shows the same trend and an unusual selectivity reversal occurs. The microscopic coadsorption mechanism obtained for the CO_2/CH_4 system (see Fig. 7.3) revealed that at low pressure, the small tetrahedron-shaped pockets are the preferential adsorption sites for both CH_4 and CO_2 molecules, which is not different from

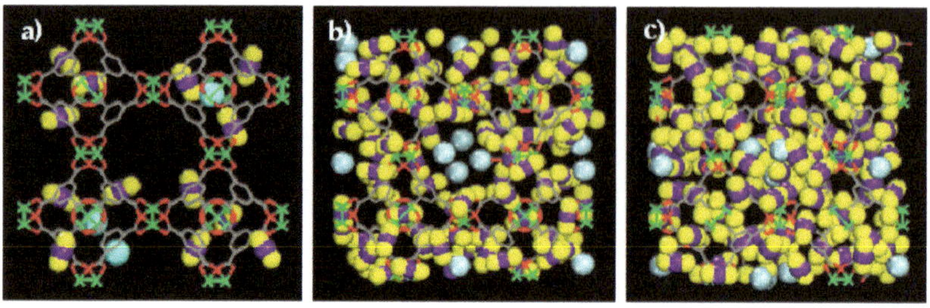

Figure 7.3 Snapshots of structures of Cu-BTC with adsorbed mixtures of CH_4 and CO_2 at various pressures. (a) $P = 0.05$ MPa, (b) $P = 1.08$ MPa, and (c) $P = 2.0$ MPa. The Cu-BTC framework is shown in line style with color: Cu, green; O, red; C, gray; methane: blue; carbon dioxide: C, violet, O, yellow. Reprinted with permission from Ref. [30]. Copyright 2006, Wiley-VCH.

those observed for the single-component adsorption. With pressure increasing, CH_4 molecules are pushed out of the pockets to the larger square-shaped pores by CO_2, followed by the filling of the channels with further increasing pressure. Later, we extended this study to compare the separation behaviors of a $CO_2/CH_4/H_2$ mixture in Cu-BTC with another well-known material, IRMOF-1 (also named MOF-5) [34], showing that both the geometry and pore size of MOFs can affect their separation efficiency largely. IRMOF-1 possesses a simple structure with large cubic pores and consequently leads to simple selectivity behaviors, while Cu-BTC with a pocket/channel pore structure can result in more complex separation behavior with different selectivity steps. The CO_2/CH_4 selectivity in Cu-BTC is significantly enhanced compared to IRMOF-1 because of the stronger electrostatic interactions for CO_2 with the former.

In another simulation study, we found that Cu-BTC also exhibits a promising application for separating CO_2 from a simulated flue gas ($CO_2/N_2/O_2$) [31]. The results also manifested that due to stronger interactions with the coordinatively unsaturated Cu sites, the larger quadrupolar CO_2 molecules are adsorbed much closer to the Cu sites than N_2 and O_2. From our systematic studies, it can be seen that the electrostatic property of MOFs plays an important role in separating CO_2 from natural or flue gases. This leads to a conclusion that strengthening the electrostatic field gradient in MOFs should be an efficient approach to improve their separation performance for the gases that are easily induced by electrostatic effects.

Ionic MOFs with charge-balancing extra-framework ions belong to a unique subset of MOFs. In contrast to zeolites, the extra-framework ions can be cationic or anionic type, depending on the framework constituents of MOFs. The first theoretical study involving ionic MOFs was published by Jiang et al. [39], where they studied the separation behavior of a CO_2/CH_4 mixture in a cationic *soc*-MOF with NO_3^- ions using molecular simulations. In this charged MOF, NO_3^- ions were found to essentially vibrate around their favorable location sites because of their strong binding with the framework and the steric restriction of the carcerand-like capsules in this material. These extra-framework ions can act as additional sites to augment the interactions between adsorbate and *soc*-MOF, particularly the electrostatic interactions for quadrupolar CO_2 molecules, and thus substantially enhance the selectivity of CO_2 over CH_4. Using a similar simulation method, they also investigated the coadsorption of a CO_2/H_2 mixture in this MOF and found that with increasing pressure, the selectivity increases sharply at low pressure and then decreases [40]. The initial increases were attributed to the strong interactions of CO_2 with *soc*-MOF and the cooperative interactions between adsorbed CO_2 molecules, whereas the entropy (packing) effect accounts for the decreasing of the selectivity at high pressure. Later, the selective CO_2 adsorptions from CO_2/H_2, CO_2/N_2, and CO_2/CH_4 gas mixtures were also investigated by them in an anionic *rho*-ZMOF with Na^+ cations [41]. They found that the selectivities for all these three mixtures decrease monotonically with increasing pressure, which was considered a consequence of the significant reduction in the electrostatic interactions between CO_2 and *rho*-ZMOF with increased loading. To quantitatively examine the effect of charges, using our strategy described in Section 7.2.2, they further investigated the adsorption behaviors of these mixtures in a neutral structure by switching off the charges of the Na^+ ions and framework. The simulation results showed there is a significant drop of the selectivities relative to those in the charged framework, especially in the low-pressure region. These studies clearly demonstrated that the incorporation of extra-framework ions into MOFs is one alternative method to enhance the electrostatic interactions between adsorbates and MOFs that can lead to exceptionally high separation capability of MOFs.

Another interesting type of MOFs are those materials with highly flexible frameworks such as Matériaux de l'Institut Lavoisier (MIL)-53(M) (M = Al, Cr, or Fe) and Cu(dhbc)$_2$(4,4'-bpy) that can exhibit unusual reversible features of "breathing" [42] and "gate opening" [43] effects upon the adsorption of guest species or the stimulation of external temperature and pressure. In addition, the structural transformation of these materials is also related to the polar nature of the probe gas molecule, which could pave the way to use them for separating strategic gases with different polarities. Llewellyn et al. [44] conducted an investigation to examine the performance of MIL-53(Cr) for separating a CO_2/CH_4 gas mixture using a combination of GCMC simulations and different experimental techniques. A hybrid Monte Carlo (HMC) simulation approach in the osmotic statistical ensemble was also employed in this work to follow the evolution of the composition of the adsorbed mixture when the MIL-53(Cr) switches from the large-pore (LP) to the narrow-pore (NP) form. Their results showed that the coadsorption of CO_2 and CH_4 leads to a similar breathing behavior of MIL-53(Cr) as with pure CO_2. This breathing effect is mainly controlled by the partial pressure of CO_2, but increasing the CH_4 content progressively decreases the transformation of LP to NP. The HMC simulation results gave clear evidence that CH_4 is excluded from the NP form, which is filled exclusively by CO_2 molecules. Except for the one and only study [45] of a CO_2/CH_4 mixture in the gate-opening material Cu(dhbc)$_2$(4,4'-bpy) using a thermodynamic analytical method (i.e., osmotic framework adsorption solution theory, OFAST), to the best of our knowledge, there is no related molecular simulation exploration reported for gas separation in such type of flexible MOFs. This OFAST method was also extended to study the coadsorption behaviors of a CO_2/CH_4 mixture in a breathing solid MIL-53(Al) [45, 46].

In reality, natural and flue gas sources usually contain a small amount of H_2O or industrial separation process might be operated under a humid condition. Consequently, an important question is how the presence of H_2O affects the performance of MOFs for CO_2 capture. Snurr et al. were the first to address the issue using molecular simulations [47]. They discovered that the presence of H_2O molecules coordinated to the copper open metal sites in Cu-BTC can increase the adsorption of CO_2, especially at low pressure, whereas the adsorption of CH_4 was not significantly affected and adsorption of N_2 was even reduced. This interesting observation was also verified by their experimental measurements. Through examining the contributions of different energies, they found the Coulombic interactions between H_2O and CO_2 are mostly responsible for the increase in CO_2 adsorption. The electrostatic interactions arising from the quadrupole moment of CO_2 interacting with the electric field gradient of the adsorbent are increased when Cu atoms are occupied by water molecules. They further simulated equimolar mixtures of CO_2/CH_4 and CO_2/N_2 in dry Cu-BTC and hydrated Cu-BTC with preadsorbing 4 wt% water. The predicted selectivities showed that the coordinated water molecules significantly improve the selectivity of CO_2 over CH_4 and N_2, as shown in Fig. 7.4. They suggested that axial ligation of the

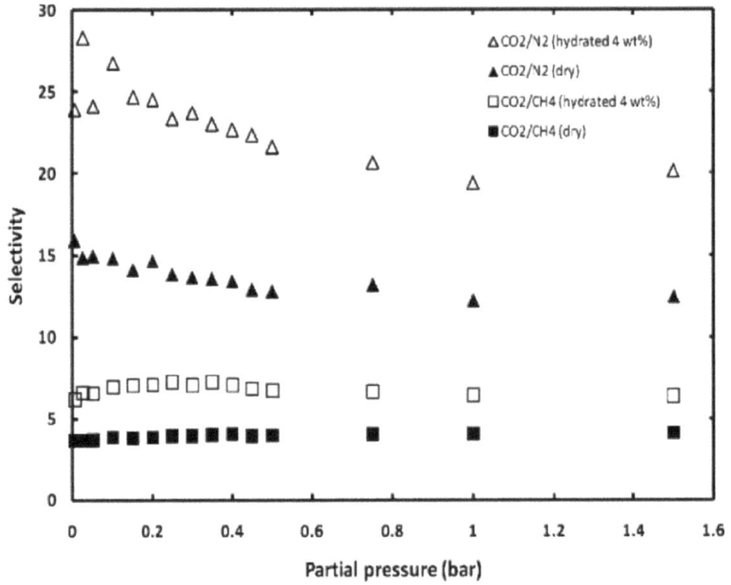

Figure 7.4 CO_2 selectivities from simulations of equimolar mixtures of CO_2/CH_4 and CO_2/N_2 in dry and hydrated Cu-BTC at 298 K.

coordinatively unsaturated metal sites by various molecules could open new possibilities for tuning the adsorption behavior of MOFs for CO_2 capture and other applications.

Liu and Smit simulated the adsorption behaviors of two three-component mixtures ($CO_2/CH_4/H_2O$ and $CO_2/N_2/H_2O$) in zeolitic imidazolate framework (ZIF)-68 and ZIF-69 [48]. They demonstrated that only if the amount of H_2O in the bulk phase is above 5%, the presence of water influences the adsorption selectivities in these ZIFs significantly. Jiang found that even the presence of a trace amount of H_2O, there is a significant influence on separation of a CO_2/H_2 mixture in the charged *soc*-MOF [40]. The CO_2/H_2 selectivity increases at low pressure due to promoted adsorption of CO_2 by the H_2O molecules strongly bound to the exposed intraframework indium atoms but decreases at high pressure as a consequence of the competitive adsorption of H_2O over CO_2. In contrast, Jiang et al. discovered [49] that the selectivity for CO_2 over CH_4 in the charged *rho*-MOF with extra-framework Na^+ ions decreases by one order of magnitude when a trace mount of H_2O presents in the bulk mixture. Their analysis showed that the interaction between CO_2 and Na^+ is substantially reduced upon adding H_2O, inducing a strong shielding on cations and thus exerting an adverse effect on CO_2 adsorption. They suggested that a pretreatment is required to remove moisture from mixtures prior to using the charged MOFs for separation.

7.3.2 *Gas Mixtures of Olefin and Paraffin*

The separation of olefin and paraffin mixtures represents a class of most important and also difficult chemical separations in the petrochemical industry [50]. Energy-efficient adsorptive separation is widely considered as a promising alternative to the traditional cryogenic distillation technologies for this application, but it requires a careful choice of suitable adsorbents. Compared to the numerous studies for CO_2 capture, to date, only a few computational investigations have been conducted to study the separation of gaseous olefin/paraffin mixtures in MOFs.

One of the industrial processes for the production of ethylene (C_2H_4) is usually accomplished by steam cracking or thermal decomposition of ethane (C_2H_6) [51]. This process results in a mixture of C_2H_4 and uncracked C_2H_6 that has to be separated to obtain high-purity C_2H_4. In 2008, we performed GCMC simulations to study the separation of an equimolar C_2H_4/C_2H_6 mixture in Cu-BTC [52]. It was found that this MOF is more favorable to the adsorption of C_2H_4 than C_2H_6 and the calculated selectivity is about 2 at low pressure. Our simulations were also consistent with the QM calculation results obtained by Nicholson and Bhatia [53]. They showed that C_2H_4 has a higher binding energy than C_2H_6, attributed to an increased interaction with the coodinatively unsaturated Cu sites and stronger hydrogen bonding with the basic framework oxygen atoms. The difference in binding energy between them leads to the selectivity factor observed by us at low pressure. Bae et al. also performed a GCMC simulation study of this mixture in MOF-74(Co) [54]. They revealed that the selectivity of C_2H_4 over C_2H_6 shows a decreasing tendency with increasing pressure, and a considerable amount of pore volume is still left after all the metal sites are occupied by C_2H_4 molecules.

As an important commercial petrochemical, the large-scale industrial production of propene (C_3H_6) requires to the separation of its mixture with propane (C_3H_8) [55]. Rodrigues et al. performed a systematic molecular simulation study combined with experiments to examine the adsorption of propane, propylene and their binary mixtures in Cu-BTC [55, 56]. They reported that unsaturated Cu^{2+} sites in this material interact preferentially with C_3H_6, due to the specific interactions between the π-orbitals of this molecule and the open metal sites. These specific interactions are the primary cause of the enhanced affinity of Cu-BTC for C_3H_6, leading to a moderate selectivity of about 4 for C_3H_6 over C_3H_8 [56]. On the basis of the results obtained from first-principles calculations and GCMC simulations [54], Snurr et al. found that, for three isostructural frameworks MOF-74(M) (M = Co, Mn, or Mg), C_3H_6 binding to the unsaturated Co^{2+} sites is the strongest and MOF-74(Co) also gives the largest difference in binding energies between C_3H_6 and C_3H_8. Interestingly, this work also showed that the C_3H_6/C_3H_8 selectivity increases with increasing pressure, in sharp contrast to the decreasing trend for the separation of a C_2H_4/C_2H_6 mixture. This unusual selectivity behavior for a C_3H_6/C_3H_8 mixture in MOF-74 was ascribed to the proper match between the pore size and the size of the molecules, that is, the pore volume is filled when each metal site is occupied by one C_3H_6 molecule, which strongly suppresses the adsorption of C_3H_8.

As can be seen from the existing studies, the adsorption-based separations of olefin/paraffin mixtures are mainly conducted in MOFs with open metal sites. It has been well-documented that the classical force fields employed in molecular simulations are inadequate for describing the adsorption behaviors of the adsorbates strongly interacting with these metal sites [57, 58]. For examining the separation performance of such systems, ideal adsorbed solution theory (IAST), developed by Myers and Prausnitz [59], was usually adopted at the moment with the experimental adsorption isotherms for single components as input [54, 60, 61]. Using this method, Long et al. [60] showed that MOF-74(Fe) has excellent performance characteristics for separating olefins from both C_2H_4/C_2H_6 and C_3H_6/C_3H_8 mixtures at 318 K, which was confirmed by their breakthrough data obtained experimentally for these two mixtures. Their simulated breakthrough data also predicted high selectivities and capacities of this MOF for the fractionation of four-component ($CH_4/C_2H_6/C_2H_4/C_2H_2$) mixtures as well as removal of C_2H_2 impurities from C_2H_4. The group of Jorge conducted GCMC simulations to calculate the selectivities of C_3H_6 over C_3H_8 in Cu-BTC on the basis of their DFT-derived force fields and compared them with those obtained from IAST calculations [61]. It was found that the IAST predicts a constant selectivity in the entire composition range, while molecular simulations predicted increasing selectivity with C_3H_8 mole fraction. For this discrepancy, they claimed that three possible reasons, inadequacy of the IAST model for describing adsorption in this system, inadequacy of the VDW force field, and inadequacy of their approach to account for the Cu–π interactions, cannot be ruled out.

MOFs have also been tested their separation performance for other important olefin/paraffin mixtures involved in the chemical industry, such as acetylene/ethylene [62, 63] and isobutene/isobutene [64] gas mixtures. These studies showed that some MOFs have large potential for the separation of these systems by selective adsorption. As the scope of this chapter is limited to focus on the scope of molecular modeling, these experimental studies are not described in detail here.

7.3.3 *Other Gas Mixtures*

Besides the systems discussed above, molecular modeling has also been used to investigate the adsorption of other gas mixtures in MOFs. For examples, Düren and Snurr [65] used GCMC simulation to study adsorption separation of an n-C_4H_{10}/CH_4 mixture in six IRMOFs with similar chemical composition but different pore size, showing that the selectivity toward n-C_4H_{10} increases with decreasing cavity size and increasing number of carbon atoms in the organic ligands. As far as we know, this is the first molecular modeling study of mixture separation in MOFs. There are also some studies to test MOFs for adsorptive separation of alkane isomer mixtures [66, 67]. Greathouse et al. [68] simulated the adsorption of noble gas mixtures IRMOF-1 and disclosed that this material selectively adsorbs Xe from Xe/Kr and Xe/Ar mixtures. Yang and Zhong [34] compared the separation performance of IRMOF-1 and Cu-BTC for a CH_4/H_2 mixture and the GCMC simulation results showed that Cu-BTC is a promising material for this purpose. Garberoglio [69] calculated the isotopic separation for T_2/H_2 and D_2/H_2 mixtures in a series of MOFs and COFs using path integral MC simulations. The selectivity shows a less pronounced dependence on the external pressure, but different behaviors are found for different materials. Liu et al. [70] also studied the separation of hydrogen isotopes separation in nine MOFs with channel-like pores, where a new concept named "quantum effective pore size" (QEPS) was proposed to correlate with the selectivity. Molecular simulations were first applied by Fischer et al. [71] to assess MOFs for C_2H_2/CO_2 separation and found that magnesium formate emerges as the most efficient one amongst the three novel materials studied by them.

7.4 Membrane-Based Separation

Due to their unique properties, MOFs are also considered as excellent candidates for fabricating membranes toward the separation of gas mixtures. Compared to the competing technologies such as pressure swing adsorption and cryogenic distillation, gas separation by membrane permeation can be very energy efficient and requires lower investment cost in many industrial-scale processes [5, 72]. Membrane separation is usually based on the size and shape of the adsorbates to be separated or on the interplay of their adsorption and diffusion properties in the membrane material [73]. As a rule, the performance of a membrane is evaluated by two characteristic parameters, permeability and selectivity. The permeability (P_i) of component i in the mixture reflects its transport rate through a given membrane, which relates the net flux of it to the membrane thickness (L) and its partial pressure (Δp_i) drop across the membrane by

$$P_i = \frac{J_i \times L}{\Delta p_i} \tag{7.3}$$

The selectivity (or permeation selectivity) estimates the separation efficiency of a membrane, which is defined as the ratio of the permeabilities of two gas components in a mixture:

$$S_{\text{perm},ij} = \frac{J_i}{J_j} \tag{7.4}$$

If the fluxes of pure components are used in Eq. 7.4, it corresponds to the ideal selectivity.

From the point view of the cost and challenge of the preparation of MOF membranes, computational methods can predict the material properties needed to characterize membrane performance and thus can greatly accelerate the development of practical MOF-containing membrane. Compared to the significant efforts devoted to characterizing MOFs for adsorption-based separations, exploration of the incorporation of MOFs into membranes using molecular modeling is still in an extremely early stage.

7.4.1 *CO_2 Related Systems*

Among the existing simulation studies testing MOF membranes for gas separation, the removal of CO_2 from various mixtures might be the most popular topic. The first information on the performance of MOFs as membranes for gas separations was reported by Keskin and Sholl [74]. They performed atomically detailed simulations to investigate the separation of a CO_2/CH_4 mixture in an ideal IRMOF-1 membrane. The calculations showed that mixture effects play a crucial in determining the properties of membranes. The predictions from single-component results suggest that IRMOF-1 membrane would give strong ideal selectivities for CH_4 in CO_2/CH_4 mixtures, while the mixture permeation study predicts that this kind of membrane shows at best weak selectivity for CO_2. The studying approach adopted in this work was subsequently extended by them to assess the separations of CO_2/H_2 and CO_2/N_2 mixtures in the IRMOF-1 membrane [75] and CO_2/CH_4 and CO_2/H_2 mixtures in the Cu-BTC membrane [76]. They further showed that it is crucial to characterize MOF membranes on the basis of their performance for mixed-gas feeds rather than extrapolating their performance from the results for pure gases. Their results also demonstrated that the Cu-BTC membrane shows higher selectivities than the IRMOF-1 membrane, as shown in Fig. 7.5.

To considerably save the computational resources and time in these calculations with atomically detailed models, the group of Sholl introduced an efficient approximate method for screening MOFs to accelerate the modeling of membrane applications [77]. In this method, they modified Krishna and van Baten's formulation [78] applied for the permeate side close to vacuum conditions and expressed the permeation selectivity of a membrane for one species over another as a product of adsorption selectivity and the diffusion selectivity. The former at an arbitrary bulk-phase composition was calculated using Eq. 7.2, and the latter was calculated from the mixture self-diffusivities at the corresponding adsorbed compositions. On the basis of the comparisons with their previous detailed calculations for CH_4/H_2, CO_2/CH_4, and CO_2/H_2 mixtures permeating through IRMOF-1 and Cu-BTC membranes, they proposed a rule of thumb for using this approximate method: if the IAST accurately predicts the mixture isotherm for the adsorbed mixture of interest, then this method is expected to give results that are accurate enough to be used in screening MOF membranes. In this study, they further studied five MOFs and a covalent organic framework COF-102 in order to examine the effect of chemical diversity and interpenetration on the performance of MOF-based

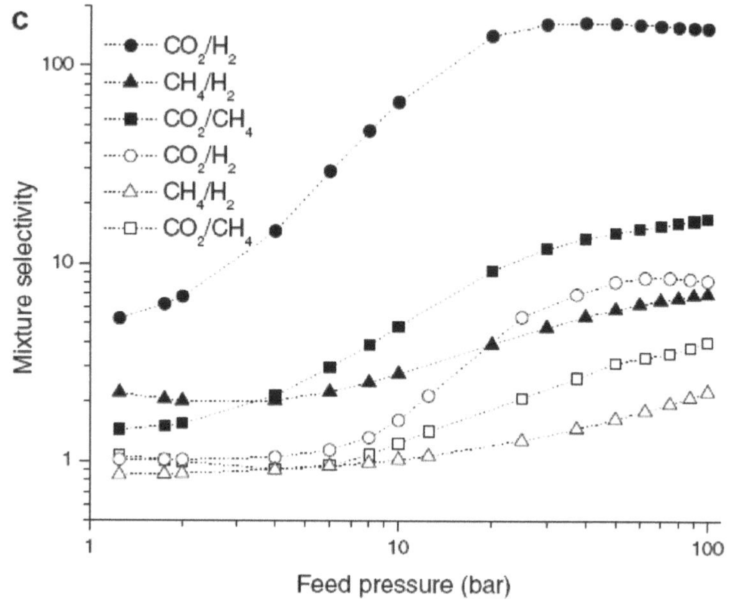

Figure 7.5 Mixture selectivity of equimolar mixtures in Cu-BTC (closed symbols) and IRMOF-1 (open symbols) membranes. The transmembrane pressure drop is 80% of feed pressure.

membranes for these three gas mixtures. On the basis of this approach, Keskin et al. also examined the permeation behaviors of CO_2/CH_4 and CO_2/H_2 mixtures in ZIFs [79, 80] and bio-MOF-11 [81] membranes.

Generally speaking, MOF membrane-based separations that rely on a combination of adsorption and diffusion selectivities would have lower permeation selectivity than the adsorption-based one. However, Sholl et al. demonstrated that this conflicting trend is not a universal property of all MOF-containing membranes [82]. In this study, they evaluated the separation performance for a CO_2/CH_4 mixture in a microporous MOF membrane, Cu(hfipbb)(H_2hfipbb)$_{0.5}$, by a composite computational method involving GCMC, MD, transition-state theory (TST), and plane-wave DFT calculations. Their predictions showed that this material exhibits excellent properties for a membrane-based separation of CO_2 from CO_2/CH_4 mixtures. Apart from the favorable CO_2 adsorption by a moderate factor, diffusion favors CO_2 by orders of magnitude larger than CH_4, indicating that an effective kinetic separation is applicable. The unprecedented net selectivity (10^4–10^5) of this MOF membrane is higher than any known polymeric membrane for this separation [83] or the high-zeolite membranes [84, 85] developed at that time. In addition, the research method used in this work should also make it possible to identify other MOF membranes with attractive properties for kinetic separation of gas mixtures.

Accompanying with a molecular-level understanding of the various diffusion coefficients that arise in the phenomenological Maxwell–Stefan formulation, Krishna et al. have published a series of computational studies to evaluate various MOF-based membranes for the separation of CO_2-related gas mixtures [58, 86–90]. On the basis of their collaboration with Bux et al., GCMC simulations jointed with time- and space-resolved infrared microscopy (IRM) measurements were employed to estimate the CO_2/CH_4 permeation selectivity of a ZIF-8 molecular sieve membrane [86]. A rather unexpected behavior was observed, that is, both the pure component and the gas mixture transport (or Fick) diffusivities of CH_4 increase much steeper than CO_2 with increasing loading. From the almost coinciding trends with increasing temperature between the membrane and GCMC-simulated adsorption selectivities (see Fig. 7.6), they concluded that permeation separation on the ZIF-8 membrane is controlled by adsorption rather than diffusion. Moreover, their findings indicates that the pore apertures of ZIF-8 are rather flexible than rigid, and hence molecular modeling should take framework flexibility into consideration. In a subsequent study [88], their results revealed that the permeances of CO_2 in a CO_2/CH_4 mixture are practically the same as for the pure component over the entire range of loading, whereas the permeances of CH_4 in the mixture are significantly lower than the unary ones at high mixture loadings. This reduction was attributable to the hindering effect caused by the preferentially location of CO_2 at the window regions of ZIF-8, which in turn can result in significant higher permeation selectivity than that obtained from unary permeation data. Krishna et al. also studied the performance of MOF-74(Mg) membranes in separating CO_2/H_2, CO_2/N_2, and CO_2/CH_4 mixtures that are important in carbon capture [89, 90]. These calculations

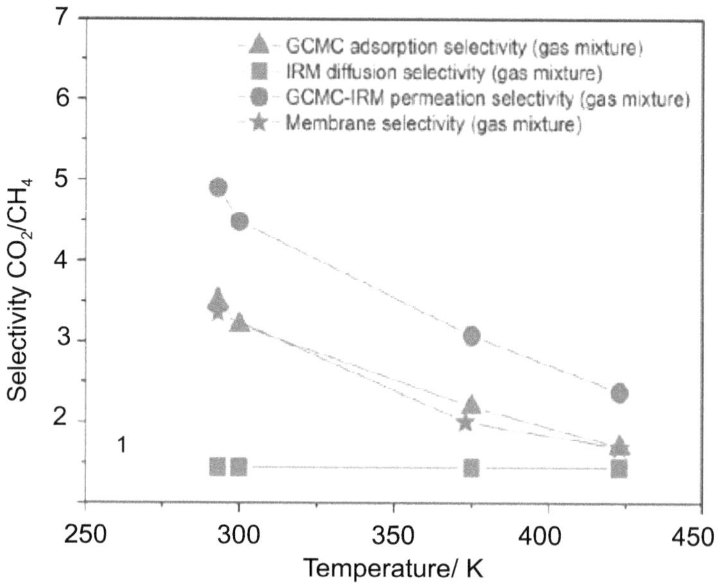

Figure 7.6 Estimated CO_2/CH_4 adsorption and diffusion selectivities as well as mixture permeation selectivities of the ZIF-8 membrane from GCMC-IRM studies.

showed that increasing upstream pressure leads to a significant enhancement of permeation selectivity, and this enhancement is directly traceable to the diffusion correlations within the 1D channel of this material. Such correlations have the effect of slowing down the more mobile partner species in the mixtures. They found that the permeation selectivity and permeability of this MOF membrane for CO_2/H_2 and CO_2/N_2 mixtures are higher than those reported in the published literature with zeolite membranes.

To boost the performance of polymer membrane, a growing research effort is underway over the last few years on the incorporation of MOF particles into a polymeric matrix to fabricate mixed-matrix membranes (MMMs). One key challenge in developing such MMMs is the requirement of efficiently screening appropriate polymer/MOF combinations toward the specific separation of interest. Theoretical models that can make quantitative predictions for the performance of new MMMs involving MOFs play an important role prior to experimental exploration [91–94]. To the best our knowledge, in 2010, Keskin and Sholl [91] reported the first quantitative information for MOF-based MMMs, using a combination of atomistic and continuum modeling. The predictions indicated that the separation property of the Cu(hfipbb)(H_2hfipbb)$_{0.5}$/Matrimid membrane exceeds Robeson's upper-bound curve for CO_2/CH_4 separation. They also demonstrated that not all MOFs yielded similar dramatic improvements in membrane performance related to a pure polymer membrane, such as both IRMOF-1 and Cu-BTC increase the permeability of CO_2 through MMMs but do not improve permeation selectivity. Following this study, Erucar and Keskin [92] examined the performance of composite membranes where Zn(bdc)(ted)$_{0.5}$ mixed with various polymer matrices for separation of CO_2 from its mixtures with CH_4 and H_2. The results indicated that using an MOF with low selectivity/high permeability as filler particles in polymers at most enhances the gas permeability and that MOFs with high selectivity and porosity should be used to increase both selectivity and permeability. Recently, by comparing with the experimental data for pure gases (CO_2, CH_4) and mixture (CO_2/CH_4) permeation through IRMOF-1/Matrimid and Cu-BTC/Matrimid MMMs, the same authors [93] discovered that Maxwell and modified Felske models give the best predicting performance among the models considering ideal (Maxwell, Bruggeman, Lewis–Nielson, Pal) and nonideal (modified Maxwell, Felske, modified Felske) morphologies, respectively. The former models assume that there are no defects in the polymer–particle interface, while the latter ones contain interface voids and polymer rigidification around particles. These two models were further employed by them to test the performance of 80 new MOF-based MMMs that were composed of 10 different MOFs and 8 different polymers for CO_2/CH_4 separation [93]. This group also used Maxwell model to assess several MMMs composed of various polymers and ZIF-65 or ZIF-90 for separation of H_2 from CO_2 [94].

7.4.2 *Gas Mixtures of Olefin and Paraffin*

At the present time, only a few molecular modeling studies have been conducted on pure MOF membranes for separating olefin/paraffin gas mixtures. Similar to their previous work [86], Caro et al. [95] studied the performance of a ZIF-8 membrane for an equimolar C_2H_4/C_2H_6 separation at room temperature and interpreted their results using GCMC-supported IRM measurements on a large ZIF-9 single crystal. They found that membrane selectivity can be predicted with sufficient accuracy as the product of adsorption and diffusion selectivity. For this cation-free nonpolar pore system, at 6 bar feed pressure, the adsorption selectivity (0.5) multiplied by the diffusion selectivity (4.8) gives an estimated permeation selectivity of 2.4 for the C_2H_4/C_2H_6 mixture. Clearly, the reduced C_2H_4 selectivity of the ZIF-8 membrane is contributed by the competition of the preferential C_2H_4 diffusion with the preferential C_2H_6 adsorption. Therefore, they suggested that for improving the C_2H_4/C_2H_6 separation performance, a membrane with improved C_2H_4 adsorption should be developed [96]. Long et al. [60] tested the potential use of the MOF-74(Fe) membrane for the separation of equimolar C_2H_4/C_2H_6 and C_3H_6/C_3H_8 mixtures using a simulation methodology combining the IAST, Maxwell–Stefan diffusion equations, and MD simulations for pure components. The calculated C_2H_4/C_2H_6 (C_3H_6/C_3H_8) permeation selectivities in the range of 13 to 20 (14 to 16) for total upstream pressures between 0.1 bar and 1.0 bar (0.05 bar and 1.0 bar), which are much greater than the above-observed ones in the ZIF-8 membrane. Their molecular-level analysis showed that the strong correlation effects in the molecular hops of the mobile and tardier species serve to bring the similar component diffusivities in the mixture and thus results in these permeation selectivities close to the adsorption ones. They also demonstrated that the diffusivities within this material are about 2 to 3 orders of magnitude greater than those in ZIF-8, conferring both selectivity and permeability advantages of the MOF-74(Fe) membrane.

As far we know, there is no related computational study available to evaluate the performance of MMMs for the separation of olefin/paraffin mixtures.

7.4.3 *Other Gas Mixtures*

Apart from above studies, molecular simulations have also been performed to investigate other gas mixtures in pure MOF or

MOF/polymer membranes [75, 97, 98]. For examples, Sholl et al. reported a computational screening method for identifying MOF membranes for kinetic separations of a CH_4/H_2 mixture [10]. This method can be used to evaluate large numbers of MOFs for kinetic separations of light gases on the basis of finding materials with large differences between the diffusion coefficients of adsorbed gas species. This group also investigated the performance of MOF-5 membrane for the separation of N_2/H_2 and N_2/CH_4 mixtures at room temperature [75]. Krishna et al. also studied the CH_4/H_2 separation in many pure MOF membranes [58, 89]. The calculations showed that MOF-74(Mg) and MOF-74(Zn) exhibits similar performance and are the "optimum" MOF structure that offers the best combination of permeation selectivity and permeability toward CH_4 among the MOFs examined by them [58]. Erucar and Keskin examined the separation performance of MMMs for separation of H_2 from CH_4 [99, 100]. The good agreement between experiments and the theoretical predictions can be obtained in IRMOF-1/Matrimid, Cu-BTC/PSF, Cu-BTC/PDMS, and Cu-BPY-HFS/Matrimd MMMs [100], which motivated them to identify several MMMs with very high H_2 selectivity and permeability relative to pure polymer membranes.

7.5 Strategy for Separation Enhancement

As described above, a wide variety of existing MOFs with different structural topologies, pore sizes, and chemical properties have been tested for potential applications in separating diverse industrial gas mixtures. On the basis of the knowledge obtained from these molecular modeling studies, several strategies have been brought out for designing MOFs with enhanced separation performance, which in parallel have been implemented by experimentalists for synthesizing new MOFs.

7.5.1 *Tailoring Pore Size and Shape*

The pore size and shape might often be the first consideration in selecting a porous adsorbent for special separation [5]. A prominent study for this topic was reported by Düren and Snurr using GCMC simulations [65]. As mentioned in Section 7.3.3 of this chapter, they showed that the selectivity for n-C_4H_{10} over CH_4 increases with decreasing cavity size. On the basis of IRMOF-1 with the smallest pore size (10.9/14.3 Å), they computationally designed a new material, IRMOF-993, with an even smaller pore size (6.3/14.5 Å), which shows dramatically higher selectivity in comparison to that in IRMOF-1. By comparing the calculation results obtained with and without blocking the small side pockets in Cu-BTC, we also proved that the presence of these pockets can largely enhance the separation capability of this material [101]. Furthermore, through properly tailing the pore size and shape, MOFs were also utilized to experimentally prepare molecular sieve membranes to kinetically separate gases [102]. Therefore, tailing MOFs with appropriate pore size and shape is a very effective approach to strengthen their separation performance for both adsorption-based and membrane-based processes.

7.5.2 *Catenation*

In experiments, it is often difficult to isolate pure phases of catenated and noncatenated versions of the same MOF, especially for those materials synthesized with expanded organic ligands [103]. However, the occurrence of such catenation in MOFs can become a beneficial factor for gas separations. In our previous computational studies [32, 104], the predictions showed that the adsorption selectivities for CH_4 over H_2 or CO_2 over H_2 are greatly enhanced in three catenated MOFs compared to their noncatenated counterparts, due to the formation of additional small pores and adsorption sites by the catenation of frameworks. Similar enhancing effects were also found for separating other mixtures [39, 66, 105]. Actually, catenation is also an alternative way to tune the pore size of MOFs for kinetic separations, which are easiest to appreciate for membrane-based separations [33]. As discussed above, Sholl et al. showed that the catenated MOF Cu(hfipbb)$(H_2hfipbb)_{0.5}$ is an attractive membrane material for CO_2/CH_4 separation due to the kinetic effect, which was later confirmed experimentally by Bao et al. [106]. Due to the same mechanism, this material was also shown able to separate normal C_2, C_3, and n-C_4 olefins and alkanes from all branched alkanes and all normal hydrocarbons above C_4

[107]. Accordingly, both the experimental and modeling studies have shown that forming rational self-catenated frameworks in MOFs is a feasible strategy for improving their separation efficiency.

7.5.3 *Chemical Modification*

Compared to other conventional adsorbents, one of the most distinguishing characteristic of MOFs is the ability to precisely predesign their structures at a molecular level. On the other hand, for the majority of MOFs, pore surface properties play a decisive role in their separation performance for a particular target. From many in silico studies of MOFs involving chemical modifications, multiple approaches for functionalizing the pore surfaces are available to achieve improved separation performance of MOFs.

The strong interactions of the open metal sites (or coordinatively unsaturated sites) with many chemicals provide us with an approach to enhance the separation performance of MOFs. For example, by replacing Zn^{2+} ions with the chemically more active Mg^{2+} ions in the framework of MOF-74, the simulation results of Krishna et al. [89, 90] showed that both adsorption and permeation selectivities are higher in MOF-74(Mg) than those in the isostructural material MOF-74(Zn) for separating CO_2/H_2, CO_2/N_2, and CO_2/N_2 mixtures. The reason can be traced to the significantly higher binding energy of CO_2 in MOF-74(Mg), as indicated by the DFT calculations of Valenzano et al. [108]. Similarly, due to stronger interaction for Fe^{2+} ions with the π-electron cloud of the "double bond" in the olefin molecules, Long et al. found that MOF-74(Fe) shows much better performance than MOF-74(Mg) for both adsorption- and membrane-based separations of C_2H_4/C_2H_6 and C_3H_6/C_3H_8 mixtures [60].

Open metal sites are usually created by removing the axially coordinated solvent molecules in the as-synthesized MOFs through activation treatment. Thus, anchoring functional organic molecules onto these active sites by postsynthesis is a useful approach to modify the chemical properties of the pore surfaces of MOFs, which can be applied to achieve the selective adsorption of guest molecules. A prototypical example is the study reported by Hupp et al. [109]. Starting from the parent solid $Zn_2(BTTB)(DMF)_2$, they synthesized a novel compound $Zn_2(BTTB)(py\text{-}CF_3)_2$ by replacing the DMF molecules coordinated to Zn^{2+} ions with a highly polar ligand py-CF_3. Their IAST calculations indicated that the py-CF_3-modified MOF exhibits larger CO_2/N_2 and CO_2/CH_4 selectivities than $Zn_2(BTTB)(DMF)_2$ and the material $Zn_2(BTTB)$ with open metal sites. The authors explained this enhancement from a combination of two factors: one is more attractive affinity of the $-CF_3$ groups for CO_2 than N_2 or CH_4, and the other is the more constricted pores of $Zn_2(BTTB)(py\text{-}CF_3)_2$ that result in more strongly adsorbed CO_2 due to the increased potential. A second example is the enhanced adsorption selectivity for a CO_2/N_2 mixture reported by Long et al. [110], where alkylamine functional group was incorporated as a ligand on the Cu^{2+} sites exposed on the pore surfaces of the parent framework $H_3[(Cu_4Cl)_3(BTTri)_8$ (CuBTTri).

As shown above, ionic MOFs can have significantly higher adsorption selectivity for mixture separations compared to other conventional MOFs. Thus, for such type materials, the exchange of counterions is an efficient strategy to tune the pore surface properties of MOFs and enhance their selective adsorption performance. To understand how different extra-framework ions would affect gas separation, Jiang et al. performed molecular simulations to investigate the separation of a CO_2/H_2 mixture in seven cation-exchanged *rho*-ZMOFs [111]. The cations include monovalent Na^+, K^+, Rb^+, Cs^+, divalent Mg^{2+}, Ca^{2+}, and trivalent Al^{3+}. As in the *rho*-ZMOF with Na^+ cations, the selectivity in other six *rho*-ZMOFs also decreases with increasing pressure. At a given pressure, the selectivity increases with a sequence of $Cs^+ < Rb^+ < K^+ < Na^+ < Ca^{2+} < Mg^{2+} \approx Al^{3+}$, which largely follows the increasing order of the charge-to-diameter ratio of cations. On the other hand, experimental studies also confirmed ion exchange is a useful approach to enhance separation performance of MOFs [112, 113].

Another widely used strategy for pore surface modifications is the functionalization of the organic ligands in MOFs. On the basis of our findings on the electrostatic characteristic of MOFs for gas separation, our group did a series of computational studies to explore the possibility for enhancing the separation performance of MOFs by grafting the organic ligands with various functional groups

[114, 115]. Through substituting the hydrogen in the aromatic rings in IRMOF-1 with four types of functional groups (–CH_3, –F, –OH, and –NH_2) [114], the calculations showed that the introduction of stronger electron-donating functional groups may induce an electrostatic potential with a larger gradient and higher absolute values around the organic linkers, leading to a higher adsorption selectivity for CO_2/CH_4 mixture separation. By grafting the organic ligands in UiO-66(Zr) with different hydrophobic and hydrophilic groups, we discovered that the materials modified using hydrophilic groups strongly outperform the parent solid and those functionalized forms with hydrophobic groups [115]. This approach has also been successfully used in other theoretical studies to enhance the selective adsorption performance of various MOFs toward different gases [116, 117]. A typical experiment conducted by Yaghi et al. [118] also proved that those ZIFs with more polar functionalities in the organic ligands can achieve higher adsorption selectivity of CO_2 over CH_4 or N_2. A second experimental exemplification is that MIL-53(Al) functionalized with amino groups showed drastically higher selectivity for CO_2/CH_4 separation compared to the nonmodified form [119]. In addition, we also proposed a route to achieve high separation performance of MOFs by physically or chemically doping metal atoms onto the organic ligands [120, 121]. It was found that the introduction of Li atoms can induce charge transfer from them to framework atoms, leading to stronger electrostatic interactions and thus higher adsorption selectivity for separating gas mixtures such as CO_2/CH_4 [120] and CO_2/N_2 [121]. The idea in this computational work has recently been confirmed by the experimental study of Hupp et al. [122], where they reported an improvement in CO_2/CH_4 selectivity by Li doping in MOFs.

7.6 Summary and Suggestions

The foregoing sections have introduced some recent advances in the studies of gas separation in MOFs using molecular modeling. These studies evidently elucidate that MOFs show a great application potential in both the adsorption-based and membrane-based separation of various industrial gas mixtures. Although the current state of knowledge for MOFs have been progressively achieved, there are still several major challenges that molecular modeling efforts need to be devoted, as outlined below.

(1) **Modeling of MOFs with highly flexible frameworks**. An increasing number of highly flexible MOFs have been synthesized and have shown very attractive advantages in both adsorption- and membrane-based separations. However, the present molecular simulations mainly focused on the adsorption and diffusion of pure gases in these MOFs [123–125]. Therefore, further systematical studies should pay attention to understanding how the cooperative actions between the components in the mixtures affect the structural transitions of highly flexible MOFs, thereby revealing the influencing factors in connection with their performance for gas separation.
(2) **Modeling of MOFs with open metal sites**. For those gases strongly interacting with the open metal sites in MOFs, their adsorption behaviors usually cannot be well captured by simulations with general force fields such as universal force field (UFF) and DREIDING, especially in the low-pressure domain. Force fields derived from high-level first-principles calculations can offer a much more accurate description of gas–adsorbent interactions, giving much better agreement with experimental observations [61, 126, 127]. This not only opens a door to gain an understanding of the competing nature of different gases, but also mixture simulation results can be used to check the applicability of the IAST, which is usually used to calculate the selectivity in MOFs with open metal sites. However, such highly desirable force fields are in extreme shortage at the moment. In addition, Leclerec et al. [128] recently showed that partial reduction of open metal sites Fe^{3+} to Fe^{2+} within MIL-100(Fe) occurs upon an appropriate thermal treatment. It appeared that C_3H_6 molecules are strongly adsorbed on Fe^{2+} sites despite its weaker Lewis acidity compared to that of Fe^{3+} ones, but the occurrence of Fe^{2+} sites does not modify the interaction with CO_2 and C_3H_8. This is a very interesting experimental phenomenon, while there is no related modeling report, to the

best of our knowledge, for gas separation in MOFs with open metal sites having mixed oxidation states.

(3) **Modeling of membrane-based separation**. Currently, it is still in the primary stage for computationally examining MOF-based membranes for gas separation, and thus many fundamental problems exist. First, the skeletons of MOFs are usually treated as rigid in simulations, while their framework flexibilities have already been shown to greatly influence the diffusivities of guest molecules [123, 129–131]. Such treatment might deviate from the intrinsic nature of MOF membranes. Hence, there is no unambiguous answer for whether the framework flexibility of MOFs is beneficial for their membrane performance. Second, some hypotheses have been adopted in MOF membrane simulations, such as perfect solids without defects and with the pores aligning in the same direction and no surface diffusion resistance. Third, for evaluating MMMs, only macroscopic mathematic models are used by a combination of the information of pure MOFs and polymer membranes, neglecting many important influencing factors, including the sizes and shapes of MOF particles and real networks formed by the mutual impacts of MOFs and polymers. Finally, the thickness of the simulated membranes is limited to the nanometer scale that is much smaller than the practical situation. Therefore, future efforts should focus on these issues using more detailed molecular modeling.

Without doubt, molecular modeling is playing an indispensable role in identifying the influencing factors and understanding the microscopic mechanisms that control the separation performance of MOFs at a molecular level. With the help of ever-growing computational source, molecular modeling is becoming an extremely powerful and efficient tool for large-scale in silico screening of MOFs to isolate best-performing materials for gas separation. In addition, detailed molecular modeling studies not only can provide very useful strategies to synthesize robust MOFs for a specific separation of interest but also can contribute the development of novel nanoporous materials in new application areas.

Acknowledgments

This work was supported by the financial support of the Natural Science Foundation of China (Nos: 21136001, 21121064, 21322603, and 21276009), the Program for New Century Excellent Talents in University (No. NCET-12-0755), the National Key Basic Research Program of China ("973") (2013CB733503), and the Specialized Research Fund for the Doctoral Program of Higher Education of China (Contract 20110010130001). The support of Beijing Nova Program (No. 2008B15) was greatly appreciated.

References

1. Férey, G. (2008). Hybrid porous solids: past, present, future, *Chem. Soc. Rev.*, **37**, 191–214.
2. Kitagawa, S., Matsuda, R. (2007). Chemistry of coordination space of porous coordination polymers, *Coord. Chem. Rev.*, **251**, 1–21.
3. Long, J. R., Yaghi, O. M. (2009). The pervasive chemistry of metal–organic frameworks, *Chem. Soc. Rev.*, **38**, 1213–1214.
4. Meek, S. T., Greathouse, J. A., Allendorf, M. D. (2011). Metal–organic frameworks: a rapidly growing class of versatile nanoporous materials, *Adv. Mater.*, **23**, 249–267.
5. Li, J.-R., Sculley, J., Zhou, H.-C. (2012). Metal–organic frameworks for separations, *Chem. Rev.*, **112**, 869–932.
6. D'Alessandro, D. M., Smit, B., Long, J. R. (2010). Carbon dioxide capture: prospects for new materials, *Angew. Chem., Int. Ed.*, **49**, 6058–6082.
7. Wu, D., Wang, C., Liu, B., Liu, D., Yang, Q., Zhong, C. (2012). Large-scale computational screening of metal–organic frameworks for CH_4/H_2 separation, *AIChE J.*, **58**, 2078–2084.
8. Vaidhyanathan, R., Iremonger, S. S., Shimizu, G. K. H., Boyd, P. G., Alavi, S., Woo, T. K. (2012). Competition and cooperativity in carbon dioxide sorption by amine-functionalized metal–organic frameworks, *Angew. Chem., Int. Ed.*, **51**, 1826–1829.
9. Wilmer, C E., Leaf, M., Lee, C. Y., Farha, O. K., Hauser, B. G., Hupp, J. T., Snurr, R. Q. (2012). Large-scale screening of hypothetical metal–organic frameworks, *Nat. Chem.*, **4**, 83–89.

10. Haldoupis, E. Nair, S., Sholl, D. (2010). Efficient calculation of diffusion limitations in metal organic framework materials: a tool for identifying materials for kinetic separations, *J. Am. Chem. Soc.*, **132**, 7528–7539.

11. Haldoupis, E. Nair, S., Sholl, D. (2012). Finding MOFs for highly selective CO_2/N_2 adsorption using materials screening based on efficient assignment of atomic point charges, *J. Am. Chem. Soc.*, **134**, 4313–4323.

12. Frenkel, D., Smit, B. (2002). *Understanding Molecular Simulation: From Algorithms to Applications*, 2nd edn. (Academic Press San Diego).

13. Allen, M. P., Tildesley, D. J. (1987). *Computer Simulation of Liquids* (Clarendon Press, Oxford).

14. David, D. S., Steckel, J. A. (2009). *Density Functional Theory: A Practical Introduction* (John Wiley & Sons).

15. Sagara, T., Klassen, J., Ganz, E. (2004). Computational study of hydrogen binding by metal-organic framework-5, *J. Chem. Phys.*, **121**, 12543–12547.

16. Watanabe, T., Manz, T. A., Sholl, D. S. (2011). Accurate treatment of electrostatics during molecular adsorption in nanoporous crystals without assigning point charges to framework atoms, *J. Phys. Chem. C*, **115**, 4824–4836.

17. Chen, Y. F. Babarao, R., Sandler, S. I., Jiang, J. W. (2010). Metal–organic framework MIL-101 for adsorption and effect of terminal water molecules: from quantum mechanics to molecular simulation, *Langmuir*, **26**, 8743–8750.

18. Ramsahye, N. A., Maurin, G., Bourrelly, S., Llewellyn, P. L., Loiseau, T., Férey G. (2007). Charge distribution in metal organic framework materials: transferability to a preliminary molecular simulation study of the CO_2 adsorption in the MIL-53(Al) system, *Phys. Chem. Chem. Phys.*, **9**, 1059–1063.

19. Young, D. C. (2001). *Computational chemistry: A Practical Guide for Applying Techniques to Real-World Problems* (John Wiley & Sons).

20. Manz, T. A., Sholl, D. S. (2010). Chemically meaningful atomic charges that reproduce the electrostatic potential in periodic and nonperiodic materials, *J. Chem. Theory Comput.*, **6**, 2455–2468.

21. Yang, Q., Zhong, C., Chen, J.-F. (2008). Computational study of CO_2 storage in metal–organic frameworks, *J. Phys. Chem. C*, **112**, 1562–1569.

22. Dubbeldam, D., Walton, K. S., Ellis, D. E., Snurr, R. Q. (2007). Exceptional negative thermal expansion in isoreticular metal–organic frameworks, *Angew. Chem., Int. Ed.*, **46**, 4496–4499.

23. Babarao, R., Jiang, J. (2008). Molecular screening of metal-organic frameworks for CO_2 storage, *Langmuir*, **24**, 6270–6278
24. Xu, Q., Zhong, C. (2010). A General approach for estimating framework charges in metal–organic frameworks, *J. Phys. Chem. C*, **114**, 5035–5042.
25. Zheng, C., Zhong, C. (2010). Estimation of framework charges in covalent organic frameworks using connectivity-based atom contribution method, *J. Phys. Chem. C*, **114**, 9945–9951.
26. Keskin, S. (2011). Atomistic simulations for adsorption, diffusion, and separation of gas mixtures in zeolite imidazolate frameworks, *J. Phys. Chem. C*, **115**, 800–807.
27. Meek, S. T., Greathouse, J. A., Allendorf, M. D. (2011). Metal–organic frameworks: a rapidly growing class of versatile nanoporous materials, *Adv. Mater.*, **23**, 249–267.
28. Rappé, A. K., Goddard III, W. A. (1991). Charge equilibration for molecular dynamics simulations, *J. Phys. Chem.*, **95**, 3358–3363.
29. Wlimer, C. E., Snurr, R. Q. (2011). Towards rapid computational screening of metal–organic frameworks for carbon dioxide capture: calculation of framework charges via charge equilibration, *Chem. Eng. J.*, **17**, 775–781.
30. Yang, Q., Zhong, C. (2006). Electrostatic-field-induced enhancement of gas mixture separation in metal–organic frameworks: a computational study, *ChemPhysChem*, **7**, 1417–1421.
31. Yang, Q. Xue, C., Zhong, C., Chen, J.-F. (2007). Molecular simulation of separation of CO_2 from flue gases in Cu-BTC metal–organic framework, *AIChE J.*, **53**, 2832–2840.
32. Yang, Q. Xu, Q., Liu, B., Zhong, C., Smit, B. (2009). Molecular simulation of CO_2/H_2 mixture separation in metal–organic frameworks: effect of catenation and electrostatic interactions, *Chin. J. Chem. Eng.*, **17**, 781–790.
33. Keskin, S., van Heest, T. M., Sholl, D. S. (2010). Can metal–organic framework materials play a useful role in large-scale carbon dioxide separations?, *ChemSusChem*, **3**, 879–891
34. Yang, Q., Zhong, C. (2006). Molecular simulation of carbon dioxide/methane/hydrogen mixture adsorption in metal–organic frameworks, *J. Phys. Chem. B*, **110**, 17776–17783.
35. Liu, Y., Liu, D., Yang, Q., Zhong, C., Mi, J. (2010). Comparative study of separation performance of COFs and MOFs for $CH_4/CO_2/H_2$ mixtures, *Ind. Eng. Chem. Res.*, **49**, 2902–2906.

36. Yang, Q. Wiersum, A. D., Jobic, H., Guillerm, V., Serre, C., Llewellyn, P. L., Maurin, G. (2011). Understanding the thermodynamic and kinetic behaviour of the CO_2/CH_4 gas mixture within the porous zirconium terephthalate UiO-66(Zr): a joint experimental and modeling approach *J. Phys. Chem. C*, **115**, 13768–13774.
37. Yang, Q., Ma, L., Zhong, C. An, X., Liu D. (2011). Enhancement of CO_2/N_2 mixture separation using the thermodynamic stepped behavior of adsorption in metal-organic frameworks, *J. Phys. Chem. C*, **115**, 2790–2797.
38. Zhang, W., Huang, H., Zhong, C., Liu, D. (2012). Cooperative effect of temperature and linker functionality on CO_2 capture from industrial gas mixtures in metal–organic frameworks: a combined experimental and molecular simulation study, *Phys. Chem. Chem. Phys.*, **14**, 2317–2325.
39. Babarao, R., Jiang, J. W., Sandler, S. I. (2009). Molecular simulations for adsorptive separation of CO_2/CH_4 mixture in metal-exposed, catenated, and charged metal-organic frameworks, *Langmuir*, **25**, 5239–5247.
40. Jiang, J. W. (2009). Charged *soc* metal-organic framework for high-efficacy H_2 adsorption and syngas purification: atomistic simulation study, *AIChE J.*, **55**, 2422–2432.
41. Babarao, R., Jiang, J. W. (2009). Unprecedentedly high selective adsorption of gas mixtures in *rho* zeolite-like metal–organic framework: a molecular simulation study, *J. Am. Chem. Soc.*, **131**, 11417–11425.
42. Bourrelly, S., Llewellyn, P. L., Serre, C., Millange, F., Loiseau, T., Férey, G. (2005). Different adsorption behaviors of methane and carbon dioxide in the isotypic nanoporous metal terephthalates MIL-53 and MIL-47, *J. Am. Chem. Soc.*, **127**, 13519–13521.
43. Kitaura, R., Seki, K., Akiyama, G., Kitagawa, S. (2003). Porous Coordination-Polymer Crystals with Gated Channels Specific for Supercritical Gases, *Angew. Chem., Int. Ed.*, **42**, 428–431.
44. Hamon, L., Llewellyn, P. L., Devic, T., Ghoufi, A. Z., Clet, G., Guillerm, V., Pirngruber, G. M., Maurin, M., Serre, C., Driver, G., van Beek, W., Jolimaître, E., Vimont, A., Daturi, M., Férey, G. (2009). Co-adsorption and separation of CO_2-CH_4 mixtures in the highly flexible MIL-53(Cr) MOF, *J. Am. Chem. Soc.*, **131**, 17490–17499.
45. Coudert, F.-X., Mellot-Draznieks, C., Fuchs, A. H., Boutin, A. (2009). Prediction of breathing and gate-opening transitions upon binary mixture adsorption in metal-organic frameworks, *J. Am. Chem. Soc.*, **131**, 11329–11331.

46. Ortiz, A. U., Springuel-Huet, M.-A., Coudert, F.-X., Fuchs, A. H., Boutin, A. (2012). Predicting mixture coadsorption in soft porous crystals: experimental and theoretical study of CO_2/CH_4 in MIL-53(Al), *Langmuir*, **28**, 494–498.
47. Yazaydın, A., Ö., Benin, A. I., Faheem, S. A., Jakubczak, P., Low, J. J., Willis, R. R., Snurr, R. Q. (2009). Enhanced CO_2 adsorption in metal-organic frameworks via occupation of open-metal sites by coordinated water molecules, *Chem. Mater.*, **21**, 1425–1430.
48. Liu, B., Smit, B. (2010). Molecular simulation studies of separation of CO_2/N_2, CO_2/CH_4, and CH_4/N_2 by ZIFs, *J. Phys. Chem. C*, **114**, 8515–8522.
49. Babarao, R., Jiang, J. W. (2009). Upgrade of natural gas in *rho* zeolite-like metal–organic framework and effect of water: a computational study, *Energy Environ. Sci.*, **2**, 1088–1093.
50. Padin, J., Yang, R. T., Munson, C. L. (1999). New sorbents for olefin/paraffin separations and olefin purification for C4 hydrocarbons, *Ind. Eng. Chem. Res.*, **38**, 3614–3621.
51. Eldridge, R. B. (1993). Olefin/paraffin separation technology: a review, *Ind. Eng. Chem. Res.*, **32**, 2208–2212.
52. Wang, S., Yang, Q., Zhong, C. (2008). Adsorption and separation of binary mixtures in a metal-organic framework Cu-BTC: a computational study, *Sep. Purif. Technol.*, **60**, 30–35.
53. Nicholson, T. M., Bhatia, S. K. (2006). Electrostatically mediated specific adsorption of small molecules in metallo-organic frameworks *J. Phys. Chem. B*, **110**, 24834–24836.
54. Bae, Y.-S., Lee, C. Y., Kim, K. C., Farha, O. K., Nickias, P., Hupp, J. T., Nguyen, S. T., Snurr, R. Q. (2012). High propene/propane selectivity in isostructural metal–organic frameworks with high densities of open metal sites, *Angew. Chem., Int. Ed.*, **51**, 1857–1860.
55. Lamia, N., Jorge, M., Granato, M. A., AlmeidaPaz, F. A., Chevreau, H., Rodrigues, A. E. (2009). Adsorption of propane, propylene and isobutene on a metal–organic framework: molecular simulation and experiment, *Chem. Eng. Sci.*, **64**, 3246–3259.
56. Jorge, M., Lamia, N., Rodrigues, A. E. (2009). Molecular simulation of propane/propylene separation on the metal–organic framework CuBTC, *Colloids Surf. A*, **357**, 27–34.
57. Yazaydın, A. Ö., Snurr, R Q., Park, T. H., Koh, K., Liu, J., LeVan, D. M., Benin, A. I., Jakubczak, P. Lanuza, M., Galloway, D. B., Low, J. J., Willis, R. R. (2009). Screening of metal–organic frameworks for carbon dioxide

capture from flue gas using a combined experimental and modeling approach, *J. Am. Chem. Soc.*, **131**, 18198–18199.

58. Krishna, R., van Baten, J. M. (2011). In silico screening of metal–organic frameworks in separation Applications, *Phys. Chem. Chem. Phys.*, **13**, 10593–10616.
59. Myers, A. L., Przusnitz, J. M. (1965). Thermodynamics of mixed-gas adsorption, *AIChE J.*, **11**, 121–127.
60. Bloch, E. D., Queen, W. L., Krishna, R., Zadrozny, J. M., Brown, C. M., Long, J. R. (2012). Hydrocarbon separations in a metal-organic framework with open iron(II) coordination sites, *Science*, **335**, 1606–1610.
61. Fischer, M., Gomes, J. R. B., Fröba, M., Jorge, M. (2012). Modeling adsorption in metal–organic frameworks with open metal sites: propane/propylene Separations, *Langmuir*, **28**, 8537–8549.
62. Xiang, S.-H., Zhang, Z., Zhao, C.-G., Hong, K., Zhao, X., Ding, D.-R., Xie, M.-H., Wu, C.-G., Das, M. C., Gill, R., Thomas, K. M., Chen, B. (2011). Rationally tuned micropores within enantiopure metal-organic frameworks for highly selective separation of acetylene and ethylene, *Nat. Commun.*, **2**, 204.
63. He, Y., Xiang, S., Chen, B. (2011). A microporous hydrogen-bonded organic framework for highly selective C_2H_2/C_2H_4 separation at ambient temperature, *J. Am. Chem. Soc.*, **133**, 14570–14573.
64. Hartmann, M., Kunz, S., Himsl, D. Tangermann, O. Ernst, S. Wagener, A. (2008). Adsorptive separation of isobutene and isobutane on $Cu_3(BTC)_2$, *Langmuir*, **24**, 8634–8642.
65. Düren, T., Snurr, R. Q. (2004). Assessment of isoreticular metal-organic frameworks for adsorption separations: a molecular simulation study of methane/*n*-butane mixtures, *J. Phys. Chem. B*, **108**, 15703–15708.
66. Babarao, R., Tong, Y. H., Jiang, J. W. (2009). Molecular insight into adsorption and diffusion of alkane isomer mixtures in metal-organic frameworks, *J. Phys. Chem. B*, **113**, 9129–9136.
67. Zhang, L., Wang, Q., Wu, T., Liu, Y.-C. (2007). Understanding adsorption and interactions of alkane isomer mixtures in isoreticular metal–organic frameworks, *Chem. Eur. J.*, **13**, 6387–6396.
68. Greathouse, J. A., Kinnibrugh, T. L., Allendorf, M. D. (2009). Adsorption and separation of noble gases by IRMOF-1: grand canonical Monte Carlo simulations, *Ind. Eng. Chem. Res.*, **48**, 3425–3431.
69. Garberoglio, G. (2009). Quantum sieving in organic frameworks, **467**, 270–275.

70. Liu, D., Wang, W., Mi, J., Zhong, C., Yang, Q., Wu, D. (2012). Quantum sieving in metal-organic frameworks: a computational study, *Ind. Eng. Chem. Res.*, **51**, 434–442.
71. Fischer, M., Hoffmann, F., Fröba, M. (2010). New microporous materials for acetylene storage and C_2H_2/CO_2 separation: insights from molecular simulations, **11**, 2220–2229.
72. Sumida, K., Rogow, D. L., Mason, J. A., McDonald, T. M., Bloch, e. D., Herm, Z. R., Bae, T.-H., Long, J. R. (2012). Carbon dioxide capture in metal–organic frameworks, *Chem. Rev.*, **112**, 724–781.
73. Gascon, J., Kapteijn, F. (2010). Metal-organic framework membranes: high potential, bright future?, *Angew. Chem., Int. Ed.*, **49**, 1530–1532.
74. Keskin, S., Sholl, D. S. (2007). Screening metal-organic framework materials for membrane-based methane/carbon dioxide separation, *J. Phys. Chem. C*, **111**, 14055–14059.
75. Keskin, S., Sholl, D. S. (2009). Assessment of a metal-organic framework membrane for gas separations using atomically detailed calculations: CO_2 CH_4 N_2 H_2 mixtures in MOF-5, *Ind. Eng. Chem. Res.*, **48**, 914–922.
76. Keskin, S. Liu, J. Rankin, R. B. Johnson, J. K., Sholl, D. S. (2009). Atomically detailed models of gas mixture diffusion through CuBTC membranes, *Microporous Mesoporous Mater.*, **125**, 101–106.
77. Keskin, S., Sholl, D. S. (2009). Efficient methods for screening of metal organic framework membranes for gas separations using atomically detailed models, *Langmuir*, **25**, 11786–11795.
78. Krishna, R., van Baten, J. M. (2007). Using molecular simulations for screening of zeolites for separation of CO_2/CH_4 mixtures, *Chem. Eng. J.*, **133**, 121–131.
79. Keskin, S. (2011). Atomistic simulations for adsorption, diffusion, and separation of gas mixtures in zeolite imidazolate frameworks *J. Phys. Chem. C*, **115**, 800–807.
80. Liu, J. Keskin, S. Sholl, D. S., Johnson, J. K. (2011). Molecular simulations and theoretical predictions for adsorption and diffusion of CH_4/H_2 and CO_2/CH_4 mixtures in ZIFs, *J. Phys. Chem. C*, **115**, 12560–12566.
81. Atci, E., Erucar, I., Keskin, S. (2011). Adsorption and transport of CH_4, CO_2, H_2 mixtures in a bio-MOF material from molecular simulation, *J. Phys. Chem. C*, **115**, 6833–6840.
82. Watanabe, T., Keskin, S., Nair S., Sholl, D. S. (2009). Computational identification of a metal organic framework for high selectivity membrane-

based CO_2/CH_4 separations: Cu(hfipbb)(H_2hfipbb)$_{0.5}$, *Phys. Chem. Chem. Phys.*, **11**, 11389–11394.

83. Robeson, L. M. (2008). The upper bound revisited, *J. Membr. Sci.*, **320**, 390–400.
84. Li, S. Falconer, J. L., Noble, R. D. (2006). Improved SAPO-34 membranes for CO_2/CH_4 separations, *Adv. Mater.*, **18**, 2601–2603.
85. Carreon, M. A. Li, S., Falconer, J. L., Noble, R. D. (2008). Alumina-supported SAPO-34 membranes for CO_2/CH_4 separation, *J. Am. Chem. Soc.*, **130**, 5412–5413.
86. Bux, H., Chmelik, C., van Baten, J. M., Krishna, R., Caro, J. (2010). Novel MOF-membrane for molecular sieving predicted by IR-diffusion studies and molecular modeling, *Adv. Mater.*, **22**, 4741–4743.
87. Krishna, R., van Baten, J. M. (2011). A simplified procedure for estimation of mixture permeances from unary permeation data, *J. Membr. Sci.*, 367, 204–210.
88. Chmelik, C., van Baten, J. M., Krishna, R. (2012). Hindering effects in CO_2/CH_4 mixture diffusion in ZIF-8 crystals, *J. Membr. Sci.*, **397–398**, 87–91.
89. Krishna, R., van Baten, J. M. (2011). Investigating the potential of MgMOF-74 membranes for CO_2 capture, *J. Membr. Sci.*, **377**, 249–260.
90. Krishna, R., van Baten, J. M. (2011). Maxwell-Stefan modeling of slowing-down effects in mixed gas permeation across porous membranes, *J. Membr. Sci.*, **383**, 289–300.
91. Keskin, S., Sholl, D. S. (2010). Selecting metal organic frameworks as enabling materials in mixed matrix membranes for high efficiency natural gas purification, *Energy Environ. Sci.*, **3**, 343–351.
92. Erucar, I., Keskin, S. (2011). Separation of CO_2 mixtures using Zn(bdc)(ted)$_{0.5}$ membranes and composites: a molecular simulation study, *J. Phys. Chem. C*, **115**, 13637–13644.
93. Erucar, I., Keskin, S. (2011). Screening metal-organic framework-based mixed-matrix membranes for CO_2/CH_4 separations, *Ind. Eng. Chem. Res.*, **50**, 12606–12616.
94. Atci, E., Keskin, S. (2012). Atomically detailed models for transport of gas mixtures in ZIF membranes and ZIF/polymer composite membranes, *Ind. Eng. Chem. Res.*, **51**, 3091–3100.
95. Bux, H., Chmelik, C., Krishna, R., Caro, J. (2011). Ethene/ethane separation by the MOF membrane ZIF-8: molecular correlation of permeation, adsorption, diffusion, *J. Membr. Sci.*, **369**, 284–289.

96. Caro, J. (2011). Are MOF membranes better in gas separation than those made of zeolites?, *Curr. Opin. Chem. Eng.*, **1**, 77–78.

97. Liu, J., Johnson, J. K. (2009). Prediction of CH_4/H_2 mixture selectivity in Zn(tbip) from computer simulations, *J. Low Temp. Phys.*, **157**, 268–276.

98. Liu, B., Sun, C., Chen, G. (2011). Molecular simulation studies of separation of CH_4/H_2 mixture in metal-organic frameworks with interpenetration and mixed-ligand, *Chem. Eng. Sci.*, **66**, 3012–3019.

99. Keskin S. (2010). Molecular simulation study of CH_4/H_2 mixture separations using metal organic framework membranes and composites, *J. Phys. Chem. C*, **114**, 13047–13054.

100. Erucar I., Keskin S. (2012). Computational screening of metal organic frameworks for mixed matrix membrane applications, *J. Membr. Sci*, **407–408**, 221–230.

101. Xue, C., Yang, Q., Zhong, C. (2009). Effects of the side pockets on gas separation in metal-organic framework Cu-BTC: a molecular simulation study, *Mol. Simulat.*, **35**, 1249–1255.

102. Li, Y., Liang, F., Bux, H., Feldhoff, A., Yang, W., Caro, J. (2010). Molecular sieve membrane: supported metal-organic framework with high hydrogen selectivity, *Angew. Chem., Int. Ed.*, **49**, 548–551.

103. Furukawa, H., Ko, N., Go, Y. B., Aratani, N, Choi, S. B. Choi, E. Yazaydin, A. Ö., Snurr, R. Q. O'Keeffe, M. Kim J. Yaghi, O. M. (2010). Ultrahigh porosity in metal-organic frameworks, *Science*, **329**, 424–428.

104. Liu, B., Yang, Q., Xue, C., Zhong, C., Chen, B., Smit, B. (2008). Enhanced adsorption selectivity of hydrogen/methane mixtures in metal-organic frameworks with interpenetration: a molecular simulation study, *J. Phys. Chem. C*, **112**, 9854–9860.

105. Liu, B., Smit, B. (2009). Comparative molecular simulation study of CO_2/N_2 and CH_4/N_2 separation in zeolites and metal-organic frameworks, *Langmuir*, **25**, 5918–9860.

106. Bao, Z., Alnemrat, S., Yu, L., Vasiliev, I., Ren, Q., Lu, X., Deng, S. (2011). Kinetic separation of carbon dioxide and methane on a copper metal–organic Framework, *J. Colloid Interface Sci*, **357**, 504–509.

107. Pan, L., Olson, D. H., Ciemnolonski, L. R., Heddy, R., Li, J. (2006). Separation of hydrocarbons with a microporous metal–organic framework, *Angew. Chem., Int. Ed.*, **45**, 616–619.

108. Valenzano, L., Civalleri, B., Sillar, K., Sauer, J. (2011). Heats of adsorption of CO and CO_2 in metal–organic frameworks: quantum mechanical study of CPO-27-M (M = Mg, Ni, Zn), *J. Phys. Chem. C*, **115**, 21777–21784.

109. Bae, Y.-S., Farha, O. K., Hupp, J. T., Snurr, R. Q. (2009). Enhancement of CO_2/N_2 selectivity in a metal-organic framework by cavity modification, *J. Mater. Chem.*, **19**, 2131–2134.

110. McDonald, T. M., D'Alessandro, D. M., Krishna, R., Long, J. R. (2011). Enhanced carbon dioxide capture upon incorporation of *NN'*-dimethylethylenediamine in the metal–organic framework CuBTTri, *Chem. Sci.*, **2**, 2022–2028.

111. Chen, Y. F., Nalaparaju, A., Eddaoudi, M., Jiang, J. W. (2012). CO_2 adsorption in mono-, di- and trivalent cation-exchanged metal–organic frameworks: a molecular simulation study, *Langmuir*, **28**, 3903–3910.

112. Procopio, E. Q., Linares, F., Montoro, C., Colombo, V., Maspero, A., Barea, E., Navarro, J. A. R. (2010). Cation-exchange porosity tuning in anionic metal–organic frameworks for the selective separation of gases and vapors and for catalysis, *Angew. Chem., Int. Ed.*, **49**, 7308–7311.

113. Maji, T. K., Matsuda, R., Kitagawa, S. (2007). A flexible interpenetrating coordination framework with a bimodal porous functionality, *Nat Mater.*, **6**, 142-148.

114. Mu, W., Liu, D., Yang, Q., Zhong, C. (2010). Computational study of the effect of organic linkers on natural gas upgrading in metal–organic frameworks, *Microporous Mesoporous Mater.*, **130**, 76–82.

115. Yang, Q., Wiersum A. D., Llewellyn P. L., Guillerm V., Serre, C., Maurin, G. (2011). Functionalizing porous zirconium terephthalate UiO-66(Zr) for natural gas upgrading: a computational exploration, *Chem. Commun.*, **47**, 9603–9605.

116. Torrisi, A., Bell, R. G., Mellot-Draznieks, C. (2010). Functionalized MOFs for enhanced CO_2 capture, *Cryst. Growth Des*, **10**, 2839–2841.

117. Amrouche, H., Aguado, S., Pérez-Pellitero, J., Chizallet, C., Siperstein, F., Farrusseng, D., Bats, N., Nieto-Draghi, C. (2011). Experimental and computational study of functionality impact on sodalite–zeolitic imidazolate frameworks for CO_2 separation, *J. Phys. Chem. C*, **115**, 16425–16432.

118. Banerjee, R., Furukawa, H., Britt, D., Knobler, C., O'Keeffe, M., Yaghi, O. M. (2009). Control of pore size and functionality in isoreticular zeolitic imidazolate frameworks and their carbon dioxide selective capture properties, *J. Am. Chem. Soc.*, **131**, 3875–3877.

119. Couck, S, Denayer, J. F. M., Baron, G. V., Rémy, T., Gascon, J., Kapteijn, F. (2009). An amine-functionalized MIL-53 metal-organic framework with large separation power for CO_2 and CH_4, *J. Am. Chem. Soc.*, **131**, 6326–6327.

120. Xu, Q., Liu, D., Yang, Q., Zhong, C., Mi, J. (2010). Li-modified metal-organic frameworks for CO_2/CH_4 separation: a route to achieving high adsorption selectivity, *J. Mater. Chem.*, **20**, 706–714.

121. Yang, Q., Ma, L., Zhong, C., An, X., Liu, D. (2011). Enhancement of CO_2/N_2 Mixture separation using the thermodynamic stepped behavior of adsorption in metal-organic frameworks, *J. Phys. Chem. C*, **115**, 2790–2797.

122. Bae, Y.-S., Hauser, B., Snurr, R. Q., Hupp, J. T. (2011). Enhancement of CO_2/CH_4 selectivity in metal-organic frameworks containing lithium cations, *Microporous Mesoporous Mater.*, **141**, 231–235.

123. Férey, G., Serre, C., Devic, T., Maurin, G., Jobic, H., Llewellyn, P. L., Weireld, G. D., Vimont, A., Daturif. M., Chang, J.-S. (2011). Why hybrid porous solids capture greenhouse gases?, *Chem. Soc. Rev.*, **40**, 550–562; and references therein.

124. Coombes, D S. Corà, F., Mellot-Draznieks, C., Bell, R. G. (2009). Sorption-Induced breathing in the flexible metal organic framework CrMIL-53: force-field simulations and electronic structure analysis, *J. Phys. Chem. C*, **113**, 544–552.

125. Stavitski, E., Pidko, E. A., Couck, S., Remy, T., Hensen, E. J. M., Weckhuysen, B. M., Denayer, J., Gascon, J., Kapteijn, F. (2011). Complexity behind CO_2 capture on NH_2-MIL-53(Al), *Langmuir*, **27**, 3970–3976.

126. Fischer, M., Kuchta, B., Firlej, L., Hoffmann, F., Fröba, M. (2010). Accurate prediction of hydrogen adsorption in metal-organic frameworks with unsaturated metal sites via a combined density-functional theory and molecular mechanics approach, *J. Phys. Chem. C*, **114**, 19116–19126.

127. Chen, L., Grajciar, L., Nachtigall, P., Düren, T., Accurate prediction of methane adsorption in a metal-organic framework with unsaturated metal sites by direct implementation of an ab initio derived potential energy surface in GCMC simulation, *J. Phys. Chem. C*, **115**, 23074–23080.

128. Leclerc, H., Vimont, A., Lavalley, J.-C., Daturi, M., Wiersum, A. D., Llwellyn, P. L., Horcajada, P., Férey, G., Serre, C. (2011). Infrared study of the influence of reducible iron(III) metal sites on the adsorption of CO, CO_2, propane, propene and propyne in the mesoporous metal–organic framework MIL-100, *Phys. Chem. Chem. Phys.*, **13**, 11748–11756.

129. Yang, Q., Jobic, H., Salles, F., Kolokolov, D., Guillerm, V., Serre, C., Maurin, G. (2011). Probing the dynamics of CO_2 and CH_4 within the porous zirconium terephthalate UiO-66(Zr): a synergic combination

of neutron scattering measurements and molecular simulations, *Chem. Eur. J.*, **17**, 8882–8889.

130. Greathouse, J. A., Allendorf, M. D. (2008). Force field validation for molecular dynamics simulations of IRMOF-1 and other isoreticular zinc carboxylate coordination polymers, *J. Phys. Chem. C*, **112**, 5795–5802.
131. Amirjalayer, S., Tafipolsky, M., Schmid, A. S. (2007). Molecular dynamics simulation of benzene diffusion in MOF-5: importance of lattice dynamics, *Angew. Chem., Int. Ed.*, **46**, 463–466.

Chapter 8

Molecular Modeling of Metal–Organic Frameworks for Carbon Dioxide Separation Applications

Seda Keskin
Chemical and Biological Engineering, Koc University, Rumeli Feneri Yolu, 34450 Sariyer, Istanbul, Turkey
skeskin@ku.edu.tr

Metal–organic frameworks (MOFs) are relatively new nanoporous materials having fascinating structural, physical, and chemical properties. MOFs, also known as porous coordination polymers (PCPs) or porous coordination networks (PCNs), are hybrid materials composed of single metal ions or polynuclear metal clusters linked by organic ligands through strong coordination bonds. MOFs have several useful properties such as very large surface areas, ultralow densities, uniformly structured pores, and good thermal and mechanical properties. The most significant advantage of MOFs over other traditional nanoporous materials is the ability to obtain pore structures with desired shape, size, and surface characteristics by variation of the organic ligands and metal clusters during synthesis. In this way, it is possible to obtain a large variety of unique materials by combining different

Metal–Organic Frameworks: Materials Modeling towards Potential Engineering Applications
Edited by Jianwen Jiang

ISBN 978-981-4613-45-3 (Hardcover), 978-981-4613-46-0 (eBook)
www.panstanford.com

metals and organic linkers. Hundreds of MOF materials with various physical and chemical characteristics have been synthesized to date. Most of the studies in the literature have focused on a few specific MOF groups such as isoreticular metal–organic frameworks (IRMOFs) [1], zeolite imidazolate frameworks (ZIFs) [2], covalent organic frameworks (COFs) [3], coordination polymers of Oslo (CPOs) [4, 5], Matériaux de l'Institut Lavoisier (MILs) [6], copper 1,3,5-benzenetricarboxylate (Cu-BTC) [7], and zinc 1,4-benzenedicarboxylic acid-triethylenediamine (Zn(bdc)(ted)$_{0.5}$) [8].

MOFs have gained significant attention from academia and industry due to their great potential in various applications, including gas storage, gas separation, chemical sensing, catalysis, and biotechnology. Among these application areas, gas storage and gas separation can be considered as the most mature areas due to the high number of studies on these topics. Considering the large number of synthesized MOFs, using purely experimental methods to identify the most appropriate MOFs for gas separation applications is not feasible. As an alternative to experiments, molecular modeling methods have played a significant role in identification, design, and development of MOFs for gas separations. Molecular simulations have become powerful tools that can complement experimental research by providing molecular-level insights for potentials of MOFs in gas separations, specifically in CO_2 separations [9–11]. Figure 8.1 shows that using molecular simulations to assess the CO_2 separation performances of MOFs is a new area and that there is a rapid growth in the number of publications featuring the terms "MOFs," "molecular simulation," and "CO_2 separation" over the past decade.

The objective of this chapter is to review the recent advances in molecular modeling of MOFs for CO_2 separation applications. Section 8.1 introduces the importance of CO_2 separation and the potential of MOFs in this area. Molecular modeling methods such as Monte Carlo and molecular dynamics to study CO_2 adsorption and diffusion in MOFs are reviewed in Section 8.2. Section 8.3 focuses on molecular modeling studies that investigated adsorption-based separation of CO_2 from other gases using MOF adsorbents, whereas Section 8.4 reviews the computational studies for MOF membranes. Molecular modeling works on CO_2 diffusion in MOFs are discussed in

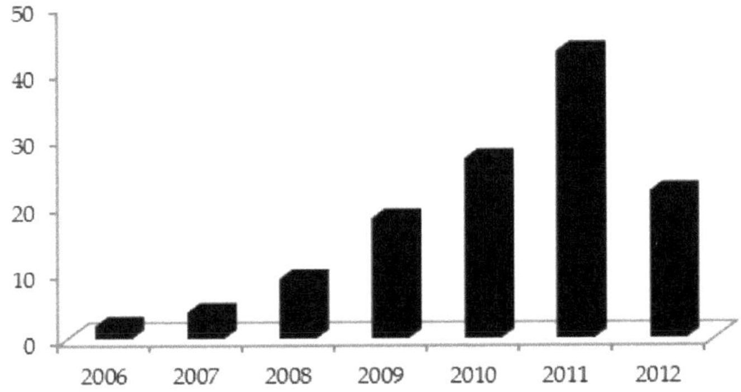

Figure 8.1 Number of publications featuring the terms "metal- organic framework," "molecular simulations," and "CO_2 separation" as a function of years. Source: *Web of Science*, data retrieved on May 8, 2012.

Section 8.5. Finally, opportunities and challenges in using molecular modeling for examining the potential of MOFs in CO_2 separations are summarized in Section 8.6.

8.1 Introduction

The capture of CO_2 emitted from industrial processes is a global challenge. In many industrial processes, CO_2 is present at low partial pressures among other gases that ideally should be recycled. Recent climate changes and global warming threats have stimulated scientific research into finding new technologies to decrease CO_2 emission to the atmosphere. The potential benefits of CO_2 separation include reduced air pollution and improvement of human health [12]. Traditional technology to separate CO_2 from other gases is absorption into amines. This technology requires regeneration of amines by heating, which is very costly; in addition to this, amine-based solvents are toxic and very reactive. Adsorption-based and membrane-based separations of CO_2 by nanoporous materials are economical and feasible alternatives to absorption due to milder operating conditions and lower energy demands. Nanoporous materials such as zeolites and carbon nanotubes have been widely investigated for adsorption- and membrane-based separations of

CO_2, and identifying new materials with high CO_2 selectivity, high capacity, high permeability, and low cost of generation is the main goal of ongoing research.

As a new class of nanoporous materials, MOFs are promising candidates for CO_2 capture since their high surface areas, tunable pore sizes, and chemical functionalities enable the selective capture of CO_2 in large quantities. What MOFs offer is a near-limitless array of porous materials that can be tuned to impart desired pore sizes, pore functionalities, and other factors that will make MOFs important materials in the future development of CO_2 separation processes [10, 13]. Considering the huge numbers of MOFs synthesized to date, it is far too time consuming to test every single MOF for CO_2 separation using experimental manners. Molecular modeling methods have been extensively used to screen materials and identify the best ones for CO_2 separation prior to experiments. These methods can provide information in the molecular level that cannot be viewed directly during an experiment. This molecular level information can be used to design a new MOF or modify an existing one to achieve a better performance for CO_2 separation. A similar type of molecular simulations has been previously used to assess gas adsorption capacities of MOFs and design MOFs with high storage capacities [14].

8.2 Molecular Modeling Methods

Molecular modeling of MOFs for CO_2 adsorption and diffusion generally includes grand canonical Monte Carlo (GCMC) simulations and equilibrium molecular dynamics (EMD) simulations, respectively. In these simulations, CO_2 molecules typically interact with the MOF atoms through dispersive, repulsive, and coulombic interactions. In the following sections, details of GCMC and EMD simulations together with the atomic models used for CO_2 and MOFs are described.

8.2.1 *Monte Carlo Simulations*

GCMC is the most widely used simulation method to predict adsorption equilibria in nanoporous materials such as zeolites and

MOFs. GCMC simulations have been compared rather successfully with the adsorption experiments for several MOFs. In GCMC simulations, chemical potential, volume, and temperature are kept constant, while the number of molecules is allowed to fluctuate. In an experimental set-up, the adsorbed gas is in equilibrium with the gas in the reservoir. Under equilibrium conditions, the temperature and chemical potential of the gas inside and outside of the adsorbent must be equal. The gas that is in contact with the adsorbent can be considered as a reservoir that imposes a temperature and chemical potential on the adsorbed gas. Knowing only the temperature and the chemical potential is enough to determine the equilibrium concentration inside the adsorbent. This is exactly what is mimicked in a GCMC simulation, allowing the fluctuation of the number of molecules during the simulation at an imposed temperature and chemical potential where the latter is related to the gas phase pressure by an equation of state. Each Monte Carlo simulation consists of millions of random insertions, deletions, translations, and rotations, which are accepted or rejected according to criteria based on a Boltzmann-type weighting [15]. The simulation results are then averaged over the runs. After running GCMC simulations at different pressures, the adsorption isotherms of a gas species can be obtained. The statistical mechanical basis of GCMC simulations and more details related to the simulation method can be found elsewhere [16].

8.2.2 *Molecular Dynamics Simulations*

Molecular dynamics (MD) is a method to simulate the motion and dynamics of molecules and atoms. The main concept in an MD simulation is to generate successive configurations of a system by integrating Newton's law of motion. Several diffusion coefficients such as self-diffusivity, corrected diffusivity, transport diffusivity can be measured from the trajectories showing how the position and velocities of gases vary with time in the system. First a short GCMC simulation is applied to generate the initial configurations of the atoms in the MOF before the MD simulation. Initial velocities are generally randomly assigned to each particle on the basis of Maxwell–Boltzmann velocity distribution [16, 17]. The number of molecules, volume of the structure, and temperature are kept constant to apply the *NVT* ensemble for equilibrating the system. The Nosé–Hoover thermostat is widely applied to keep the desired temperature, and the integration of the system dynamics is based on the explicit $N-V-T$ chain integrator by Martyna et al. [18, 19]. After the equilibration, Newton's equation is integrated and the positions of each particle in the system are recorded at a prespecified rate. The MD algorithm of Ciccotti et al. [20] is widely used to describe the dynamics of rigid-linear CO_2 molecules. The so-called order N algorithm [16] is implemented to calculate the diffusivities from the saved trajectories. More details of MD simulations for computing diffusivities of gas molecules in nanoporous structures can be found elsewhere [16].

8.2.3 *Force Fields for CO_2 and MOFs*

To perform GCMC and/or MD simulations to compute CO_2 adsorption and/or diffusion in MOFs' pores, one needs to define interactions between CO_2 molecules and between CO_2 molecules and MOF atoms. The combination of the functional form with a set of parameters completely describes the interactions in the model and is known as *force field* [14]. There are several functional forms defining the interactions such as Lennard–Jones. (LJ) [21], Morse. [22], and harmonic potentials. Among these functions, LJ is the most widely used one for gas adsorption and gas diffusion simulations in MOFs.

$$U_{\mathrm{LJ}} = 4\varepsilon_{ij}\left[\left(\frac{\sigma_{ij}}{r_{ij}}\right)^{12} - \left(\frac{\sigma_{ij}}{r_{ij}}\right)^{6}\right] \tag{8.1}$$

Parameters ε and σ represent the energy and size parameters of the LJ potential energy (U_{LJ}), whereas r_{ij} is the distance between atoms i and j. It is important to note that this potential computes the total energy as the sum of interaction energy between every pair of atoms, which is an approximation. The nonbonded interactions between CO_2 molecules and MOF atoms are represented as the sum of LJ interactions and coulombic interactions. The coulomb potential (U_{c}) is represented as following where q is the charge of atoms:

$$U_{\mathrm{C}} = \frac{q_i q_j}{4\pi\varepsilon_0 r_{ij}^2} \tag{8.2}$$

Force field parameters play an important role in determining the accuracy of molecular modeling studies. In almost all of the molecular simulation studies for MOFs, CO_2 is represented as a rigid linear triatomic molecule with three charged LJ interaction sites located at each atom. The LJ potential parameters of CO_2 is taken from the force field developed by Potoff and Siepmann [23] in which potential parameters of C and O and the C–O bond length were determined as follows: $\sigma_C = 0.280$ nm $\varepsilon_C/k_B = 27$ K, $\sigma_O = 0.305$ nm, $\varepsilon_O/k_B = 79$ K, and $l = 0.116$ nm. Partial point charges centered at each LJ site approximately represent the first-order electrostatic and second-order induction interactions, and these electrostatic interactions are generally handled by the Ewald summation technique.

Molecular modeling studies for CO_2 separations in MOFs have used general-purpose force fields such as the universal force field (UFF) [24], the DREIDING force field [25], and optimized potential for the liquid simulations all-atom (OPLS-AA) [26] force field for representing MOF atoms. A few studies have used quantum mechanical (QM) properties to develop new potentials for specific MOF-adsorbate interactions [27, 28]. There are studies where the parameters of the force fields are refined to match the predictions of simulations with the experimental measurements (gas adsorption isotherm data) or using first principles calculations [28–32]. Refining force field parameters to match the results of simulations with the experimental data is a not strong strategy because the accuracy of the experiments can be significantly affected by the defects of as-synthesized MOFs or trapped residual solvent molecules present in the samples [9]. Once the force field parameters for CO_2 and MOF atoms are determined, one can compute the LJ parameters for CO_2-MOF interactions using standard mixing rules. One of the most widely used mixing rule is known as Lorentz–Berthelot [17] rule, and it computes arithmetic averages for sigma and geometric averages for epsilon.

8.2.4 *Modeling of MOFs*

The early molecular modeling studies have assumed rigid MOF structures and kept the framework atoms fixed at their crystallographic positions obtained from x-ray diffraction experiments. In rigid framework simulations, only the nonbonding parameters describing the pairwise interactions between the CO_2 molecules and the MOF atoms of the particular force field were used. Of course, assumption of a rigid framework brings a huge computational efficiency and therefore current molecular modeling studies continue to use rigid MOF structure in GCMC and EMD simulations. On the other hand, recent research have shown that the inclusion of the framework motion and deformation is crucial for an accurate description of diffusion of large gas molecules in narrow-pore MOFs since large molecules fit tightly in the MOF pores, forcing the MOF to deform in order to allow a migration from pore to pore [33].

To carry out GCMC and EMD simulations for adsorption and diffusion of CO_2 in MOFs, the electrostatic interactions between MOFs and CO_2 molecules must be taken into account, which requires the assignment of partial charges to the MOF atoms. Since CO_2 molecules possess a significant quadrupole moment, it is expected that framework charge-quadrupole interactions (CQI) contributes to the adsorption and diffusion of CO_2 within the framework. Recent studies showed that the CQI are crucial at low loadings, and if the CQI are not taken into account in molecular simulations then the adsorption isotherms or diffusivities of CO_2 molecules can be significantly overestimated or underestimated [34]. One way to assign fixed partial charges is to use QM calculations; however, multiple methods exist for partitioning the net electron density determined in a QM calculation [9], and none of these methods give an unambiguous definition of the resulting point charges. These methods generally do not agree with each other nor are they expected to because atomic charges are not experimentally observable, and the methods used to derive them depend on the physical phenomenon the charges are intended to be reproduced [35].

Keskin et al. [9] reviewed the partial charges assigned to IRMOF-1 on the basis of QM calculations and showed that there is a significant variation in the charge values, depending on the method used. This variation may have a significant impact on the outcome of classical force field calculations in examples where electrostatic interactions are important. A new approach called connectivity-based atom contribution (CBAC) was recently developed as an alternative to time consuming QM calculations [36]. In CBAC method, it is assumed that the partial charge of an atom in a framework is determined by its bonding connectivity, and the atoms with the same connectivity have identical charges. The results of CBAC method on 43 different MOF structures showed that the CBAC charges give nearly identical results to those from QM calculations for adsorption isotherms of CO_2, CO, and N_2 molecules in these MOFs. Another recent study on two different MOFs (ZIF-3 and ZIF-10) showed that CO_2 adsorption isotherms and diffusivities computed using the charges from QM methods based on the ChelpG [37] density functional theory (DFT) calculations are very similar to the ones computed using charges from CBAC method [38]. Later on, Zheng and Zhong [39] extended this charge assignment method to COFs and showed that it is possible to apply this method to other types of MOFs for predicting CO_2 adsorption. The same research group [40] also concluded that the contribution of framework charges into CO_2 adsorption in MOFs cannot be ignored in computational screening of MOFs for CO_2 capture under low pressure conditions, whereas at moderate pressures, the contribution can be ignored in large-scale prescreening for certain processes such as natural gas upgrading (CO_2/CH_4 separations).

Wilmer and Snurr [35] investigated the feasibility of a method that bypasses the computationally expensive QM calculations to compute the partial charges for MOFs using charge equilibration method and showed that this new method is sufficiently accurate for screening purposes to study CO_2 adsorption and diffusion in MOFs. Watanabe et al. [41] used electrostatic potential energy surface (EPES) computed from plane-wave DFT calculations, which includes information from the full electron density of periodic material and demonstrated that it is not necessary to assign point charges to framework atoms for simulating CO_2 adsorption and diffusion. Haldoupis et al. [42] recently described a semiemprical charge equilibration method suitable for periodic materials that can be used to efficiently assign charges and allow molecular simulations for a large number of MOFs for CO_2/N_2 separations.

8.3 Adsorption-Based Carbon Dioxide Separation

Adsorption-based selectivity ($S_{\text{adsorption}}$) is a measure to evaluate the adsorption-based separation performance of a nanoporous material using the following expression:

$$S_{\text{adsorption}(i/j)} = \frac{x_i \cdot y_j}{x_j \cdot y_i} \tag{8.3}$$

In this definition x and y are the mole fractions of the components in the adsorbed and bulk phases, respectively. If i is CO_2 and j is the other component in a binary mixture, the CO_2 selectivities greater than unity imply that CO_2 is more strongly adsorbed in the pores of the material compared to the other component. A significant amount of published research on MOFs is dedicated to CO_2 capture by selective adsorption. The separation of CO_2 from CH_4, N_2, H_2, and CO has been widely studied in the literature since these gas separations have industrial importance. The molecular modeling studies for adsorption-based CO_2 separations are reviewed in the following sections.

8.3.1 *CO_2/CH_4 Separations*

Natural gas is mainly composed of CH_4 and thought to be one of the cleanest carbon fuels due to the low emissions of carbonaceous greenhouse gas. The presence of CO_2 within CH_4 reduces the energy content of the natural gas and brings the risk of pipeline corrosion due to the acidic structure of CO_2. Pressure swing adsorption (PSA) is known to be an efficient and economical process to remove CO_2 from CH_4 for natural gas upgrading. There has been a huge effort to identify the best material for adsorption-based separation of CO_2/CH_4 mixtures, and significant emphasis was given to nanoporous materials due to their high surface areas and large pore volumes, which are all desired for an adsorption-based separation process. The critical point is to design and develop nanoporous materials that show high selectivity and high capacity for CO_2. For example, traditional zeolites such as NaX and NaY are known to exhibit high separation performance for CO_2/CH_4 mixtures when used as adsorbents; however, it is difficult to regenerate these materials without significant heating, which causes high costs and low productivity. MOFs have emerged as great alternatives to traditional zeolites due to their adjustable pore sizes and the flexibility to control the chemical affinity of the framework atoms for CO_2.

Several research groups have computed adsorption-based separation performance of MOFs for CO_2/CH_4 mixtures using GCMC simulations. Babarao et al. [43] used GCMC simulations to compute the CO_2 selectivity of IRMOF-1, the prototype of isoreticular MOFs (IRMOFs), from CO_2/CH_4 mixtures and reported that the selectivity of IRMOF-1 ($\sim$2) is similar to that of zeolites MFI ($\sim$2.0–2.4) and less than C_{168} schwarzite ($\sim$5). Yang and Zhong [44] studied CO_2/CH_4 separation in Cu–benzene-1,3,5-tricarboxylate (Cu-BTC) and reported a higher CO_2 selectivity ($\sim$6–10) than IRMOF-1 for CO_2. Gutierrez-Sevillano et al. [45] performed molecular simulations and found that Cu-BTC is strongly CO_2 selective due to the differences in polarizability of CO_2 and CH_4 molecules. They also used EMD simulations and showed that diffusion selectivity favors CH_4 since it is a smaller and lighter molecule than CO_2. Martin-Calvo et al. [46] studied the same MOF and reported that the CO_2 selectivity of Cu-BTC obtained by molecular simulations of different research groups show similar trends.

Several researchers have used doping strategy to increase the affinity of MOFs for CO_2 molecules. Mu et al. [47] combined GCMC simulations with DFT calculations to study the effect of doping different metal atoms on CO_2/CH_4 separation in IRMOF-1. Computational results showed that alkali metal doping can enhance the CO_2 selectivity from 4 to 40 at 5 bar, 298 K if Li or Na is used as the doping metal. DFT studies indicated that doping materials with weak first-ionization energies, low LJ potential parameters, and a large length from the aromatic ring in the linker of IRMOF-1 would enhance electrostatic interactions and consequently improve the adsorption selectivity for CO_2. Xu et al. [48] studied adsorption-based selectivity of Li-modified IRMOF-1 using GCMC simulations and DFT calculations and showed that adsorption selectivity of Li-modified IRMOF-1 for CO_2 can increase up to 40 at 5 bar, 298 K due to strong binding of CO_2 to Li.

Molecular simulations have been also used to examine the effects of organic linkers on the selectivity of MOFs. Mu et al. [49] studied the effects of chemical properties of eight organic linkers (–CH_3, –F, –OH, –NH_2, –$(OH)_2$, –$(NH_2)_2$, –$(NH_2)_3$, and –$(NH_2)_4$) on IRMOF-1 for CO_2/CH_4 separations. Their results showed that the organic linkers decorated with electron donating groups can strengthen the distribution of the electrostatic field in the pores of MOF and enhance the adsorption selectivity of CO_2/CH_4 separation. Zhu et al. [50] replaced the ligand of COF-102 by extended aromatic moieties to study the effect of ligands on CO_2/CH_4 separation. They observed that CO_2 selectivity of COF-102 decreases from 10 to 3 after moiety replacement, and this was attributed to the enlargement of the pores by the ligand, which reduces the confinement of CO_2 in the pores and hence decreases selectivity. Amrouche et al. [51] studied CO_2/CH_4 selectivity of ZIF-8, ZIF-90, ZIF-Cl, and two hypothetical ZIFs (ZIF-COOH and ZIF-NO_2) designated using DFT calculations to understand the effects of ligands on CO_2 selectivity. The hypothetical ZIF-NO_2 exhibited higher CO_2/CH_4 selectivity ($\sim$14) compared to other ZIFs, indicating that choosing the chemical function can be a strategy for enhancing the CO_2 selectivity of the material.

Babarao and Jiang [52] reported a molecular simulation study for separation of CO_2/CH_4 mixtures in *rho*-ZMOF (*rho*-zeolite like MOF) that contains charge balancing extra framework Na^+ ions and showed that this MOF has significantly higher CO_2 selectivity than IRMOF-1, Cu-BTC, carborane-based MOFs due to additional CO_2 adsorption sites provided by Na^+ ions and strong electrostatic interactions. He et al. [53] carried out GCMC simulations to investigate the effects of extra framework cations (Na, Li, NH_4, Mg, Sr, Al) in *rho*-ZMOF on the separation of CO_2/CH_4 mixtures. Results showed that under typical conditions (1 atm, 298 K) of a PSA process, Al-exchanged *rho*-ZMOF exhibits an unprecedented high CO_2 selectivity ($\sim$200) over CH_4. This result suggested that variation of extra framework cations in MOFs is an efficient way to improve the separation capacity of MOFs.

Huang et al. [54] combined experiments with molecular simulation studies to investigate the effect of temperature on separation of CO_2/CH_4 mixtures in ZIF-8. At 1 bar and 300 K, CO_2 selectivity of ZIF-8 from an equimolar mixture was found to be $\sim$2, and this number increased (decreased) with increasing pressure (temperature) at constant temperature (pressure). Yang et al. [55] investigated

CO_2/CH_4 selectivity of UiO-66(Zr) using GCMC simulations and calculated a CO_2 selectivity of 5.5, which is comparable to selectivity of MIL-53 and ZIF-8 (4-7). They also calculated working capacity of CO_2 in a CO_2/CH_4 mixture at 10 bar and 1 bar and showed that UiO-66(Zr) can be a promising material for a PSA process since it exhibits almost three times higher working capacity than NaX and offers easy regeneration, which is very desirable for a PSA process.

Chen et al. [56] examined adsorption-based separation of CO_2/CH_4 mixtures in MIL-101 and hydrated MIL-101 using GCMC simulations. They concluded that adsorption selectivity of hydrated MIL-101 for CO_2 is higher than that of dehydrated MOF since terminal water molecules in the former act as additional adsorption sites that enhance adsorption of CO_2 and CH_4. The enhancement is stronger for CO_2 due to the electrostatic interactions between CO_2 and water molecules. Adsorption-based CO_2 selectivity of several MOFs predicted by molecular modeling studies for CO_2/CH_4 separations is given in Table 8.1.

8.3.2 *CO_2/N_2 Separations*

The removal of CO_2 from flue gas is important to reduce greenhouse gas emission. Relatively low CO_2 concentration and low pressure of the flue gas make CO_2/N_2 separation a difficult one. Cryogenic distillation, although widely used for other gas separations, is generally not considered as a practical means to separate CO_2 from flue gas because of the high energy costs involved [68]. The separation of CO_2/N_2 mixture has been widely studied using nanoporous materials. For example, at room temperature and slightly above atmospheric pressure, the CO_2 selectivities are 19 in Na-4A, 16 in activated carbon Norit-R1, 30 in silicalite, and 100 in ITQ-3 [69–71]. Yang et al. [72] studied CO_2/N_2 separation in Cu-BTC using GCMC simulations and reported CO_2 selectivities in the range of 20–40, depending on the bulk gas composition. Starting from the existing framework of Cu-BTC, Wells et al. [73] proposed two new modified scandium-based MOFs and simulated them for CO_2/N_2 separations. GCMC simulations showed that these new structures have higher CO_2/N_2 selectivity and a CO_2 working capacity compared to original Cu-BTC both in PSA and

Table 8.1 Adsorption-based CO_2 selectivity of MOFs for CO_2/CH_4 separations

MOF	*P*, *T*, composition (CO_2/CH_4)	CO_2 selectivity	References
Bio-MOF-11	10 bar, 298 K, 10/90	10	[57]
COF-102	10 bar, 298 K	12	[50]
COF-102	10 bar, 298 K, 50/50	3.8	[58]
Cu-BTC	1 bar, 298 K 50/50 hydrated 4 wt%	7	[59]
Cu-BTC	1 bar, 298 K, 50/50	4	[59]
Diphenyl-COF-102	10 bar, 298 K	3	[50]
IRMOF-$(NH_2)_2$	10 bar, 298 K, 50/50	5.5	[49]
IRMOF-$(OH)_2$	10 bar, 298 K, 50/50	4.5	[49]
IRMOF-1	10 bar, 298 K, 50/50	2.8	[60]
IRMOF-1	10 bar, 298 K, 50/50	3	[49]
IRMOF-10	10 bar, 298 K, 50/50	1.8	[60]
IRMOF-14	10 bar, 298 K, 50/50	0.5	[60]
IRMOF-1-Li	10 bar, 298 K, 10/90	30	[47]
IRMOF-1-Mg	10 bar, 298 K, 10/90	10	[47]
IRMOF-1-Na	10 bar, 298 K, 10/90	40	[47]
IRMOF-1-Rb	10 bar, 298 K, 10/90	20	[47]
IRMOF-8	10 bar, 298 K, 50/50	1.8	[60]
IRMOF-9	10 bar, 298 K, 50/50	4	[60]
IRMOF-CH3	10 bar, 298 K, 50/50	3.5	[49]
IRMOF-F	10 bar, 298 K, 50/50	3	[49]
IRMOF-NH_2	10 bar, 298 K, 50/50	4	[49]
IRMOF-OH	10 bar, 298 K, 50/50	4	[49]
MMIF	10 bar, 298 K, 50/50	8	[60]
MOF-177	1 bar, 300 K, 50/50	5	[61]
pyrene-COF-102	10 bar, 298 K	8	[50]
rho-ZMOF	1 bar, 300 K, 50/50	100	[61]
rho-ZMOF-Al	1 bar, 298 K, 50/50	100	[53]
UHM-6	1 bar, 298 K, 50/50	5	[62]
Uio-66	1 bar, 298 K, 50/50	6	[63]
UiO-66-Br	1 bar, 298 K, 50/50	4	[63]
UiO-66-NH_2	1 bar, 298 K, 50/50	4.2	[63]
UMCM-1	10 bar, 298 K, 50/50	2	[64]
UMCM-2	10 bar, 298 K, 50/50	3	[64]
ZIF-10	10 bar, 298 K, 50/50	5	[60]
ZIF-2	1 bar, 298 K	1.5	[65]
ZIF-4	1 bar, 298 K	1.6	[65]
ZIF-5	1 bar, 298 K	3.5	[65]
ZIF-68	10 bar, 298 K, 50/50	5.5	[66]
ZIF-69	10 bar, 298 K, 50/50	4	[66]
ZIF-78	10 bar, 298 K 50/50	5	[67]

MOF	*P*, *T*, composition (CO_2/CH_4)	CO_2 selectivity	References
ZIF-79	10 bar, 298 K 50/50	5	[67]
ZIF-8	1 bar, 298 K, 50/50	2.3	[54]
ZIF-8	1 bar, 298 K	1.5	[65]
ZIF-9	1 bar, 298 K	2	[65]
ZIF-CH_3	10 bar, 303.15 K, 25/75	2	[51]
ZIF-Cl	10 bar, 303.15 K, 25/75	4	[51]
ZIF-COOH	10 bar, 303.15 K, 25/75	3.5	[51]
ZIF-HOO	10 bar, 303.15 K, 25/75	9	[51]
ZIF-NO_2	10 bar, 303.15 K, 25/75	10	[51]

vacuum swing adsorption (VSA) processes. They also performed MD simulations and reported that the stability of new MOFs when loaded with water is better than Cu-BTC. Yazaydin et al.' [59] simulations showed that hydrated Cu-BTC has a CO_2 selectivity of 20 from equimolar CO_2/N_2 mixtures, whereas the dry Cu-BTC has a selectivity of 15 at 298 K and 2 bar. This result was later validated by experiments. A detailed analysis of simulation data revealed that CO_2 adsorption is enhanced due to the interaction between CO_2's quadrupole moment and the electric field created by water molecules.

Chen et al. [74] examined the microscopic properties of a composite of ionic liquid (IL) supported on IRMOF-1 for separation of CO_2/N_2 mixtures using GCMC. Compared to pure IRMOF-1, the IL/IRMOF-1 composite has a substantially higher selectivity for CO_2/N_2. At ambient conditions, CO_2/N_2 selectivity of the composite was calculated to be 70 for a weight ratio of IL/IRMOF-1 = 1.5, whereas a pure IRMOF can only have a CO_2 selectivity of 10. Yang et al. [75] examined CO_2/N_2 separation in IRMOFs and demonstrated that CO_2 selectivities from N_2 show remarkable steps below certain temperatures, and these stepped behaviors can greatly enhance the separation performance of IRMOFs for CO_2 up to selectivity of 500 at 10 bar, 195 K.

Liu and Smit [66] performed GCMC simulations to evaluate the separation performance of two typical ZIFs, ZIF-68 and ZIF-69, for CO_2/N_2 separations. Their results showed that due to the presence of chlorine atoms in imidazolate linkers in ZIF-69, the adsorption selectivity for CO_2 over N_2 is higher in ZIF-69 compared to ZIF-68. This work also showed that including electrostatic interactions between gas molecules and the framework is very important to achieve high adsorption selectivities for CO_2. Li et al. [67] investigated adsorption selectivity of CO_2/N_2 and CO_2/CH_4 mixtures in ZIF-78 and ZIF-79 with different CO_2 fractions using GCMC simulations. The selectivity of the CO_2/N_2 mixture is higher in ZIF-78 (20–25) than ZIF-79 (15–20) under the same conditions, and this was attributed to the electronic structure and polarity of the imidazolate linkers that affect the adsorption of CO_2. Both ZIFs exhibit higher CO_2 selectivity (15-25) for CO_2/N_2 mixtures compared to CO_2 selectivity (5-10) from CO_2/CH_4 mixtures.

Wells and Chaffee [76] simulated CO_2/N_2 selectivity and the CO_2 working capacity of Cu-BTC, MOF-5, Sc-MOF, COF-102, ZIF-69, MOF-74, MOF-177, and Zn(bdc)(ted)$_{0.5}$ for PSA and VSA processes. Results showed that Sc-MOF has a very high CO_2 selectivity (20) but a very low working capacity (<0.2 mol/kg) due to its small pores. Cu-BTC and MOF-74 show promising selectivity/working capacity combinations for both PSA and VSA processes. Adsorption-based CO_2 selectivity of several MOFs for CO_2/N_2 separations predicted from molecular modeling works is summarized in Table 8.2.

8.3.3 *CO_2/H_2 Separations*

Separation of CO_2 from H_2 is important in methane steam reforming and water–gas shift reactions since 95% of the H_2 used in fuel cells is now produced in these ways. Yang and Zhong [81] studied Cu-BTC and IRMOF-1 for CO_2/H_2 separations and reported selectivities of $\sim$150 and $\sim$25 at 30 bar, 298 K for these MOFs, respectively. The high selectivity of Cu-BTC was attributed to the side pockets of the material. Xue et al. [79] investigated the effects of the side pockets on CO_2/H_2 separation in Cu-BTC with and without blocking the side pockets in their GCMC simulations. The CO_2/H_2 selectivity in the side pockets was found to be much larger than in other regions, directly showing the positive contribution of the side pockets to separation. This was attributed to the stronger confinement of CO_2 in narrow pores of side pockets. Yang et al. [82] studied CO_2/H_2

Table 8.2 Adsorption-based CO_2 selectivity of MOFs for CO_2/N_2 separations

MOF	*P*, *T*, composition (CO_2/N_2)	CO_2 selectivity	References
COF-102	15 bar, 393 K	2.5	[76]
Cu-BTC	10 bar, 298 K, 50/50	32	[60]
Cu-BTC	15 bar, 393 K	7	[76]
Cu-BTC	10 bar, 298 K, 50/50	30	[77]
Cu-BTC	10 bar, 298 K, 50/50	33	[78]
Cu-BTC	1 bar, 298 K, 50/50	13	[59]
Cu-BTC	1 bar, 298 K, 50/50 hydrated 4 wt%	20	[59]
Cu-BTC	10 bar, 298 K, 50/50	20	[79]
Cu-BTC	1 bar, 298 K, 10/90	17	[46]
IL/IRMOF-1	1 bar w(IL/MOF-1)=1.5	70	[74]
IRMOF-1	10 bar, 298 K, 50/50	3	[75]
IRMOF-1	10 bar, 298 K, 50/50	5	[77]
IRMOF-1	10 bar, 298 K, 50/50	5	[78]
IRMOF-1	1 bar, 298 K, 10/90	5	[46]
IRMOF-10	10 bar, 213 K, 15/85	100	[75]
IRMOF-11	10 bar, 298 K, 50/50	15	[78]
IRMOF-12	10 bar, 213 K, 15/85	100	[75]
IRMOF-12	10 bar, 298 K, 50/50	5	[78]
IRMOF-13	10 bar, 298 K, 50/50	15	[78]
IRMOF-14	10 bar, 213 K, 15/85	130	[75]
IRMOF-3	10 bar, 298 K, 50/50	6	[77]
IRMOF-8	10 bar, 213 K, 15/85	125	[75]
MgMOF-74	1 bar, 300 K	90	[80]
MIL-47	15 bar, 393 K	4.5	[76]
MMIF	10 bar, 298 K, 50/50	45	[60]
MOF-177	15 bar, 393 K	2.5	[76]
MOF-5	15 bar, 393 K	3	[76]
MOF-74	15 bar, 393 K	11.5	[76]
Sc-MOF	15 bar, 393 K	20	[76]
UMCM-1	10 bar, 298 K, 50/50	4	[64]
UMCM-2	10 bar, 298 K, 50/50	5	[64]
ZIF-2	1 bar, 298 K	4	[65]
ZIF-4	1 bar, 298 K	7	[65]
ZIF-5	1 bar, 298 K	17	[65]
ZIF-68	10 bar, 298 K, 15/85	13	[66]
ZIF-69	15 bar, 393 K	7.5	[76]
ZIF-69	10 bar, 298 K, 15/85	20	[66]
ZIF-78	10 bar, 298 K, 50/50	35	[67]
ZIF-79	10 bar, 298 K, 50/50	30	[67]

(Contd.)

Table 8.2 (*Contd.*)

MOF	*P*, *T*, composition (CO_2/N_2)	CO_2 selectivity	References
ZIF-8	1 bar, 298 K, 50/50	6	[54]
ZIF-8	1 bar, 298 K	3	[65]
ZIF-9	1 bar, 298 K	15	[65]
ZIF-CH_3	10 bar, 303.15 K, 50/50	5	[51]
ZIF-Cl	10 bar, 303.15 K, 50/50	5	[51]
ZIF-COOH	10 bar, 303.15 K, 50/50	10	[51]
ZIF-HOO	10 bar, 303.15 K, 50/50	20	[51]
ZIF-NO_2	10 bar, 303.15 K, 50/50	25	[51]
Zn-MOF	10 bar, 298 K, 50/50	15	[77]

separation performances of catenated IRMOFs using GCMC and showed that catenation can greatly enhance the CO_2 selectivity from CO_2/H_2 mixtures in MOFs due to presence of small pores with different sizes. Although significant attention was given to IRMOFs and Cu-BTC, new MOFs were also examined for CO_2/H_2 separation. Frahm et al. [62] synthesized a new MOF called UHM-6 and used GCMC simulations to explore CO_2/H_2 selectivity of this material. The CO_2 selectivity of UHM-6 for an equimolar CO_2/H_2 mixture was found to be higher (~90) than selectivities of IRMOFs and in the same range with the selectivity of Cu-BTC at 1 bar and 298 K.

Erucar and Keskin [83] examined CO_2/H_2 selectivity of Zn(bdc)(ted)$_{0.5}$ and showed that the CO_2 selectivity of this MOF (~200) is significantly higher than that of IRMOFs (~10) under the same conditions. Peng et al. [64] used computer simulations to predict CO_2/H_2 selectivity of UMCM-1 and UMCM-2 in addition to separation of CO_2 from N_2 and CH_4. Results showed that UMCM-2 exhibits higher CO_2/H_2 selectivity (~10–25) compared to UMCM-1 (~10–15) since the former has two types of micropores in contrast to one micropore of the latter, and these additional micropores contribute to the more adsorption of CO_2. Battisti et al. [65] computed zero pressure adsorption selectivity of ZIF-i (i=2–10) and showed that ZIF-7 and ZIF-9 show a very high CO_2/H_2 selectivity (~284 and ~275, respectively) due to their narrow pores that strongly confine CO_2 molecules. Chen and Jiang [84] computed CO_2 selectivity of bioMOF-11 from CO_2/H_2:15/85 mixture as 325 at

10 bar, 298 K. To examine the effect of H_2O, the adsorption of a CO_2/H_2 mixture was simulated in the presence of 0.1% H_2O, and results showed that H_2O has a negligible effect on the adsorption of both CO_2 and H_2 and that the CO_2/H_2 selectivity is slightly enhanced due to the co-adsorption between CO_2 and adsorbed H_2O.

The effect of cation exchange on CO_2/H_2 separation was examined by two studies: Babarao and Jiang [85] used a hierarchical approach combining first-principles modeling and molecular simulations to investigate separation of CO_2/H_2 mixtures in Li^+-exchanged MOFs. The selectivity of Li^+-exchanged MOF for CO_2 was computed as 550, whereas the CO_2 selectivity of the neutral MOF (without framework charges and cations) was 60 at 298 K, 1 bar, showing that framework charges and cations have a substantial effect on CO_2 selectivity. Chen et al. [86] studied CO_2/H_2 separation in cation-exchanged *rho*-ZMOF and found that selectivity at a given pressure increases following the increasing order of the charge to diameter ratio. Adsorption-based CO_2 selectivity of several MOFs predicted from molecular simulations for CO_2/H_2 separations is given in Table 8.3.

8.3.4 *CO_2/CO Separations*

The CO_2/CO mixtures are produced from a large variety sources including metallurgic plants, synthesis gas from steam reforming and coal gasification. Separation of CO_2 from CO is important because pure CO is a valuable raw material for the synthesis of a variety of chemicals such as phosgene for polyurethanes and polycarbonates, and acetic acid. The number of studies on adsorption-based separation of CO_2/CO mixtures is less compared to other CO_2 mixtures. Karra and Walton [87] used GCMC simulations for IRMOFs and Cu-BTC to understand the interplay factors such as pore size, open metal sites, electrostatic interactions, ligand functionalization contributing to separation of CO_2/CO mixtures. Simulation results showed that CO_2/CO selectivities increases as the pore size of the MOF decreases, and MOFs having open metal sites (such as Cu-BTC) exhibit higher CO_2 selectivity than MOFs without open metal sites. Zhang et al. [63] investigated the effects of temperature and linker functionality in MOFs on CO_2 selectivity from CO_2/CO,

Table 8.3 Adsorption-based CO_2 selectivity of MOFs for CO_2/H_2 separations

MOF	*P*, *T*, composition (CO_2/H_2)	CO_2 selectivity	References
BioMOF-11	10 bar, 298 K, 10/90	500	[57]
UMCM-1	10 bar, 298 K, 50/50	4	[64]
UMCM-2	10 bar, 298 K, 50/50	5	[64]
MgMOF-74	1 bar, 300 K	700	[80]
ZIF-2	1 bar, 298 K	20	[65]
ZIF-4	1 bar, 298 K	70	[65]
ZIF-5	1 bar, 298 K	70	[65]
ZIF-8	1 bar, 298 K	13	[65]
ZIF-9	1 bar, 298 K	275	[65]
IRMOF-9	10 bar, 298 K, 50/50	50	[82]
IRMOF-10	10 bar, 298 K, 50/50	10	[82]
IRMOF-11	10 bar, 298 K, 50/50	80	[82]
IRMOF-12	10 bar, 298 K, 50/50	10	[82]
IRMOF-13	10 bar, 298 K, 50/50	90	[82]
IRMOF-14	10 bar, 298 K, 50/50	10	[82]
rho-ZMOF	1 bar, 300 K, 15/85	1000	[61]
ZnMOF-74	1 bar, 300 K, 15/85	100	[61]

CO_2/N_2, and CO_2/H_2 mixtures. They functionalized the linkers of UiO-66 to obtain UiO-66-Br, UiO-66-NO_2, and UiO-66-NH_2, and GCMC simulations showed that CO_2 selectivities all decrease with increasing temperature and for a given binary gas mixture; and the larger the affinity of functional group for CO_2, the quicker the decrease of selectivity with temperature. Adsorption-based CO_2 selectivity of several MOFs for CO_2/CO separations predicted from molecular modeling studies is in Table 8.4.

8.3.5 *Other CO_2 Separations*

Although significant attention has been given to separation of CO_2/CH_4 and CO_2/N_2 mixtures, there have been a few molecular modeling studies on separation of CO_2 from other gases. Wang et al. [88] conducted GCMC simulations to evaluate adsorption-based separation of CO_2/C_2H_4 mixtures in Cu-BTC and reported low C_2H_4 selectivity (~2–4). Yang et al. [72] studied CO_2/O_2 separation in Cu-BTC using GCMC simulations and reported CO_2 selectivities

Table 8.4 Adsorption-based CO_2 selectivity of MOFs for CO_2/CO separations

MOF	*P*, *T*, composition (CO_2/CO)	CO_2 selectivity	References
Cu-BTC	10 bar, 298 K, 50/50	10	[77]
Cu-BTC	10 bar, 298 K, 50/50	15	[88]
IRMOF-1	10 bar, 298 K, 50/50	5	[77]
IRMOF-3	10 bar, 298 K, 50/50	6	[77]
ZIF-8	1 bar, 298 K, 50/50	5.5	[54]
Zn-MOF	10 bar, 298 K, 50/50	10	[77]

in the range of 30–40 depending on the pressure. Fischer et al. [89] studied C_2H_2/CO_2 separation in Cu-BTC and concluded that in spite of the high C_2H_2 storage capacity, Cu-BTC exhibits modest C_2H_2/CO_2 selectivity. Wu et al. [90] studied CO_2 capture from CO_2/O_2 and CO_2/C_2H_4 in a Li-modified chem-4Li MOF using GCMC simulations. The results showed that this modified MOF shows exceptional CO_2 capture capability due to the enhancement of the electrostatic potential in the presence of lithium. Wells and Chaffee [76] simulated CO_2/H_2O selectivity and CO_2 working capacity of Cu-BTC, MOF-5, Sc-MOF, COF-102, ZIF-69, MOF-74, MOF-177, and Zn(bdc)(ted)$_{0.5}$ for PSA and VSA processes. Results showed that MIL-47 exhibits a high CO_2 selectivity over H_2O (6-8) compared to other MOFs studied due to its hydrophobicity.

8.3.6 *Comparing MOFs with Other Nanoporous Materials for CO_2 Separations*

It is important to compare the performance of MOFs with other nanoporous materials to assess the potential of MOFs in adsorption-based separations. Krishna and van Baten [61] performed one of the largest in silico screening of MOFs for CO_2 separation applications using configurational bias Monte Carlo (CBMC) and EMD techniques. They calculated the adsorption selectivity and working capacity of various MOFs and compared these with traditional zeolites such as NaX and NaY for CO_2/CH_4, CO_2/N_2, and CO_2/H_2 separations. Results showed that MgMOF-74, ZnMOF-74, Cu-BTC, MIL-47, and MIL-53 exhibit higher adsorption selectivity and working capacity for CO_2 than traditional zeolites DDR, MFI, ITQ-29, and TON in CO_2/CH_4 separations. Similarly, MgMOF-74 and *rho*-ZMOF outperform LTA, MFI, and DDR in CO_2/H_2 and CO_2/N_2 separations. It was concluded that among the MOFs they investigated, MgMOF-74 shows the best CO_2 capture performance due to strong electrostatic interactions between CO_2 molecules and exposed metal cation sites of the framework [61, 91]. Traditional zeolites NaX and NaY have higher CO_2 selectivity than CO_2/CH_4, CO_2/N_2, and CO_2/H_2 mixtures; however, they suffer from relatively low working capacity, which is important in PSA processes. Krishna and van Baten [92] also compared adsorption-based CO_2 selectivity of MOF-177, Cu-BTT, and MgMOF-74 with zeolites MFI, JBW, AFX, and NaX using molecular simulations. They concluded that MgMOF-74 outperforms all the materials considered in their study in terms of working capacity due to high free volume and strong affinity of CO_2 to the open metal sites of MOF. Liu and Smit [78] performed a systematic simulation study to compare the separation performance of CO_2/N_2 mixtures in MOFs (Cu-BTC, MIL-47(V), IRMOF-1, IRMOF-11, IRMOF-12, IRMOF-13, and IRMOF-14) and zeolites (MFI, LTA, and DDR). This study reiterated the conclusion that MOFs perform much better for CO_2 storage but their CO_2 separation performance is only comparable to zeolites. For the systems where the preferable component has a larger quadrupole moment, both zeolites and MOFs can enhance the separation selectivity.

8.4 MOFs for Membrane-Based Carbon Dioxide Separation

Permeation through CO_2-selective membranes is another potential means of separating CO_2 from other gases. Adsorption-based separation processes rely on equilibrium, whereas membrane-based separations depend on both equilibrium and kinetic properties of gases in the material. Membrane-based separations are energetically efficient, environmentally friendly, and continuous processes. Polymeric membranes have been widely used for CO_2/CH_4, CO_2/N_2, and CO_2/H_2 separations in industry, yet these membranes suffer from a trade-off between selectivity and permeability. Polymeric membranes having high selectivity generally possess low gas permeability, which requires large surface areas therefore high capital cost for a membrane-based separation process. Robeson [93] used large collections of experimental data to demonstrate the inverse relationship between selectivity and permeability of polymer membranes. Identification and development of membranes that can exceed Robeson's upper bound for CO_2 selectivity and CO_2 permeability has been the central focus of research in high performance membranes for the past two decades.

MOF membranes for CO_2 separation is an emerging field. A few pure MOF membranes and several mixed-matrix membranes (MMMs) with MOF fillers have been fabricated to date, and only a small fraction of these membranes have been tested for CO_2 separations [94–105]. Considering the large number of MOFs, experimental screening of MOF membranes is very time consuming. Molecular modeling methods that predict the performance of MOF membranes for gas separations of interest has been very useful to direct experimental efforts to more promising membrane materials. To model MOF membranes, both equilibrium and kinetic data of gases in MOFs' pores is required, and this information can be obtained using molecular simulations.

Keskin and Sholl [106] used molecular simulations to make the first prediction about the performance of a MOF membrane for CO_2/CH_4 separations. Their results revealed that the performance of MOF membranes cannot be evaluated only using the single component gas properties because ideal and mixture selectivities of MOF membranes can be enormously different due to multicomponent effects. Later on Keskin and Sholl [107, 108] predicted the selectivity of MOF-5 and Cu-BTC membranes for CO_2/CH_4, CO_2/H_2, and CO_2/N_2 mixtures and showed that Cu-BTC can exhibit higher selectivities than the MOF-5 membrane, indicating the importance of the structure on the membrane's performance. Modifying Krishna and van Baten's formulation [109], Keskin and Sholl [58] suggested an approximate method to predict the selectivity of a MOF membrane for separation of binary gas mixtures under arbitrary conditions:

$$S_{\text{permeation(i/j)}} = S_{\text{adsorption}_{(i/j)}} \cdot \frac{D_{i,\text{self}}(x_i, x_j)}{D_{j,\text{self}}(x_i, x_j)} \tag{8.4}$$

This method can predict a MOF membrane's mixture selectivity at a specified feed pressure and bulk gas composition on the basis of a single-mixture GCMC simulation and an EMD simulation performed at the loadings determined from this GCMC calculation. After this method was validated by comparing the predictions with full mixture calculations, it was used to estimate the selectivities of IRMOFs, Cu-BTC, Zn(bdc)(ted)$_{0.5}$ for CO_2/CH_4 and CO_2/H_2 separations [38, 58].

Chmelik et al. [110] studied CO_2/CH_4 separation in ZIF-8 and demonstrated that the membrane selectivity for the separation of a gas mixture can be well described by the interplay of mixture adsorption and mixture diffusion according to the simple relation. The mixed-gas adsorption and diffusion data necessary for this type of calculation can be derived by sorption/desorption studies on giant single crystals of MOFs with IR microscopy detection.

Watanabe et al. [111] used GCMC and EMD simulations to estimate the separation performance of a Cu(hfipbb)(H_2hfipbb)$_{0.5}$ membranes, and their results showed that due to the narrow pore structure of this MOF, CH_4 diffuses several orders of magnitude slower than CO_2, leading to a very highly CO_2 selective ($\sim 10^4$–10^5) MOF membrane. Molecular simulations in a similar structure, MMIF, showed diffusion of CO_2 is rapid enough to be characterized, whereas CH_4 cannot diffuse due to the large energy barrier experienced at the narrow pore openings of MMIF. This work concluded that MMIF can be great candidate for kinetic separation of CO_2 from CH_4 [60].

Krishna and van Baten [61] combined CBMC and MD simulations to predict the performance of MOFs as membranes in separation of CO_2/CH_4, CO_2/N_2, and CO_2/H_2 mixtures. They calculated CO_2 selectivity and CO_2 permeability of MOF membranes and compared these values with zeolites membranes. Results showed that MgMOF-74 and ZnMOF-74 membranes are on the lower end of the Robeson's upper bound for CO_2/CH_4 separations, whereas DDR, CHA, and ITQ-29 membranes are well above the upper bound due to their high CO_2 selectivities caused by narrow pore windows that control molecular transport. Similarly, Cu-BTC is on the Robeson's upper bound for CO_2/N_2 separations but NaX, NaY, and CHA membranes show higher CO_2 selectivity than MOF membranes. Krishna and van Baten [80] used molecular simulations of adsorption and diffusion to investigate permeation selectivity and permeability of CO_2 in zeolites and Mg-MOF-74. This MOF was found to exhibit higher CO_2/H_2 permeation selectivity (~200) than MFI, CHA, and LTA-Si membranes, higher CO_2/CH_4 permeation selectivity (~20) than MFI and FAU-Si membranes, and higher CO_2/N_2 permeation selectivity (~60) than CHA, LTA-Si, and MFI membranes. These high selectivities were attributed to the multicomponent correlation effects in one dimensional channel structure of MgMOF-74. This MOF is also a promising membrane material for CO_2/N_2 separation with a permselectivity of 40 at 10 bar, 300 K.

Battisti et al. [65] computed permeation selectivity of ZIF-2, ZIF-4, and ZIF-8 using GCMC and MD simulations for separation of CO_2/H_2, CO_2/CH_4, and CO_2/N_2 mixtures in the zero pressure limit and found that ZIF-4 can be a promising membrane material with a CO_2/N_2 permeation selectivity of 10.4, followed by ZIF-8 with a selectivity of 5.72. Liu et al. [112] computed CO_2/CH_4 selectivity of ZIF-68 and ZIF-70 membranes and found selectivity values between 2 and 6, showing that those materials are not very promising candidates for CO_2 separations. Predicted permeation selectivities from molecular simulations for CO_2/CH_4 and CO_2/H_2 are shown in Figs. 8.2 and 8.3, respectively [58, 61, 80, 83, 92, 113–115].

Atci et al. [57] studied CO_2 separation performance of a bioMOF and concluded that bioMOF-11 can outperform several isoreticular MOFs, traditional zeolites, and ZIFs in membrane-based separations of CO_2/CH_4 and CO_2/H_2 mixtures due to its high gas permeability and permeation selectivity. Atci and Keskin [116] later studied CO_2/H_2 separation performance of ZIF-65 and ZIF-90 after validating their atomistic models with experimental measurements of fabricated ZIF-90 membrane. The results showed that ZIF-65 and ZIF-90 are H_2-selective membranes in contrast to DDR, MFI, and CHA membranes in CO_2/H_2 separations.

Another potential application of MOFs in membrane-based gas separations is using MOFs as filler particles in polymers to make MMMs. The high gas permeability of MOFs and low cost and easy fabrication of polymers can be combined to achieve high-performance composite membranes for CO_2 separations. Several MMMs have been already synthesized, and the experimental

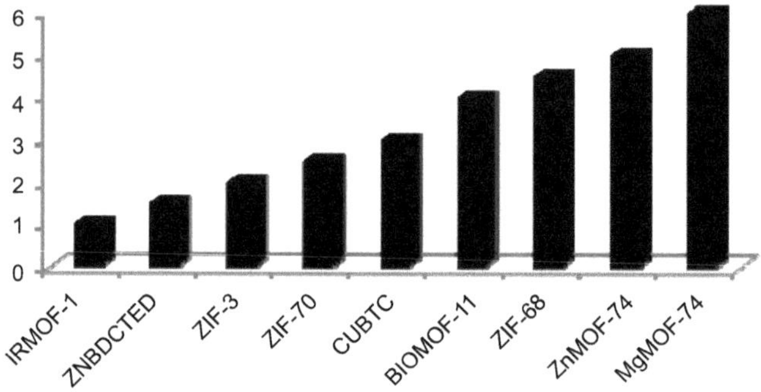

Figure 8.2 Predicted permeation selectivity for CO_2/CH_4 mixtures at 10 bar and 298 K. The composition of the bulk gas mixtures is CO_2/CH_4:10/90 for Zn(bdc)(ted), ZIF-3, and bioMOF-11, and the adsorbed composition of the CO_2/CH_4 mixture is 25/75 for ZIF-68 and ZIF-70 and equimolar for others.

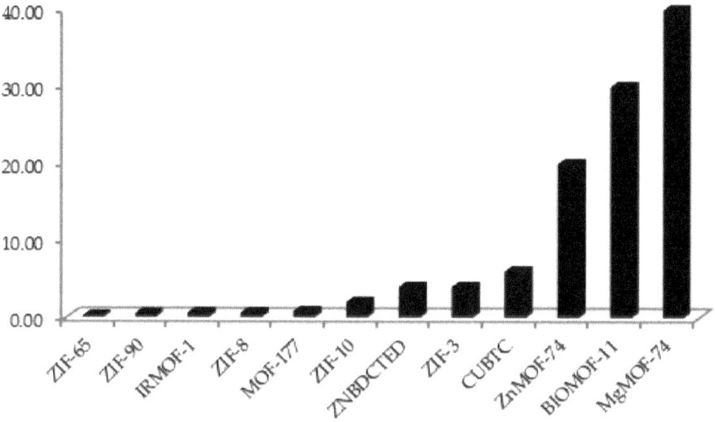

Figure 8.3 Predicted permeation selectivity for CO_2/H_2 mixtures at 10 bar and 298 K. The composition of the bulk gas mixtures is CO_2/H_2:1/99 for ZIF-3, ZIF-10, Zn(bdc)(ted), and bioMOF-11, 10/90 for IRMOF-1 and Cu-BTC, and equimolar for the others.

studies showed that incorporation of MOFs can enhance both CO_2 permeability and selectivity of pure polymer membranes [94, 102–104]. Considering the high number of available polymers and MOFs, it is a challenge to match polymers with MOFs for specific gas separation applications. Keskin and Sholl introduced a modeling approach, which combines molecular simulations with continuum modeling to predict the performance of MOFs as filler particles in polymers [117]. Erucar and Keskin [83] tested the validity of various theoretical permeation models to estimate the gas permeability of several MOF-based MMMs for CO_2/CH_4 separations. These studies showed that it is possible to enhance the CO_2 selectivity and CO_2 permeability of pure polymers by incorporating highly selective MOFs as filler particles.

8.5 Diffusion of Carbon Dioxide in MOFs

Besides thermodynamics, kinetics is an important factor that can significantly affect the separation performances of adsorbents and membranes. Understanding the dynamics of CO_2 in MOFs is essential to gain insights for the design and development of MOF adsorbents and membranes. Although the number of studies related to CO_2 adsorption in MOFs is increasing very fast, studies on CO_2 diffusion in MOFs is still limited.

Skoulidas and Sholl [118] reported CO_2 diffusion in a MOF for the first time using EMD simulations. Their calculations showed that the self- and transport diffusivities of light gases follow the order of $H_2 > N_2 = CH_4 = Ar > CO_2$ in IRMOF-1 at room temperature. Yang et al. [119] examined self-diffusivity of CO_2 in IRMOF-10, IRMOF-14, and MOF-177 at room temperature and reported diffusivities in the order of $\sim 4 \times 10^{-4} cm^2/s$. Yang et al. [120] used MD simulations to understand concentration dependence of self-diffusivities of CO_2 in UiO-66(Zr) and estimated the self-diffusivity of CO_2 as 10^{-5}–10^{-6} cm^2/s. Salles et al. [121] combined quasi-elastic neutron scattering (QENS) measurements with MD simulations, for the first time, to examine the transport diffusivity of CO_2 in a flexible MOF, MIL-53(Cr). The same research group [122] later used the same method to determine the self-diffusivity, corrected diffusivity, and transport diffusivity of CO_2 in MIL-47 (V) over a wide range of loadings. The QENS-MD combined approach introduced in these works became helpful to address the dynamics of a probe molecule in a MOF host, which undergoes guest-assisted structural transitions.

Diffusion of CO_2 in ZIFs has been studied by different research groups. Pantatosaki et al. [123] computed corrected and transport diffusivities of CO_2 in ZIF-8 as a function of adsorbed molecules and predicted diffusivities of CO_2 in the order of $\sim 10^{-6} cm^2/s$. Rankin et al. [124] used MD simulations to compute self- and transport diffusivities of CO_2 in ZIF-68 and ZIF-70 and examined the effect of CQI on diffusivities. Their results showed that CQI have a moderate impact on diffusion barriers at high adsorbate loadings. Liu et al. [125] studied self-diffusivity of CO_2 in ZIF-68 and ZIF-69 at 273 K and showed that diffusion in ZIF-69 is slower than ZIF-68 due to the relatively small pore size and the steric hindrance effects of chlorine atoms in the linkers of the former. Sirjoosingh et al. [126] used MD simulations to obtain mean square displacement (MSD) of CO_2 in the pores and channels of ZIF-68 and ZIF-69 at 273 K and ambient pressure. Babarao et al. [127] demonstrated that the pore topology and accessibility must be analyzed carefully so that the assumptions in the simulations reflect the reality. They performed MD studies for CO_2 diffusion in ZIF-68 and ZIF-69 and showed that small channels of these ZIFs are inaccessible to tightly bound CO_2 molecules due to the strong interactions between CO_2 molecules and the nitro groups forming the small cages.

8.6 Summary and Outlook

Due to the recent developments in reticular chemistry, a large number of MOF materials having various metal atoms and organic linkers have been synthesized. Molecular modeling methods provide a very important tool to screen a large variety of MOFs to identify the most promising ones for CO_2 separation applications. These simulations also provide a molecular-level understanding of the material's structure–performance relationships, which is crucial to design the ideal material prior to experiments. Once the structure–performance relationships are examined using molecular simulations and the structural parameters that have significant effects on the material's performance are identified, it can be possible to use reverse engineering and design the material for a specific gas separation application.

Choice of force field parameters plays a significant role in the accuracy of molecular simulations. The lack of a proper force field limits the investigation of gas separation performance of MOFs at the molecular level. At that point, multiscale simulation methods can be very helpful. In the multiscale simulation approach [128], accurate force fields can be developed for new materials using first-principles calculations and ab initio calculations can provide additional information for CO_2–MOF interactions. For example, Getman et al. [14] recently discussed that generic force fields are less successful at capturing adsorption in MOFs with open metal sites, and new force fields most likely from quantum chemical calculations are needed to describe the interaction of gases with open metal sites of MOFs.

As discussed before, a significant amount of the molecular modeling studies for MOFs assumed rigid frameworks. However, recent research identified several MOFs exhibiting unusual flexibility [121, 129]. Snurr et al. [130] developed a computational tool to efficiently calculate the unit cell shape and size for flexible MOFs loaded with adsorbates and demonstrated that this method can predict the structural change of MIL-53 in the presence of water. Development of molecular modeling methods that include framework flexibility in adsorption and diffusion simulations of MOFs will be very helpful to study a large group of flexible MOFs.

As can be seen from the literature review given in the previous section, the quantitative information about molecular diffusion is limited. Detailed characterization of molecular transport in MOFs is required to consider MOFs in membrane-based CO_2 separations. Further studies to compute molecular diffusion in MOFs will be helpful to visualize these materials in practical gas separation applications.

It is also important to note that molecular simulation methods typically do not make predictions about the long-term stability of MOFs. Stability is an important issue especially for membrane-based separations. A material with high CO_2 selectivity cannot find place in practical applications if the material is not stable under the presence of water vapor, and some earlier studies demonstrated that this can be a problem for MOFs [131, 132]. The stability of MOFs in water has received little attention in the literature, perhaps because they can be prepared and characterized in other solvents. However, the technological applications of MOFs would be greatly increased if new water-resistant frameworks were developed. It is likely that this issue is more straightforward to address experimentally; however, molecular simulations that can predict the stability of MOFs in CO_2 separations would be very useful.

References

1. Eddaoudi, M., Kim, J., Rosi, N., Vodak, D., Wachter, J., O'Keeffe, M., Yaghi, O. M. (2002). Systematic design of pore size and functionality in isoreticular MOFs and their application in methane storage, *Science*, **295**, 469–472.
2. Park, K. S., Ni, Z., Cote, A. P., Choi, J. Y., Huang, R. D., Uribe-Romo, F. J., Chae, H. K., O'Keeffe, M., Yaghi, O. M. (2006). Exceptional chemical and thermal stability of zeolitic imidazolate frameworks, *Proc. Natl. Acad. Sci. U. S. A.*, **103**, 10186–10191.
3. El-Kaderi, H. M., Hunt, J. R., Mendoza-Cortés, J. L., Côté, A. P., Taylor, R. E., O'Keeffe, M., Yaghi, O. M. (2007). Designed synthesis of 3D covalent organic frameworks, *Science*, **316**, 268–272.
4. Dietzel, P. D. C., Morita, Y., Blom, R., Fjellvag, H. (2005). An in situ high-temperature single-crystal investigation of a dehydrated metal-organic framework compound and field-induced magnetization of one-dimensional metaloxygen chains, *Angew. Chem., Int. Ed.*, **44**, 6354–6358.
5. Dietzel, P. D. C., Panella, B., Hirscher, M., Blom, R., Fjellvag, H. (2006). Hydrogen adsorption in a nickel based coordination polymer with open metal sites in the cylindrical cavities of the desolvated framework, *Chem. Commun.*, 959–961.
6. Loiseau, T., Serre, C., Huguenard, C., Fink, G., Taulelle, F., Henry, M., Bataille, T., Ferey, G. (2004). A rationale for the large breathing of the porous aluminum terephthalate (MIL-53) upon hydration, *Chem. Eur. J.*, **10**, 1373–1382.
7. Chui, S. S. Y., Lo, S. M. F., Charmant, J. P. H., Orpen, A. G., Williams, I. D. (1999). A chemically functionalizable nanoporous material [Cu-$_3$(TMA)($_2$)(H_2O)($_3$)](n), *Science*, **283**, 1148–1150.
8. Li, H., Eddaoudi, M., Groy, T. L., Yaghi, O. M. (1998). Establishing micro-porosity in open metal-organic frameworks: gas sorption isotherms

for zn(bdc) (bdc = 1,4-benzenedicarboxylate), *J. Am. Chem. Soc.*, **120**, 8571–8572.

9. Keskin, S., Liu, J., Rankin, R. B., Johnson, J. K., Sholl, D. S. (2009). Progress, opportunities, and challenges for applying atomically detailed modeling to molecular adsorption and transport in metal-organic framework materials, *Ind. Eng. Chem. Res.*, **48**, 2355–2371.
10. Keskin, S., Heest, T. M. V., Sholl, D. S. (2010). Can metal–organic framework materials play a useful role in large-scale carbon dioxide separations? *ChemSusChem*, **3**, 879–891.
11. Li, J. R., Kuppler, R. J., Zhou, H. C. (2009). Selective gas adsorption and separation in metal-organic frameworks, *Chem. Soc. Rev.*, **38**, 1477–1504.
12. Ferey, G., Serre, C., Devic, T., Maurin, G., Jobic, H., Llewellyn, P. L., De Weireld, G., Vimont, A., Daturi, M., Chang, J. S. (2011). Why hybrid porous solids capture greenhouse gases? *Chem. Soc. Rev.*, **40**, 550–562.
13. Jones, C. W. (2011). CO_2 capture from dilute gases as a component of modern global carbon management, *Annu. Rev. Chem. Biomol.*, **2**, 31–52.
14. Getman, R. B., Bae, Y. S., Wilmer, C. E., Snurr, R. Q. (2012). Review and analysis of molecular simulations of methane, hydrogen, and acetylene storage in metal-organic frameworks, *Chem. Rev.*, **112**, 703–723.
15. Düren, T., Bae, Y. S., Snurr, R. Q. (2009). Using molecular simulation to characterize metal organic frameworks for adsorption applications, *Chem. Soc. Rev.*, **38**, 1237–1247.
16. Frenkel, D., Smit, B. (2002). *Understanding Molecular Simulation: From Algorithms to Applications*, 2nd ed. (Academic Press, San Diego).
17. Allen, M. P., Tildesley, D. J. (1987). *Computer Simulation of Liquids* (Oxford University Press, New York).
18. Martyna, G. J., Tuckerman, M. E., Klein, M. L. (1992). Nose-hoover chains: the canonical ensemble via continuous dynamics, *J. Chem. Phys.*, **97**, 2635–2643.
19. Martyna, G. J., Tuckerman, E., Tobias, D. J., Klein, M. L. (1996). Explicit reversible integration algorithms for extended systems, *Mol. Phys.*, **87**, 1117–1157.
20. Ciccotti, G., Ferrario, M., Ryckaert, J.-P. (1982). Molecular-dynamics of rigid systems in cartesian coordinates: a general formulation, *Mol. Phys.*, **47**, 1253–1264.
21. Lennard-Jones, J. E. (1931). Cohesion, *Proc. Phys. Soc.*, **43**, 461–482.

22. Morse, P. M. (1929). Diatomic molecules according to the wave mechanics. II. Vibrational levels, *Phys. Rev.*, **34**, 57–64.
23. Potoff, J. J., Siepmann, J. I. (2001). Vapor-liquid equilibria of mixtures contaning alkanes, carbon dioxide, and nitrogen, *AIChE J.*, **47**, 1676–1682.
24. Rappe, A. K., Casewit, C. J., Colwell, K. S., Goddard, W. A., Skiff, W. M. (1992). UFF, a full periodic table force field for molecular mechanics and molecular dynamics simulations, *J. Am. Chem. Soc.*, **114**, 10024–10035.
25. Mayo, S. L., Olafson, B. D., Goddard, W. A. (1990). Dreiding: a generic force field for molecular simulations, *J. Phys. Chem. C*, **94**, 8897–8909.
26. Jorgensen, W. L., Maxwell, D. S., Tirado-Rives, J. (1996). Development and testing of the OPLS all-atom force field on conformational energetics and properties of organic liquids, *J. Am. Chem. Soc.*, **118**, 11225–11236.
27. Bordiga, S., Vitillo, J. G., Ricchiardi, G., Regli, L., Cocina, D., Zecchina, A., Bj.Arstad, Bjørgen, M., Hafizovic, J., Lillerud, K. P. (2005). Interaction of hydrogen with MOF-5, *J. Phys. Chem. B*, **109**, 18237–18242.
28. Sagara, T., Klassen, J., Ganz, E. (2004). Computational study of hydrogen binding by metal-organic framework-5, *J. Chem. Phys.*, **121**, 12543–12547.
29. Sagara, T., Ortony, J., Ganz, E. (2005). New isoreticular metal-organic framework materials for high hydrogen storage capacity, *J. Chem. Phys.*, **123**, 214707–214712.
30. Sagara, T., Klassen, J., Ortony, J., Ganz, E. (2005). Binding energies of hydrogen molecules to isoreticular metal-organic framework materials, *J. Chem. Phys.*, **123**, 014701–014705.
31. Yang, Q., Zhong, C. (2005). Molecular simulation of adsorption and diffusion of hydrogen in metal-organic frameworks, *J. Phys. Chem. B*, **109**, 11862–11864.
32. Yang, Q., Zhong, C. (2006). Understanding hydrogen adsorption in metal-organic frameworks with open metal sites: a computational study, *J. Phys. Chem. B*, **110**, 655–658.
33. Tafipolsky, M., Amirjalayer, S., Schmid, R. (2007). Ab initio parametrized MM3 force field for the metal-organic framework MOF-5, *J. Comput. Chem.*, **28**, 1169–1176.
34. Rankin, R. B., Liu, J., Kulkarni, A. D., Johnson, J. K. (2009). Adsorption and diffusion of light gases in ZIF-68 and ZIF-70: a simulation study, *J. Phys. Chem. C*, **113**, 16906–16914.

35. Wilmer, C. E., Snurr, R. Q. (2011). Towards rapid computational screening of metal-organic frameworks for carbon dioxide capture: calculation of framework charges via charge equilibration, *Chem. Eng. J.*, **171**, 775–781.
36. Xu, Q., Zhong, C. (2010). A general approcah for estimating framework charges in metal organic frameworks, *J. Phys. Chem. C*, **114**, 5035–5042.
37. Francl, M. M., Carey, C., Chirlian, L. E., Gange, D. M. (1996). Charges fit to electrostatic potentials. II. Can atomic charges be unambiguously fit to electrostatic potentials? *J. Comput. Chem.*, **17**, 367–383.
38. Keskin, S. (2011). Adsorption, diffusion, and separation of gas mixtures in zeolite imidazolate frameworks, *J. Phys. Chem. C*, **115**, 800–807.
39. Meng, C. C., Zhong, C. L. (2010). Estimation of framework charges in covalent organic frameworks using connectivity-based atom contribution method, *J. Phys. Chem. C*, **114**, 9945–9951.
40. Zheng, C. C., Liu, D. H., Yang, Q. Y., Zhong, C. L., Mi, J. G. (2009). Computational study on the influences of framework charges on CO_2 uptake in metal-organic frameworks, *Ind. Eng. Chem. Res.*, **48**, 10479–10484.
41. Watanabe, T., Manz, T. A., Sholl, D. S. (2011). Accurate treatment of electrostatics during molecular adsorption in nanoporous crystals without assigning point charges to framework atoms, *J. Phys. Chem. C*, **115**, 4824–4836.
42. Haldoupis, E., Nair, S., Sholl, D. S. (2012). Finding MOFs for highly selective CO_2/N_2 adsorption using materials screening based on efficient assignment of atomic point charges, *J. Am. Chem. Soc.*, **134**, 4313–4323.
43. Babarao, R., Hu, Z. Q., Jiang, J. W., Chempath, S., Sandler, S. I. (2007). Storage and separation of CO_2 and CH_4 in silicalite, C-168 schwarzite, and IRMOF-1: a comparative study from monte carlo simulation, *Langmuir*, **23**, 659–666.
44. Yang, Q. Y., Zhong, C. L. (2006). Electrostatic-field-induced enhancement of gas mixture separation in metal-organic frameworks: a computational study, *ChemPhysChem*, **7**, 1417–1421.
45. Gutierrez-Sevillano, J. J., Caro-Perez, A., Dubbeldam, D., Calero, S. (2011). Molecular simulation investigation into the performance of Cu-BTC metal-organic frameworks for carbon dioxide-methane separations, *Phys. Chem. Chem. Phys.*, **13**, 20453–20460.

46. Martin-Calvo, A., Garcia-Perez, E., Castillo, J. M., Calero, S. (2008). Molecular simulations for adsorption and separation of natural gas in IRMOF-1 and Cu-BTC metal-organic frameworks, *Phys. Chem. Chem. Phys.*, **10**, 7085–7091.
47. Mu, W., Liu, D. H., Zhong, C. L. (2011). A computational study of the effect of doping metals on CO_2/CH_4 separation in metal-organic frameworks, *Micropororous Mesoporous Mater.*, **143**, 66–72.
48. Xu, Q., Liu, D. H., Yang, Q. Y., Zhong, C. L., Mi, J. G. (2010). Li-modified metal-organic frameworks for CO_2/CH_4 separation: a route to achieving high adsorption selectivity, *J. Mater. Chem.*, **20**, 706–714.
49. Mu, W., Liu, D. H., Yang, Q. Y., Zhong, C. L. (2010). Computational study of the effect of organic linkers on natural gas upgrading in metal-organic frameworks, *Microporous Mesoporous Mater.*, **130**, 76–82.
50. Zhu, Y. J., Zhou, J. H., Hu, J., Liu, H. L., Hu, Y. (2011). Computer simulation of adsorption and separation of CO_2/CH_4 in modified COF-102, *Chin. J. Chem. Eng.*, **19**, 709–716.
51. Amrouche, H., Aguado, S., Perez-Pellitero, J., Chizallet, C., Siperstein, F., Farrusseng, D., Bats, N., Nieto-Draghi, C. (2011). Experimental and computational study of functionality impact on sodalite-zeolitic imidazolate frameworks for CO_2 separation, *J. Phys. Chem. C*, **115**, 16425–16432.
52. Babarao, R., Jiang, J. W. (2009). Unprecedentedly high selective adsorption of gas mixtures in rho zeolite-like metal-organic framework: a molecular simulation study, *J. Am. Chem. Soc.*, **131**, 11417–11425.
53. He, P., Liu, H., Li, Y. F., Lei, Z. G., Huang, S. P., Wang, P., Tian, H. P. (2012). Enhanced selectivity and capacity of adsorption of CO_2 and CH_4 in zeolite-like metal-organic frameworks with different extra-framework cations: a molecular simulation study, *Mol. Simul.*, **38**, 72–83.
54. Huang, H. L., Zhang, W. J., Liu, D. H., Liu, B., Chen, G. J., Zhong, C. L. (2011). Effect of temperature on gas adsorption and separation in ZIF-8: a combined experimental and molecular simulation study, *Chem. Eng. Sci.*, **66**, 6297–6305.
55. Yang, Q. Y., Wiersum, A. D., Jobic, H., Guillerm, V., Serre, C., Llewellyn, P. L., Maurin, G. (2011). Understanding the thermodynamic and kinetic behavior of the CO_2/CH_4 gas mixture within the porous zirconium terephthalate UiO-66(Zr): a joint experimental and modeling approach, *J. Phys. Chem. C*, **115**, 13768–13774.
56. Chen, Y. F., Babarao, R., Sandler, S. I., Jiang, J. W. (2010). Metal organic framework MIL-101 for adsorption and effect of terminal

water molecules: from quantum mechanics to molecular simulation, *Langmuir*, **26**, 8743–8750.

57. Atci, E., Erucar, I., Keskin, S. (2011). Adsorption and transport of CH_4, CO_2, H_2 mixtures in a bio-MOF material from molecular simulations, *J. Phys. Chem. C*, **115**, 6833–6840.
58. Keskin, S., Sholl, D. S. (2009). Efficient methods for screening of metal organic framework membranes for gas separations using atomically-detailed models, *Langmuir*, **25**, 11786–11795.
59. Yazaydin, A. O., Benin, A. I., Faheem, S. A., Jakubczak, P., Low, J. J., Willis, R. R., Snurr, R. Q. (2009). Enhanced CO_2 adsorption in metal-organic frameworks via occupation of open-metal sites by coordinated water molecules, *Chem. Mater.*, **21**, 1425–1430.
60. Keskin, S. (2011). High CO_2 selectivity of a microporous metal-imidazolate framework: a molecular simulation study, *Ind. Eng. Chem. Res.*, **50**, 8230–8236.
61. Krishna, R., van Baten, J. M. (2011). In silico screening of metal-organic frameworks in separation applications, *Phys. Chem. Chem. Phys.*, **13**, 10593–10616.
62. Frahm, D., Fischer, M., Hoffmann, F., Froba, M. (2011). An interpenetrated metal-organic framework and its gas storage behavior: simulation and experiment, *Inorg. Chem.*, **50**, 11055–11063.
63. Zhang, W. J., Huang, H. L., Zhong, C. L., Liu, D. H. (2012). Cooperative effect of temperature and linker functionality on CO_2 capture from industrial gas mixtures in metal-organic frameworks: a combined experimental and molecular simulation study, *Phys. Chem. Chem. Phys.*, **14**, 2317–2325.
64. Peng, X., Cheng, X., Cao, D. P. (2011). Computer simulations for the adsorption and separation of $CO_2/CH_4/H_2/N_2$ gases by UMCM-1 and UMCM-2 metal organic frameworks, *J. Mater. Chem.*, **21**, 11259–11270.
65. Battisti, A., Taioli, S., Garberoglio, G. (2011). Zeolitic imidazolate frameworks for separation of binary mixtures of CO_2, CH_4, N_2 and H_2: a computer simulation investigation, *Microporous Mesoporous Mater.*, **143**, 46–53.
66. Liu, B., Smit, B. (2010). Molecular simulation studies of separation of CO_2/N_2, CO_2/CH_4, and CH_4/N_2 by ZIFs, *J. Phys. Chem. C*, **114**, 8515–8522.
67. Li, B., Wei, S. H., Chen, L. (2011). Molecular simulation of CO_2, N_2 and CH_4 adsorption and separation in ZIF-78 and ZIF-79, *Mol. Simul.*, **37**, 1131–1142.

68. Choi, S., Drese, J. H., Jones, C. W. (2009). Adsorbent materials for carbon dioxide capture from large anthropogenic point sources, *ChemSusChem*, **2**, 796–854.

69. Akten, E. D., Siriwardane, R., Sholl, D. S. (2003). Monte carlo simulation of single- and binary-component adsorption of CO_2, N_2, and H_2 in zeolite Na-4A, *Energy Fuels*, **17**, 977–983.

70. Dreisbach, F., Staudt, R., Keller, J. U. (1999). High pressure adsorption data of methane, nitrogen, carbon dioxide and their binary and ternary mixtures on activated carbon, *Adsorption*, **5**, 215–227.

71. Goj, A., Sholl, D. S., Akten, E. D., Kohen, D. (2002). Atomistic simulations of CO_2 and N_2 adsorption in silica zeolites: the impact of pore size and shape, *J. Phys. Chem. B*, **106**, 8367–8375.

72. Yang, Q. Y., Xue, C. Y., Zhong, C. L., Chen, J. F. (2007). Molecular simulation of separation of CO_2 from flue gases in Cu-BTC metal-organic framework, *AIChE J.*, **53**, 2832–2840.

73. Wells, B. A., Webley, P. A., Chaffee, A. L. (2011). Simulations of model metal-organic frameworks for the separation of carbon dioxide, *Energy Proc.*, **4**, 568–575.

74. Chen, Y. F., Hu, Z. Q., Gupta, K. M., Jiang, J. W. (2011). Ionic liquid/metal-organic framework composite for CO_2 capture: a computational investigation, *J. Phys. Chem. C*, **115**, 21736–21742.

75. Yang, Q. Y., Ma, L. L., Zhong, C. L., An, X. H., Liu, D. H. (2011). Enhancement of CO_2/N_2 mixture separation using the thermodynamic stepped behavior of adsorption in metal-organic frameworks, *J. Phys. Chem. C*, **115**, 2790–2797.

76. Wells, B. A., Chaffee, A. L. (2011). Modeling gas separation in metal-organic frameworks, *Adsorption*, **17**, 255–264.

77. Morris, W., Leung, B., Furukawa, H., Yaghi, O. K., He, N., Hayashi, H., Houndonougbo, Y., Asta, M., Laird, B. B., Yaghi, O. M. (2010). A combined experimental-computational investigation of carbon dioxide capture in a series of isoreticular zeolitic imidazolate frameworks, *J. Am. Chem. Soc.*, **132**, 11006–11008.

78. Liu, B., Smit, B. (2009). Comparative molecular simulation study of CO_2/N_2 and CH_4/N_2 separation in zeolites and metal-organic frameworks, *Langmuir*, **25**, 5918–5926.

79. Xue, C. Y., Yang, Q. Y., Zhong, C. L. (2009). Effects of the side pockets on gas separation in metal-organic framework Cu-BTC: a molecular simulation study, *Mol. Simul.*, **35**, 1249–1255.

80. Krishna, R., van Baten, J. M. (2011). Investigating the potential of MgMOF-74 membranes for CO_2 capture, *J. Membr. Sci.*, **377**, 249–260.

81. Yang, Q. Y., Zhong, C. L. (2006). Molecular simulation of carbon dioxide/methane/hydrogen mixture adsorption in metal-organic frameworks, *J. Phys. Chem. B*, **110**, 17776–17783.

82. Yang, Q. Y., Xu, Q., Liu, B., Zhong, C. L., Smit, B. (2009). Molecular simulation of CO_2/H_2 mixture separation in metal-organic frameworks: effect of catenation and electrostatic interactions, *Chinese J. Chem. Eng.*, **17**, 781–790.

83. Erucar, I., Keskin, S. (2011). Separation of CO_2 mixtures using Zn(bdc)(ted)$_{0.5}$ membranes and composites: a molecular simulation study, *J. Phys. Chem. C*, **15**, 13637–13644.

84. Chen, Y. F., Jiang, J. W. (2010). A bio-metal-organic framework for highly selective CO_2 capture: a molecular simulation study, *ChemSusChem*, **3**, 982–988.

85. Babarao, R., Jiang, J. W. (2011). Cation characterization and CO(2) capture in Li(+)-exchanged metal-organic frameworks: from first-principles modeling to molecular simulation, *Ind. Eng. Chem. Res.*, **50**, 62–68.

86. Chen, Y. F., Nalaparaju, A., Eddaoudi, M., Jiang, J. W. (2012). CO_2 adsorption in mono-, di- and trivalent cation-exchanged metal-organic frameworks: a molecular simulation study, *Langmuir*, **28**, 3903–3910.

87. Karra, J. R., Walton, K. S. (2010). Molecular simulations and experimental studies of CO_2, CO, and N_2 adsorption in metal-organic frameworks, *J. Phys. Chem. C*, **114**, 15735–15740.

88. Wang, S. Y., Yang, Q. Y., Zhong, C. L. (2008). Adsorption and separation of binary mixtures in a metal-organic framework Cu-BTC: a computational study, *Sep. Purif. Technol.*, **60**, 30–35.

89. Fischer, M., Hoffmann, F., Froba, M. (2010). New microporous materials for acetylene storage and C_2H_2/CO_2 separation: insights from molecular simulations, *ChemPhysChem*, **11**, 2220–2229.

90. Wu, D., Xu, Q., Liu, D. H., Zhong, C. L. (2010). Exceptional CO_2 capture capability and molecular-level segregation in a Li-modified metal-organic framework, *J. Phys. Chem. C*, **114**, 16611–16617.

91. Herm, Z. R., Krishna, R., Long, J. R. (2012). CO_2/CH_4, CH_4/H_2 and $CO_2/CH_4/H_2$ separations at high pressures using Mg-2(dobdc), *Microporous Mesoporous Mater.*, **151**, 481–487.

92. Krishna, R., van Baten, J. M. (2012). A comparison of the CO_2 capture characteristics of zeolites and metal-organic frameworks, *Sep. Purif. Technol.*, **87**, 120–126.

93. Robeson, L. M. (1991). Correlation of separation factor versus permeability for polymeric membranes, *J. Membr. Sci.*, **62**, 165–185.
94. Zhang, C., Dai, Y., Johnson, J. R., Karvan, O., Koros, W. J. (2012). High performance ZIF-8/6FDA-DAM mixed matrix membrane for propylene/propane separations, *J. Membr. Sci.*, **389**, 34–42.
95. Jin, W. Q., Nan, J. P., Dong, X. L., Wang, W. J., Xu, N. P. (2011). Step-by-step seeding procedure for preparing HKUST-1 membrane on porous alpha-alumina support, *Langmuir*, **27**, 4309–4312.
96. Basu, S., Cano-Odena, A., Vankelecom, I. F. J. (2011). MOF-containing mixed-matrix membranes for CO_2/CH_4 and CO_2/N_2 binary gas mixture separations, *Sep. Purif. Technol.*, **81**, 31–40.
97. Venna, S. R., Carreon, M. A. (2010). Highly permeable zeolite imidazolate framework-8 membranes for CO_2/CH_4 separation, *J. Am. Chem. Soc.*, **132**, 76.
98. Li, Y.-S., Liang, F.-Y., Bux, H., Feldhoff, A., Yang, W.-S., Caro, J. (2010). Molecular sieve membrane: supported metal–organic framework with high hydrogen selectivity, *Angew. Chem., Int. Ed.*, **49**, 548–551.
99. Jeong, H. K., Guerrero, V. V., Yoo, Y., McCarthy, M. C. (2010). HKUST-1 membranes on porous supports using secondary growth, *J. Mater. Chem.*, **20**, 3938–3943.
100. Huang, A., Dou, W., Caro, J. (2010). Steam stable zeolitic imidazolate framework ZIF-90 membrane with hydrogen selectivity through covalent functionalizaiton, *J. Am. Chem. Soc.*, **132**, 15562–15564.
101. Huang, A., Bux, H., Steinbach, F., Caro, J. (2010). Molecular-sieve membrane with hydrogen permselectivity: ZIF-22 in LTA topology prepared with 3-aminopropyltriethoxysilane as covalent linker, *Angew. Chem., Int. Ed.*, **49**, 4958–4961.
102. Hu, J., Cai, H., Ren, H., Wei, Y., Xu, Z., Liu, H., Hu, Y. (2010). Mixed-matrix membrane hollow fibers of $Cu_3(btc)_2$ MOF and polyimide for gas separation and adsorption, *Ind. Eng. Chem. Res.*, **49**, 12605–12612.
103. Basu, S., Cano-Odena, A., Vankelecom, I. F. J. (2010). Asymmetric Matrimid®/$[Cu_3(BTC)_2]$ mixed matrix membranes for gas separations, *J. Memb. Sci.*, **362**, 478–487.
104. Perez, E. V., Balkus, K. J., Ferraris, J. P., Musselman, I. H. (2009). Mixed-matrix membranes containing MOF-5 for gas separations, *J. Membr. Sci.*, **328**, 165–173.
105. Liu, Y., Ng, Z., Khan, E. A., Jeong, H.-K., Ching, C.-B., Lai, Z. (2009). Synthesis of continous MOF-5 membranes on porous alpha-alumina substrates, *Microporous Mesoporous Mater.*, **118**, 296–301.

106. Keskin, S., Sholl, D. S. (2007). Screening metal-organic framework materials for membrane-based methane/carbon dioxide separations, *J. Phys. Chem. C*, **111**, 14055–14059.
107. Keskin, S., Liu, J., Johnson, J. K., Sholl, D. S. (2009). Atomically-detailed models of gas mixture diffusion through CuBTC membranes, *Microporous Mesoporous Mater.*, **125**, 101–106.
108. Keskin, S., Sholl, D. S. (2009). Assessment of a metal-organic framework membrane for gas separations using atomically detailed calculations: CO_2, CH_4, N_2, H_2 mixtures in MOF-5, *Ind. Eng. Chem. Res.*, **48**, 914–922.
109. Krishna, R., van Baten, J. M. (2007). Using molecular simulations for screening of zeolites for separation of CO_2/CH_4 mixtures, *Chem. Eng. J.*, **133**, 121–131.
110. Chmelik, C., Bux, H., Voss, H., Caro, J. (2011). Adsorption and diffusion: basis for molecular understanding of permeation through molecular sieve membranes, *Chem. Ing. Tech.*, **83**, 104–112.
111. Watanabe, T., Keskin, S., Nair, S., Sholl, D. S. (2009). Computational identification of a metal organic framework for high selectivity membrane-based gas separations, *Phys. Chem. Chem. Phys.*, **11**, 11389–11394.
112. Liu, J., Keskin, S., Sholl, D., Johnson, J. K. (2011). Molecular simulations and theoretical predictions for adsorption and diffusion of CH_4/H_2 and CO_2/CH_4 mixtures in ZIF-68 and ZIF-70, *J. Phys. Chem. C*, **115**, 12560–12566.
113. Keskin, S. (2010). Comparing performance of CPO and IRMOF membranes for gas separations using atomistic models, *Ind. Eng. Chem. Res.*, **48**, 11689–11696.
114. Keskin, S. (2011). Atomistic simulations for adsorption, diffusion, and separation of gas mixtures in zeolite imidazolate frameworks, *J. Phys. Chem. C*, **115**, 800–807.
115. Keskin, S. (2012). Adsorption, diffusion and separation of CH4/H2 mixtures in covalent organic frameworks: molecular simulations and theoretical predictions, *J. Phys. Chem. C*, **116**, 1772–1779.
116. Atci, E., Keskin, S. (2012). Atomically detailed models for transport of gas mixtures in ZIF membranes and ZIF/polymer composite membranes, *Ind. Eng. Chem. Res.*, **51**, 3091–3100.
117. Keskin, S., Sholl, D. S. (2010). Selecting metal organic frameworks as enabling materials in mixed matrix membranes for high efficiency natural gas purification, *Energy Environ. Sci.*, **3**, 343–351.

118. Skoulidas, A. I., Sholl, D. S. (2005). Self-diffusion and transport diffusion of light gases in metal-organic framework materials assessed using molecular dynamics simulations, *J. Phys. Chem. B*, **109**, 15760–15768.
119. Yang, Q. Y., Zhong, C. L., Chen, J. F. (2008). Computational study of CO_2 storage in metal-organic frameworks, *J. Phys. Chem. C*, **112**, 1562–1569.
120. Yang, Q. Y., Jobic, H., Salles, F., Kolokolov, D., Guillerm, V., Serre, C., Maurin, G. (2011). Probing the dynamics of CO_2 and CH_4 within the porous zirconium terephthalate UiO-66(Zr): a synergic combination of neutron scattering measurements and molecular simulations, *Chem. Eur. J.*, **17**, 8882–8889.
121. Salles, F., Jobic, H., Ghoufi, A., Llewellyn, P. L., Serre, C., Bourrelly, S., Ferey, G., Maurin, G. (2009). Transport diffusivity of CO_2 in the highly flexible metal-organic framework MIL-53(cr), *Angew. Chem., Int. Ed.*, **48**, 8335–8339.
122. Salles, F., Jobic, H., Devic, T., Llewellyn, P. L., Serre, C., Ferey, G., Maurin, G. (2010). Self and transport diffusivity of CO_2 in the metal-organic framework MIL-47(V) explored by quasi-elastic neutron scattering experiments and molecular dynamics simulations, *ACS Nano*, **4**, 143–152.
123. Pantatosaki, E., Megariotis, G., Pusch, A. K., Chmelik, C., Stallmach, F., Papadopoulos, G. K. (2012). On the impact of sorbent mobility on the sorbed phase equilibria and dynamics: a study of methane and carbon dioxide within the zeolite imidazolate framework-8, *J. Phys. Chem. C*, **116**, 201–207.
124. Rankin, R. B., Liu, J. C., Kulkarni, A. D., Johnson, J. K. (2009). Adsorption and diffusion of light gases in ZIF-68 and ZIF-70: a simulation study, *J. Phys. Chem. C*, **113**, 16906–16914.
125. Liu, D. H., Zheng, C. C., Yang, Q. Y., Zhong, C. L. (2009). Understanding the adsorption and diffusion of carbon dioxide in zeolitic imidazolate frameworks: a molecular simulation study, *J. Phys. Chem. C*, **113**, 5004–5009.
126. Sirjoosingh, A., Alavi, S., Woo, T. K. (2010). Grand-canonical monte carlo and molecular-dynamics simulations of carbon-dioxide and carbon-monoxide adsorption in zeolitic imidazolate framework materials, *J. Phys. Chem. C*, **114**, 2171–2178.
127. Babarao, R., Dai, S., Jiang, D. E. (2011). Effect of pore topology and accessibility on gas adsorption capacity in zeolitic-imidazolate frameworks: bringing molecular simulation close to experiment, *J. Phys. Chem. C*, **115**, 8126–8135.

128. Xiang, Z. H., Cao, D. P., Lan, J. H., Wang, W. C., Broom, D. P. (2010). Multiscale simulation and modelling of adsorptive processes for energy gas storage and carbon dioxide capture in porous coordination frameworks, *Energy Environ. Sci.*, **3**, 1469–1487.

129. Boutin, A., Coudert, F. X., Springuel-Huet, M. A., Neimark, A. V., Ferey, G., Fuchs, A. H. (2010). The behavior of flexible MIL-53(al) upon CH_4 and CO_2 adsorption, *J. Phys. Chem. C*, **114**, 22237–22244.

130. Dubbeldam, D., Krishna, R., Snurr, R. Q. (2009). Method for analyzing structural changes of flexible metal-organic frameworks induced by adsorbates, *J. Phys. Chem. C*, **113**, 19317–19327.

131. Greathouse, J. A., Allendorf, M. D. (2006). Molecular dynamics simulations of metal-organic framework-5 and its interactions with water, *Abstr. Pap. Am. Chem. S*, **232**, 835–835.

132. Greathouse, J. A., Allendorf, M. D. (2006). The interaction of water with MOF-5 simulated by molecular dynamics, *J. Am. Chem. Soc.*, **128**, 10678–10679.

Chapter 9

Modeling of Zeolitic-Like Hybrid Materials for Gas Separation

Carlos Nieto-Draghi[a] **and Javier Pérez-Pellitero**[b]
[a]*IFPEN, 1-4 avenue de Bois Préau, 92500 Rueil-Malmaison, France*
[b]*IFPEN, Rond-point de l'échangeur de Solaize, BP3, 69360 Solaize, France*
carlos.nieto@ifpen.fr, javier.perez-pellitero@ifpen.fr

The separation of CO_2 and other greenhouse emissions from flue gases is one of the most important industrial challenges nowadays. The use of hybrid materials as molecular sieves to achieve this task has received great interest in the scientific community. The possibility of tailoring these materials in terms of pore size or group functionalities provides a huge variety of combinations. The use of molecular simulation techniques becomes a useful tool to select the optimum solid for a particular application. In this chapter, we provide an overview of the most recent advances in the area of modeling the adsorption and diffusion phenomena of gases using zeolitic imidazolate framework (ZIF) materials. The key point in the prediction of those properties is the development of accurate intermolecular potentials to reproduce solid–gas interactions. This fact makes feasible the use of modern simulation techniques for an in silico screening of adsorption and diffusion of several gases in ZIFs at a large scale.

Metal–Organic Frameworks: Materials Modeling towards Potential Engineering Applications
Edited by Jianwen Jiang

ISBN 978-981-4613-45-3 (Hardcover), 978-981-4613-46-0 (eBook)
www.panstanford.com

9.1 Introduction

The capture of carbon dioxide gas is one of the most important challenges facing the scientific community today. In recent decades, human activity has been responsible for the release of a huge quantity of CO_2 into the atmosphere. Due to its stability, CO_2 is usually the final product of combustion reactions between a fossil fuel (such as coal or heating oil) and oxygen. The N_2/CO_2 separation of CO_2 from combustion gases appears to be the main problem. CO_2 capture is also an issue in the natural gas industry, as the presence of carbon dioxide in natural gas, which is mainly composed of methane, reduces its caloric power and also induces pipeline corrosion. In this case, a gas separation process is necessary. Among the various solutions we present here two examples for gas purification, pressure swing adsorption (PSA), which is known to be one of the most efficient and affordable processes for removing CO_2 from natural gas [1, 2], and membrane separation. The first technology could be extended to separate gas mixtures obtained from combustion reactions. PSA is a three-step process: First, the gas mixture is injected into a separation column; then, during a separation step, CO_2 is adsorbed by a porous adsorbent while the other gases are evacuated; and finally, the adsorbent is regenerated to restart the process. This separation process requires an adsorbent of sufficient capacity and selectivity. Although zeolite materials are recognized to be effective for gas separation, their renewability is energy intensive [3]. Research for new materials is therefore strongly recommended for the successful design of adsorption separation processes that are competitive with respect to the liquid separation processes used today. The second technology, membranes, is used in separation processes without additional heating. This separation process is purely physical, and it is based on the difference of relative mobility (diffusion) of the species involved. The main challenge of this approach is to produce a material having sufficient affinity (selectivity) for the target gas as well as an acceptable permeability to reduce excessive pressure losses.

The recent appearance of metal–organic frameworks (MOFs) has generated a vast volume of studies and attracted the attention of an important part of the scientific community. The possibility of tailoring these materials in terms of pore size [4] or group functionalities makes them extremely attractive for a wide variety of applications such as CO_2 storage, alkylation catalysis, or alkyl-benzene separation among many others. Theoretically, there are a huge number of possible combinations of organic linker and metallic oxides to form MOFs with different topological and chemical characteristics. In this chapter, we will focus our attention to the so-called zeolitic imidazolate frameworks (ZIFs) [5–8], which are a particularly interesting class of materials. ZIFs are constructed from tetrahedral $M^{2+}(Im^{-})_{4/2}$ units, where M^{2+} represents a bivalent cation (usually, but not exclusively Zn^{2+}) and Im^{-} a (substituted) imidazolate anion. Because of the strong bonding between the imidazolate linker and the metal center many ZIFs have high thermal (>673 K) and moisture stability compared to many other MOF structures [9]. As can be seen in Fig. 9.1 the $M^{2+}(Im^{-})_{4/2}$ units are analogous to the $SiO_{4/2}$ tetrahedra in zeolites and the Im–M–Im bond angle is similar to the Si–O–Si bond angle [11]. It is therefore possible to form zeolite-like networks from these building units. Gmelinite (GME), sodalite (SOD), Linde Type A (LTA), and rho (RHO) (http://izasc.biw.kuleuven.be/fmi/xsl/IZA-SC/ft.xsl) are some of the zeolite topologies for which ZIF analogues have been found. A nonexhaustive list of known ZIFs is reported in the work of Phan et al. [12].

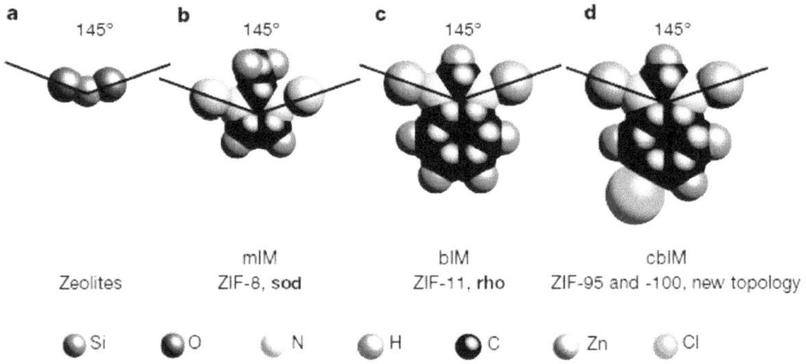

Figure 9.1 Comparison of the angle formed by O–Si–O and the angle formed by imidazolates bridging with transition metals in some ZIFs. Reproduced with permission from Ref. [11]. Copyright 2008.

The possibility of tuning the linker enables creating a huge variety of candidates for separation applications [10]. Nevertheless, it is unreasonable to explore all possible linker–structure combinations, and efficient scanning methods are strongly desirable to guide the development of structures [13]. Such methods must provide a direct connection between the topological and chemical nature of materials and the thermodynamic properties that describe the adsorption and diffusion behavior of gases. Molecular simulation appears to be an appropriate tool to scan MOF materials for CO_2 capture [14]. To achieve this titanic goal a reliable description of solid–gas interactions must be available. Such description relies on the development of accurate intermolecular force-fields capable of reproducing the experimental adsorption isotherms and diffusion phenomena of CO_2 and other gases on ZIFs. This is the subject of this chapter.

9.2 Development of Force Fields for ZIFs

As previously mentioned the appearance of ZIFs in early 2000 opened the door for the tailored synthesis of a wide variety of materials and consequently attracted the attention of the molecular simulation community. The need for reliable and transferable force fields able to accurately reproduce and predict the behavior of such materials for different types of applications has motivated numerous works. In the following sections we introduce the different approaches proposed to model ZIFs previously to the consideration of their different applications to the calculation of adsorptive and diffusive properties.

9.2.1 *Standard Force Fields*

Different generic force fields are available in the literature [15–17]. The majority of these force fields were primarily designed to reproduce structural properties. Many of the molecular modeling force fields used today can be interpreted in terms of a simple four-term description of the inter- and intramolecular forces governing the system:

$$V\left(r^{N}\right)=\sum_{\text{bonds}}\frac{k_b}{2}\left(l_b-l_{b,0}\right)^2+\sum_{\text{angles}}\frac{k_a}{2}\left(\theta_a-\theta_{a,0}\right)^2$$
$$+\sum_{\text{torsions}}\frac{k_t}{2}\left(1+\cos(n\omega-\gamma)\right)$$
$$+\sum_{i=1}^{N}\sum_{j=i+1}^{N}\left(4\varepsilon_{ij}\left[\left(\frac{\sigma_{ij}}{r_{ij}}\right)^{12}-\left(\frac{\sigma_{ij}}{r_{ij}}\right)^{6}\right]+\frac{q_i q_j}{4\pi\varepsilon_0 r_{ij}}\right) \tag{9.1}$$

where $V(r^N)$ is the potential energy, which is a function of the positions r of N particles (usually atoms) in the system. The three first terms introduce energy penalties accounting for the deviations from the equilibrium values of the different bond lengths ($\mathbf{l}_{\mathbf{b},\mathbf{0}}$), angles ($\theta_{\mathbf{b},\mathbf{0}}$) and torsions ($\omega$). The last term accounts for the intermolecular contributions arising from dispersion–repulsion and electrostatic forces. These contributions being modeled through a Lennard-Jones (LJ) and a coulombic term, respectively. According to the consideration of the intramolecular terms, force fields can be classified into rigid or flexible types.

9.2.2 *Dedicated Rigid Force Fields for ZIFs*

It is well accepted that the adsorption properties of small guest molecules in ZIFs can be in general modeled by means of force fields where the intramolecular energy contributions are neglected, that is, rigid models. Under this assumption, these kinds of force fields are mainly governed by the dispersion–repulsion and electrostatic intermolecular contributions. An appropriate repartition of these two contributions is necessary to correctly reproduce the behavior of adsorptive molecules bearing and not bearing significant electrostatic contributions. To obtain such a correct distribution, force field parameterization is in general performed in a sequential way where, in a first step, dispersion–repulsion parameters are determined or validated with experimental data of nonpolar molecules (noble gases or alkanes). Then, in a second step, the electrostatic contribution is added ant the total behavior validated with experimental data of polar molecules.

The first approaches to model ZIFs were done by Liu et al. [18]. They proposed the combination of the universal force field (UFF) with density functional theory (DFT)-derived electrostatic charges to study the adsorption and diffusion of CO_2 in ZIF-68 and ZIF-69. They obtained a good level of agreement for the CO_2 adsorption isotherms. In another study [19] the adsorption of hydrogen in ZIF-8 was modeled by means of a combination of DREIDING and OPLS-AA force fields. Nevertheless, the transferability of the UFF was not tested to other guest molecules and in particular to those not bearing a significant electrostatic contribution. Although the behavior of generic force fields to model the adsorption properties of nonpolar gases in other families of MOFs has been shown to be satisfactory [19–23], no theoretical evidence confirming the transferability of the generic LJ parameters to adsorption systems has been given.

Recent works [24–26] revealed the difficulties of conventional force fields to accurately reproduce the adsorption behavior of gases bearing and not bearing significant electrostatic contributions. These works showed an important overestimation of methane isotherms in different ZIF structures when using generic force fields. In particular, Pérez-Pellitero et al. proposed a modification of the UFF by rescaling the LJ parameters in order to match the experimental isotherm of methane. A good level of agreement was then obtained when considering electrostatic interactions in the case of CO_2 and N_2. In the same way, good agreement was obtained when considering other structures involving different types of topology. A similar approach was adopted by Liu and Smit [27] to study binary mixtures of different gases in ZIF-68 and ZIF-69. A compilation of the LJ parameters used in the different force fields is given in Table 9.1. Concerning the electrostatic description of the different materials, different approaches can be adopted. The most extended consists of the determination of electrostatic point charges, which are usually determined from quantum calculations applied to clusters representative of the material. Different methodologies are available in the literature for the determination of electrostatic point charges. They can be classified into two general groups, one based on the partition of the

Table 9.1 LJ potential parameters of different force fields

	UFF		DREIDING		Pérez-Pellitero et al. [24]		Fairen et al. [26]		Liu and Smit [27]	
	σ (Å)	ε/k_B (K)	σ (Å)	ε/k_B (K)	σ (Å)	ε/k_B (K)	σ (Å)	ε/k_B (K)	σ (Å)	ε/k_B (K)
Zn	2.462	62.40	4.045	27.69	2.338	43.08	2.462	36.928	2.462	53.40
O	3.118	30.19	3.033	48.19	3.118	20.85	-	-	3.118	21.75
C	3.431	52.84	3.473	47.89	3.259	36.48	3.431	31.270	3.431	40.27
N	3.261	34.72	3.263	38.98	3.097	23.97	3.261	20.549	3.261	22.72
H	2.571	22.14	2.846	7.654	2.440	15.29	2.571	13.103	2.571	14.14
Cl	3.516	114.23	3.519	142.66	3.516	78.87	-	-	3.516	99.23

wavefunction or the electron density (i.e., Mulliken, Hirshfeld) and one based on the electrostatic potential (ESP).

More recently, periodic methodologies permit to overcome the problems associated to the finite-size character of the cluster calculations have been developed [28, 29]. Despite the many advantages of using electrostatic point charges the main inconvenience of this approach is precisely that no unique definition or methodology exists to determine their values [30]. For instance, one of the particular problems of the methods based on the electrostatic surface potential is the fact that in the presence of buried atoms (as is the case of metals in ZIFs) they can provide sets of charges, which are not chemically meaningful, that is, a low or negative values of the point charges is obtained for the metallic centers [31]. This fact evidences the importance of parameterizing the LJ and electrostatic contributions of the force field in an independent way if the transferability wants to be improved. To overcome this problem, very recent works by Mc Daniel and Schmidt [32, 33] had dealt with the development of ab initio derived force fields for ZIFs. In this case the parameterization does not follow the conventional potential energy functional described in Eq. 9.1. Instead, an inherent energy decomposition consisting of exchange, electrostatic, induction, dispersion, and a small δ-Hartree–Fock term is proposed. They were able to obtain a set of parameters able to correctly reproduce the adsorption isotherms of CO_2 in ZIFs presenting RHO, SOD, and GME topologies. Despite the high level of accuracy and transferability obtained with this methodology, the main inconvenience of this approach is the lack of accuracy of DFT calculations for the prediction of the dispersion–repulsion term and the lack of Monte Carlo (MC) and molecular dynamics (MD) codes adapted to the use of the proposed energy decomposition.

9.2.3 *Dedicated Flexible Force Fields for ZIFs*

In simulating adsorption isotherms in microporous materials, it is customary to assume that the framework is rigid. This is usually a very good approximation, except for the materials, which are known to exhibit a structural phase transition upon adsorption [34].

Although ZIFs do not fall generally into this category, framework flexibility might have a non-negligible impact on the diffusion properties of adsorbed molecules particularly when the size of guest molecules gets closer to that of the pore windows of the microporous solids. For instance, in the case of a study devoted to the isoreticular metal–organic framework (IRMOF)-1 material [35], it was found that the diffusion coefficient of benzene was significantly affected when framework flexibility was taken into account. Besides, it has been experimentally shown that molecules exhibiting kinetic diameters higher than the size of the windows can access the microporosity of ZIF-8 (3.4 Å). Methane (3.8 Å) [24], ethane (3.8 Å) [36], propane (4.0 Å) [37], or even xylene isomers can pass through the six-membered ring of the ZIF-8 SOD topology. To account for this phenomenon, flexible force fields have been recently proposed [38–41]. In a first approach Battisti et al. [38] used the UFF-ZIF [29] to predict the diffusivities of different gases in ZIF-2, ZIF-4, ZIF-5, ZIF-6, ZIF-7, ZIF-8, ZIF-9, and ZIF-10. Zheng et al. [39] derived a flexible force field based on the general AMBER [17] force field developed for proteins. Additional parameterization based on quantum mechanics calculations was performed to better mimic the interactions of the tetrahedral ZnN_4 due to the relevance of this interaction to reproduce the lattice behavior [40]. The force field was validated with experimental x-ray diffraction data before carrying out CO_2 self-diffusion by means of MD calculations. Finally the impact of different simulation details over the CO_2 self-diffusivity was studied using the experimental lattice parameter. As can be seen in Table 9.2, the self-diffusion coefficient is overestimated by a factor of 6 when the simulation is performed in absence of electrostatics and keeping the crystallographic positions. On the other hand, when considering the electrostatic contributions and the average positions obtained after relaxation of the structure, in absence of sorbate molecules, the self-diffusivity is overestimated by only 30% compared to the fully flexible model. According to the obtained results, increasing the level of detail in the description of the system (electrostatics + flexibility) resulted in a decrease of the self-diffusivity of CO_2. On the other hand, for the case of molecules like methane exhibiting kinetic diameters larger than that of the window size, they expected an

Table 9.2 Influence of framework flexibility and electrostatics on the values of self-diffusion coefficients of CO_2 in ZIF-8

Model	D_s (10^{-10} m^2/s)	Self-diffusivity ratio with respect to flexible framework and electrostatic
Crystallographic without electrostatic charges	10.46 ± 0.36	5.8
Crystallographic with electrostatic charges	3.87 ± 0.65	2.2
Relaxed structure with electrostatic charges	2.31 ± 0.33	1.3
Flexible with electrostatic charges	1.79 ± 0.10	1.0

increase of the self-diffusivity since the introduction of the flexibility should result in allowing motion between neighboring cages.

In the same way, Hertag et al. [40] proposed an AMBER inspired force field to study the diffusive properties of CO_2 and H_2 in ZIF-8. Finally Hu et al. [41] also proposed an AMBER inspired force field, which was shown to be able to reproduce crystalline, mechanical, thermophysical, and diffusive properties of ZIF-8. In agreement with the results of Battisti et al. [38], they showed that in the absence of framework flexibility, methane vibrated in the ZIF-8 cavity and was not able to go through the aperture. In the case of CO_2 and contrary to the behavior observed in Ref. [39] the introduction of flexibility in the model was traduced in a significant increase of the self-diffusion coefficient.

9.3 Molecular Modeling of Adsorption in ZIFs

The high thermal and moisture stability of different ZIF structures together with the possibility of tailoring their porosity makes this type of materials good candidates for adsorption applications. The fact that M-Im-M is longer than Si-O-Al units in zeolites allows obtaining in ZIFs larger pore volumes than those found in the analogous zeolites. The combination of a large pore volume with pore sizes in the subnanometer range translates into a potentially high adsorption capacity. In addition the possibility of tuning the affinity of ZIFs for adsorbate molecules by modifying the imidazolate linkers allows us to obtain a very large variety of potential energy landscapes resulting on different adsorption behaviors and selectivities. As a consequence and to identify the best candidates for a given separation, the molecular modeling techniques can help to analyze how the complex interplay of linker, pore size, and topology affects the adsorption behavior of ZIFs. Molecular simulation separates and quantifies the contributions of the three parameters cited above; thereby it provides a possible explanation and may lead to predictions of the adsorption behavior of ZIFs. Many studies in molecular simulation have been published in the last years concerning the adsorption of different gases in ZIFs. Before describing such works it is interesting to quickly review the concepts involving the adsorption in solids. Then the general features of ZIF adsorbents and their particular behavior for different given applications is considered.

9.3.1 *Basic Concepts of Adsorption*

In molecular simulation the accessible volumes of the different solids can be evaluated using the following volume integral:

$$V_{\text{free}} = \int_V \exp\left(-U_{\text{ads}}^{\text{He}}(r)/k_{\text{B}}T\right) dr \tag{9.2}$$

where $U_{\text{ads}}^{\text{He}}$ is the LJ interaction between a single helium atom and the adsorbent (or any other gas supposed to exhibit a negligible interaction with the solid). The porosity ϕ can be then calculated as the ratio between the free volume (V_{free}) and the total volume of the unit cell (V_{UC}).

The fugacities corresponding to the different pressure conditions chosen to calculate the adsorption isotherms can be determined either from MC simulations in the isobaric–isothermal (NPT) ensemble combined with the particle insertion method [42] or from an equation of state (EoS) when available. In the same way, the bulk densities needed to transform the absolute number of molecules adsorbed in the solid into excess quantities can be obtained. This transformation was done according to

$$N_{\text{ex}} = N_{\text{abs}} - \rho_{\text{bulk}} V_{\text{free}} \tag{9.3}$$

where N_{ex} and N_{abs} are the excess and absolute number of adsorbed particles respectively, ρ_{bulk} is the density of the bulk gas calculated from the NPT simulations or an EoS, and V_{free} is the free volume calculated from the interaction of a helium particle with the solid. Isosteric heats of adsorption q_{st} can be calculated according to

$$q_{st} = H_b - \left[\frac{\partial(U_a)}{\partial N}\right]_{T,V} \tag{9.4}$$

where H_b is the enthalpy of the bulk phase and U_{ads} is the energy of the adsorbed phase. Using a fluctuations method with ideal gas assumptions under Henry's regime, the isosteric heats can be readily calculated from grand canonical Monte Carlo (GCMC) simulations according to

$$q_{st} = RT - \frac{\left\langle U^{s}_{ext} N\right\rangle - \left\langle U^{s}_{ext}\right\rangle \langle N\rangle}{\left\langle N^2\right\rangle - \langle N\rangle^2} \tag{9.5}$$

where U^{s}_{ext} is the intermolecular energy of the adsorbed phase and N is the number of adsorbed molecules. The angle brackets denote averages in the grand canonical ensemble.

9.3.2 *General Behavior of Adsorption in ZIFs*

Although adsorption properties depend both on adsorbate and adsorbent properties, general behavior can be associated to the intrinsic properties of ZIFs because of the common $ZnIm_2$ tetrahedral structure. In particular the importance of metallic centers and electrostatic interaction into the overall adsorption properties is considered in the next sections.

9.3.2.1 Metal sites and ligand adsorption

One important question arises when analyzing the adsorption on ZIF materials: what role do metals play in adsorption? Since ZIFs exhibit zeolitic topologies we can guess that adsorption should not be strongly affected by metals as far as van der Waals (vdW) interactions are concerned. The influence of the vdW interactions of metal sites has been investigated by removing these energy contributions [24]. As can be seen in Fig. 9.2 for different ZIFs, less than 10% of relative deviation is obtained for the different

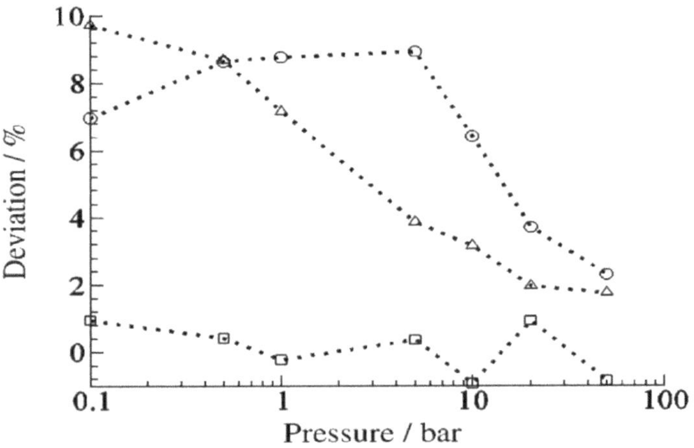

Figure 9.2 Deviation observed on absolute adsorption of CO_2 for ZIF-8 (circles), ZIF-69 (triangles), and ZIF-76 (squares) when no vdW interactions are considered for Zn metals. Reproduced with permission from Ref. [24]. Copyright © 2009, John Wiley and Sons.

materials considered when not considering the vdW contribution of the metal sites. This contribution becomes even less important at higher pressures, confirming that the behavior of ZIFs is similar to that of normal zeolite structures, in which guest-Si and guest-Al vdW contributions are not explicitly taken into account in adsorption modeling by molecular simulations.

9.3.2.2 Importance of electrostatics in adsorption

As in the case of metal sites, different studies have considered the importance of electrostatic interactions in ZIFs. [18, 24]. As it is shown in Fig. 9.3, relative deviations of up to 80% in the case of ZIF-76 were found. This effect is in particular important at low loadings, having a strong impact in the determination of Henry constants and isosteric heats of adsorption. At high pressure, fluid–fluid interactions become important and the absence of electrostatic host–guest interactions is not as important as in the low-pressure regime, under which the energy of the system is practically dominated by host–guest interactions.

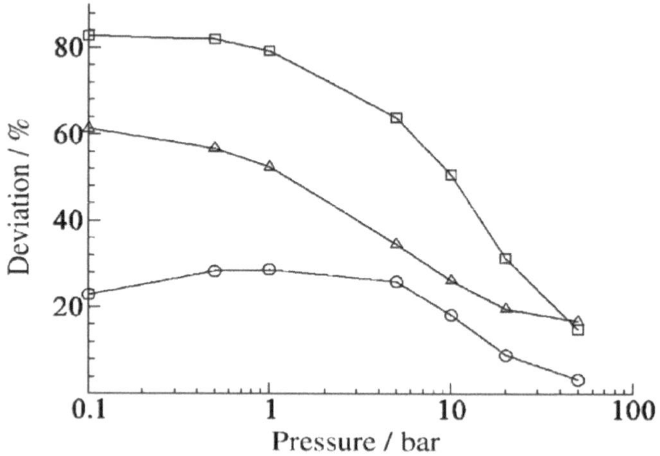

Figure 9.3 Deviation observed on absolute adsorption of CO_2 for ZIF-8 (circles), ZIF-69 (triangles), and ZIF-76 (squares) when no electrostatic interactions are considered. Reproduced with permission from Ref. [24]. Copyright © 2009, John Wiley and Sons.

9.3.3 *Adsorption of Alkanes in ZIFs*

Studying the adsorption of small hydrocarbons in ZIFs allows a better understanding of the adsorption mechanism in this family of adsorbents. Since in the case of alkanes the importance of electrostatics can be neglected, the dissociation of the electrostatic a dispersion–repulsion effect becomes evident. Several works have modeled the adsorption of small alkanes in ZIFs. A nonexhaustive list is provided in Table 9.3. Molecular simulation techniques allow obtaining very detailed information about the adsorption mechanism of guest molecules in the adsorbents. As can be seen in Fig. 9.4 [24] for the case of the prototypical ZIF-8 structure, by examining the locations of the centers of mass of the CH_4 adsorbed molecules, the most favorable locations from the dispersion–repulsion point of view can be identified.

In this case, the guest molecules adsorb close to the imidazolic rings contrary to the behavior previously observed in other families of MOFs where the metallic centers seem to play a more important role in the adsorption. This mechanism is in agreement with the analysis reported in the previous section where it was anticipated

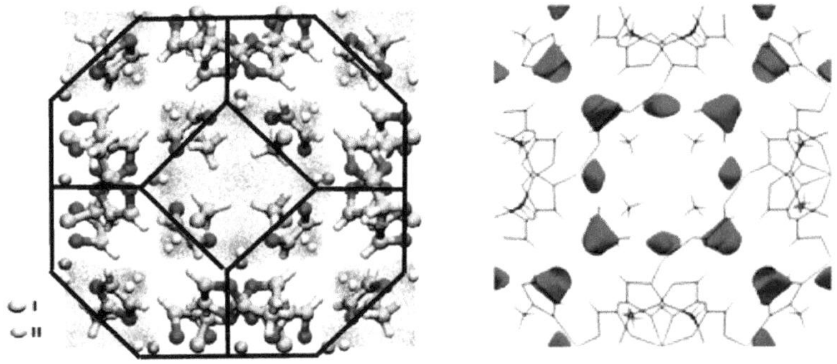

Figure 9.4 (Left) Representation of the center of mass of adsorbed CH_4 molecules on ZIF-8 at 303 K and 5 bar. (Right) Loci of sites for maximum vdW interaction in ZIF-8. Main adsorption sites I and II are highlighted. Reproduced with permission from Ref. [24]. Copyright © 2009, John Wiley and Sons.

that the partial contribution of the metallic centers to the potential energy appears to not to be relevant. As expected, when analyzing the location of sites for maximum vdW interaction in ZIF-8 the same information is obtained. The preferential sites for vdW interactions are inside the sodalite cage, close to the methyl groups of the mIm linkers (site I) and on both sides of the center of the six-ring windows (site II). More recently, in a combined molecular simulation/experimental approach, Zhang et al. [43] found the same behavior for ethane and propane and concluded that for the case of ZIF-8 there was almost no competition effect from the size or energetic point of view for these short alkanes. Nevertheless, the agreement between the identified adsorption sites and the loci of maximal dispersion–repulsion energy could disappear when considering longer or branched alkanes.

The fact that many ZIFs exhibit window openings close to the kinetic diameters of alkanes and alkenes have focused the attention of different recent works. In particular, van den Bergh et al. [44] explained by means of DFT calculations the ability of ZIF-7 to preferentially adsorb C2 and C3 alkanes from mixtures over the corresponding alkenes. This anomalous selectivity behavior confirms the need for flexible molecular force fields when considering adsorbates commensurable to the window openings.

Table 9.3 Recent simulation works on hydrocarbon adsorption in ZIFs

Material	Framework	Solid–gas FF	Properties	T (K)	Ref.
ZIF-8, ZIF-69, ZIF-76	Rigid	UFF inspired	• CH_4 isotherms • Q_{st} • Henry constant • Adsorption sites	(273, 303)	[24]
ZIF-78, ZIF-79	Rigid	UFF, UFF inspired, DREIDING	• CO_2, CH_4, N_2 isotherms • Q_{st}	298	[45]
ZIF-8	Rigid	OPLS-AA/DREIDING	• Methane, ethane, ethene, propane, propene adsorption isotherms • Adsorption sites of methane	300	[46]
ZIF-7	–	DFT	• Ethane, ethene, propane, propene, butane, butene isomers adsorption isotherms • Energy profiles along the window	(298, 373)	[44]
ZIF-8, ZIF-10	Rigid	DREIDING	Methane/ethane/propane equimolar mixtures	298	[43]
ZIF-8	Rigid (structures at low pressure and high pressure)	UFF inspired	• Methane, ethane, propane, butane adsorption isotherms • Q_{st}	(125, 300)	[26]

9.3.4 CO_2 Adsorption in ZIFs

In the last years a growing interest in lowering the concentration of greenhouse gases in the atmosphere has been stated [13]. Carbon capture and storage (CCS) is one option for reducing these harmful CO_2 emissions. While a variety of technologies and methods have been developed, the separation of CO_2 from gas streams is still a critical issue. The exploration of capture materials with high separation abilities and low cost is consequently a subject of capital importance. In this framework, adsorption of CO_2 in ZIFs has been studied in several works by means of molecular simulation techniques. A nonexhaustive list of these studies is provided in Table 9.4.

Concerning the CO_2 adsorption mechanism, as in the case of alkanes, it was found in general [24] that CO_2 molecules adsorb close to the imidazolate ligands in ZIF-8. As can be seen in Fig. 9.5, the preferential adsorption sites for electrostatic interactions are on both sides of six-ring windows (similar to site II observed for CH_4), and in the middle of the faces of the accessible surface area of ZIF-8 (site III), close to the aromatic rings of the mIm linker, above the center of the four-ring windows. Site III cannot be described by the vdW energy but corresponds to the zones where electrostatic interactions are maximal as can be observed in the projection of the electrostatic interaction of a positive charge (+1)

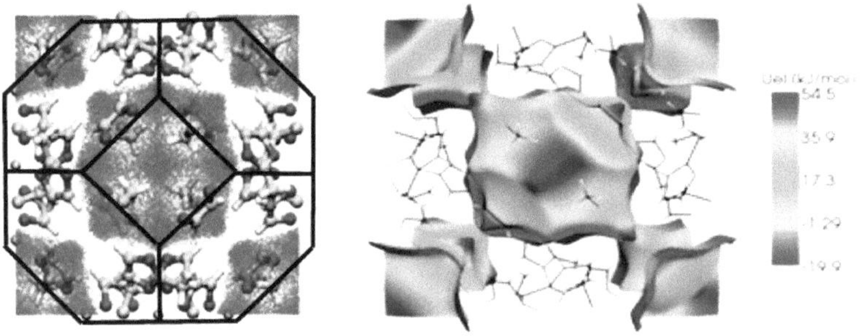

Figure 9.5 (Left) Representation of the center of mass of adsorbed CO_2 molecules on ZIF-8 at 303 K and 5 bar. (Right) Projection of the electrostatic interactions of a positive charge (+1) over the accessible surface area. Reproduced with permission from Ref. [24]. Copyright © 2009, John Wiley and Sons.

Table 9.4 Recent simulation works on CO_2 adsorption in ZIFs

Material	Framework	Solid–gas FF	Properties	T (K)	Ref.
ZIF-68, ZIF-69	Rigid	UFF	• CO_2 isotherms • Q_{st} • Adsorption sites	273	[18]
ZIF-8, ZIF-69, ZIF-76	Rigid	UFF inspired	• CO_2, CO, CH_4, N_2 isotherms • Q_{st} • Henry constant • Adsorption sites	(273, 303)	[24]
ZIF-68, ZIF-69	Rigid	UFF	CO_2, CO isotherms	273	[47]
ZIF-25, ZIF-71, ZIF-93, ZIF-96, ZIF-97	Rigid	UFF	CO_2 isotherms	298	[31]
ZIF-78, ZIF-79	–	UFF, UFF inspired, DREIDING	• CO_2, CH_4, N_2 isotherms • Q_{st}	298	[48]
ZIF-8	Rigid (structures at low pressure and high pressure)	UFF inspired	• CO_2 adsorption isotherms • Q_{st} • Henry constant	(125, 300)	[26]

over the accessible surface area of ZIF-8. Finally, particularly for the case of CO_2 and placed in the middle of the ZIF-8 cavity, is also present at high loadings an extra adsorption site. However, this site cannot be explained by guest-host interactions and is the result of a preferential arrangement between guest adsorbed molecules.

More in particular, for those topologies presenting more than one type of cage, CO_2 adsorbs first in the smaller cavities due to the confinement effect. As expected and due to the important quadrupolar interaction exhibited by CO_2, this molecule was found to strongly interact with the adsorbent. This fact confirms the non-negligible contribution of the electrostatic interactions in ZIFs.

9.3.5 *Impact of Functionality on Adsorption*

The possibility of tailoring ZIFs by modifying the metal or imidazolate ligand opens the door to the research and screening of optimal adsorbents for separation applications. Since the choice of the metal has been shown to have a moderate impact on the adsorption properties, understanding the impact of the different ligand functionalities becomes relevant to link the chemical composition of the adsorbent with the correspondent performance for a given separation. A nonexhaustive list of works concerning this point is given in Table 9.5. Due to the fact that in some cases the considered functionalized structures has not been yet synthesized, the DFT or hybrid DFT/force field methodologies are used to predict the crystallographic structure of the hypothetical materials. For instance, in a first approach, Hou et al. [49] showed that although CO_2 and CO are adsorbed in a first stage in the small pores of the GME topology of ZIF-68 and ZIF-69, the strong interaction of the adsorbate molecules with the nitro functionality of the small pore linkers causes the blocking of the entrance of additional molecules. As a consequence the larger pore cbIM linkers become responsible of the adsorption ability. On that basis, they proposed the substitution of electron-withdrawing groups in bIM linkers in order to improve the CO_2 adsorption. More recently, Amrouche et al. [50] studied the functionalization at the −2 position of the IM linkers of the SOD topology (ZIF-8, ZIF-90). For the case of this single cage porosity it was found that the adsorption of CO_2

Table 9.5 List of simulation works focused on the functionalization of ZIFs

Material/topology	Framework	Solid–gas FF	Properties	T (K)	Ref.
ZIF-68, ZIF-69 (GME)		ONIOM2(B3LYP/ 6-31G**:UFF)	CO_2, CO interaction energies and adsorption sites	–	[49]
ZIF-8, ZIF-90 and other SOD materials	–	DFT/UFF inspired	• CO_2, CH_4 isotherms • CO_2/CH_4, CO_2/N_2 selectivities • Q_{st}	303	[50]
ZIF-68, ZIF-69, ZIF-78 (GME)	–	DFT	CO_2 binding energies	–	[51]
ZIF-25, ZIF-71, ZIF-93, ZIF-96, ZIF-97	Rigid	UFF	CO_2 isotherms	298	[31]

was mainly governed by the polarity of the ligand at low and moderate pressures. According to that observation, they proposed a correlation between the CO_2 isosteric heat of adsorption at zero coverage and the ligand dipole moment. In the same way they studied the impact of the functionalization over different gas mixtures showing that the separation properties can be enhanced if the solid linker is judiciously chosen.

9.4 Membrane Separation in ZIFs

Before describing the recent works on the diffusion of gas molecules in ZIF materials it is interesting to quickly review the concepts involving the diffusion mechanism in solids.

9.4.1 *Basic Concepts of Diffusion and Permeation*

The self-diffusivity $D_{s,i}$ describes the motion of individual particles of an ith species inside an isotropic three-dimensional material. It can be obtained in molecular simulation by the calculation of the mean-squared displacement of each particle after time t though the Einstein relation

$$D_{s,i}(c) = \lim_{t\to\infty} \frac{1}{6t} \left\langle |\vec{r}(t) - \vec{r}(0)|^2 \right\rangle \tag{9.6}$$

In this expression $\vec{r}(t)$ is the position of a tagged particle at time t, whereas the angular brackets stand for an ensemble average. This coefficient depends on the concentration (c) of molecules and can be computed for any particle/species in the system [52]. The transport diffusivity $D_t(c)$ can be directly derived from the Fick's law by means of the proportionality constant multiplying the spatial concentration gradient to obtain the macroscopic flux of one species. The transport diffusivity is also commonly referred to as Fickian diffusivity [53, 54], the chemical diffusivity [52], or the collective diffusivity [55], and it is related to the Maxwell–Stefan mutual diffusion coefficient $D^{MS}(c)$ [56] by the following expression:

$$D_t(c) = D^{MS}(c) \left(\frac{d \ln f}{d \ln c} \right)_T \tag{9.7}$$

where f is the fugacity of the adsorbed gas. This expression can also be computed for each type of molecule present in the system (in this case both coefficients D_t and D^{MS} depend on the concentration of all species). The last term in Eq. 9.7 is known as the thermodynamic correction factor Γ. This correction factor can be directly estimated by adsorption isotherms by means of fluctuation formula [57]. In the case of binary mixtures (components 1 and 2), one can define the diffusion selectivity S_{diff} as

$$S_{\text{diff}} = \frac{D_{s,1}}{D_{s,2}} \tag{9.8}$$

The most important factor to estimate the permeability of a particular membrane is the ideal permeation selectivity S_{perm} [58], defined by

$$S_{\text{perm}} = S_{\text{ads}} \times S_{\text{diff}} \tag{9.9}$$

In cases where a negligible downstream loading is assumed; a more rigorous treatment of the membrane permeation (Π) can be determined from MD simulations by using the following expression [59]

$$\Pi_i = \frac{\varphi D_{s,i} c_i}{f_i} \tag{9.10}$$

where ϕ represents the fractional pore volume of the microsporus crystalline material and the concentration c_i is defined in terms of concentration of species i at the upstream face of the membrane. In this line, Keskin and Sholl have recently proposed a methodology for in silico screening of MOF multilayer membranes for natural gas purification [60].

9.4.2 *Diffusion Behavior of CO_2 and Other Gases in ZIFs*

The study of gas diffusion and permeability on MOF-based materials/membranes has been experimentally performed on several systems (see the Li et al. review [13] and references therein). However, relatively few works (compared to those published for adsorption simulations) have been devoted to model the diffusion mechanism of gases in MOFs from the scope of molecular simulation. The first reported study off MD simulation to estimate diffusion coefficients of alkanes [61] and Ar [62] in MOF-5 and Cu-BTC, respectively, were published on 2004. In the case of ZIF materials, the lack of this kind of studies may come from two facts: 1) only few experimental works on diffusion and permeation are available [40, 63–73], and 2) accurate force-fields adapted for ZIF materials are relatively recent (see Section 9.2). The first MD calculations of the self-diffusion coefficient of CO_2 in ZIFs were those of Liu et al. [18] and Rankin et al. [25]. A nonexhaustive list of simulation works is presented in Table 9.6.

In view of the literature about simulation of diffusion of gas molecules in ZIFs, it is interesting to see that most works have adopted a rigid framework representation. Some works included the effect of flexibility, but they are exclusively devoted to ZIF-8 (a normal situation, since most of the available experimental data concerns ZIF-8). The permeation of gases in ZIF structures is difficult to predict due to the different degree of flexibility of the framework (i.e., a priori the degree of flexibility of ZIFs is not necessarily the same as the one observed for ZIF-8). The lack of flexibility in some approaches allows the analysis of diffusion mechanism of small gas molecules (H_2, CO, CO_2) since their kinetic diameters are smaller than the usual window apertures (d_a) of most common ZIFs [10]. The window aperture of some ZIF materials is shown in Fig. 9.6.

Some gases, such as CH_4 (also Ar, Kr, or Xe) can diffuse inside ZIF-8, even with a rigid model, since the CH_4 kinetic diameter is close enough to the d_a. This fact is in agreement with experiments where both gases CO_2 and CH_4 yield an S_{diff} value of the order of unity in ZIF-8 [63]. Battisti et al. performed a systematic test of the adsorption and diffusion of gas mixtures (containing CO_2, CH_4, N_2, and H_2) in different ZIFs structures using MD simulations [38]. They showed that in the materials with the smaller pores (ZIF-5, ZIF-7 and ZIF-9); the diffusion coefficient of H_2 is orders of magnitude larger than that of other gases. There are, however, open questions concerning the diffusion mechanism of small gases in ZIF materials. For instance, CO_2 and CH_4 molecules can diffuse inside the pores of ZIF-20, which has a $d_a \approx 2.8$ Å but not inside ZIF-11, where the d_a is about 3 Å [9, 10]. This fact reveals the presence of complex permeation mechanisms with a nonideal influence of the chemical nature of the organic ligand.

Table 9.6 Recent simulation works on ZIFs permeation and diffusion

Material	Framework	Solid–gas FF	Property	*T* (K)	System	Ref.
ZIF-68, ZIF-69	Rigid	UFF	D_s	273	CO_2	[18]
ZIF-68, ZIF-70	Rigid	UFF, DREIDING	D_s, D_t	298	CO_2, CH_4, H_2	[25]
ZIF-8	Rigid	Zhou et al. [19]	D_t	300	CO_2, CH_4, H_2	[46]
ZIF-8, ZIF-68	Rigid	UFF, DREIDING, Zhou et al. [19]	D^{MS}, Π, S_{diff}, S_{perm}	300	CO_2+H_2, CO_2+CH_4, CO_2+N_2	[59]
Na-*rho*-ZMOF	Rigid	UFF	D_s, S_{diff}, S_{perm}	323, 373	Water+ methanol	[74]
ZIF-2 to ZIF-10	Rigid	UFF-ZIF [24]	D_s, S_{diff}, S_{perm}	298	CO_2+H_2, CO_2+CH_4, CO_2+N_2, CH_4+H_2, CH_4+N_2	[38]
ZIF-8	Rigid	UFF, DREIDING	D_s, Π, S_{diff}, S_{perm}	300	CO_2+H_2, CO_2+CH_4, CO_2+N_2, CH_4+H_2	[75]
ZIF-8	Flexible	DREIDING, AMBER	D_s, S_{diff}, S_{perm}	295	CH_4, H_2, CH_4+H_2	[40]
ZIF-8	Rigid	UFF	J_w^a	(280–310)	Water+NaCl	[76]
ZIF-68, ZIF-70	Rigid	UFF	D_s, S_{diff}, S_{perm}	298	CO_2+CH_4, CH_4+H_2	[77]
ZIF-8	Flexible	AMBER	D_s	300	CO_2	[39]
ZIF-8	Flexible	AMBER+	D_s	300	CO_2, CH_4	[41]
ZIF-8	Rigid, flexible	UFF	D_s, D_t	298	CO_2 CH_4	[78]
ZIF-1, ZIF-2, ZIF-3, ZIF-10	Rigid	DREIDING	D_s, Π, S_{diff}, S_{perm}	298	Xe, Kr, Ar, Xe+Kr, Xe+Ar	[79]
ZIF-7	Rigid	UFF	J_i^a, Π	300	$H_2+CO_2+CO+CH_4+H_2O$	[80]
ZIF-90, ZIF-65, ZIF-90 +Matrimid, ZIF-65 +Ultem	Rigid	UFF	D_s, D_t, Π, S_{diff}, S_{perm}	298, 473	CH_4, CO_2, H_2, N_2, H_2+CO_2, H_2+N_2, H_2+CH_4	[81]

[a] Mass flow

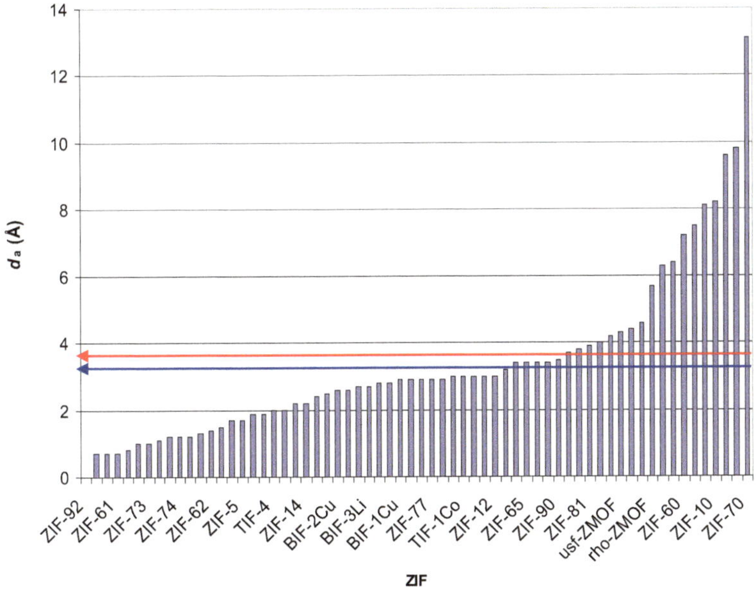

Figure 9.6 Pore window diameter (d_a) of some ZIF materials (information extracted from Ref [12]). The blue and red lines correspond to the kinetic diameter of CO_2 (~3.3 Å) and CH_4 (~3.8 Å), respectively.

Gas permeation of small hydrocarbon molecules (ethane, ethene, propane, propene, etc.) inside ZIF-8 [73] or ZIF-7 [67] was reported experimentally. The prediction of these phenomena in simulation is a real challenge, even using flexible force field models. van der Berg et al. attempt to understand the diffusion mechanism of alkane/alkene separation in ZIF-7 by means of DFT calculations and a thermodynamic analysis of the adsorption–desorption isotherms [44]. The authors explain the difference in the behavior of alkenes and alkanes toward adsorption on ZIF-7 by the ability of these molecules to form adsorption complexes at the solid's external surface. In addition, the difference in the pressure at which alkenes and alkanes are taken up (gate opening) is due to the more favorable adsorption of the alkenes on the external surfaces of ZIF-7 as a result of more efficient containment of the flat, unsaturated hydrocarbon between the benzimidazole linker of the molecules of the six-member ring (6MR) pore.

It is clear that the concept of window aperture, as traditionally conceived for the more rigid zeolite structures, is no longer valid for ZIFs [82, 83]. A systematic analysis of the influence of the topology and the chemical nature of the organic linker (including polarity, the ability to form hydrogen bonding with guest molecules) on the diffusion mechanism of gases in ZIF materials is still lacking. The problem is even worst for liquid phases where entropic effects may play a major role.

9.5 New Perspectives on Modeling and Applications

A number of reviews mention the recent advances in the areas of molecular simulation concerning adsorption separation [13, 84–86] and sensors [87] using MOF materials. We will discuss in particular some original works where molecular simulation techniques have been used for screening purposes, aiming at selecting the most appropriate material for a particular application. The development of optical sensors [88, 89] and purification of gaseous fission products [90], among others, are only two examples of new possible alternative applications for ZIFs. In general, the objective of this section is to mention new possible areas of research where molecular simulation techniques may contribute to enhance the comprehension of complex physical phenomena involving ZIF materials as well as other types of MOFs.

9.5.1 *Correlative Models and in silico Screening*

As we mention in previous section, the possibility of tuning the chemical nature of the organic linker in MOFs provides a huge research space to obtain the ideal solid for a particular application. Most of the recent works on screening of materials are based on a large set of simulations predicting adsorption and diffusion properties on MOFs and/or ZIFs in order to rank solids according to some kind of "efficiency criteria." Thornton et al. have estimated the selectivity and membrane permeability of a series of MOFs and ZIFs for hydrogen purification, the authors conclude that ZIF-11 has exceptional characteristics for this application [91].

The use of a limited number of variables (descriptors) allows the classification and analysis of a reduced database of materials. If a vast amount of information is required, the development of correlative models requires the use of highly efficient statistical approaches, such as quantitative structure–property relationship (QSPR). Amrouche et al. developed a QSPR approach proposing a series of structural and molecular descriptors (e.g., organic linker dipole moment, pore curvature, gas polarity, and boiling temperature) that are useful to predict the isosteric heat of adsorption at zero coverage (q_{st}°) of polar and nonpolar gases in ZIFs [92]. The authors also propose a simple relationship between q_{st}° and water vaporization enthalpy to estimate the hydrophobic/hydrophilic nature of ZIF materials. One can consider q_{st}° as a crucial variable to model PSA adsorption processes and one that has been previously used to classify MOFs solids for CO_2 capture [14]. In the same line, Wu et al. propose a QSPR model that can account for the CO_2/N_2 [93] and CH_4/H_2 [94] mixture selectivity of an extended series of MOFs/ZIFs database with a very large chemical and topological diversity under industrial pressure condition. The authors base their approach combining solid (porosity) and solid–gas descriptors (zero-coverage heat of adsorption difference of adsorbed gases $\Delta Q_{st}^{\circ} = q_{st}^{\circ}{}_{CO2} - q_{st}^{\circ}{}_{N2}$).

Lin et al. have recently present an in silico approach to rank absorbents according to their performance in carbon dioxide capture and storage [95]. The authors propose a special descriptor called "parasitic energy," which takes into account the easiness of regeneration of a solid in PSA process after a compression step. The idea behind is to quantify the electric load imposed on power plants to achieve a given CO_2 purity, including more realistic process constraints. The parasitic energy was computed for different ZIFs structures in Fig. 9.7.

Such kind of massive screening approach is impossible without having accurate intermolecular potentials (see Section 9.2), highly efficient MC calculation of adsorption thermodynamic data [96], and efficient calculation of structural and topological descriptors of pores in solids [97].

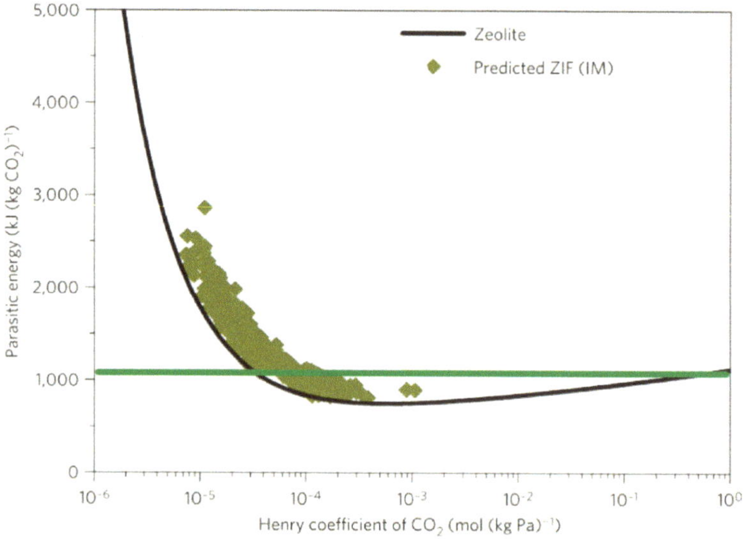

Figure 9.7 The parasitic energy as a function of the CO_2 Henry coefficient for ZIFs is shown. The green lines give the parasitic energy of the current MEA technology, and the black line is the minimal parasitic energy calculated for a given value of the Henry coefficient in the all-silica structures. Reprinted by permission from Macmillan Publishers Ltd: *Nature Materials*, Ref. [95], copyright 2012.

9.5.2 *Prediction of New Solid Structures*

One of the most interesting aspects of using molecular simulation applied to crystalline ZIFs solid material is the possibility to conceive not-yet synthesized structures. In this line we can mention the pioneering work of Baburin et al., where 21 ZIF materials based on Zn connectors were derived on the basis of DFT calculation and compared to known ZIFs networks [98]. This kind of DFT approach has a clear advantage in the comprehension of the role played by the metal sites, the organic linker chemical nature and topological aspects in the structural stability of ZIFs [99, 100]. These approaches, however, only provide a relative stability metrics that may explain why only a certain number of zeolite-like topologies have been synthesized. They don't predict at all if one structure will be formed starting from the initial chemical species in solution (including reactants, templates, and other important synthesis variables such as pH, for instance). Several attempts have been done by including reactive force fields in MOFs [101] or by using large time-land-scale simulations to study the kinetically controlled crystal growth of zeolites [102]. This approach of in silico synthesis represents a real challenge for actual and future modeling techniques.

9.6 Conclusions

In the last years a few hundred of ZIF materials have been synthesized and their structure was more or less well characterized; however, only a few dozens of structures were actually tested experimentally for separation applications. Apart from ZIF-8, where a set of experimental and simulation works provides a good overview of its mechanical, structural, and adsorptive capabilities, most of the ZIF solids are still badly characterized. This problem arises from the difficulties on synthesizing enough amounts of solid crystals to perform adsorption and membrane permeation experiments. This fact opens the door for an exhaustive use of molecular simulation techniques for screening purposes. There are several works devoted to the implementation and optimization of accurate force field to describe solid–gas interactions on such family of MOFs. Though generic force fields have been adapted and they are accurate enough for screening purposes on adsorption applications, the development of more accurate force fields for ZIFs is still required. In particular the inclusion of framework flexibility is a key point to assess the potential use of this family of materials for membrane separation. Even considering ZIF materials as being not as flexible as other classes of MOFs, the concept of pore aperture and permeability is a fundamental issue that is not well understood. Finally, simulation techniques can be used not only for selecting useful materials but also for evaluating/predicting new structure–function–property relationships able to guide the design and optimization of new ZIF materials. It is then evident that molecular simulation will, in fact, play a key role in this research.

References

1. Gomes, V. G., Yee, K. W. K. (2002). Pressure swing adsorption for carbon dioxide sequestration from exhaust gases, *Sep. Purif. Technol.*, **28**, 161–171.
2. Tagliabue, M., Farrusseng, D., Valencia, S., Aguado, S., Ravon, U., Rizzo, C., Corma, A., Mirodatos, C. (2009). Natural gas treating by selective adsorption: material science and chemical engineering interplay, *Chem. Eng. J.*, **155**, 553–566.
3. Bourrelly, S., Llewellyn, P. L., Serre, C., Millange, F., Loiseau, T., Ferey, G. (2005). Different adsorption behaviors of methane and carbon dioxide in the isotypic nanoporous metal terephthalates MIL-53 and MIL-47, *J. Am. Chem. Soc.*, **127**, 13519–13521.
4. Eddaoudi, M., Kim, J., Rosi, N., Vodak, D., Wachter, J., O'Keeffe, M., Yaghi, O. M. (2002). Systematic design of pore size and functionality in isoreticular MOFs and their application in methane storage, *Science*, **295**, 469–472.
5. Huang, X. C., Lin, Y. Y., Zhang, J. P., Chen, X. M. (2006). Ligand-directed strategy for zeolite-type metal–organic frameworks: zinc(II) imidazolates with unusual zeolitic topologies, *Angew. Chem., Int.*, **45**, 1557–1559.
6. Liu, Y., Kravtsov, V. C., Larsen, R., Eddaoudi, M. (2006). Molecular building blocks approach to the assembly of zeolite-like metal–organic frameworks (ZMOFs) with extra-large cavities, *Chem. Commun.*, 1488–1490.
7. Tian, Y. Q., Zhao, Y. M., Chen, Z. X., Zhang, N. G., Weng, L. H., Zhao, D. Y. (2007). Design and generation of extended zeolitic metal–organic frameworks (ZMOFs): synthesis and crystal structures of zinc(II) imidazolate polymers with zeolitic topologies, *Chem. Eur. J.*, 13, 4146–4154.
8. Wu, T., Bu, X.: Liu, R., Lin, Z., Zhang, J., Feng, P. (2008). A new zeolitic topology with sixteen-membered ring and multidimensional large pore channels, *Chem. Eur. J.*, **14**, 7771–7773.
9. Park, K. S., Ni, Z., Côté, A. P., Choi, J. Y., Huang, R., Uribe-Romo, F. J., Chae, H. K., O'Keeffe, M., Yaghi, O. M. (2006). Exceptional chemical and thermal stability of zeolitic imidazolate frameworks, *Proc. Natl. Acad. Sci.*, **103**, 10186–10196.
10. Banerjee, R., Phan, A., Wang, B., Knobler, C., Furukawa, H., O'Keeffe, M., Yaghi, O. M. (2008). High-throughput synthesis of zeolitic imidazolate frameworks and application to CO2 capture, *Science*, **319**, 939–943.

11. Wang, B., Côté, A. P., Furukawa, H., O'Keeffe, M., Yaghi, O. M. (2008). Colossal cages in zeolitic imidazolate frameworks as selective carbon dioxide reservoirs, *Nature*, **453**, 207–211.
12. Phan, A., Doonan, C. J., Uribe-Romo, F. J., Knobler, C. B., O'Keeffe, M., Yaghi, O. M. (2009). Synthesis, structure, and carbon dioxide capture properties of zeolitic imidazolate frameworks, *Acc. Chem. Res.*, **43**, 58–67.
13. Li, J.-R., Ma, Y., McCarthy, M. C., Sculley, J., Yu, J., Jeong, H. K., Balbuena, P. B., Zhou, H-C. (2011). Carbon dioxide capture-related gas adsorption and separation in metal-organic frameworks, *Coord. Chem. Rev.*, **255**, 1791–1823.
14. Yazaydın, A. Ö., Snurr, R. Q., Park, T.-H., Koh, K., Liu, J., LeVan, M. D., Benin, A. I., Jakubczak, P., Lanuza, M., Galloway, D. B., Low J. J., Willis, R. R. (2009). Screening of metal-organic frameworks for carbon dioxide capture from flue gas using a combined experimental and modeling approach, *J. Am. Chem. Soc.*, **131**, 18198–18199.
15. Mayo, S. L., Olafson, B. D., Goddard, W. A. (1990). DREIDING: a generic force field for molecular simulations, *J. Phys. Chem.*, **94**, 8897–8909.
16. Jorgensen, W. L., Maxwell, D. S., Tirado-Rives, J. (1996). Development and testing of the OPLS all-atom force field on conformational energetics and properties of organic liquids, *J. Am. Chem. Soc.*, **118**, 11225–11236.
17. Duan, Y., Wu, C., Chowdhury, S., Lee, M. C., Xiong, G., Zhang, W., Yang, R., Cieplak, P., Luo, R., Lee, T., Caldwell, J., Wang, J., Kollman, P. (2003). A point-charge force field for molecular mechanics simulations of proteins based on condensed-phase quantum mechanical calculations, *J. Comp. Chem.*, **24**, 1999–2012.
18. Liu, D., Zheng, C., Yang, Q., Zhong, C. (2009). Understanding the adsorption and diffusion of carbon dioxide in zeolitic imidazolate frameworks: a molecular simulation study, *J. Phys. Chem. C*, **113**, 5004–5009.
19. Zhou, M., Wang, Q., Zhang, L., Liu, Y.-C., Kang, Y. (2009). Adsorption sites of hydrogen in zeolitic imidazolate frameworks, *J. Phys. Chem. B*, **113**, 11049–11053.
20. Düren, T., Snurr, R. Q. (2004). Assessment of isoreticular metal–organic frameworks for adsorption separations: a molecular simulation study of methane/n-butane mixtures, *J. Phys. Chem. B*, **108**, 15703–15708.
21. Düren, T., Sarkisov, L., Yaghi, O. M., Snurr, R. Q. (2004). Design of new materials for methane storage, *Langmuir*, **20**, 2683–2689.

22. Jiang, J., Sandler, S. I. (2006). Monte Carlo simulation for the adsorption and separation of linear and branched alkanes in IRMOF-1, *Langmuir*, **22**, 5702–5707.
23. Babarao, R., Hu, Z., Jiang, J., Chempath, S., Sandler, S. I. (2007). Storage and separation of CO2 and CH4 in silicalite, C168 schwarzite, and IRMOF-1: a comparative study from Monte Carlo simulation, *Langmuir*, **23**, 659–666.
24. Pérez-Pellitero, J., Amrouche, H., Siperstein, F. R., Pirngruber, G., Nieto-Draghi, C., Chaplais, G., Simon-Masseron, A., Bazer-Bachi, D., Peralta, D., Bats, N. (2010). Adsorption of CO(2), CH(4), and N(2) on zeolitic imidazolate frameworks: experiments and simulations, *Chem. Eur. J.*, **16**, 1560–1571.
25. Rankin, R. B., Liu, J., Kulkarni, A. D., Johnson, J. K. (2009). Adsorption and diffusion of light gases in ZIF-68 and ZIF-70: a simulation study, *J. Phys. Chem. C*, **113**, 16906–16914.
26. Fairen-Jimenez, D., Galvelis, R., Torrisi, A., Gellan, A. D., Wharmby, M. T., Wright, P., Mellot-Draznieks, C., Düren, T. (2012). Flexibility and swing effect on the adsorption of energy-related gases on ZIF-8: combined experimental and simulation study, *Dalton Trans.*, **41**, 10752–10762.
27. Liu, B., Smit, B. (2010). Molecular simulation studies of separation of CO2/N2, CO2/CH4, and CH4/N2 by ZIFs, *J. Phys. Chem. C*, **114**, 8515–8522.
28. Campañá, C., Mussard, B., Woo, T. K. (2009). Electrostatic potential derived atomic charges for periodic systems using a modified error functional, *J. Chem. Theory Comput.*, **5**, 2866–2878.
29. Manz, T. A., Sholl, D. S. (2010). Chemically meaningful atomic charges that reproduce the electrostatic potential in periodic and nonperiodic materials, *J. Chem. Theory Comput.*, **6**, 2455–2468.
30. Rana, M. K., Pazzona, F. G., Suffritti, G. B., Demontis, P., Masia, M. (2011). Estimation of partial charges in small zeolite imidazolate frameworks from density functional theory calculations, *J. Chem. Theory Comput.*, **7**, 1575–1582.
31. Morris, W., Leung, B., Furukawa, H., Yaghi, O. K., He, N., Hayashi, H., Houndonougbo, Y., Asta, M., Laird, B. B., Yaghi, O. M. (2010). A combined experimental-computational investigation of carbon dioxide capture in a series of isoreticular zeolitic imidazolate frameworks, *J. Am. Chem. Soc.*, **132**, 11006–11008.
32. McDaniel, J. G., Yu, K., Schmidt, J. R. (2012). Ab initio, physically motivated force fields for CO2 adsorption in zeolitic imidazolate frameworks, *J. Phys. Chem. C*, **116**, 1892–1903.

33. McDaniel, J. G., Schmidt, J. R. (2012). Robust, transferable, and physically motivated force fields for gas adsorption in functionalized zeolitic imidazolate frameworks, *J. Phys. Chem. C*, **116**, 14031–14039.
34. Salles, F., Ghoufi, A., Maurin, G., Bell, R. G., Mellot-Draznieks, C., Férey, G. (2008). Molecular dynamics simulations of breathing MOFs: structural transformations of MIL-53(Cr) upon thermal activation and CO2 adsorption, *Angew. Chem., Int. Ed.*, **47**, 8487–8491.
35. Amirjalayer, S., Tafipolsky, M., Schmid, R. (2007). Molecular dynamics simulation of benzene diffusion in MOF-5: importance of lattice dynamics, *Angew. Chem., Int. Ed.*, **46**, 463–466.
36. Bux, H., Chmelik, C., Krishna, R., Caro, J. (2011). Ethene/ethane separation by the MOF membrane ZIF-8: molecular correlation of permeation, adsorption, diffusion, *J. Membr. Sci.*, **369**, 284–289.
37. Bux, H., Feldhoff, A., Cravillon, J., Wiebcke, M., Li, Y., Caro, J. (2011). Oriented zeolitic imidazolate framework-8 membrane with sharp H2/C3H8 molecular sieve separation, *Chem. Mater.*, **23**, 2262–2269.
38. Battisti, A., Taioli, S., Garberoglio, G. (2011). Zeolitic imidazolate frameworks for separation of binary mixtures of CO2, CH4, N2 and H2: a computer simulation investigation, *Microporous Mesoporous Mater.*, **143**, 46–53.
39. Zheng, B., Sant, M., Demontis, P., Suffritti, G. B. (2012). Force field for molecular dynamics computations in flexible ZIF-8 framework, *J. Phys. Chem. C*, **116**, 933–938.
40. Hertäg, L., Bux, H., Caro, J., Chmelik, C., Remsungnen, T., Knauth, M., Fritzsche, S. (2011). Diffusion of CH4 and H2 in ZIF-8, *J. Membr. Sci.*, **377**, 36–41.
41. Hu, Z., Zhang, L., Jiang, J. (2012). Development of a force field for zeolitic imidazolate framework-8 with structural flexibility, *J. Chem. Phys.*, **136**, 244703.
42. Frenkel D., Smit B. (2001). *Understanding Molecular Simulation: From Algorithms to Applications* (Academic Press, San Francisco).
43. Zhang, L., Li, Q., Lu, Z., Wang, X. (2011). Understanding adsorption and separation behavior of shorter chain alkane mixtures in zeolitic imidazolate frameworks by molecular simulations, *Proc. Comp. Sci.*, **4**, 1193–1202.
44. van den Bergh, J., Gücüyener, C., Pidko, E. A., Hensen, E. J. M., Gascon, J., Kapteijn, F. (2011). Understanding the anomalous alkane selectivity of ZIF-7 in the separation of light alkane/alkene mixtures, *Chem. Eur. J.*, **17**, 8832–8840.

45. Li, B., Wei, S., Chen, L. (2011). Molecular simulation of CO2, N2 and CH4 adsorption and separation in ZIF-78 and ZIF-79, *Mol. Simul.*, **37**, 1131–1142.
46. Bux, H., Chmelik, C., van Baten, J. M., Krishna, R., Caro, J. (2010). Novel MOF-membrane for molecular sieving predicted by IR-diffusion studies and molecular modeling, *Adv. Mater.*, **22**, 4741–4743.
47. Sirjoosingh, A., Alavi, S., Woo, T. K. (2010). Grand-canonical Monte Carlo and molecular-dynamics simulations of carbon-dioxide and carbon-monoxide adsorption in zeolitic imidazolate framework materials, *J. Phys. Chem. C*, **114**, 2171–2178.
48. Li, B., Wei, S., Chen, L. (2011). Molecular simulation of CO2, N2 and CH4 adsorption and separation in ZIF-78 and ZIF-79, *Mol. Simul.*, **37**, 1131–1142.
49. Hou, X.-J., Li, H. (2010). Unraveling the high uptake and selectivity of CO2 in the zeolitic imidazolate frameworks ZIF-68 and ZIF-69, *J. Phys. Chem. C*, **114**, 13501–13508.
50. Amrouche, H., Aguado, S., Pérez-Pellitero, J., Chizallet, C., Siperstein, F., Farrusseng, D., Bats, N., Nieto-Draghi, C. (2011). Experimental and computational study of functionality impact on sodalite–zeolitic imidazolate frameworks for CO2 separation, *J. Phys. Chem. C*, **115**, 16425–16432.
51. Liu, Y., Liu, J., Chang, M., Zheng, C. (2012). Effect of functionalized linker on CO2 binding in zeolitic imidazolate frameworks: density functional theory study, *J. Phys. Chem. C*, **116**, 16985–16991.
52. Uebing, C., Gomer, R. (1994). Determination of surface diffusion coefficients by Monte Carlo methods: comparison of fluctuation and Kubo–Green methods, *J. Chem. Phys.*, **100**, 7759.
53. Auerbach, S. M. (2000). Theory and simulation of jump dynamics, diffusion and phase equilibrium in nanopores, *Int. Rev. Phys. Chem.*, **19**, 155–198.
54. Sholl, D. S. (2000). Predicting single-component permeance through macroscopic zeolite membranes from atomistic simulations, *Ind. Eng. Chem. Res.*, **39**, 3737–3746.
55. Mak, C. H., Andersen, H. C., George, S. M. (1988). Monte Carlo studies of diffusion on inhomogeneous surfaces, *J. Chem. Phys.*, **88**, 4052.
56. Krishna, R. (1990). Multicomponent surface diffusion of adsorbed species. A description based on the generalized Maxwell-Stefan diffusion equation, *Chem. Eng. Sci.*, **45**, 1779–1791.

57. Reed, D. A., Ehrlich, G. (1981). Surface diffusion, atomic jump rates and thermodynamics, *Surf. Sci.*, **102**, 588–609.
58. Keskin, S., Sholl, D. S. (2009). Efficient methods for screening of metal organic framework membranes for gas separations using atomically detailed models, *Langmuir*, **25**, 11786–11795.
59. Krishna, R., van Baten, J. M. (2010). In silico screening of zeolite membranes for CO2 capture, *J. Membr. Sci.*, **360**, 323–333.
60. Keskin, S., Sholl, D. S. (2010). Selecting metal organic frameworks as enabling materials in mixed matrix membranes for high efficiency natural gas purification, *Energy Environ. Sci.*, **3**, 343–351.
61. Sarkisov, L., Düren, T., Snurr, R. Q. (2004). Molecular modelling of adsorption in novel nanoporous metal–organic materials, *Mol. Phys.*, **102**, 211–221.
62. Skoulidas, A. I. (2004). Molecular dynamics simulations of gas diffusion in metal–organic frameworks: argon in CuBTC, *J. Am. Chem. Soc.*, **126**, 1356–1357.
63. Bux, H., Liang, F., Li, Y., Cravillon, J., Wiebcke, M., Caro, J. (2009). Zeolitic imidazolate framework membrane with molecular sieving properties by microwave-assisted solvothermal synthesis, *J. Am. Chem. Soc.*, **131**, 16000–16001.
64. Huang, A., Bux, H., Steinbach, F., Caro, J. (2010). Molecular-sieve membrane with hydrogen permselectivity: ZIF-22 in LTA topology prepared with 3-aminopropyltriethoxysilane as covalent linker, *Angew. Chem., Int. Ed.*, **49**, 4958–4961.
65. Huang, A., Dou, W., Caro, J. (2010). Steam stable zeolitic imidazolate framework ZIF-90 membrane with hydrogen selectivity through covalent functionalization, *J. Am. Chem. Soc.*, **132**, 15562–15564.
66. Venna, S. R., Carreon, M. A. (2010). Highly permeable zeolite imidazolate framework-8 membranes for CO_2/CH_4 separation, *J. Am. Chem. Soc.*, **132**, 76–78.
67. Gücüyener, C., Bergh, J. van den, Gascon, J., Kapteijn, F. (2010). Ethane/ethene separation turned on its head: selective ethane adsorption on the metal-organic framework ZIF-7 through a gate-opening mechanism, *J. Am. Chem. Soc.*, **132**, 17704–17706.
68. Li, Y., Liang, F., Bux, H., Yang, W., Caro, J. (2010). Zeolitic imidazolate framework ZIF-7 based molecular sieve membrane for hydrogen separation, *J. Membr. Sci.*, **354**, 48–54.
69. Aguado, S., Nicolas, C.-H., Moizan-Baslé, V., Nieto, C., Amrouche, H., Bats, N., Audebrand, N., Farrusseng, D. (2011). Facile synthesis of an

ultramicroporous MOF tubular membrane with selectivity towards CO_2, *New J. Chem.*, **35**, 41–44.

70. Chmelik, C., Voß, H., Bux, H., Caro, J. (2011). Adsorption and diffusion: basis for molecular understanding of permeation through molecular sieve membranes, *Chem. Ing. Tech.*, **83**, 104–112.
71. Díaz, K., López-González, M., Castillo, L. F. del, Riande, E. (2011). Effect of zeolitic imidazolate frameworks on the gas transport performance of ZIF8-poly(1,4-phenylene ether-ether-sulfone) hybrid membranes, *J. Membr. Sci.*, **383**, 206–213.
72. Chmelik, C., Baten, J. V., Krishna, R. (2012). Hindering effects in diffusion of CO_2/CH_4 mixtures in ZIF-8 crystals, *J. Membr. Sci.*, **397–398**, 87–91.
73. Chmelik, C., Freude, D., Bux, H., Haase, J. (2012). Ethene/ethane mixture diffusion in the MOF sieve ZIF-8 studied by MAS PFG NMR diffusometry, *Microporous Mesoporous Mater.*, **147**, 135–141.
74. Nalaparaju, A., Zhao, X. S., Jiang, J. W. (2011). Biofuel purification by pervaporation and vapor permeation in metal–organic frameworks: a computational study, *Energy Environ. Sci.*, **4**, 2107–2116.
75. Krishna, R., van Baten, J. M. (2011). In silico screening of metal-organic frameworks in separation applications, *Phys. Chem. Chem. Phys.*, **13**, 10593–10616.
76. Hu, Z., Chen, Y., Jiang, J. (2011). Zeolitic imidazolate framework-8 as a reverse osmosis membrane for water desalination: insight from molecular simulation, *J. Chem. Phys.*, **134**, 134705.
77. Liu, J., Keskin, S., Sholl, D. S., Johnson, J. K. (2011). Molecular simulations and theoretical predictions for adsorption and diffusion of CH_4/H_2 and CO_2/CH_4 mixtures in ZIFs, *J. Phys. Chem. C*, **115**, 12560–12566.
78. Pantatosaki, E., Megariotis, G., Pusch, A.-K., Chmelik, C., Stallmach, F., Papadopoulos, G. K. (2012). On the impact of sorbent mobility on the sorbed phase equilibria and dynamics: a study of methane and carbon dioxide within the zeolite imidazolate framework-8, *J. Phys. Chem. C*, **116**, 201–207.
79. Gurdal, Y., Keskin, S. (2012). Atomically detailed modeling of metal organic frameworks for adsorption, diffusion, and separation of noble gas mixtures, *Ind. Eng. Chem. Res.*, **51**, 7373–7382.
80. Hu, Z., Chen, Y. (2012). Zeolitic imidazolate framework-7 as an ultra-selective nanofilter for H2 purification: atomistic simulation study, *Mol. Simul.*, **38**, 830–837.

81. Atci, E., Keskin, S. (2012). Atomically detailed models for transport of gas mixtures in ZIF membranes and ZIF/polymer composite membranes. *Ind. Eng. Chem. Res.*, **51**, 3091–3100.
82. Krishna, R., van Baten, J. M. (2010). Modeling adsorption and self-diffusion of methane in LTA zeolites: the influence of framework flexibility, *J. Phys. Chem. C*, **114**, 18017–18021.
83. Krishna, R., van Baten, J. M. (2011). A molecular dynamics investigation of the diffusion characteristics of cavity-type zeolites with 8-ring windows, *Microporous Mesoporous Mater.*, **137**, 83–91.
84. Keskin, S., Heest, T. M. van, Sholl, D. S. (2010). Can metal-organic framework materials play a useful role in large-scale carbon dioxide separations? *ChemSusChem*, **3**, 879–891.
85. Liu, J., Thallapally, P. K., McGrail, B. P., Brown, D. R., Liu, J. (2012). Progress in adsorption-based CO2 capture by metal-organic frameworks, *Chem. Soc. Rev.*, **41**, 2308–2322.
86. Mani-Biswas, M., Cagin, T. (2012). Insights from theoretical calculations on structure, dynamics, phase behavior and hydrogen sorption in nanoporous metal organic frameworks, *Comp. Theor. Chem.*, **987**, 42–56.
87. Liu, B. (2012). Metal–organic framework-based devices: separation and sensors, *J. Mater. Chem.*, **22**, 10094–10101.
88. Liu, S., Xiang, Z., Hu, Z., Zheng, X., Cao, D. (2011). Zeolitic imidazolate framework-8 as a luminescent material for the sensing of metal ions and small molecules, *J. Mater. Chem.*, **21**, 6649–6653.
89. Hinterholzinger, F. M., Ranft, A., Feckl, J. M., Rühle, B., Bein, T., Lotsch, B. V. (2012). One-dimensional metal–organic framework photonic crystals used as platforms for vapor sorption, *J. Mater. Chem.*, **22**, 10356–10362.
90. Sava, D. F., Rodriguez, M. A., Chapman, K. W., Chupas, P. J., Greathouse, J. A., Crozier, P. S., Nenoff, T. M. (2011). Capture of volatile iodine, a gaseous fission product, by zeolitic imidazolate framework-8, *J. Am. Chem. Soc.*, **133**, 12398–12401.
91. Thornton, A. W., Dubbeldam, D., Liu, M. S., Ladewig, B. P., Hill, A. J., Hill, M. R. (2012). Feasibility of zeolitic imidazolate framework membranes for clean energy applications, *Energy Environ. Sci.*, **5**, 7637–7646.
92. Amrouche, H., Creton, B., Siperstein, F., Nieto-Draghi, C. (2012). Prediction of thermodynamic properties of adsorbed gases in zeolitic imidazolate frameworks, *RSC Adv.*, **2**, 6028–6035.

93. Wu, D., Yang, Q., Zhong, C., Liu, D., Huang, H., Zhang, W., Maurin, G. (2012). Revealing the structure-property relationships of metal-organic frameworks for CO_2 capture from flue gas, *Langmuir*, **28**, 12094–12099.

94. Wu, D., Wang, C., Liu, B., Liu, D., Yang, Q., Zhong, C. (2012). Large-scale computational screening of metal-organic frameworks for CH4/H2 separation, *AIChE J.*, **58**, 2078–2084.

95. Lin, L.-C., Berger, A. H., Martin, R. L., Kim, J., Swisher, J., Jariwala, K., Rycroft, C. H., Bhown, A. S., Deem, M. W., Haranczyk, M., Smit, B. (2012). In silico screening of carbon-capture materials, *Nat. Mater.*, **11**, 633–641.

96. Kim, J., Smit, B. (2012). Efficient Monte Carlo simulations of gas molecules inside porous materials, *J. Chem. Theor. Comp.*, **8**, 2336–2343.

97. Martin, R. L., Smit, B., Haranczyk, M. (2012). Addressing challenges of identifying geometrically diverse sets of crystalline porous materials, *J. Chem. Inf. Mod.*, **52**, 308–318.

98. Baburin, I. A., Leoni, S., Seifert, G. (2008). Enumeration of not-yet-synthesized zeolitic zinc imidazolate MOF networks: a topological and DFT approach, *J. Phys. Chem. B*, **112**, 9437–9443.

99. Lewis, D. W., Ruiz-Salvador, A. R., Gómez, A., Rodriguez-Albelo, L. M., Coudert, F.-X., Slater, B., Cheetham, A. K., Mellot-Draznieks, C. (2009). Zeolitic imidazole frameworks: structural and energetics trends compared with their zeolite analogues, *CrystEngComm*, **11**, 2272–2276.

100. Galvelis, R., Slater, B., Cheetham, A. K., Mellot-Draznieks, C. (2012). Comparison of the relative stability of zinc and lithium-boron zeolitic imidazolate frameworks, *CrystEngComm*, **14**, 374–378.

101. Han, S. S., Choi, S.-H., van Duin, A. C. T. (2010). Molecular dynamics simulations of stability of metal–organic frameworks against H_2O using the ReaxFF reactive force field, *Chem. Commun.*, **46**, 5713–5715.

102. Cubillas, P., Stevens, S. M., Blake, N., Umemura, A., Chong, C. B., Terasaki, O., Anderson, M. W. (2011). AFM and HRSEM investigation of zeolite a crystal growth. Part 1: in the absence of organic additives, *J. Phys. Chem. C*, **115**, 12567–12574.

Chapter 10

Modeling Adsorptive Separations Using Metal–Organic Frameworks

Alexandre F. P. Ferreira, Ana M. Ribeiro, João C. Santos, Marta C. Campo, and Alírio E. Rodrigues

Laboratory of Separation and Reaction Engineering, Associate Laboratory LSRE/LCM, Department of Chemical Engineering, Faculty of Engineering, University of Porto, Rua Dr. Roberto Frias s/n, 4200-465 Porto, Portugal

aferreir@fe.up.pt

This chapter aims to offer a review on recent advances and applications of metal–organic frameworks (MOFs) in modeling of cyclic adsorptive separations. Adsorption processes, such as pressure swing adsorption (PSA) or simulated moving bed (SMB) present opportunities for energy savings, process intensification, and costs reduction. In recent years MOFs have been pointed out as promising adsorbents to carry out several different separations, either in gas or in liquid phase. The number of microporous/mesoporous MOFs proposed for adsorptive separations, exploiting molecular sieving or specific interactions, is clearly growing since more and more MOFs are being synthesized every day. However, new approaches will be required in the future in terms of material design to overcome the bottlenecks in the most *challenging* separations. In this chapter we will address two of those cases, paraffin/olefin

Metal–Organic Frameworks: Materials Modeling towards Potential Engineering Applications
Edited by Jianwen Jiang

ISBN 978-981-4613-45-3 (Hardcover), 978-981-4613-46-0 (eBook)
www.panstanford.com

separation (propane/propylene) in the gas phase and the separation of isomeric mixtures (xylene isomers) in the liquid phase. The most studied material for propane/propylene separation is Cu-BTC (also known as HKUST-1), which selectively adsorbs the olefin in detriment of the paraffin. In the liquid phase separation of xylene isomers there are several materials being explored (e.g., Matériaux de l'Institut Lavoisier [MIL]-53(Al), MIL-47(V), UiO-66(Zr)). However, the most promising materials are the ones with the topology of MIL-125(Ti), such as MIL-125(Ti) itself, MIL-125(Ti)_NH_2, and CAU-1, which are *p*-xylene selective. After a brief introduction of MOF- and adsorption-based cyclic processes, this review will focus on adsorptive gas-phase separations, with the example of propane/propylene separation. The present review will be concluded with a short summary and future challenges.

10.1 Introduction

The early years of the 1970s brought about significant advances in materials science, bringing in a large family of materials, the metal–organic materials (MOMs). Within this class of materials is an interesting subclass for adsorption-based separation processes, which is microporous/mesoporous metal–organic frameworks (MOFs) (Fig. 10.1). However, their application to adsorptive separations only started to be studied in the late 1990s and early 2000s (Fig. 10.2). The first studies were focused in adsorption properties such as adsorption equilibrium, adsorption kinetics, and gas mixture adsorption selectivity. These early studies were mainly experimental reports like the work by Yaghi et al. [1–3] and the work by several different other authors [4–9]. Nevertheless, some efforts on molecular simulations of gas adsorption in MOFs were already being developed by a few authors, for example, the works reported by Vishnyakov et al. [10], in 2003, and by Skoulidas [11], in 2004. MOFs are showing great potential for several different gas separations from the energetic, biological, and environmental standpoints [12]. The most studied gas-phase separations are light gases (e.g., H_2, N_2, O_2, CO_2, and CH_4), light olefins and paraffins (e.g., ethane/ethylene and propane/propylene), harmful gases (e.g., H_2S, SO_2, and CO),

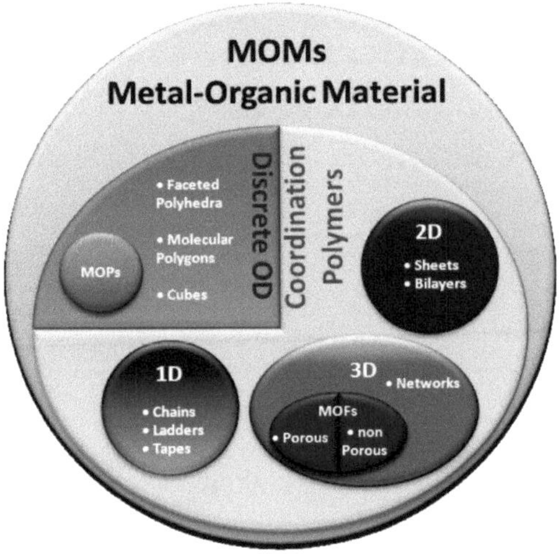

Figure 10.1 Euler diagram of the metal–organic materials family.

noble gases (e.g., Ar, He, and Xe), and vapors (e.g., solvents, xylene isomers, and hexane isomers). There are various extensive reviews in this topic. The most comprehensive ones are by Li et al. [12], Ferey [13], Kitagawa et al. [14], and Rowsell et al. [15]. It is not the objective of this manuscript to extensively present all the studied materials and separations; for that the readers should refer to the previously mentioned reviews. In liquid-phase separations the focus has been made on organic molecules with different properties and/or functional groups (e.g., alcohols, amine, acids, and aromatics), organic molecules with different sizes and shapes (e.g., short-chain alcohols, aromatic amines, large aromatics), organosulfur compounds (e.g., thiophene, benzothiophene, and mercaptans), structural isomers (e.g., xylene isomers, diolefins, and linear/branched alkanes), and stereoisomers (e.g., enantiomers and *cis-/trans*-isomers). An extensive review in liquid-phase separations is presented by Li et al. [12].

Like in zeolites separations in microporous/mesoporous MOFs are based mainly in two phenomena, molecular sieving and specific interactions (enthalpy effects), yet there are also reported some separations based on entropic effects. The narrow micropores

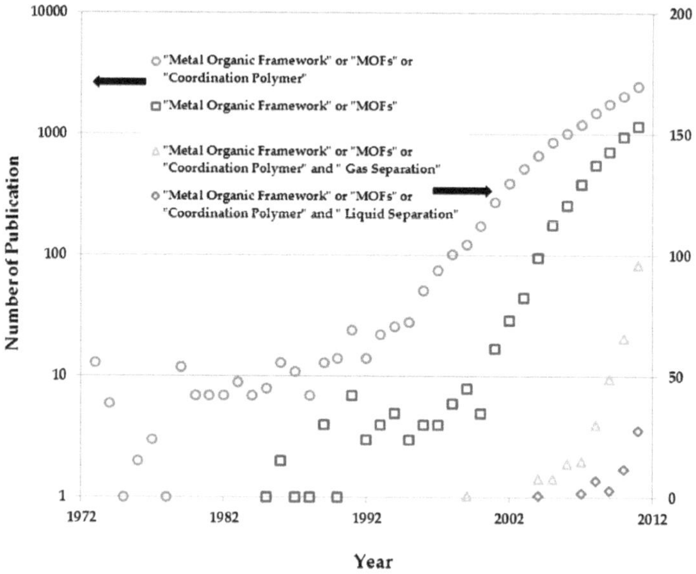

Figure 10.2 Number of publications containing the keywords "metal–organic framework," "MOFs," "coordination polymer," "gas separation," and "liquid separation" from 1972 until 2012 (data obtained from the SCOPUS database). One should note that the obtained numbers are just indicative of the trends and magnitude orders, since the used keywords might not be broad enough to enclose all published works in the area.

(4.5 Å) of Mn formate hinder the adsorption of molecules with a large kinetic diameter, such as Ar, N_2, and CH_4, but allow the adsorption of CO_2 and H_2 [1]. This is a classic example of molecular sieving in gas-phase adsorption with MOFs. The selectivity of Cu-BTC for propylene in detriment of propane is conventionally attributed to the π-complexation between the double bond of propylene and the coordinatively unsaturated sites of Cu^{+2}. This case is the most studied example of gas-phase separation based on specific interactions between adsorbent and adsorbate [16–23]. Barcia et al. claim that the observed reverse shape selectivity for hexane and xylene isomers, vapor phase, in UiO-66(Zr) is a result of entropic effects [24]. The selectivity toward bulkier isomers suggests that the retention in small cavities is dictated by steric constraints, that is, the rotational freedom in those cavities [24].

Regarding liquid-phase adsorption, the adsorption of xylene isomers in the analogous structures MIL-47(V) and MIL-53(Al) presents itself as an interesting case [25]. Although the structures are analogous, present similar selectivity and adsorption preference, the adsorption mechanisms are very different. The preference for styrene in MIL-47(V) is based on differences in entropy, due to its unique packing, favoring π–π stacking [25]. On the other hand, for ethylbenzene, the methyl protons of the conformationally flexible ethyl group interact with the oxygen atoms in the pore wall, resulting in lower packing density [25]. For MIL-53(Al) the styrene selectivity is due to enthalpic effects. Molecular sieving in liquid-phase adsorption finds a good example in the separation of benzene, naphthalene, anthracene, phenanthrene, pyrene, 1,3,5-triphenylbenzene, and 1,3,5-tris(4-bromophenyl)benzene in MOF-5 and Cu-BTC [26]. For Cu-BTC the sieving effect starts for kinetic diameters equal or larger than pyrene molecular size (pyrene, 1,3,5-triphenylbenzene, and 1,3,5-tris(4-bromophenyl)benzene). The smaller molecules (benzene, naphthalene, anthracene, and phenanthrene) are retained and have elution times proportional to their size [26]. In MOF-5, which has larger pores, the retention times are lower for all retained components, and pyrene is already retained in this structure, in contrast to 1,3,5-tris(4-bromophenyl)benzene that still elutes without being retained [26].

In spite of recent progresses in MOF synthesis, and that several different MOF structures have been studied to test their potential on different separations, the key issues, like pilot-scale testing and process design, remain substantially unexplored [31]. However, few examples can be found in the literature of testing at the pilot scale and process design, the separation of propane/propylene using Cu-BTC [19, 21], the separation of a C3/C4 mixture with MIL-100(Fe) [28], the separation of acidic gas in MIL-53(Al) [27], the separation of Kr/Xe [32], and methane purification from natural gas [33]. In the liquid phase there are even fewer pilot-scale tests and process designs; one example is the separation of xylene isomers by simulated moving bed (SMB) using MIL-53(Al) [29] and UiO-66(Zr) [30] (see Table 10.1). The restricted number of publications/studies at the pilot scale, and the subsequent process development, finds its main reasons in the costs of MOF production,

Table 10.1 Examples of pilot tests and process simulations found in the literature

Separation	MOF	mads (g)	Column (l × *d*) (cm × cm)	Process	Ref.
Propane/propylene	Cu-BTC	177	83 × 2.1	-	[21]
Propane/propylene	Cu-BTC	152	81 × 2.1	PSA	[19]
CH_4/CO_2	MIL-53(Al)	200	50 × 5.0	PSA	[27]
C3/C4	MIL-100(Fe)	132	81 × 2.1	-	[28]
Xylene isomers	MIL-53(Al)	53.1	35 × 1.9	SMB	[29]
Xylene isomers	UiO-66(Zr)	48.7	35 × 1.9	-	[30]

their up-scale synthesis, and their shaping. The large-scale synthesis and shaping have been mainly developed by BASF (Basolite Z1200–Zn(II)/2-methylimidazole, Basolite M050–Mg(II)/formic acid, Basolite A520 Al(III)/fumaric acid, Basolite A100–Al(III)/terephthalate, Basolite C300–Cu(II)/benzene-1,3,5-tricarboxylate, Basolite F300 Fe(III)/benzene-1,3,5-tricarboxylate) [31, 34, 35] and the Korea Research Institute of Chemical Technology (KRICT) (Cu-BTC [19, 36, 37], MIL-100(Fe) [28, 37], UiO-66(Zr) [30, 37], MIL-100(Fe) [37, 38], and MIL-101 [37]).

10.2 Adsorptive Gas-Phase Separation Processes

The application of adsorption to industrial separation processes dates from the 1960s due to the combined appearance of pressure swing adsorption (PSA) processes [39, 40] and molecular sieves [41, 42]. Nowadays, adsorptive gas mixture separation processes are widely used for various separations. They compete with several other processes such as distillation, membrane-based processes, gas extraction with organic amines/bases, etc. The most important adsorption-based processes, for gas-phase separations, are PSA and temperature swing adsorption (TSA). In general they differ mostly in the regeneration step. In PSA, short production/regeneration cycles are possible, processing large flow rates. Adsorbent regeneration is achieved by lowering the partial pressure of the adsorbed components in the gas phase. This is performed by reducing the total pressure, even to below atmospheric pressures (vacuum pressure swing adsorption [VPSA]), or by rinsing the adsorbent bed with an inert gas (lower the partial pressure of the components) or a portion of the product. The regeneration step in TSA is attained by heating the adsorbent bed with a preheated purge gas. As each heating/cooling cycle takes from a few hours to over a day, TSA is mainly used for purification, in which small quantities of adsorptive gases are processed. The various applications of these processes are related to biogas/natural gas upgrading, flue gas, CO_2 capture, air separation, O_2 and/or N_2 production, syngas/H_2 production, and volatile organic compounds (VOCs) removal. Although adsorptive separations by PSA/TSA are already a well-established field, there are a few challenges to face, such as the light olefin/paraffin separation and the implementation of SMB technology to gas-phase separations.

10.2.1 *Propylene/Propane Separation by PSA*

Propylene is the feedstock of a series of chemical products, and it is also used in refinery operations; however, its demand is mainly driven by polypropylene production [43]. The raw material for propylene manufacture is usually a C_n fraction with varying composition depending on its origin (by-product from petroleum cracking or ethylene manufacture) [44]. Propane/propylene separation is a difficult separation due to the close volatilities of the compounds, which result in an energy-intensive distillation. The purity requirements for the propylene product will depend on the intended final use: polymer production (ca. 99%), chemical grade (ca. 95%), or refinery grade (ca. 65%) [45]. A wide variety of adsorbents have been studied: silica-based adsorbent [46], carbon molecular sieves [47], zeolites [48–52], chemical adsorbents with complexing metals (i.e., copper or silver ions) [53–55], and, most recently, MOFs [20, 56–61]. Among these, Cu-BTC (or HKUST-1) has shown to present high adsorption capacity and equilibrium selectivity for propylene over propane [18–21, 56, 57, 59, 62, 63] (see Fig. 10.3). Cu-BTC is an MOF with a large surface area (Brunauer–Emmett–Teller [BET] surface area up to 1600 m^2 g^{-1}), high pore volume, high chemical stability, and Lewis acidity with high potential for applications in the field of adsorption, such

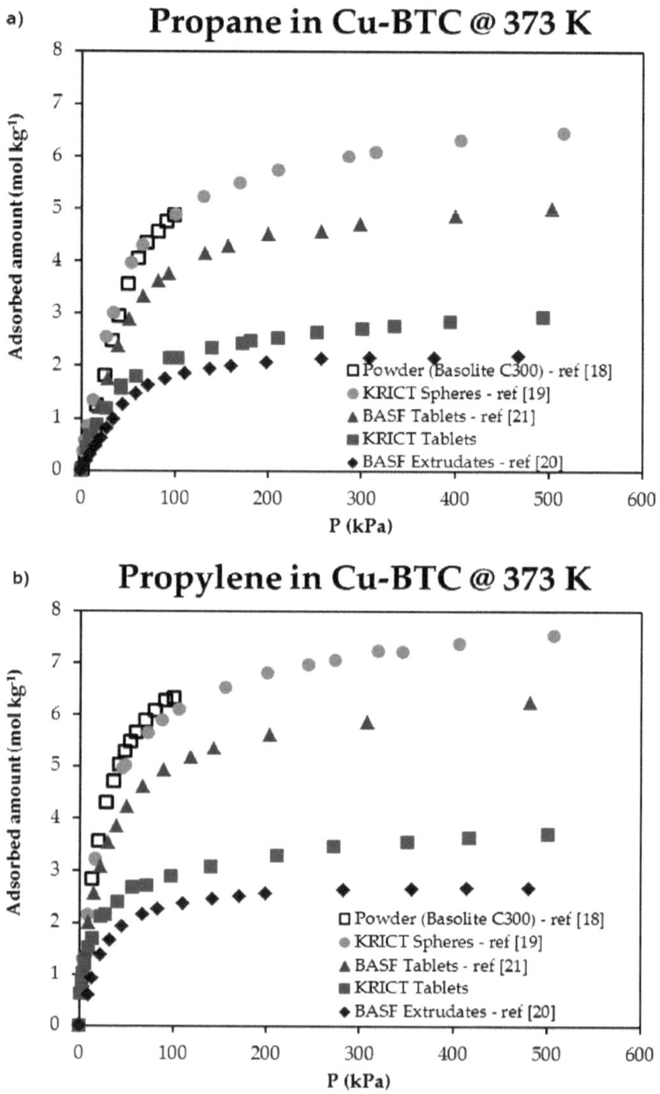

Figure 10.3 Comparison of the adsorption isotherms at 373 K of (a) propane and (b) propylene over several different Cu-BTC materials (powder and shaped).

as gas separation and gas storage [4, 8, 64–67]. However, the strong interaction of Cu-BTC with propylene, resulting in a steep adsorption isotherm, requires a desorption/regeneration step at subatmospheric pressure.

Proper modeling of the equilibrium data is of critical importance for process design. The most used models to describe adsorption isotherms are the Langmuir model (Eq. 10.1), the Toth model (Eq. 10.2), the Sips model (Eq. 10.3), the dual Langmuir model (Eq. 10.4), and the dual Sips model (Eq. 10.5).

$$n^{\mathrm{a}} = n^{\mathrm{a}}_{\mathrm{sat}} \frac{kp}{1 + kp} \tag{10.1}$$

$$n^{\mathrm{a}} = n^{\mathrm{a}}_{\mathrm{sat}} \frac{p}{(b + p^{m})^{1/m}} \tag{10.2}$$

$$n^{\mathrm{a}} = n^{\mathrm{a}}_{\mathrm{sat}} \frac{(kp)^{1/m}}{1 + (kp)^{1/m}} \tag{10.3}$$

$$n^{\mathrm{a}} = n^{\mathrm{a}}_{1,\mathrm{sat}} \frac{k_1 p}{1 + k_1 p} + n^{\mathrm{a}}_{2,\mathrm{sat}} \frac{k_2 p}{1 + k_2 p} \tag{10.4}$$

$$n^{\mathrm{a}} = n^{\mathrm{a}}_{1,\mathrm{sat}} \frac{(k_1 p)^{1/m_1}}{1 + (k_1 p)^{1/m_1}} + n^{\mathrm{a}}_{2,\mathrm{sat}} \frac{(k_2 p)^{1/m_2}}{1 + (k_2 p)^{1/m_2}} \tag{10.5}$$

For the hydrogen adsorption on Cu-BTC, Liu et al. attributed three different adsorption sites; this was later corroborated by García-Pérez et al. [68, 69]. However, the assessment of the total adsorbed amount by macroscopic techniques (gravimetric and/or manometric) lacks detailed information on the adsorbed amount per site. Therefore, to describe the adsorption equilibrium of propane and propylene in Cu-BTC, simpler models, which take into consideration only one site or a maximum of two sites, such as Langmuir [19] and dual Sips [18, 20, 21, 70] and are easier to handle in process simulation, are preferred. In Table 10.2 are presented the adsorption equilibrium parameter values for the adsorption of propane and propylene in Cu-BTC found in the literature. Although shaping can decrease the adsorbent capacity by several reasons, such as dilution with an inert material/binder, pore blocking with a binder, and/or destruction of the crystalline material by extrusion/pressure, the adsorbate–adsorbent interactions are not affected since the isosteric heats of adsorption are the same for the different adsorbent forms (see Fig. 10.4).

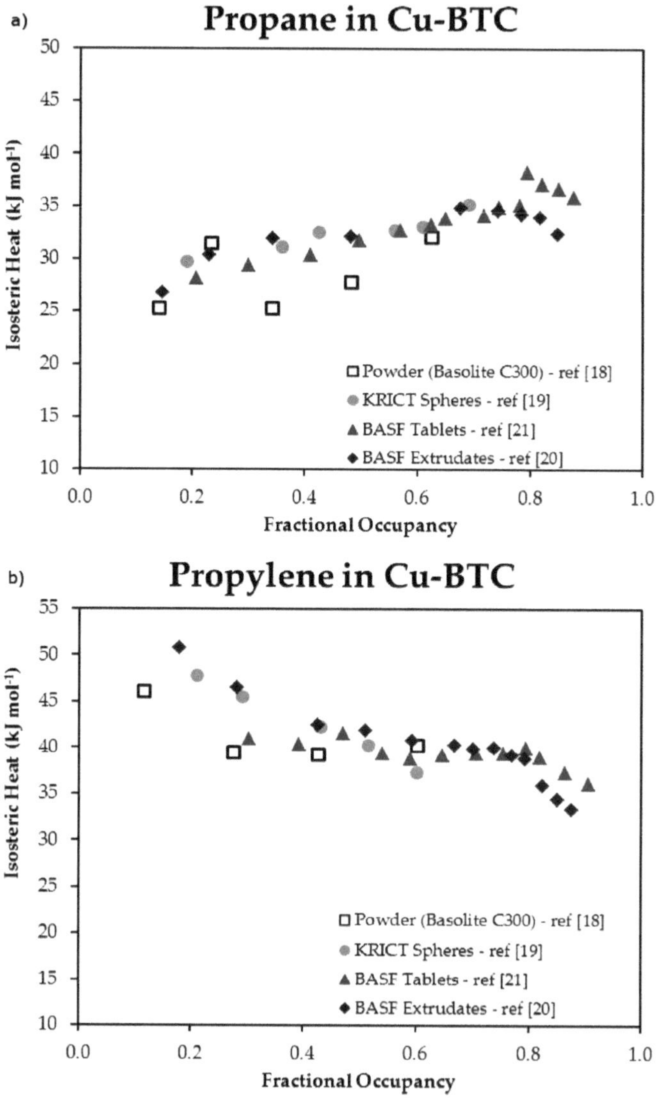

Figure 10.4 Comparison of the isosteric heats of adsorption of (a) propane and (b) propylene over several different Cu-BTC materials (powder and shaped).

Table 10.2 Equilibrium parameters for the adsorption of propane and propylene on different-shaped Cu-BTC materials

Adsorbate	$n^a_{1,sat}/n^a_{2,sat}$ (mol kg^{-1})	$k_{1,0}/k_{2,0}$ (10^{-8}kPa^{-1})	$-\Delta H^0_{1,ads}/-\Delta H^0_{2,ads}$ (kJ mol^{-1})	n_1/n_2 (-)	Ref.
Propane	6.21/1.14	0.07/0.16	28.7/34.1	0.82/0.32	[18][1]
	2.12/0.72	0.90/1.46	31.9/36.5	0.58/0.88	[20][1]
	4.90/0.90	0.07/0.16	28.7/34.1	0.82/0.32	[21][1]
	7.06	0.09	31.0	-	[19][2]
Propylene	7.09/1.22	0.42/0.06	41.9/47.0	1.00/0.80	[18][1]
	2.75/0.54	3.42/5.54	43.6/20.2	0.98/1.20	[20][1]
	5.60/0.96	0.42/0.06	41.9/47.0	1.00/0.80	[21][1]
	7.73	0.01	42	-	[19][2]

[1] Dual-site Sips model
[2] Langmuir model

To design an adsorption-based process an adequate dynamic model is required for an adsorption fixed bed. The selected model is commonly validated by single-component and multicomponent (coadsorption and counteradsorption) breakthrough experiments, usually at the pilot scale. The most used model is a nonisothermal, nonadiabatic, bidisperse mass transfer control model [50]. The bimodal pore size distribution of the adsorbent (crystal micropores surrounded by macropores formed upon shaping) is accounted for by double linear driving force (LDF) approximation. For an equilibrium-controlled process the LDF approximation proves adequate for most operating conditions [71]. The multicomponent adsorption equilibrium is usually predicted by the extended form of the chosen single-component adsorption equilibrium model (e.g., extended Langmuir, Eq. 10.6; extended dual-site Sips, Eq. 10.7), with different saturation capacities for each adsorbate.

$$n_i^a = n_{i,\text{sat}}^a \frac{k_i\, p_i}{1 + \sum\limits_{i=1}^{n} k_i\, p_i} \tag{10.6}$$

$$n_i^a = n_{i,1,\text{sat}}^a \frac{(k_{i,1}\, p_i)^{1/m_{i,1}}}{1 + \sum\limits_{i=1}^{n} (k_{i,1}\, p_i)^{1/m_{i,1}}} + n_{i,2,\text{sat}}^a \frac{(k_{i,2}\, p_i)^{1/m_{i,2}}}{1 + \sum\limits_{i=1}^{n} (k_{i,2}\, p_i)^{1/m_{i,2}}} \tag{10.7}$$

Although these models are not thermodynamically consistent, they are extremely useful to predict the multicomponent adsorption equilibrium and not too demanding computationally. Ideal adsorption solution theory (IAST) [72] can also be used to predict the multicomponent adsorption equilibrium and implemented in commercial packages, usually used to perform process simulation, such as *gPROMS*.

The fixed-bed adsorption process can be described through a mathematical model that includes the mass, energy, and momentum balances that govern the process. To build the mathematical model the following assumptions are made: ideal gas behavior for the gas phase, axially dispersed plug flow, constant void fraction, local pressure drop following the Ergun equation, absence of temperature gradients within the solid particle, and the column wall only interchanges energy with the gas phase and the external environment. When all the previous assumptions are made, the adsorption fixed-bed model follows as presented next:

Mass balance to the gas phase:

$$\varepsilon_{\mathrm{b}}\frac{\partial C_i}{\partial t} = \frac{\partial}{\partial z}\left(\varepsilon_{\mathrm{b}} D_{zm}\frac{\partial C_i}{\partial z}\right) - \frac{\partial\left(uC_i\right)}{\partial z} - N_i \tag{10.8}$$

Mass transfer to the solid:

$$N_i = \left(1-\varepsilon_{\mathrm{b}}\right)\frac{a_{\mathrm{p}}}{Bi_{m,i}+1}\left(C_i - C_{m,i}\right); \quad a_{\mathrm{p}} = \frac{3}{R_{\mathrm{p}}} \tag{10.9}$$

Mass balance to the macroporous phase:

$$\frac{\partial C_{m,i}}{\partial t} = 15\frac{D_{\mathrm{p},i}}{R_{\mathrm{p}}^2}\frac{Bi_{m,i}}{Bi_{m,i}+1}\left(C_i - \overline{C_i}\right) - \frac{\rho_{\mathrm{p}}}{\varepsilon_{\mathrm{p}}}\frac{\partial n_i^{\mathrm{a}}}{\partial t} \tag{10.10}$$

Mass balance to the crystals:

$$\frac{\partial n_i^{\mathrm{a}}}{\partial t} = 15\frac{D_{C,i}}{r_{\mathrm{c}}^2}\left(n_i^{\mathrm{a},*} - n_i^{\mathrm{a}}\right) \tag{10.11}$$

Ideal gas law:

$$P = \sum_{i=1}^{n} C_i R_{\mathrm{g}} T \tag{10.12}$$

Ergun equation:

$$\frac{\partial P}{\partial t} = 150\frac{\mu\left(1-\varepsilon_{\mathrm{b}}\right)^2}{\varepsilon_{\mathrm{b}}^3 d_{\mathrm{h}}^2}u + 1.75\frac{\left(1-\varepsilon_{\mathrm{b}}\right)^2\rho_{\mathrm{g}}}{\varepsilon_{\mathrm{b}}^3 d_h}\left|u\right|u \tag{10.13}$$

Energy balance to the gas phase:

$$\frac{\partial T_g}{\partial t} = \frac{1}{\varepsilon_b C\tilde{C}_v}\frac{\partial}{\partial z}\left(\lambda\frac{\partial T_g}{\partial z}\right) - u\frac{\tilde{C}_p}{\tilde{C}_v}\frac{\partial T_g}{\partial z} + \frac{R_g T_g}{C\tilde{C}_v}\frac{\partial C}{\partial t}$$
$$-\frac{(1-\varepsilon_b)}{\varepsilon_b}\frac{ah_f}{C\tilde{C}_v}\left(T_g - T_s\right) - \frac{2h_w}{\varepsilon_b C\tilde{C}_v R_w}\left(T_g - T_w\right) \quad (10.14)$$

Energy balance to the solid:

$$(1-\varepsilon_b)\left[\varepsilon_p\sum_{i=1}^{n}\left(C_{m,i}\tilde{C}_{v,i}\right) + \rho_p\sum_{i=1}^{n}\left(n^a_{i,\text{total}}\tilde{C}_{v,\text{ads},i}\right) + \rho_p\hat{C}_{ps}\right]$$
$$\frac{\partial T_s}{\partial t} = (1-\varepsilon_b)\,\varepsilon_p R T_s\frac{\partial \bar{C}}{\partial t}$$
$$+\rho_b\sum_{j=1}^{ns}\left(\sum_{i=1}^{n}\left(\left(-\Delta H^0_{\text{ads},i,j}\right)\frac{\partial n^a_{i,j}}{\partial t}\right)\right)$$
$$+(1-\varepsilon_b)\,a_p h_f\left(T_g - T_s\right) \quad (10.15)$$

Energy balance to the wall:

$$\frac{\partial T_w}{\partial t} = \frac{\alpha_w h_w}{\rho_w\hat{C}_{pw}}\left(T_g - T_w\right) - \frac{\alpha_{wl}U}{\rho_w\hat{C}_{pw}}\left(T_w - T_\infty\right);$$
$$\alpha_w = \frac{D_w}{e\left(D_w + e\right)}; \quad \alpha_{wl} = \frac{2}{\left(D_w + e\right)\ln\left(\frac{D_w+2e}{D_w}\right)} \quad (10.16)$$

where the following parameters can be calculated by:

Mass axial dispersion:

$$\frac{\varepsilon_b D_{zm}}{D_m} = 20 + 0.5\text{ScRe}$$

External mass transfer coefficient:

$$\frac{K_{m,i}d_h}{D_{m,i}} = 2.0 + 1.1\text{Re}^{0.6}\text{Sc}_i^{1/3}$$

Heat axial dispersion:

$$\frac{\lambda}{k_g} = 7 + 0.5\,\text{Pr}\,\text{Re}$$

Particle film heat transfer coefficient:

$$\frac{h_f D_h}{k_g} = 2.0 + 1.1\text{Re}^{0.6}\text{Pr}_i^{1/3}$$

Pore diffusivity:

$$\frac{1}{D_{\mathrm{p},i}} = \tau_{\mathrm{p}} \left(\frac{1}{D_{\mathrm{m},i}} + \frac{1}{D_{k,i}} \right)$$

Knudsen diffusivity:

$$D_{k,i} = 97 r_{\mathrm{p}} \sqrt{\frac{T}{M_i}} \qquad ; [r] = m$$

Crystal diffusivity:

$$D_{\mathrm{c},i} = D_{\mathrm{c0},i} e^{\left(-E_{\mathrm{a}}/R_{\mathrm{g}}T\right)}$$

The several required parameters can be found in the literature or calculated from known equations/correlations. For example, crystal diffusivity parameters for propane and propylene over Cu-BTC are given by Whering et al. [73]. The heat capacity of the solid can be considered to be similar to that of another Cu-MOF [74]. Mass and heat axial dispersion terms (D_{zm} and λ) and mass and heat transfer coefficients ($K_{m,i}$ and h_f) can be calculated using Wakao and Funazkri correlations [75]. The molecular diffusivity can be calculated from the Champman–Enskog equation [76]. The diffusivity in the macroporous phase ($D_{v,i}$) can be derived with the Bonsanquet equation [76]. The viscosity of the gas mixture can be calculated with Wilke's method and the viscosity of individual components with Chung's method [77] at the reference temperature (373 K) and then corrected locally for temperature [76]. The conductivity of the individual components was calculated making use of the Eucken equation, and the conductivity of the gas mixture was also calculated by the Wilke method. The required characteristic parameters of the gases under study were taken from the literature [78]. Plaza et al. [19] presented simulations of breakthrough experiments of propane, propylene, isobutane, and binary mixtures on Cu-BTC spheres at 373 K and 150 kPa. However, to describe well the fixed-bed breakthrough experimental behavior a correction with a factor of 0.8 had to be made to the material capacity. This was attributed to deterioration of the adsorbent material, which could be due to several factors, such as exposure to ambient conditions, "coke" formation, or chemisorption (waster sensitivity) [19].

Figures 10.5 and 10.6 represent the mass front (at the exit of the column) and the temperature front (in the middle of the column)

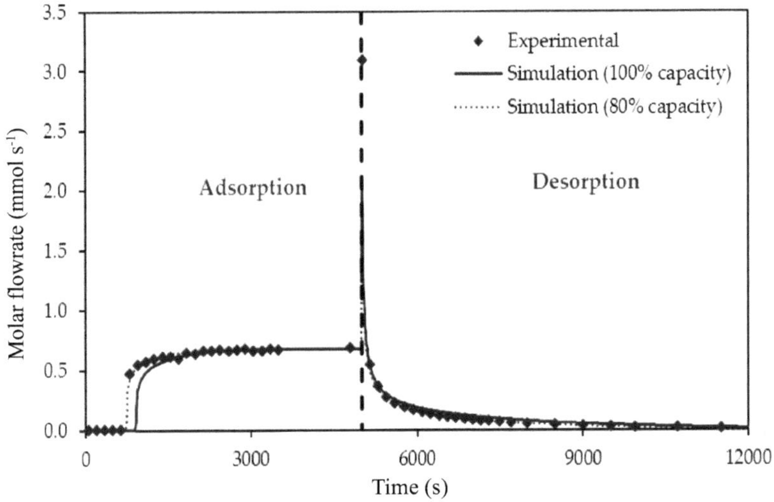

Figure 10.5 Mass front of the adsorption and desorption of propylene (76% in helium) over a bed initially full of helium at 373 K and 150 kPa. Adapted from Plaza et al. [19].

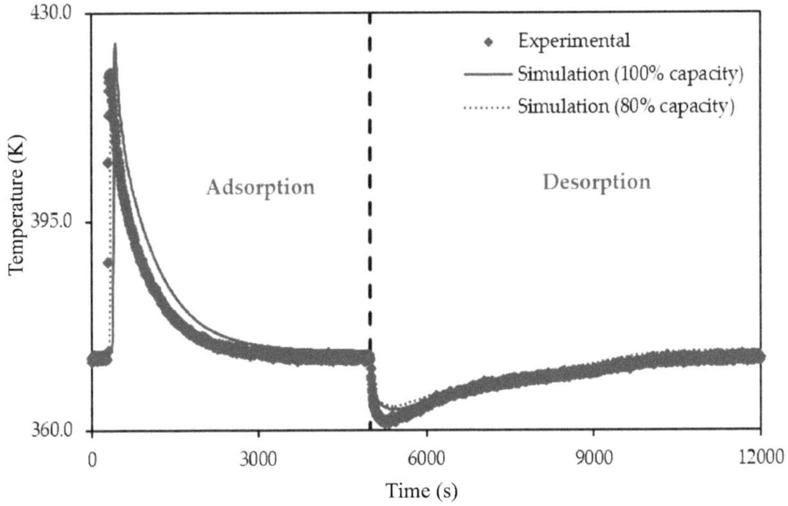

Figure 10.6 Temperature front of the adsorption and desorption of propylene (76% in helium) over a bed initially full of helium at 373 K and 150 kPa. Adapted from Plaza et al. [19].

of a breakthrough experiment and their simulation [19]. As can be seen the model described above can predict fairly well the observed experimental behavior of the fixed-bed in an adsorption/desorption steps. Following the validation of the model and its parameters by simulation of different breakthrough experiments, with different operation conditions the next step is to simulate the overall process, in this case the Pressure Swing Adsorption (PSA) process. The different breakthrough experiments, used for model's validation, should be designed in a way to mimic the different steps existing in the process, e.g., co-adsorption/binary adsorption to mimic the feed step, counter-adsorption/pseudo-binary to mimic the rinsing with light or heavy component, and the desorption with inert to mimic depressurization step. Although, validation of the model requires a large set of experimental data, process simulation is still a powerful tool that can save time and money, and aids to reduce the number of required experiments. The accuracy of the simulator will depend on the mathematical model used, and thus the complexity of the model should be large enough to account for all the main mechanisms involved in the process, while keeping the computational effort necessary to simulate as low as possible [19].

In gas separation by adsorptive processes the PSA process is the most industrialized technology [79]. PSA has been used since 1960, when the first patent in air separation by PSA was filled [40]. The Skarstrom cycle uses four basic steps, in a two-bed unit, as shown in Fig. 10.7.

Regarding propane/propylene separation Da Silva and Rodrigues proposed a five-step vacuum swing adsorption (VSA) cycle to split an equimolar mixture of propylene and propane diluted in nitrogen using zeolite 13X pellets as the adsorbent [50]. Cheng and Wilson proposed a VSA for separating propylene from a feed stream comprising propylene and propane using AlPO-14 as the adsorbent [80, 81]. More recently, Grande, Poplow, and Rodrigues proposed a two-stage VPSA unit to produce polymer-grade propylene with high recovery using zeolite 4A as the adsorbent [82]. On the basis of previous knowledge, a VSA process with five steps, for three columns, was proposed by Plaza et al. to separate a 30% propane/70% propylene mixture, using Cu-BTC MOF spheres as the adsorbent (see Fig. 10.8).

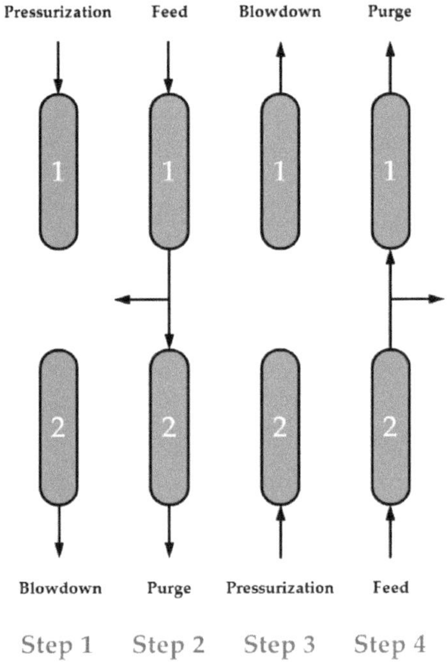

Figure 10.7 Four steps in the Skarstrom cycle.

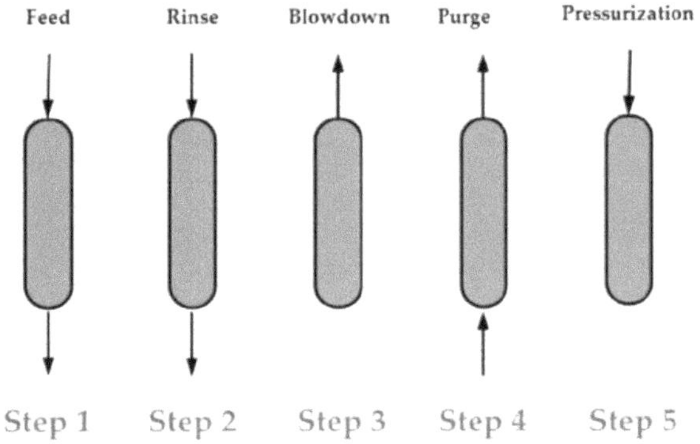

<table>
<tr><td colspan="2">Feed</td><td colspan="2">Rinse</td><td>Blowdown</td><td>Purge</td><td>Pressurization</td></tr>
<tr><td>Purge</td><td>Pressurization</td><td colspan="2">Feed</td><td>Rinse</td><td colspan="2">Blowdown</td></tr>
<tr><td colspan="2">Blowdown</td><td>Purge</td><td>Pressurization</td><td>Feed</td><td colspan="2">Rinse</td></tr>
</table>

Figure 10.8 Five-step VSA cycle configuration proposed by Plaza et al. [19] for a high-purity propylene unit.

	$z=0$	$z=L$
Step I: Countercurrent pressurization 	$u\|_{z^+}=0$ if $P<P_{\mathrm{H}}$ $\left.\frac{\partial C_i}{\partial z}\right\|_{z^+}=0$ $\left.\frac{\partial T_{\mathrm{g}}}{\partial z}\right\|_{z^+}=0$	$uC\|_{z^-}=uC\|_{z^+}$ $-\varepsilon_{\mathrm{b}}D_{\mathrm{zm},i}\left.\frac{\partial C_i}{\partial z}\right\|_{z^-}+u\,C_i\|_{z^+}-u_0C_{i,0}=0$ $-\lambda\left.\frac{\partial T_{\mathrm{g}}}{\partial z}\right\|_{z^-}+u\,C\tilde{C}_{\mathrm{pg}}T_{\mathrm{g}}\big\|_{z^-}-u\,C\tilde{C}_{\mathrm{pg}}T_{\mathrm{g}}\big\|_{z^+}=0$
Steps II and III: Feed/rinse 	$uC\|_{z^-}=uC\|_{z^+}$ $-\varepsilon_{\mathrm{b}}D_{\mathrm{zm},i}\left.\frac{\partial C_i}{\partial z}\right\|_{z^+}+u\,C_i\|_{z^+}-u\,C_i\|_{z^-}=0$ $-\lambda\left.\frac{\partial T_{\mathrm{g}}}{\partial z}\right\|_{z^+}+u\,C\tilde{C}_{\mathrm{pg}}T_{\mathrm{g}}\big\|_{z^+}-u\,C\tilde{C}_{\mathrm{pg}}T_{\mathrm{g}}\big\|_{z^-}=0$	$P\|_{z^-}=P\|_{z^+}$ $\left.\frac{\partial C_i}{\partial z}\right\|_{z^-}=0$ $\left.\frac{\partial T_{\mathrm{g}}}{\partial z}\right\|_{z^-}=0$
Step IV: Blowdown 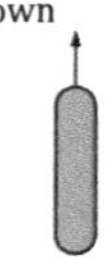	$P(t)\|_{z^+}=P(0)\|_{z^+}+\left(P_{\mathrm{L}}-P(0)\|_{z^+}\right)\left(1-e^{-Mt_{\mathrm{BD}}}\right)$ $\left.\frac{\partial C_i}{\partial z}\right\|_{z^+}=0$ $\left.\frac{\partial T_{\mathrm{g}}}{\partial z}\right\|_{z^+}=0$	$u\|_{z^+}=0$ $\left.\frac{\partial C_i}{\partial z}\right\|_{z^-}=0$ $\left.\frac{\partial T_{\mathrm{g}}}{\partial z}\right\|_{z^-}=0$
Step V: Purge 	$P\|_{z^-}=P\|_{z^+}$ $\left.\frac{\partial C_i}{\partial z}\right\|_{z^+}=0$ $\left.\frac{\partial T_{\mathrm{g}}}{\partial z}\right\|_{z^+}=0$	$uC\|_{z^-}=uC\|_{z^+}$ $-\varepsilon_{\mathrm{b}}D_{\mathrm{zm},i}\left.\frac{\partial C_i}{\partial z}\right\|_{z^-}+u\,C_i\|_{z^-}-u\,C_i\|_{z^+}=0$ $-\lambda\left.\frac{\partial T_{\mathrm{g}}}{\partial z}\right\|_{z^-}+u\,C\tilde{C}_{\mathrm{pg}}T_{\mathrm{g}}\big\|_{z^-}-u\,C\tilde{C}_{\mathrm{pg}}T_{\mathrm{g}}\big\|_{z^+}=0$

To simulate such a process it is necessary to define the appropriate boundary conditions for each step of the designed cycle.

By coupling the previously described model with the boundary conditions, it is possible to simulate the PSA process. Plaza et al. designed a PSA cycle feasible to validate experimentally with a one-column PSA unit [19]. The main results of this study are presented in Fig. 10.9. It should be noticed that after seven cycles the cyclic steady state (CSS) is reached, as predicted by the simulations.

The performance of a PSA process is commonly evaluated according to product purity, product recovery, and adsorbent productivity [83]. Plaza et al. [19] defined product purity, recovery, and productivity parameters by the following equations:

$$C_3H_8\text{ purity }(\%) \Rightarrow \frac{\int_0^{t_{\text{feed}}} F_{C_3H_8,\text{out}}dt + \int_0^{t_{\text{rinse}}} F_{C_3H_8,\text{out}}dt}{\left(\int_0^{t_{\text{feed}}} F_{C_3H_8,\text{out}}dt + \int_0^{t_{\text{feed}}} F_{C_3H_6,\text{out}}dt\right) + \left(\int_0^{t_{\text{rinse}}} F_{C_3H_8,\text{out}}dt + \int_0^{t_{\text{rinse}}} F_{C_3H_6,\text{out}}dt\right)} \times 100$$

$$C_3H_8\text{ rec }(\%) \Rightarrow \frac{\left(\int_0^{t_{\text{feed}}} F_{C_3H_8,\text{out}}dt + \int_0^{t_{\text{rinse}}} F_{C_3H_8,\text{out}}dt\right) - \left(F_{C_3H_8,\text{in press}} \times t_{\text{press}} + F_{C_3H_8,\text{in purge}} \times t_{\text{purge}}\right)}{\left(F_{C_3H_8,\text{in feed}} \times t_{\text{feed}}\right)} \times 100$$

$$C_3H_6\text{ rec }(\%) \Rightarrow \frac{\left(\int_0^{t_{\text{blowdown}}} F_{C_3H_6,\text{out}}dt + \int_0^{t_{\text{purge}}} F_{C_3H_6,\text{out}}dt\right) - \left(F_{C_3H_6,\text{in rinse}} \times t_{\text{rinse}}\right)}{\left(F_{C_3H_6,\text{in feed}} \times t_{\text{feed}}\right)} \times 100$$

$$\text{Productivity (mol kg}^{-1}\text{ h}^{-1}) \Rightarrow \frac{\left(\int_0^{t_{\text{blowdown}}} F_{C_3H_6,\text{out}}dt + \int_0^{t_{\text{purge}}} F_{C_3H_6,\text{out}}dt\right) - \left(F_{C_3H_6,\text{in rinse}} \times t_{\text{rinse}}\right)}{\text{Mass of dry adsorbent} \times t_{\text{cycle}}}$$

Plaza et al. [19] reported a propylene purity of 99.8%; however, the propylene recovery was low, and the reported values were attained by the process simulation. Therefore, the authors suggested recycling part of the raffinate stream, produced during the feed and rinsing steps, for use in the pressurization and purging steps in order to increase the recovery. After the complete validation of the model (fixed-bed model + PSA boundary conditions) it is possible to design an industrial unit with the industrially required purity, recovery, and productivity benchmarks.

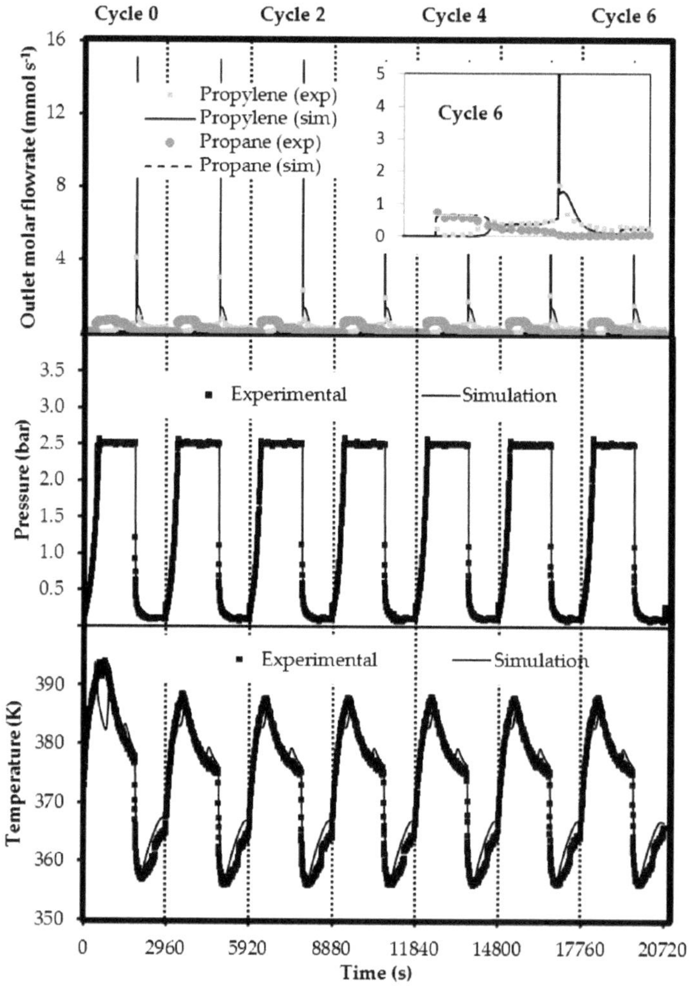

Figure 10.9 VSA experimental results and simulation for the separation of propane/propylene on Cu-BTC spheres at 373 K and 2.5 bar. Adapted from Plaza et al. [19].

10.3 Future Chalenges

Several topics are of critical importance for adsorptive processes based on MOF materials as adsorbents to become a major player in the field of (gas) separation. The most important one is the development of inexpensive MOFs, which are thermo-, chemo-, and mechanically stable. Their up-scale synthesis and shaping are challenges. However, the development of these MOFs should follow industrial needs in the most relevant separation process. Up to now, the application of MOFs in the separation field has followed the opposite route, that is, to fit the existing/synthesized MOFs to possible interesting separations. This trial-and-error route has proven highly inefficient, resulting in a scarce number of separations taken from the fundamentals studies of adsorption properties to the process development. One of the few successes is the recent patent application for the separation of propylene/nitrogen in MIL-100(Fe) [81]. Even the interesting feature of ZIF-7 being propane selective, as reported by van den Bergh et al., was found as a result of experimental work and not of its potential behavior forecast [61, 84]. However, the window effect that promotes the preference of ZIF-7 to uptake more propane than propylene at certain pressures results in its framework flexibility, which can be destroyed or limited upon shaping.

The multicomponent adsorption equilibrium, measuring/prediction, still presents itself as a rather complex task. The quite heterogeneous structure (organic linkers, metal sites, coordinated water, etc.) of MOFs creates an environment with multiple adsorption sites. For example, Cu-BTC presents itself with at least three significantly different sites [68, 69]. The competition between different adsorbates, on those sites by the standard multicomponent model, is still one of the major drawbacks in modeling adsorptive processes based on MOFs. In the liquid phase, measuring single-component isotherms is rather complex, since it is almost impossible to find an "inert eluent" that can be used. Consequently, untangling the obtained multicomponent equilibrium data, to get the necessary adsorption equilibrium parameters, is not a trivial task.

The application of MOFs in other adsorptive processes, e.g., SMB, can be an alternative to the advance of MOFs as adsorbents in industry. Recently, Rodrigues et al. proposed SMB technology for propane/propylene separation using zeolite 13X as the adsorbent in the gas phase [85]. Following the same line of reasoning, Ferreira et al. and Plaza et al. proposed the pair Cu-BTC/*iso*-butane as the adsorbent/eluent for the separation of propane/propylene with

SMB technology in the gas phase [20, 21]. However, the experimental proof of the concept (in any type of adsorbent) still requires further work. Moreira et al. studied the possible application of MIL-53(Al), UiO-66(Zr), MIL-125(Ti), and MIL-125(Ti)_NH_2 MOFs in liquid-phase SMB for the separation of *para*-xylene from its isomers [29, 30, 86, 87].

10.4 Summary

While there is an impressive number of MOF structures, very few have their way through to practical applications at the pilot level, and as far as our knowledge, none have made it to industrial/large-scale application. In the adsorption-based process field, their main application has been in gas-phase separations. Yet, the only three separations that can be found in the literature with reported process results/performance are the separation of propane/propylene on Cu-BTC [19], acid gas separation on MIL-53(Al) [27], and propylene/nitrogen separation on MIL-100(Fe) [81]. Modeling adsorptive separation processes using MOFs is still a very novel field, presenting ample space for groundbreaking solutions either by innovation in the material front or by advancements in the technology/process design (e.g., hybrid process).

Nomenclature

a	macroparticle specific area
b	Toth adsorption equilibrium constant
Bi	Biot number
C	concentration in bulk phase
$\overline{C}$	average concentration
$\tilde{C}_v$	gas mixture molar specific heat at constant volume
$\tilde{C}_p$	gas mixture molar specific heat at constant pressure
$\tilde{C}_{v,ads,i}$	molar specific heat of the adsorbed phase at constant volume
$\hat{C}_{ps}$	particle specific heat at constant pressure

$\hat{C}_{pw}$	wall specific heat at constant pressure
D	axial dispersion coefficient
d	particle diameter
D_c	micropore diffusivity
D_k	Knudsen diffusivity
D_m	molecular diffusivity
D_p	macropore diffusivity
E_a	activation energy
h	film heat transfer coefficient
k	Langmuir/Sips adsorption equilibrium constant
M	molecular mass
N	molar flux from gas phase to the particle surface
n^a	loading (adsorbate concentration in the particles)
p	pressure (adsorbate concentration in the gas phase)
Pr	Prandtl number
R	radius
Re	Reynolds number
R_g	Ideal gas constant
R_w	bed radius
S_c	Schmidt number
T	temperature
t	time
U	overall heat transfer coefficient
u	velocity
z	axial dimension

Greek letters:

α_w	ratio of the internal surface area to the volume of the column wall
α_{wl}	ratio of the log mean surface area to the volume of the column wall
ε	porosity
ρ	density
λ	heat axial dispersion coefficient
τ	tortuosity

Subscripts:

b	bulk
c	crystal
f	fluid
g	gas
h	hydraulic
i	component
m	mass parameter
p	particle
s	solid
sat	saturation
w	wall
z	axial direction parameter
1	site 1
2	site 2

References

1. Dybtsev, D. N., Chun, H., Yoon, S. H., Kim, D., Kim, K. (2003). Microporous manganese formate: A simple metal-organic porous material with high framework stability and highly selective gas sorption properties, *J. Am. Chem. Soc.*, **126**, 32–33.
2. Eddaoudi, M., Li, H., Yaghi, O. M. (2000). Highly porous and stable metal-organic frameworks: Structure design and sorption properties, *J. Am. Chem. Soc.*, **122**, 1391–1397.
3. Eddaoudi, M., Moler, D. B., Li, H., Chen, B., Reineke, T. M., O'Keeffe, M., Yaghi, O. M. (2001). Modular chemistry: secondary building units as a basis for the design of highly porous and robust metal-organic carboxylate frameworks, *Acc. Chem. Res.*, **34**, 319–330.
4. Millward, A. R., Yaghi, O. M. (2005). Metal-organic frameworks with exceptionally high capacity for storage of carbon dioxide at room temperature, *J. Am. Chem. Soc.*, **127**, 17998–17999.
5. Seki, K., Mori, W. (2002). Syntheses and characterization of microporous coordination polymers with open frameworks, *J. Phys. Chem. B*, **106**, 1380–1385.

6. Pan, L., Liu, H., Lei, X., Huang, X., Olson, D. H., Turro, N. J., Li, J. (2003). RPM-1: a recyclable nanoporous material suitable for ship-in-bottle synthesis and large hydrocarbon sorption, *Angew. Chem., Int. Ed.*, **42**, 542–546.

7. Fletcher, A. J., Cussen, E. J., Prior, T. J., Rosseinsky, M. J., Kepert, C. J., Thomas, K. M. (2001). Adsorption dynamics of gases and vapors on the nanoporous metal organic framework material Ni2(4,4'-bipyridine)3(NO3)4: guest modification of host sorption behavior, *J. Am. Chem. Soc.*, **123**, 10001–10011.

8. Min Wang, Q., Shen, D., Bülow, M., Ling Lau, M., Deng, S., Fitch, F. R., Lemcoff, N. O., Semanscin, J. (2002). Metallo-organic molecular sieve for gas separation and purification, *Microporous Mesoporous Mater.*, **55**, 217–230.

9. Eddaoudi, M., Kim, J., Rosi, N., Vodak, D., Wachter, J., O'Keeffe, M., Yaghi, O. M. (2002). Systematic design of pore size and functionality in isoreticular mofs and their application in methane storage, *Science*, **295**, 469–472.

10. Vishnyakov, A., Ravikovitch, P. I., Neimark, A. V., Bülow, M., Wang, Q. M. (2003). Nanopore structure and sorption properties of Cu-BTC metal-organic framework, *Nano Lett.*, **3**, 713–718.

11. Skoulidas, A. I. (2004). Molecular dynamics simulations of gas diffusion in metal-organic frameworks: Argon in CuBTC, *J. Am. Chem. Soc.*, **126**, 1356–1357.

12. Li, J.-R., Sculley, J., Zhou, H.-C. (2011). Metal–organic frameworks for separations, *Chem. Rev.*, **112**, 869–932.

13. Ferey, G. (2008). Hybrid porous solids: past, present, future, *Chem. Soc. Rev.*, **37**, 191–214.

14. Kitagawa, S., Kitaura, R., Noro, S.-I. (2004). Functional porous coordination polymers, *Angew. Chem., Int. Ed.*, **43**, 2334–2375.

15. Rowsell, J. L. C., Yaghi, O. M. (2004). Metal–organic frameworks: A new class of porous materials, *Microporous Mesoporous Mater.*, **73**, 3–14.

16. Wagener, A., Schindler, M., Rudolphi, F., Ernst, S. (2007). Metal-organic coordination polymers for the adsorptive separation of propane/propylene compounds, *Chem-Ing-Tech.*, **79**, 851–855.

17. Lamia, N., Granato, M. A., Gomes, P. S. A., Grande, C. A., Wolff, L., Leflaive, P., Leinekugel-le-Cocq, D., Rodrigues, A. E. (2009). Propane/propylene separation by simulated moving bed II. Measurement and prediction of binary adsorption equilibria of propane, propylene, isobutane, and 1-butene on 13x zeolite, *Separ. Sci. Technol.*, **44**, 1485–1509.

18. Lamia, N., Jorge, M., Granato, M. A., Paz, F. A. A., Chevreau, H., Rodrigues, A. E. (2009). Adsorption of propane, propylene and isobutane on a metal-organic framework: molecular simulation and experiment, *Chem. Eng. Sci.*, **64**, 3246–3259.
19. Plaza, M. G., Ribeiro, A. M., Ferreira, A., Santos, J. C., Lee, U. H., Chang, J.-S., Loureiro, J. M., Rodrigues, A. E. (2012). Propylene/propane separation by vacuum swing adsorption using Cu-BTC spheres, *Sep. Purif. Technol.*, **90**, 109–119.
20. Ferreira, A. F. P., Santos, J. C., Plaza, M. G., Lamia, N., Loureiro, J. M., Rodrigues, A. E. (2011). Suitability of Cu-BTC extrudates for propane-propylene separation by adsorption processes, *Chem. Eng. J.*, **167**, 1–12.
21. Plaza, M. G., Ferreira, A. F. P., Santos, J. C., Ribeiro, A. M., Müller, U., Trukhan, N., Loureiro, J. M., Rodrigues, A. E. (2012). Propane/propylene separation by adsorption using shaped copper trimesate mof, *Microporous Mesoporous Mater.*, **157**, 101–111.
22. Yoon, J. W., Jang, I. T., Lee, K. Y., Hwang, Y. K., Chang, J. S. (2010). Adsorptive separation of propylene and propane on a porous metal-organic framework, copper trimesate, *B Korean Chem. Soc.*, **31**, 220–223.
23. Hartmann, M., Kunz, S., Himsl, D., Tangermann, O., Ernst, S., Wagener, A. (2008). Adsorptive separation of isobutene and isobutane on Cu3(BTC)2, *Langmuir*, **24**, 8634–8642.
24. Barcia, P. S., Guimaraes, D., Mendes, P. A. P., Silva, J. A. C., Guillerm, V., Chevreau, H., Serre, C., Rodrigues, A. E. (2011). Reverse shape selectivity in the adsorption of hexane and xylene isomers in mof UiO-66, *Microporous Mesoporous Mater.*, **139**, 67–73.
25. Maes, M., Vermoortele, F., Alaerts, L., Couck, S., Kirschhock, C. E. A., Denayer, J. F. M., De Vos, D. E. (2010). Separation of styrene and ethylbenzene on metal-organic frameworks: analogous structures with different adsorption mechanisms, *J. Am. Chem. Soc.*, **132**, 15277–15285.
26. Ahmad, R., Wong-Foy, A. G., Matzger, A. J. (2009). Microporous coordination polymers as selective sorbents for liquid chromatography, *Langmuir*, **25**, 11977–11979.
27. Heymans, N., Vaesen, S., De Weireld, G. (2012). A complete procedure for acidic gas separation by adsorption on MIL-53 (Al), *Microporous Mesoporous Mater.*, **154**, 93–99.
28. Plaza, M. G., Ribeiro, A. M., Ferreira, A., Santos, J. C., Hwang, Y. K., Seo, Y. K., Lee, U. H., Chang, J. S., Loureiro, J. M., Rodrigues, A. E. (2012). Separation of C3/C4 hydrocarbon mixtures by adsorption using a mesoporous iron mof: MIL-100(Fe), *Microporous Mesoporous Mater.*, **153**, 178–190.

29. Moreira, M. A., Santos, J. C., Ferreira, A. F. P., Müller, U., Trukhan, N., Loureiro, J. M., Rodrigues, A. E. (2011). Selective liquid phase adsorption and separation of ortho-xylene with the microporous MIL-53(Al), *Sep. Sci. Technol.*, **46**, 1995–2003.

30. Moreira, M. A., Santos, J. C., Ferreira, A. F. P., Loureiro, J. M., Ragon, F., Horcajada, P., Shim, K.-E., Hwang, Y.-K., Lee, U. H., Chang, J.-S., Serre, C., Rodrigues, A. E. (2012). Reverse shape selectivity in the liquid-phase adsorption of xylene isomers in zirconium terephthalate mof UiO-66, *Langmuir*, **28**, 5715–5723.

31. Czaja, A. U., Trukhan, N., Müller, U. (2009). Industrial applications of metal-organic frameworks, *Chem. Soc. Rev.*, **38**, 1284–1293.

32. Mueller, U., Hesse, M., Puetter, H., Kamieth, M., Schierle-Arndt, K., Patent No. WO/2006/072573 (2006).

33. Müller, U., Hesse, M., Puetter, H., Schubert, M., Mirsch, D. D., Patent No. EP1674555 (2006).

34. Yilmaz, B., Trukhan, N., Müller, U. (2012). Industrial outlook on zeolites and metal organic frameworks, *Chinese J. Catal.*, **33**, 3–10.

35. Mueller, U., Schubert, M., Teich, F., Puetter, H., Schierle-Arndt, K., Pastre, J. (2006). Metal-organic frameworks-prospective industrial applications, *J. Mat. Chem.*, **16**, 626–636.

36. Seo, Y.-K., Hundal, G., Jang, I. T., Hwang, Y. K., Jun, C.-H., Chang, J.-S. (2009). Microwave synthesis of hybrid inorganic–organic materials including porous Cu3(BTC)2 from Cu(II)-trimesate mixture, *Microporous Mesoporous Mater.*, **119**, 331–337.

37. Hwang, Y. K., Seo, Y. K., Lee, J. S., Chang, J. S., Horcajada, P., Serre, C., Férey, G., in *MOF 2010: 2nd International Conference on Metal-Organic Frameworks and Open Framework Compounds* (Marseille, France, 2010).

38. Seo, Y.-K., Yoon, J. W., Lee, J. S., Lee, U. H., Hwang, Y. K., Jun, C.-H., Horcajada, P., Serre, C., Chang, J.-S. (2012). Large scale fluorine-free synthesis of hierarchically porous iron(III) trimesate MIL-100(Fe) with a zeolite mtn topology, *Microporous Mesoporous Mater.*, **157**, 137–145.

39. Guerin de Montgareuil, P., Domine, D., U.S. Patent No. US Patent 3,155,468 (1964).

40. Skarstrom, C. W., US Patent 2,944,627 (1960).

41. Milton, R. M., US Patent No. 2,882,243 (1959).

42. Yang, R. T. (1997) *Gas Separation by Adsorption Processes* (Imperial College Press, London).

43. Aitani, A. M. (2005). Propylene production, in *Encyclopedia of Chemical Processing*, 2461–2466.

44. Sherred, J. A., Fair, J. R. (1959). High purity propylene: a new problem, *Ind. Eng. Chem.*, **51**, 249–252.
45. Bryan, P. F. (2004). Removal of propylene from fuel-grade propane, *Sep. Purif. Rev.*, **33**, 157–182.
46. Zhu, W., Kapteijn, F. A., Moulijn, J. (1999). Shape selectivity in the adsorption of propane/propene on the all-silica DD3R, *Chem. Commun.*, **1999**, 2453–2454.
47. Grande, C. A., Cavenati, S., Da Silva, F. A., Rodrigues, A. E. (2005). Carbon molecular sieves for hydrocarbon separations by adsorption, *Ind. Eng. Chem. Res.*, **44**, 7218–7227.
48. da Silva, F. A., Rodrigues, A. E. (1999). Adsorption equilibria and kinetics for propylene and propane over 13X and 4A zeolite pellets, *Ind. Eng. Chem. Res.*, **38**, 2051–2057.
49. da Silva, F. A., Rodrigues, A. E. (2001). Vacuum swing adsorption for propylene/propane separation with 4A zeolite, *Ind. Eng. Chem. Res.*, **40**, 5758–5774.
50. da Silva, F. A., Rodrigues, A. E. (2001). Propylene/propane separation by vacuum swing adsorption using 13x zeolite, *AIChE J.*, **47**, 341–357.
51. Grande, C. A., Gascon, J., Kapteijn, F., Rodrigues, A. E. (2010). Propane/propylene separation with Li-exchanged zeolite 13x, *Chem. Eng. J.*, **160**, 207–214.
52. Ruthven, D. M., Reyes, S. C. (2007). Adsorptive separation of light olefins from paraffins, *Microporous Mesoporous Mater.*, **104**, 59–66.
53. Rege, S. U., Padin, J., Yang, R. T. (1998). Olefin/paraffin separations by adsorption: Π-complexation vs. kinetic separation, *AIChE J.*, **44**, 799–809.
54. Rege, S. U., Yang, R. T. (2002). Propane/propylene separation by pressure swing adsorption: sorbent comparison and multiplicity of cyclic steady states, *Chem. Eng. Sci.*, **57**, 1139–1149.
55. Chen, J. P., Yang, R. T. (1995). A molecular orbital study of the selective adsorption of simple hydrocarbon molecules on Ag+- and Cu+-exchanged resins and cuprous halides, *Langmuir*, **11**, 3450–3456.
56. Wagener, A., Rudolphi, F., Schindler, M., Ernst, S. (2006). Trennung von propan/propen-gemischen durch adsorption an metallorganischen koordinationspolymeren, *Chem-Ing-Tech.*, **78**, 1328–1329.
57. Wagener, A., Schindler, M., Rudolphi, F., Ernst, S. (2007). Metallorganische koordinationspolymere zur adsorptiven trennung von propan/propen-gemischen, *Chem-Ing-Tech.*, **79**, 851–855.

58. Plaza, M. G., Ferreira, A. F. P., Santos, J. C., Ribeiro, A. M., Müller, U., Trukhan, N., Loureiro, J. M., Rodrigues, A. E. (2012). Propane/propylene separation by adsorption using shaped copper trimesate MOF, *Microporous Mesoporous Mater.*, **157**, 101–111.

59. Yoon, J. W., Jang, I. T., Lee, K.-Y., Hwang, Y. K., Chang, J.-S. (2010). Adsorptive separation of propylene and propane on a porous metal-organic framework, copper trimesate, *B Korean Chem. Soc.*, **31**, 220–223.

60. Yoon, J., Seo, Y. K., Hwang, Y., Chang, J. S., Leclerc, H., Wuttke, S., Bazin, P., Vimont, A., Daturi, M., Bloch, E., Llewellyn, P., Serre, C., Horcajada, P., Grenèche, J. M., Rodrigues, A., Férey, G. (2010). Controlled reducibility of a metal–organic framework with coordinatively unsaturated sites for preferential gas sorption, *Angew. Chem., Int. Ed.*, **49**, 5949–5952.

61. van den Bergh, J., Gücüyener, C., Pidko, E. A., Hensen, E. J. M., Gascon, J., Kapteijn, F. (2011). Understanding the anomalous alkane selectivity of ZIF-7 in the separation of light alkane/alkene mixtures, *Chem. Eur. J.*, **17**, 8832–8840.

62. Jorge, M., Lamia, N., Rodrigues, A. E. (2010). Molecular simulation of propane/propylene separation on the metal-organic framework CuBTC, *Colloids Surf. A*, **357**, 27–34.

63. Lamia, N., Jorge, M., Granato, M. A., Almeida Paz, F. A., Chevreau, H., Rodrigues, A. E. (2009). Adsorption of propane, propylene and isobutane on a metal-organic framework: molecular simulation and experiment, *Chem. Eng. Sci.*, **64**, 3246–3259.

64. Alaerts, L., Séguin, E., Poelman, H., Thibault-Starzyk, F., Jacobs, P. A., De Vos, D. E. (2006). Probing the lewis acidity and catalytic activity of the metal–organic framework [Cu3(BTC)2] (BTC=benzene-1,3,5-tricarboxylate), *Chem. Eur. J.*, **12**, 7353–7363.

65. Dathe, H., Peringer, E., Roberts, V., Jentys, A., Lercher, J. A. (2005). Metal organic frameworks based on Cu2+ and benzene-1,3,5-tricarboxylate as host for SO2 trapping agents, *C. R. Chim.*, **8**, 753–763.

66. Schlichte, K., Kratzke, T., Kaskel, S. (2004). Improved synthesis, thermal stability and catalytic properties of the metal-organic framework compound $Cu_3(BTC)_2$, *Microporous Mesoporous Mater.*, **73**, 81–88.

67. Seo, Y.-K., Hundal, G., Jang, I. T., Hwang, Y. K., Jun, C.-H., Chang, J.-S. (2009). Microwave synthesis of hybrid inorganic-organic materials including porous $Cu_3(BTC)_2$ from Cu(II)-trimesate mixture, *Microporous Mesoporous Mater.*, **119**, 331–337.

68. García-Pérez, E., Gascón, J., Morales-Flórez, V. C., Castillo, J. M., Kapteijn, F., Calero, S. A. (2009). Identification of adsorption sites in Cu-BTC by experimentation and molecular simulation, *Langmuir*, **25**, 1725–1731.
69. Liu, Y., Brown, C. M., Neumann, D. A., Peterson, V. K., Kepert, C. J. (2007). Inelastic neutron scattering of H2 adsorbed in HKUST-1, *J. Alloys Compd.*, **446–447**, 385–388.
70. Chmelik, C., Karger, J., Wiebcke, M., Caro, J., van Baten, J. M., Krishna, R. (2009). Adsorption and diffusion of alkanes in CuBTC crystals investigated using infra-red microscopy and molecular simulations, *Microporous Mesoporous Mater.*, **117**, 22–32.
71. Ruthven, D. M., Farooq, S., Knaebel, K. S. (1994). *Pressure Swing Adsorption* (John Wiley & Sons).
72. Do, D. D. (1998). *Adsorption Analysis: Equilibria and Kinetics* (Imperial College Press, London).
73. Wehring, M., Gascon, J., Dubbeldam, D., Kapteijn, F., Snurr, R. Q., Stallmach, F. (2010). Self-diffusion studies in CuBTC by PFG NMR and MD simulations, *J. Phys. Chem. C*, **114**, 10527–10534.
74. Song, L.-F., Jiang, C.-H., Zhang, J., Sun, L.-X., Xu, F., You, W.-S., Zhao, Y., Zhang, Z.-H., Wang, M.-H., Sawada, Y., Cao, Z., Zeng, J.-L. (2010). Heat capacities and thermodynamic properties of a novel mixed-ligands mofs, *J. Ther. Anal. Calorim.*, **100**, 679–684.
75. Wakao, N., Funazkri, T. (1978). Effect of fluid dispersion coefficients on particle-to-fluid mass transfer coefficients in packed beds : correlation of sherwood numbers, *Chem. Eng. Sci.*, **33**, 1375–1384.
76. Bird, R. B., Stewart, W. E., Lightfoot, E. N. (1960). *Transport Phenomena* (John Wiley & Sons, New York).
77. Chung, T. H., Lee, L. L., Starling, K. E. (1984). Applications of kinetic gas theories and multiparameter correlation for prediction of dilute gas viscosity and thermal-conductivity, *Ind. Eng. Chem. Fund.*, **23**, 8–13.
78. Poling, B. E., Prausnitz, J. M., O'Connell, J. P. (2001). *The Properties of Gases and Liquids*, 5th ed. (McGraw Hill).
79. Sircar, S. (2002). Pressure swing adsorption, *Ind. Eng. Chem. Res.*, **41**, 1389–1392.
80. Cheng, L. S., Wilson, S. T., US Patent No. 6,293,999 B1 (Sep. 25, 2001).
81. Luck, F., Leroi, C., Fernandez, A., Ribeiro, A. M., Santos, J. C., Rodrigues, A. E., Chang, J.-S., Lee, U.-H., Hwang, Y.-K., Bonne, J.-P., France Patent No. PCT/FR2012/051623 (July 6, 2012).

82. Grande, C. A., Poplow, F., Rodrigues, A. E. (2010). Vacuum pressure swing adsorption to produce polymer-grade propylene, *Sep. Sci. Technol.*, **45**, 1252–1259.

83. Rota, R., Wankat, P. C. (1990). Intensification of pressure swing adsorption processes, *AIChE J.*, **36**, 1299–1312.

84. Gucuyener, C., van den Bergh, J., Gascon, J., Kapteijn, F. (2010). Ethane/ethene separation turned on its head: selective ethane adsorption on the metal-organic framework ZIF-7 through a gate-opening mechanism, *J. Am. Chem. Soc.*, **132**, 17704–17706.

85. Sá Gomes, P., Lamia, N., Rodrigues, A. E. (2009). Design of a gas phase simulated moving bed for propane/propylene separation, *Chem. Eng. Sci.*, **64**, 1336–1357.

86. Moreira, M. A., Santos, J. C., Ferreira, A. F. P., Loureiro, J. M., Ragon, F., Horcajada, P., Yot, P. G., Serre, C., Rodrigues, A. E. (2012). Effect of ethylbenzene in p-xylene selectivity of the porous titanium amino terephthalate MIL-125(Ti)_NH2, *Microporous Mesoporous Mater.*, **158**, 229–234.

87. Moreira, M. A., Santos, J. C., Ferreira, A. F. P., Loureiro, J. M., Ragon, F., Horcajada, P., Yot, P. G., Serre, C., Rodrigues, A. E. (2012). Toward understanding the influence of ethylbenzene in p-xylene selectivity of the porous titanium amino terephthalate MIL-125(Ti): adsorption equilibrium and separation of xylene isomers, *Langmuir*, **28**, 3494–3502.

Chapter 11

Computer Simulations of Ionic Metal–Organic Frameworks

Anjaiah Nalaparaju and Jianwen Jiang
Department of Chemical and Biomolecular Engineering, National University of Singapore, Singapore 117576, Singapore
chejj@nus.edu.sg

Metal–organic frameworks (MOFs) have emerged as new hybrid nanoporous materials. As a subfamily, ionic MOFs contain nonframework cations or anions in crystalline structures and possess unique properties for potential applications. While most studies in this field are currently focused on nonionic counterparts, there has been growing interest in ionic MOFs. This chapter reviews the recent computer simulation studies of ionic MOFs, from identifying the locations of nonframework ions to examining their performances for gas adsorption and separation, water treatment, biofuel purification, and drug loading. The microscopic insights from simulations are indispensable to elucidate mechanistic properties in ionic MOFs and provide bottom-up guidelines for the rational design of new MOFs.

Metal–Organic Frameworks: Materials Modeling towards Potential Engineering Applications
Edited by Jianwen Jiang

ISBN 978-981-4613-45-3 (Hardcover), 978-981-4613-46-0 (eBook)
www.panstanford.com

11.1 Introduction

As new hybrid nanoporous materials, metal–organic frameworks (MOFs) can be synthesized by judicious selection of inorganic and organic building blocks or via post-synthetic modification [1, 2]. The degree of diversity and multiplicity in MOF structures is substantially more extensive than any other porous materials. With these salient features, MOFs have been considered versatile materials for diverse potential applications [3]. Over the past decade, thousands of different MOFs with varying structure and functionality have been reported. While some MOFs were synthesized by rationally designed predictable topologies, others were produced fortuitously or accidentally [4, 5].

Most MOFs possess neutral frameworks, but there is a subfamily composed of ionic frameworks along with charge-balancing non-framework ions. Thus, there exist electric fields in ionic MOFs, and the interactions with guest molecules are enhanced to facilitate sorption, ion exchange, and catalytic activity. A number of ionic MOFs have been synthesized and characterized. Notably, Eddaoudi and coworkers produced two zeolite-like MOFs (ZMOFs), *rho*-ZMOF and *sod*-ZMOF, by the molecular building blocks approach [6]. The structures (Fig. 11.1a,b) are anionic and neutralized by doubly protonated 1,3,4,6,7,8-hexahydro-2H-pyrimido[1,2-a]pyrimidine (HPP) cations. These ZMOFs represent the first examples of anionic MOFs based on *rho* and *sod* topologies, respectively. Subsequently, they tested H_2 storage in several ionic MOFs such as square-octahedral (*soc*)-MOF [7,8], *usf*-ZMOF [9], *rht*-MOF [10], Li^+- and Mg^{2+}-exchanged *rho*-ZMOFs [11], and porph@MOM-10 [12]; the improved H_2 adsorption was attributed to the electric fields generated. In addition, the catalytic oxidation of cyclohexane and controlled drug release in *rho*-ZMOFs were also examined [13, 14].

Using adenine as a biomolecular ligand, Rosi and coworkers synthesized bio-MOF-1 (Fig. 11.1c), which contains an anionic framework and dimethylammonium (DMA) cations [15]. By ion exchange, a cationic drug, namely, procainamide, was introduced into bio-MOF-1; approximately 2.5 procainamide molecules per formula unit were loaded. This is the first study for the potential biomedical application of bio-MOFs. By postsynthetic exchange of DMA with other cations,

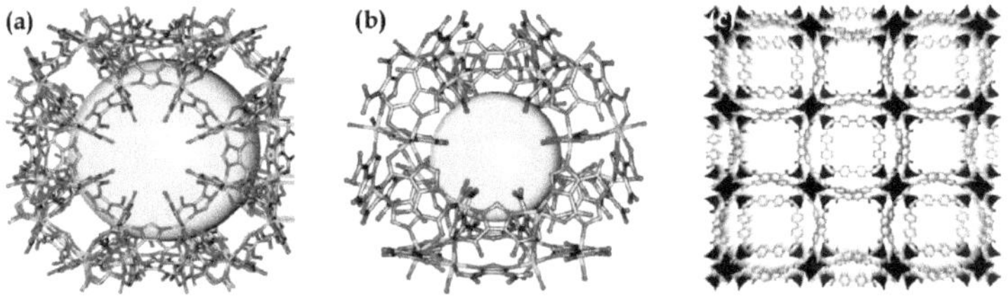

Figure 11.1 Crystal structures of (a) *rho*-ZMOF, (b) *sod*-ZMOF [6], and (c) bio-MOF-1 [15]. The nonframework ions are not shown for clarity. Reprinted with permission from Ref. [6], copyright 2006 Royal Society of Chemistry. Reprinted with permission from Ref. [15]. Copyright 2009 American Chemical Society.

such as tetramethylammonium (TMA), tetraethylammonium (TEA), tetrabutylammonium (TBA), and guanidinium (GND), the pore volume and surface area of bio-MOF-1 were systematically tuned for CO_2 adsorption [16, 17]. They also exchanged DMA with lanthanide cations and tested sensing capability for small molecules [18]. Recently, they further reported a promising strategy using large metal–biomolecule clusters to prepare mesoporous bio-MOF-100 [19].

Other research groups have also reported various ionic MOFs and demonstrated their potential applications. For example, Lu et al. synthesized three-dimensional lanthanide anionic MOFs under hydrothermal conditions. The K^+ ions within the 1D channel could be easily exchanged with various cations, leading to tunable luminescent emissions [20]. A flexible, anionic open-framework $NaLa(H_4L)$ produced by Plabst et al. was found to exhibit an exceptional ion exchange selectivity for monovalent cations [21]. On the basis of triazamacrocycle substituted with carboxylic groups as a linker, Ortiz et al. prepared a cationic MOF, namely, Zn-CBTACN. Consisting of interconnected 1D channels, the framework is charge-balanced by Cl^- ions. A high adsorption capacity was found for CO_2, with a high selectivity over CO, CH_4, N_2, and O_2 [22]. Feng and coworkers synthesized an exceptional MOF, namely, CPF-13, which integrates nanoscopic polyhedral cages, an anionic framework, and amino functionalization. Under ambient conditions, CPF-13 exhibits the highest CO_2 adsorption capacity among reported ionic MOFs [23]. They further synthesized a series of nanogated cationic MOFs and observed size- or charge-selective ion exchange of organic dyes. These ionic MOFs were suggested to be utilized as a platform for anion exchange–based separation [24].

The experiments described above demonstrate that ionic MOFs offer advantages over traditional aluminum-substituted zeolites, which are the inorganic counterparts of ionic MOFs and possess only anionic frameworks [25]. In contrast, ionic MOFs may have anionic or cationic frameworks and hence either cationic or anionic exchange capability. Furthermore, cations or anions in ionic MOFs are readily tunable; thus the electric fields generated can be manipulated for desired functionality. To achieve this target, however, an in-depth microscopic understanding of ionic MOFs is indispensable.

With ever-increasing computational resources and advances in mathematical algorithms, computer simulation has become an invaluable tool in materials characterization, screening, and design [26]. At a molecular level, simulation can provide microscopic insights from the bottom up and establish structure–function relationships. Numerous simulations have been reported on cation-exchanged zeolites for adsorption, diffusion, and catalysis [27, 28]. Comparatively, there are substantially fewer simulations on ionic MOFs. Nevertheless, the number of ionic MOFs increases rapidly, leading to growing interest. In this chapter, we review the recent simulation studies in several areas for ionic MOFs. First, a few studies identifying the locations of nonframework ions are presented. Then, those examining the potential applications of ionic MOFs for gas adsorption and separation are discussed, particularly H_2 adsorption and separation of CO_2-containing gas mixtures. This is the area where most simulation studies of MOFs have focused on. Finally, the utilization of ionic MOFs for water treatment, biofuel purification, and drug delivery is demonstrated.

11.2 Simulation Studies

11.2.1 *Locations of Nonframework Ions*

Several experiments reported the synthesis of cationic MOFs such as *soc*-MOF [7] and *rht*-MOF [10]. In these MOFs, the locations of nonframework NO_3^- ions were successfully determined by X-ray crystallography. In anionic MOFs, however, the determination of small cations is formidable. For example, the locations of Na^+ ions in Na-*rho*-ZMOF [6], DMA cations in bio-MOF-100 [19], and Li^+ ions in Li-MOF [29] were not experimentally provided. Nevertheless, this information is crucial to unravel the intrinsic properties of ionic MOFs.

The *rho*-ZMOF synthesized by Eddaoudi and coworkers contains a very open structure with large truncated cuboctahedra (α-cages) 18.2 Å in diameter. Unlike *rho*-zeolite and other *rho*-aluminosilicate or aluminophosphate, the *rho*-ZMOF possesses twice as many positive charges (48 vs. 24) in a unit cell to neutralize the anionic framework. The HPP cations in the *rho*-ZMOF could be fully exchanged with other cations (e.g., Na^+) [6]. Using Monte Carlo (MC) simulation, Babarao and Jiang identified the locations of Na^+ ions in Na-*rho*-ZMOF. On the basis of coordination environment and binding energy, two types of sites were found, as shown in Fig. 11.2. Site I is in the single eight-member ring (S8MR) near the entrance to the α-cage. The distance from site I to the nearest In atoms in the S8MR is 5.0–5.3 Å and approximately 7.8 Å to the next-to-nearest In atoms. Site II is in the α-cage and proximal to the moiety of the organic link.

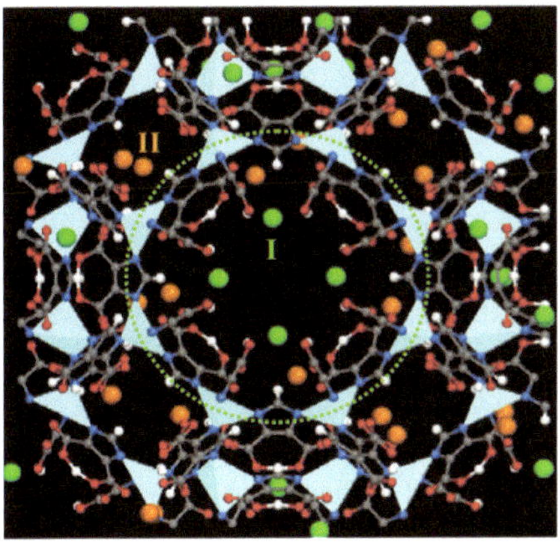

Figure 11.2 Locations of Na^+ ions in Na-*rho*-ZMOF. Site I is in the single eight-member ring, and site II is in the α-cage. Color code: In, cyan; N, blue; C, grey; O, red; and H, white. Reprinted with permission from Ref. [30]. Copyright 2009 American Chemical Society.

In one unit cell, 26 Na^+ ions are located at site I and the remaining at site II. As revealed by radial distribution functions, site I ions have a larger coordination number with neighboring atoms than site II ions and thus a stronger interaction with the framework. Interestingly, the two types of sites in the *rho*-ZMOF resemble those in *rho*-zeolite and would act as preferential sites for adsorption. Furthermore, the mobility of Na^+ ions was estimated from molecular dynamics (MD) simulation. The displacement at site I is nearly flat, but initially increases and then approaches a constant at site II. The greater mobility at site II is attributed to the weaker interaction with the framework and the larger void space around site II. Due to strong binding energy, ion hopping is largely prohibited in the Na-*rho*-ZMOF and thus the Na^+ ions only vibrate locally at their favorable sites [30].

Through ion exchange in $[(CH_3)_2NH_2]^+[In(C_{16}H_6O_8)]^-$, Schröder and coworkers produced a Li-MOF with light Li^+ ions. Half of the cations in the Li-MOF are Li^+ and the other half are H_3O^+. A dehydrated structure of the Li-MOF could be obtained by degassing for 24 hours at 180°C and 10^{-10} bar. The locations of cations in the hydrated and dehydrated Li-MOFs were not available experimentally. Nevertheless, it was underlined that precisely identifying the locations of cations would be very useful for further investigation [29]. Towards this end, Babarao and Jiang performed density functional theory (DFT) calculations to identify cation locations. As shown in Fig. 11.3a, the H^+ ion in the dehydrated Li-MOF forms a covalent bond with one carboxylic O atom, while the Li^+ ion is located between two carboxylic O atoms. The distance is approximately 0.98 Å between the H^+ ion and the O atom, and 1.87 Å between the Li^+ ion and the O atom. Both the Li^+ and H^+ ions are strongly coordinated to the framework. Similarly, the Li^+ and H_3O^+ ions in the hydrated counterpart shown in Fig. 11.3b are also located proximally to the carboxylic O atoms. The distance from the Li^+ ion to the carboxylic O atom is around 1.86 Å, while the distance between the H atom of H_3O^+ and the carboxylic O atom ranges from 1.55 Å to 1.59 Å, implying the formation of a hydrogen bond [31].

Using large Zn-adeninate vertices, Rosi and coworkers constructed mesoporous bio-MOF-100. With an extremely low crystal

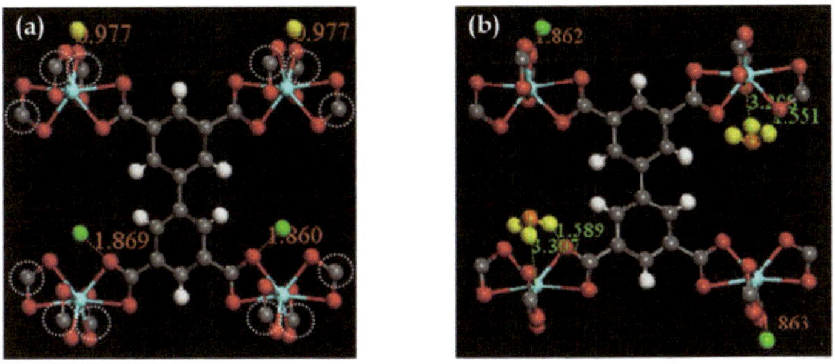

Figure 11.3 Locations of cations in (a) dehydrated Li-MOF and (b) hydrated Li- MOF. The distances are in angstrom. Color code: Li^+, green; H^+, yellow; In, cyan; C, grey; O, red; and H, white. Reprinted with permission from Ref. [31]. Copyright 2011 American Chemical Society.

density (0.302 g/cm^3), bio-MOF-100 possesses a high surface area (4033 m^2/g) and the largest pore volume (4.3 cm^3/g) among reported MOFs. Intriguingly, bio-MOF-100 contains exclusively mesopores (2.8 nm in diameter), which would allow for facile transport of large guest molecules. The framework is anionic and charge-balanced by DMA cations [19]. Chen and Jiang conducted MC simulation to identify the locations of DMA cations. As illustrated in Fig. 11.4, the DMA cations are located near the surface of mesopores. These locations are not unexpected due to the preferential interactions with the mesopores [32].

Toni et al. identified the locations of anions in cationic Zn-CBTACN synthesized by Ortiz et al. [22]. With a purpose to analyze the influence of anion size and polarizability, three anions Cl^-, Br^-, and F^- were examined. The distributions of all three anions are discrete, implying the anions are located individually in Zn-CBTACN. As illustrated in Fig. 11.5, Cl^- ions prefer to locate in between two channel intersections rather than inside the intersections. This was interpreted as a consequence to maximize the dispersive interactions with surrounding atoms. In the same channel, two Cl^- ions are separated by at least 6 Å. The sites occupied by Br^- ions are very close to Cl^-, but slightly shifted away from the framework surface due to the larger size of Br^-. In contrast, the smaller F^- ions are located in a very narrow region near Zn atoms, and they are much less mobile [33].

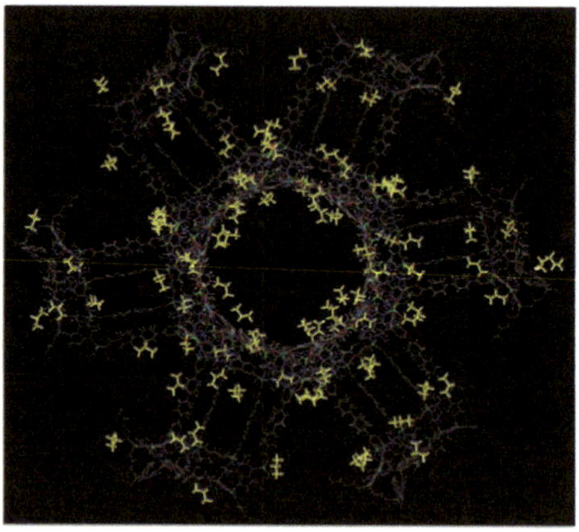

Figure 11.4 Locations of DMA cations (yellow) in bio-MOF-100. Color code: Zn, green; C, grey; O, red; N, blue; and H, white [32].

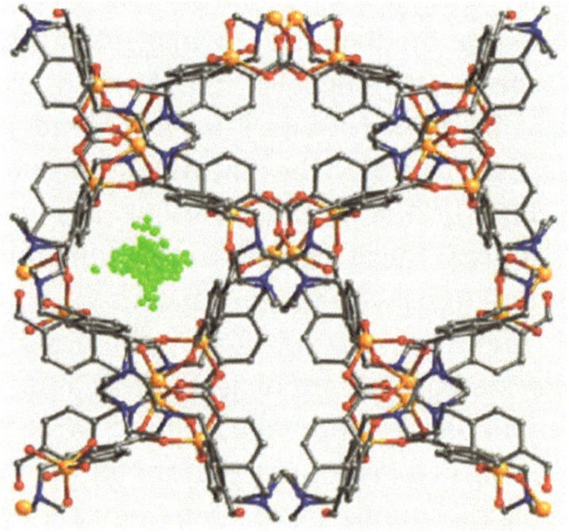

Figure 11.5 Distribution of Cl^- ions along the (100) channel in Zn-CBTACN. Reprinted with permission from Ref. [33]. Copyright 2012 American Chemical Society.

11.2.2 *Gas Adsorption and Separation*

Numerous simulation studies have been reported for gas adsorption and separation in MOFs [34–39]. Nevertheless, the majority of these simulations were focused on nonionic MOFs, and only a small number examined ionic MOFs.

11.2.2.1 Adsorption of pure gases

With a rare *soc* topology, the *soc*-MOF consists of a cationic framework and nonframework NO_3^- ions. In the structure, there exist carcerand-like capsules and narrow channels (approximately 1 nm in diameter). H_2 adsorption in the *soc*-MOF was measured by Eddaoudi and coworkers to be 2.61% at 78 K and 1.2 atm, and this high capacity was attributed to the narrow channels and the strong interactions under localized charge densities [7]. Following the experiment, they reported a simulation study to examine H_2 adsorption in the *soc*-MOF. The role of polarization was found to be a dominant contribution to H_2 adsorption. As demonstrated in Fig. 11.6, the distribution of H_2 molecular dipoles is bimodal. The low dipoles correspond to the regions near open metal sites, and

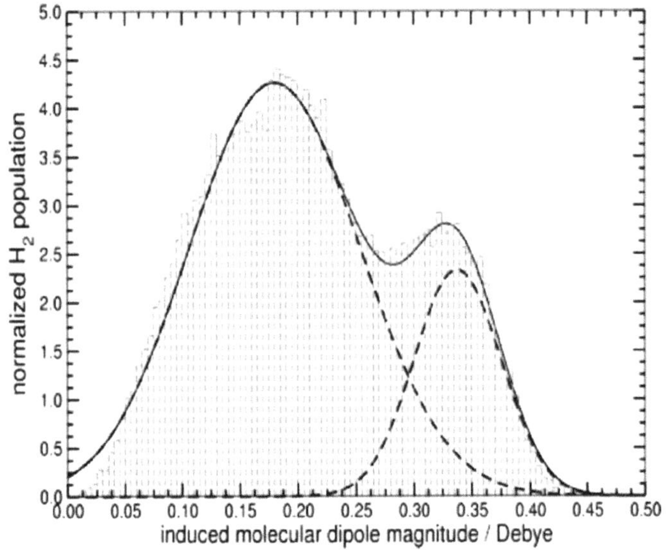

Figure 11.6 Bimodal distribution of H_2 molecular dipoles in *soc*-MOF. Reprinted with permission from Ref. [40]. Copyright 2007 American Chemical Society.

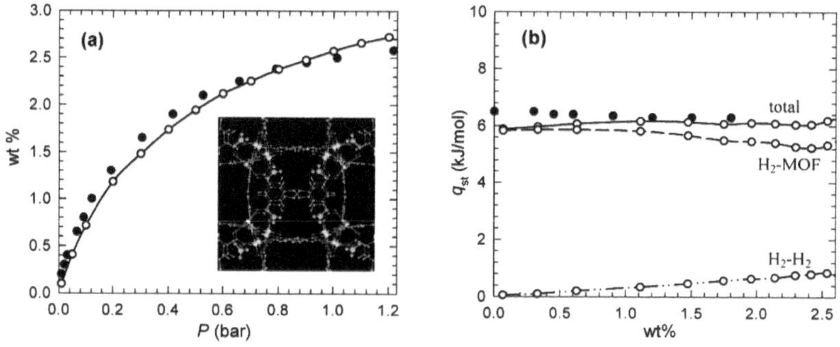

Figure 11.7 H_2 adsorption in *soc*-MOF (a) isotherm and (b) isosteric heat. Filled and open circles are experimental and simulation results, respectively. From Ref. [45]. Copyright 2009 American Institute of Chemical Engineers.

the high dipoles are associated with organic linkers. The isosteric heats were predicted to be 14.354 kJ/mol and 10.645 kJ/mol at unsaturated and saturated states, respectively. Nevertheless, these values are substantially higher than the experimentally measured 6.5 kJ/mol [40].

Jiang further investigated H_2 adsorption, as well as the separation of CO_2/H_2 mixture, in the *soc*-MOF. At low pressure, adsorption occurs concurrently at multiple sites near exposed indium atoms and organic components. At high pressure, the capsules are accessible through surrounding channels. As shown in Fig. 11.7, the simulated isotherm and isosteric heat q_{st} of H_2 adsorption agree fairly well with experimental data. The magnitude of q_{st} is comparable to the case in other ion-exchanged MOFs and zeolites [41–44]. At infinite dilution, the simulated q_{st} is about 6.0 kJ/mol and slightly lower than the measured value (6.5 kJ/mol). With increasing loading, q_{st} remains nearly a constant. The contribution from H_2–MOF interaction to q_{st} drops slightly with loading. This is because adsorption occurs initially at the most favorable sites and then in less favorable sites; consequently, the H_2–MOF interaction becomes weaker when more H_2 is adsorbed. Nevertheless, the contribution from H_2–H_2 interaction rises with loading due to the increasingly greater cooperative attractions between adsorbed H_2 molecules. Overall, q_{st} is insignificantly affected by loading, which implies the relatively homogenous feature of adsorption sites in the *soc*-MOF [45].

In another cationic MOF, namely, *rht*-MOF that is also charge-balanced by NO_3^- ions, Babarao et al. simulated the adsorption of H_2 and CO_2. The framework consists of three types of unique open cages: rhombicuboctahedral, tetrahedral, and cuboctahedral cages. At low pressure, gas is preferentially adsorbed in the vicinity of NO_3^- ions. With increasing pressure, gas enters the tetrahedral and cuboctahedral cages. No measured data of CO_2 adsorption was available for comparison; however, the predicted H_2 isotherm was found to be consistent with the experiment. The H_2 capacity was estimated to be 2.4 wt% at 1 bar and 6.2 wt% at 50 bar, among the highest in reported MOFs [46]. Recently, Space and coworkers also conducted a simulation study for H_2 adsorption in the *rht*-MOF. The NO_3^- ions were identified as the initial adsorption sites, corresponding to the high initial isosteric heat. Good agreement with the experiment was achieved for the simulated isotherm and isosteric heat. The quantum rotational levels of H_2 were calculated and compared with experimental inelastic neutron-scattering data [47].

Chen et al. systematically examined CO_2 adsorption in *rho*-ZMOFs exchanged with a series of cations (Na^+, K^+, Rb^+, Cs^+, Mg^{2+}, Ca^{2+}, and Al^{3+}). At infinite dilution, the isosteric heat Q_{st}^{o} and Henry's constant K_H are much larger than those in neutral MOFs because of strong electrostatic interaction between CO_2 and the cation/ionic framework. In addition, Q_{st}^{o} and K_H vary drastically with different cations. For monovalent cations (Na^+, K^+, Rb^+, and Cs^+), Q_{st}^{o} and K_H decrease in the order of $Na^+ > K^+ > Rb^+ > Cs^+$, that is, with increasing cation diameter. For Mg^{2+} and Ca^{2+}, Q_{st}^{o} and K_H are significantly larger than in monovalent counterparts. Among the seven *rho*-ZMOFs, Al-*rho*-ZMOF exhibits the highest Q_{st}^{o} and K_H. A similar trend was previously found for the enthalpy of CO_2 adsorption in alkali cation-exchanged zeolite-Y [48]. If the electric field generated is quantified by the charge-to-diameter ratio of the cation, as shown in Fig. 11.8, Q_{st}^{o} and K_H increase monotonically with increasing charge-to-diameter ratio [49].

To examine the favorable sites for CO_2 adsorption in Na-*rho*-ZMOF, Chen et al. analyzed the density distributions of CO_2 at 10 kPa, 100 kPa, and 1000 kPa. As indicated by Fig. 11.9, CO_2 molecules at 10 kPa are located proximately to the Na^+ ions and in six-member rings. Therefore, the Na^+ ions are the preferential site for CO_2 adsorption

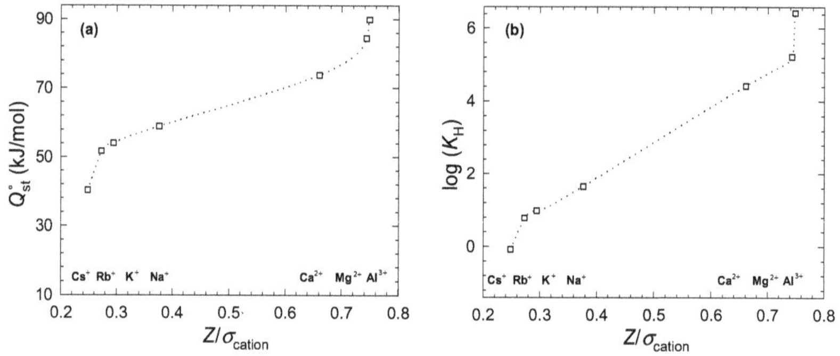

Figure 11.8 (a) Isosteric heat and (b) Henry's constant of CO_2 adsorption in *rho*-ZMOFs versus the charge-to-diameter ratio of cation. Reprinted with permission from Ref. [49]. Copyright 2012 American Chemical Society.

due to strong electrostatic interactions. With increasing pressure, more CO_2 molecules are located in the α-cage and away from the Na^+ ions; thus further adsorbed CO_2 molecules have relatively weaker interactions with the Na^+ ions. The distance between CO_2 molecules becomes shorter with increasing pressure, suggesting an enhancement in cooperative interactions between CO_2 molecules. Moreover, the locations of Na^+ ions are slightly shifted upon CO_2 adsorption, as observed in Fig. 11.9 [49].

Toni et al. conducted simulation and experimental measurement for CO_2 adsorption in Cl-Zn-CBTACN. Using a commonly used force field to describe CO_2 molecules, the agreement between simulation

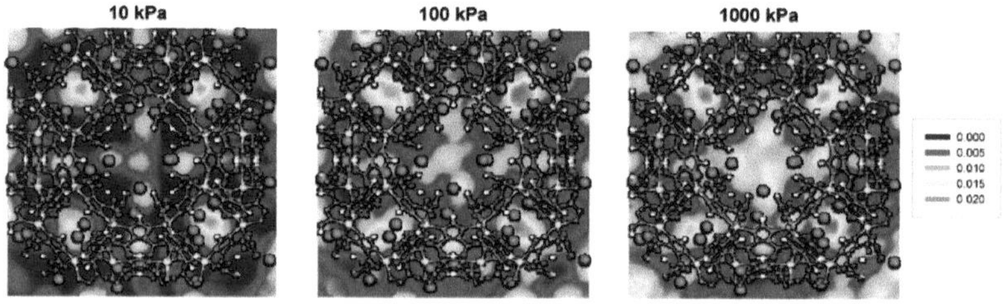

Figure 11.9 Density distributions of CO_2 adsorption in Na-*rho*-ZMOF at 10 kPa, 100 kPa, and 1000 kPa. The locations of Na^+ ions are indicted by the large spheres. The density scale is the number of CO_2 molecules per Å^3. Reprinted with permission from Ref. [49]. Copyright 2012 American Chemical Society.

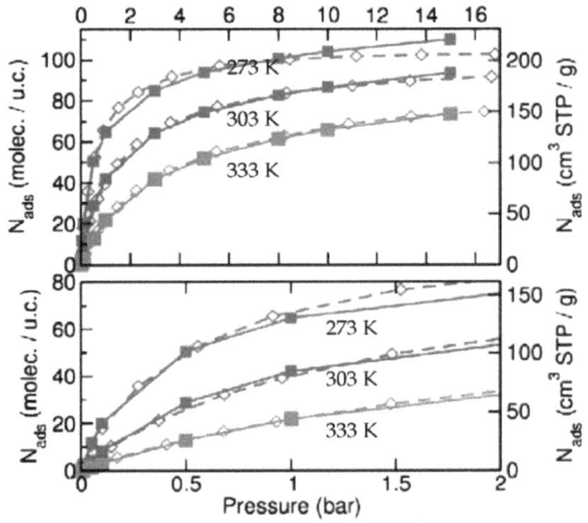

Figure 11.10 CO_2 adsorption isotherms in Cl-Zn-CBTACN at 273 K, 303 K, and 333 K. Simulation and experimental data are donated by the solid and dashed lines, respectively. Reprinted with permission from Ref. [33]. Copyright 2012 American Chemical Society.

and experiment was unsatisfactory. By turning off electrostatic interactions and adjusting dispersive potential parameters, the simulation could well reproduce measured data, as shown in Fig. 11.10, at three temperatures (273 K, 303 K, and 333 K) and both low- and high-pressure ranges. The isosteric heat calculated was approximately 24 kJ/mol, close to the measured value (21.5 kJ/mol). This study highlights that the force field optimized in the bulk phase may not be appropriate to mimic the adsorption of polar adsorbates in ionic MOFs [33].

11.2.2.2 Separation of CO_2-containing mixtures

Babarao and Jiang presented a simulation study on the separation of CO_2/CH_4, CO_2/N_2, and CO_2/H_2 mixtures in Na-*rho*-ZMOF. CO_2 is predominantly adsorbed over other small gases due to the strong electrostatic interactions with the Na^+ ions and ionic framework. The inset of Fig. 11.11 shows typical locations of CO_2 molecules in the S8MR for a CO_2/CH_4 equimolar mixture at 500 kPa. CO_2 molecules are observed to bind preferentially with Na^+ ions. The

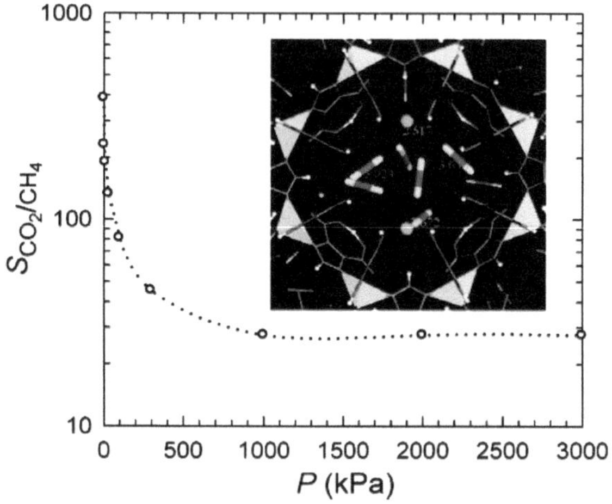

Figure 11.11 CO_2/CH_4 selectivity in Na-*rho*-ZMOF. Inset: Locations of CO_2 molecules in the S8MR. Reprinted with permission from Ref. [30]. Copyright 2009 American Chemical Society.

predicted CO_2/CH_4 selectivity at infinite dilution is 3800 and reduces to 80 at 1 atm, which is significantly higher than in nonionic MOFs. With increasing pressure, selectivity decreases monotonically as a consequence of reduction in electrostatic interaction between CO_2 and the Na-*rho*-ZMOF. Furthermore, it was found that selectivity decreases substantially by artificially switching off the charges in the Na-*rho*-ZMOF. This distinctly implies that electrostatic interaction is crucial for high selectivity in the Na-*rho*-ZMOF [30].

Adsorptive separation is usually conducted in a fixed bed packed with an adsorbent [50]. The breakthrough curves from the fixed bed provide valuable information, for example, the adsorption-regeneration cycle time, for the rational choice of the adsorbent as well as for the design and scale-up of the fixed bed. Using the adsorption isotherms of pure CO_2 and N_2 calculated from MC simulations, Nalaparaju and Jiang predicted the breakthrough curves of CO_2 and N_2 through a fixed bed packed with the Na-*rho*-ZMOF. The inlet CO_2/N_2 mixture (0.15:0.85) is at 298 K and 100 kPa, and the time is nondimensionalized as $t/(L_b/v_0)$. As illustrated in Fig. 11.12, three distinct regimes are seen in the breakthrough curves. First, nearly pure N_2 exits from the bed,

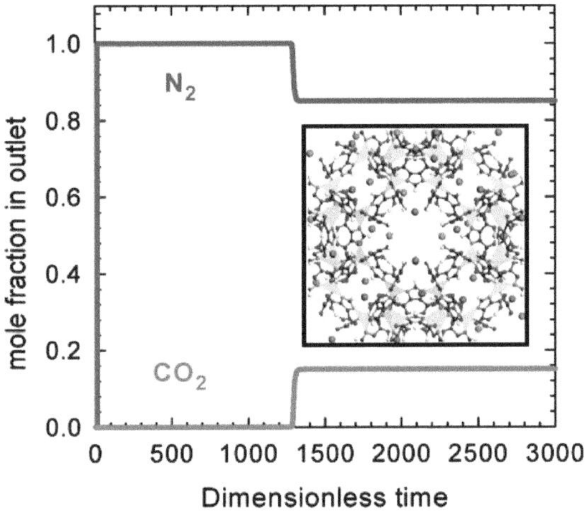

Figure 11.12 Breakthrough curves of CO_2 and N_2 through a fixed bed packed with Na-*rho*-ZMOF [51].

whereas CO_2 is adsorbed. After a certain time, CO_2 breakthrough occurs and gradually reaches its feed composition; meanwhile, N_2 concentration at the outlet drops to its feed composition. To quantity, the breakthrough time τ is defined as the time when the outlet composition is 10^{-4}. With this definition, the breakthrough times for N_2 and CO_2 in the Na-*rho*-ZMOF are 12 and 1278, respectively. Such a large difference in the breakthrough time suggests the effectiveness of the Na-*rho*-ZMOF for the separation of CO_2/N_2 mixture [51].

Babarao and Jiang also investigated the separation of CO_2/N_2 mixture in Li-MOF. Figure 11.13 shows the distributions of CO_2 in the dehydrated Li-MOF at 10 kPa, 100 kPa, and 1000 kPa. At 10 kPa, CO_2 is adsorbed proximally to the cations and metal clusters due to strong electrostatic interactions. Nevertheless, these regions are small and saturated rapidly. Therefore, CO_2 starts to occupy the square channels when pressure rises to 100 kPa. At a high pressure of 1000 kPa, adsorption occurs mostly in the square channels. Over the entire pressure range, CO_2 is overwhelmingly adsorbed over N_2. Under ambient conditions, the selectivity is approximately 60 for a CO_2/N_2 mixture, higher than that in nonionic MOFs and other nanoporous adsorbents. The charges of cations and

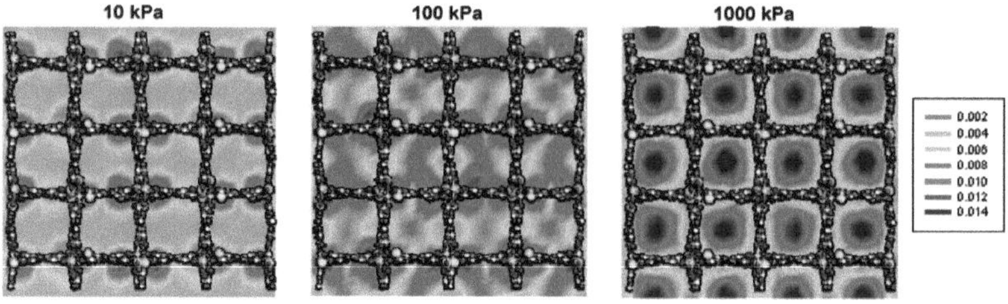

Figure 11.13 Distributions of CO_2 in Li-MOF at 10 kPa, 100 kPa, and 1000 kPa. Reprinted with permission from Ref. [31]. Copyright 2011 American Chemical Society.

the framework have a significant effect on selectivity. By switching off the charges, the selectivity was found to decrease by 1 order of magnitude. Furthermore, the adsorption capacity and selectivity in the dehydrated Li-MOF are slightly decreased upon hydration because of reduced free volume [31].

It is interesting to compare the separation performance in different ionic MOFs. For CO_2/H_2 mixtures in three ionic MOFs (Na-*rho*-ZMOF, *soc*-MOF, and *rht*-MOF), selectivity behaves differently [30, 45, 46]. In the Na-*rho*-ZMOF, selectivity decreases monotonically with increasing pressure, as a consequence of significant reduction in the electrostatic interactions between CO_2 and the Na-*rho*-ZMOF. In the *soc*-MOF, selectivity increases with pressure, reaches a maximum, and finally decreases. The increase is caused by the strong interactions of CO_2 molecules with multiple binding sites and by the cooperative interactions of CO_2 molecules; the decrease is attributed to the entropy effect. In the *rht*-MOF, selectivity decreases slightly, then increases, and finally tends to decrease. This is attributed to the counterbalance between energetic and entropy effects. Furthermore, the fractional free volume increases in the order of Na-*rho*-ZMOF <*soc*-MOF <*rht*-MOF, but the charge density increases in the opposite order. Therefore, the Na-*rho*-ZMOF has the smallest porosity and the largest charge density, thus exhibits the highest selectivity, followed by the *soc*-MOF and *rht*-MOF.

A few other simulation studies also reported the separation of CO_2/CH_4 mixture in ionic MOFs. Toni et al. examined the performance of Zn-CBTACN with three different anions: Cl^-, Br^-,

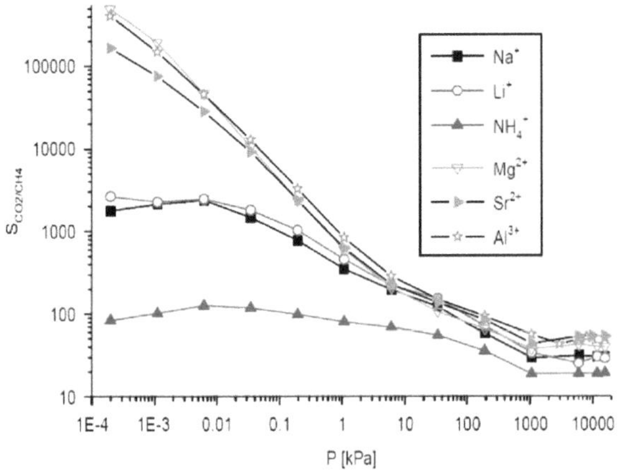

Figure 11.14 CO_2/CH_4 selectivities in *rho*-ZMOFs. Reprinted with permission from Ref. [52]. Copyright 2012 Taylor & Francis.

and F^-. At 303 K and 70 bar, the CO_2/CH_4 selectivities were predicted to be 9.2, 10.2, and 12.3 for Cl^-, Br^-, and F^-, respectively. The increase in selectivity was found to be independent of the force field used [33]. Among cation-exchanged (Li^+, Na^+, NH_4^+, Mg^{2+}, Sr^{2+}, and Al^{3+}) *rho*-ZMOFs, He et al. predicted that the Al-*rho*-ZMOF exhibits the highest CO_2/CH_4 selectivity, up to 112 at 298 K and 1 atm. As shown in Fig. 11.14, the selectivities drop monotonically with increasing pressure. From the predictions of ideal adsorbed solution theory (IAST), they found that the IAST overestimates CO_2 loading but underestimates CH_4 loading. These deviations were attributed to the increased extent of adsorbent heterogeneity induced by nonframework ions [52]. Demir and Ahunbay considered seven *sod*-ZMOFs with various nonframework ions (Li^+, Na^+, and K^+) and different ion exchange ratios. The adsorption sites for CO_2 and CH_4 in *sod*-ZMOFs are identical, but CO_2 is more strongly adsorbed in the mixture due to favorable electrostatic interaction [53].

It is noteworthy that moisture usually exists in adsorption processes, and thus capacity and selectivity could be affected. This has been, indeed, observed in zeolitic materials. Henry's constant of CO_2 in zeolite-X was found to decline exponentially with H_2O loading [54]. With a small amount of H_2O, the adsorption capacities of N_2, O_2, and Ar in Li-LSX zeolite dropped significantly [55]. In other cases, however, the presence of H_2O is beneficial. The selectivity of *p*-xylene/*m*-xylene in BaX increased with pre-adsorbed H_2O [56].

From a series of simulation studies, Jiang and coworkers systematically examined moisture effects on the adsorption and separation of CO_2 in ionic MOFs. Depending on the nature of the framework, different effects were found. For a CO_2/CH_4 mixture in the Na-*rho*-ZMOF, CO_2 adsorption drops significantly, even with only 0.1% of H_2O, while CH_4 adsorption is not discernibly influenced [57]. This is because H_2O is highly polar and interacts with Na^+ ions far more strongly than CO_2; thus H_2O is preferentially adsorbed proximal to Na^+ ions. As a result, CO_2/CH_4 selectivity is reduced approximately by 1 order of magnitude upon addition of H_2O. A similar significant effect on the separation of a CO_2/H_2 mixture was observed in the *rht*-MOF [46]. Nevertheless, the *soc*-MOF exhibits different behavior from the Na-*rho*-MOF and *rht*-ZMOF [45]. The nonframework NO_3^- ions in the *soc*-MOF are located in carcerand-like capsules. In the presence of a small amount H_2O, CO_2/H_2 selectivity in the *soc*-MOF increases at low pressures. The reason is that H_2O interacts preferentially with the readily accessible indium atoms rather than NO_3^- ions constrained in the capsules, and thus the bound H_2O molecules act as additional sites for CO_2 adsorption. With increasing H_2O at high pressures, however, H_2O competitively replaces CO_2 and selectivity decreases. It should be noted that moisture effects were predicted to be marginal for CO_2 capture in nonionic MOFs such as Matériaux de l'Institut Lavoisier (MIL)-101 [58], Zn(bdc)(ted)$_{0.5}$ [59], and bio-MOF-11 [60].

11.2.3 *Water Treatment*

With an ever-increasing world population and economic development, a large amount of contaminants such as metal ions, Cu^{2+}, Cd^{2+}, and Pb^{2+}, have been introduced into water [61]. These ions can accumulate in living organisms and lead to dysfunction in nervous, circulatory, and immune systems [62]. It is, therefore, important to remove them from water, thus minimizing health and environmental risks.

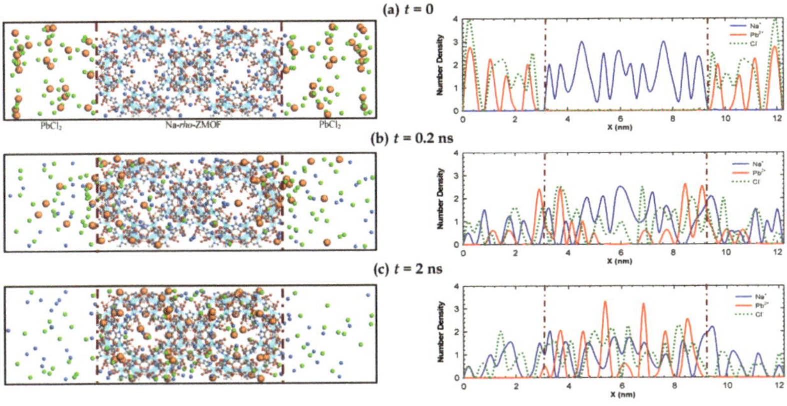

Figure 11.15 Removal of Pb^{2+} by ion exchange in Na-*rho*-ZMOF. (a) $t = 0$, (b) $t = 0.2$, and (c) $t = 2$ ns. Color code: Pb^{2+}, orange; Cl^{-}, green; and Na^{+}, blue. Left: Simulation snapshots. Right: Density profiles. Reprinted with permission from Ref. [63]. Copyright 2012 American Chemical Society.

Nalaparaju and Jiang conducted a simulation study for the removal of Pb^{2+} ions by exchange with Na^{+} ions in the Na-*rho*-ZMOF. Figure 11.15 shows the simulation system at different time durations. Initially, Pb^{2+} and Cl^{-} ions reside in solution and Na^{+} ions are in the framework. Upon starting simulation, Pb^{2+} ions rapidly move into the framework, while Na^{+} ions are exchanged out. At $t = 0.2$ ns, most Pb^{2+} ions have moved into the framework, particularly near the solution/framework interface. Meanwhile, Na^{+} ions move out and stay in the solution. A few Cl^{-} ions also move into the framework during ion exchange. Once exchanged, Pb^{2+} ions prefer staying in the framework without moving back into the solution. At $t = 2$ ns, all Pb^{2+} ions are exchanged and reside in the framework. The density profiles of Pb^{2+}, Na^{+}, and Cl^{-} ions at three stages are also depicted in Fig. 11.15. At $t = 0$, Na^{+} ions are in the framework, whereas Pb^{2+} and Cl^{-} ions are in the solution. Na^{+} ions in the framework exhibit several distinct peaks due to preferential locations at different sites. At $t = 0.2$ ns after ion exchange occurs, Na^{+} ions move out from the framework into the solution; thus the density decreases in the framework but increases in the solution. The situation for Pb^{2+} and Cl^{-} ions is opposite, and their densities

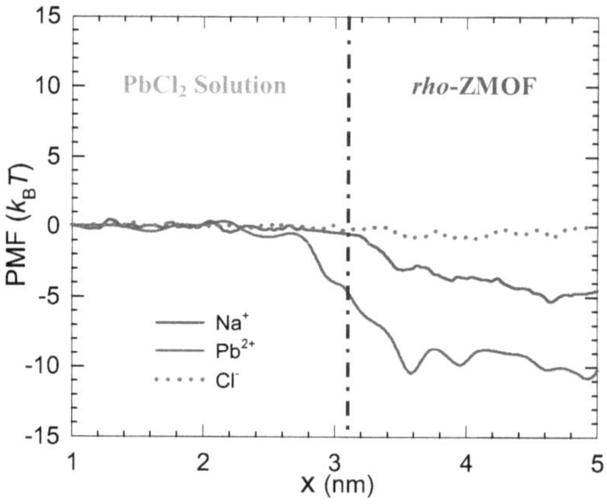

Figure 11.16 Potentials of mean force for Na^+, Pb^{2+}, and Cl^- ions moving from the solution into the *rho*-ZMOF framework. Reprinted with permission from Ref. [63]. Copyright 2012 American Chemical Society.

in the framework increase from zero. These variations are observed till dynamic equilibrium is reached at approximately 2 ns [63].

To provide fundamental insight into the observed ion exchange, the potentials of mean force (PMFs) were estimated from the simulation. As shown in Fig. 11.16, the PMFs for Pb^{2+} and Na^+ ions decrease upon moving toward the *rho*-ZMOF framework due to the attractive interactions exerted by the anionic framework. In the interior of the framework, Pb^{2+} has a larger PMF ($-10k_BT$) than Na^+ ($-5k_BT$). Thus, Pb^{2+} interacts more strongly than Na^+ with the framework, which is the driving force for the exchange between Pb^{2+} and Na^+. On the other hand, the PMFs in the framework can be considered as an energy barrier for ions to move from the framework to the solution. With a large PMF and hence a large barrier, Pb^{2+} ions cannot move back to the solution once exchanged. However, Na^+ ions in the framework and the solution exist in dynamic equilibrium and can be exchange due to the smaller PMF. There is no significant free-energy gain for Cl^- ions moving into the framework. Furthermore, from residence-time distributions and mean-squared displacements, all the exchanged Pb^{2+} ions were found to continuously stay in the framework. Nevertheless, Na^+ ions possess a shorter residence time and a larger mobility. The exchanged Pb^{2+} ions in the framework are located at the eight-, six-, and four-member rings and exhibit different dynamics. A quantitative understanding is provided by this simulation study for the mechanism of ion exchange. It is instructive to note that Pb^{2+} ions are exchanged with Na^+ ions in the Na-*rho*-ZMOF. If a cationic MOF is used, then toxic anions (e.g., F^-) could be removed [63].

11.2.4 *Biofuel Purification*

Biofuel is environmentally benign and carbon neutral with less emission of gaseous pollutants compared to fossil fuels. Fermented biofuel contains water along with alcohol (e.g., ethanol). Thus, it is crucial to efficiently separate water/alcohol mixtures to produce fuel-grade biofuel [64]. While a handful of experimental studies have been reported on biofuel purification in MOFs, simulation exploration is scarce.

As the first step, Nalaparaju et al. simulated the adsorption of water in the Na-*rho*-ZMOF. At low pressure, water is preferentially adsorbed around Na^+ ions. Interestingly, water adsorbs preferentially near Na^+ ions at site II rather than at site I. This is attributed to the steric effect as the void space near site II is larger than at site I. With increasing pressure, adsorption occurs near the framework surface. At high pressure, the α-cage is gradually filled by water and Na^+ ions appear to be solvated. From low to high pressure, the locations of Na^+ ions are shifted upon water adsorption. Moreover, water dynamics in the Na-*rho*-ZMOF were also analyzed. As revealed by the vibrational spectra in Fig. 11.17, water in the Na-*rho*-ZMOF exhibits distinct bands of librational motion, bending, and stretching. Due to confinement, the librational motion shows a frequency higher than that of bulk water. When loading increases, there is a blue shift in the bending frequency. As a consequence of strong water–ion interaction, symmetric and asymmetric modes are observed in the stretching. For all three bands, the spectral intensity and width exhibit a similar trend with increasing pressure. At low pressure, the intensity is large because water is adsorbed onto individual Na^+ ions and scattered. Thus, the distance of the H-bond is long, the vibration has great amplitude, and the width is narrow due to the limited number of H-bonds. At high pressure,

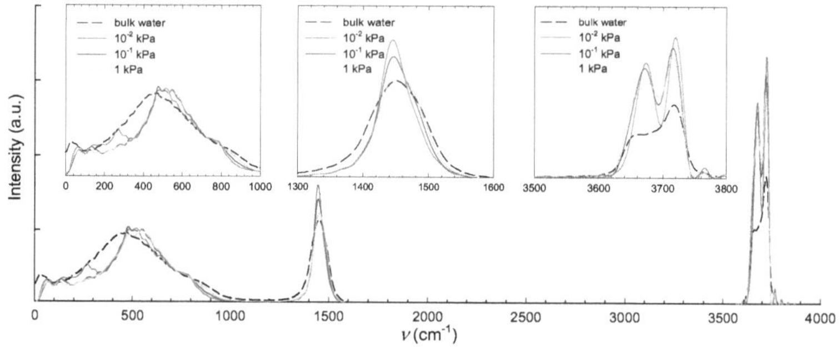

Figure 11.17 Vibrational spectra of water in Na-*rho*-ZMOF. Reprinted with permission from Ref. [65]. Copyright 2009 American Chemical Society.

water stays closely to one another and the number of H-bonds increases; consequently, the spectra have smaller amplitude and wider bandwidth. This study provides a fundamental microscopic insight into the static and dynamic properties of water in the Na-*rho*-MOF [65].

The adsorption of ethanol in the Na-*rho*-ZMOF was then examined and compared to water. As shown in Fig. 11.18, the isotherms of both ethanol and water are of type I. At low pressure, the difference between ethanol and water is small but becomes pronounced at high pressure. With increasing pressure, water adsorption increases by a much larger extent than ethanol adsorption. In addition, water exhibits a higher saturated capacity. The reason is that the substitution of hydrogen in water by the ethyl group leads to a less polar and bigger molecule. Thus, water is more strongly adsorbed than ethanol in the Na-*rho*-ZMOF [66].

With a fairly good understanding of the adsorption of pure water and ethanol, Nalaparaju et al. simulated biofuel purification in the Na-*rho*-ZMOF. The biofuel was mimicked as liquid mixtures of ethanol and water, and both pervaporation (PV) and vapor permeation (VP) were considered. Water competitively adsorbs over ethanol at all the feed compositions. This is due to the stronger interaction of highly polar water with Na^+ ions and the ionic framework. Particularly, water is preferentially adsorbed near six-member rings; however, ethanol is primarily localized in open α-cages. The adsorption selectivity of water/ethanol is high at a

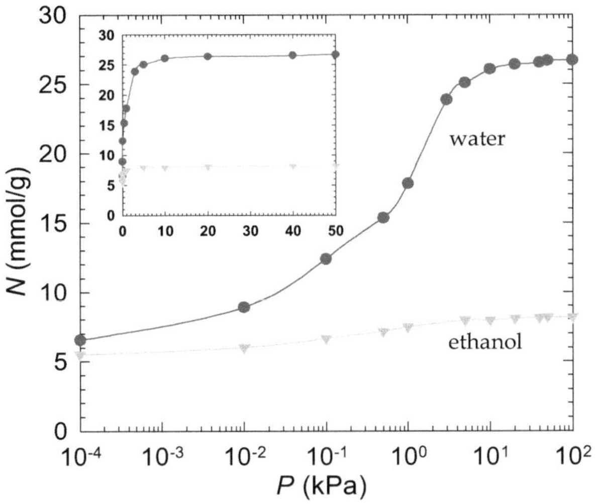

Figure 11.18 Adsorption isotherms of ethanol and water in Na-*rho*-ZMOF. Inset: Isotherms in the linear scale of pressure. Reprinted with permission from Ref. [66]. Copyright 2010 American Chemical Society.

low composition of water. With increasing water composition, the adsorption selectivity drops, whereas the diffusion selectivity rises. Figure 11.19 shows the permselectivity of water/ethanol mixtures in the Na-*rho*-ZMOF. Apparently, the Na-*rho*-ZMOF is preferable for biofuel dehydration when water composition in the feed solution is low and VP seems to be more efficient than PV. This simulation study provides useful microscopic insight into biofuel separation in ionic MOFs [67].

11.2.5 *Drug Loading*

Efficient drug delivery is critical for therapeutic treatment. While conventional materials (e.g., inorganic silica and polymeric matrices) have been tested as drug carriers, their loading capacities are usually not sufficiently high, and encapsulated drugs are difficult to be specifically released [68]. In this context, porous MOFs with readily tunable and well-defined pore structures are considered as promising candidates for drug delivery.

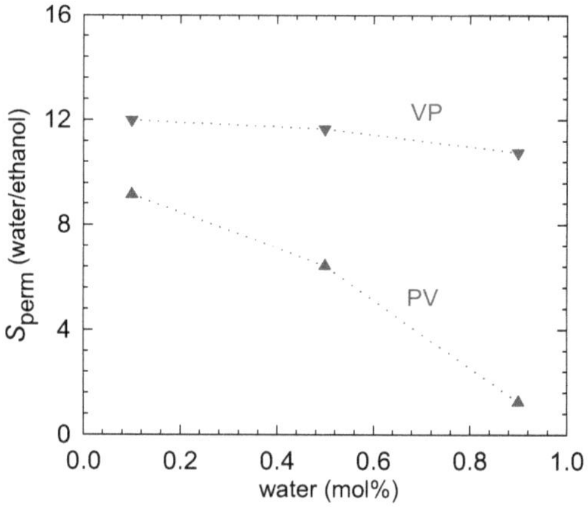

Figure 11.19 Permselectivity of water/ethanol mixtures in a Na-*rho*-ZMOF [67]. Reprinted with permission from Ref. [67], copyright 2011 Royal Society of Chemistry.

Several experimental groups reported that high loading and controlled release could be achieved in some MOFs [69]. Currently, only a few computational studies have been conducted on drug delivery in MOFs, including MIL-53 [70], MIL-101 and UMCM-1 [71], MIL-100, CDMOF-1, and bio-MOF-100 [72]. Among these, bio-MOF-100 is anionic with charge-balanced DMA cations and contains exclusively mesopores. Bernini et al. predicted that the loading of ibuprofen in the bio-MOF-100 could reach 1975 mg/g, which is remarkably higher than in silicas and zeolites. As shown by the simulation snapshots in Fig. 11.20, ibuprofen gradually forms a monolayer on the framework surface of the bio-MOF-100 at low loading; then a multilayer is formed at intermediate loading; finally, condensation occurs filling the mesopores. Structural analysis suggests that ibuprofen interacts strongly with DMA cations and the oxygen atoms of the bio-MOF-100 framework [72].

11.3 Summary

In contrast to most experimental and simulation studies focusing on nonionic MOFs, we presented an overview of the recent simulation

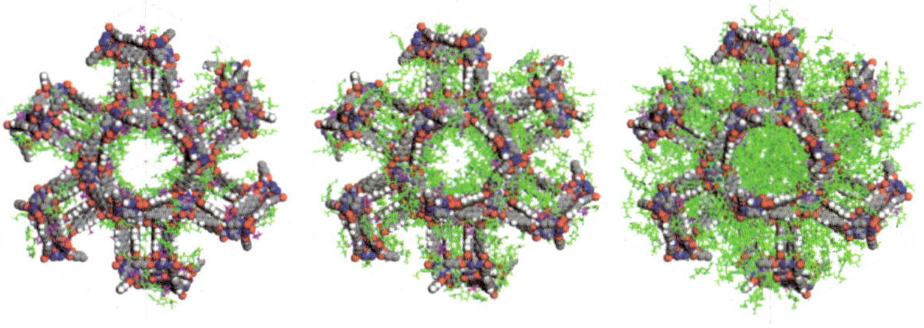

Figure 11.20 Ibuprofen in bio-MOF-100 at different loadings: 329 mg/g, 1095 mg/g, and 1975 mg/g. Ibuprofen molecules are in green, and DMA cations are in pink stick mode. Reprinted with permission from Ref. [72]. Copyright 2014 Royal Society of Chemistry.

studies on ionic MOFs. Notable progress has been achieved towards a quantitative understanding of ionic MOFs, including locations of nonframework ions, potential applications for gas adsorption and separation, water treatment, biofuel purification, and drug loading. The atomic resolution and time-resolved insights provided by simulations are extremely crucial to unravel fundamental mechanisms, complement experimental endeavors, and facilitate new developments in this burgeoning field.

Acknowledgments

The authors are grateful to the National University of Singapore, the Ministry of Education of Singapore, and the Singapore National Research Foundation for financial support.

References

1. Long, J. R., Yaghi, O. M. (2009). The pervasive chemistry of metal-organic frameworks, *Chem. Soc. Rev.*, **38**, 1213–1214.
2. Zhou, H. C., Long, J. R., Yaghi, O. M. (2012). Introduction to metal-organic frameworks, *Chem. Rev.*, **112**, 673–674.
3. Férey, G. (2008). Hybrid porous solids: past, present, future, *Chem. Soc. Rev.*, **37**, 191–214.

4. Stock, N., Biswas, S. (2012). Synthesis of metal-organic frameworks: routes to various MOF topologies, morphologies, and composites, *Chem. Rev.*, **112**, 933–969.
5. Cohen, S. M. (2012). Postsynthetic methods for the functionalization of metal-organic frameworks, *Chem. Rev.*, **112**, 970–1000.
6. Liu, Y. L., Kravtsov, V. C., Larsen, R., Eddaoudi, M. (2006). Molecular building blocks approach to the assembly of zeolite-like metal-organic frameworks with extra-large cavities, *Chem. Commun.*, **42**, 1488–1490.
7. Liu, Y. L., Eubank, J. F., Cairns, A. J., Eckert, J., Kravtsov, V. C., Luebke, R., Eddaoudi, M. (2007). Assembly of metal–organic frameworks based on indium-trimer building blocks: a porous mof with *soc* topology and high hydrogen storage, *Angew. Chem., Int. Ed.*, **46**, 3278–3283.
8. Moellmer, J., Celer, E. B., Luebke, R., Cairns, A. J., Staudt, R., Eddaoudi, M., Thommes, M. (2010). Insights on adsorption characterization of metal-organic frameworks: a benchmark study on the novel *soc*-MOF, *Microporous Mesoporous Mater.*, **129**, 345–353.
9. Liu, Y. L., Kravtsov, V. C., Eddaoudi, M. (2008). Template-directed assembly of zeolite-like metal–organic frameworks: a *usf*-ZMOF with an unprecedented zeolite topology, *Angew. Chem., Int. Ed.*, **47**, 8446–8449.
10. Nouar, F., Eubank, J. F., Bousquet, T., Wojtas, L., Zaworotko, M. J., Eddaoudi, M. (2008). Supermolecular building blocks for the design and synthesis of highly porous metal-organic frameworks, *J. Am. Chem. Soc.*, **130**, 1833–1835.
11. Nouar, F., Eckert, J., Eubank, J. F., Forster, P., Eddaoudi, M. (2009). Zeolite-like metal-organic frameworks as hydrogen storage platform: lithium and magnesium ion-exchange and H_2-*rho*-ZMOF interaction studies, *J. Am. Chem. Soc.*, **131**, 2864–2875.
12. Zhang, Z., Zhang, L., Wojtas, L., Nugent, P., Eddaoudi, M., Zaworotko, M. J. (2012). Templated synthesis, postsynthetic metal exchange, and properties of a porphyrin-encapsulating metal-organic material, *J. Am. Chem. Soc.*, **134**, 924–927.
13. Alkordi, M. H., Liu, Y. L., Larsen, R. W., Eubank, J. F., Eddaoudi, M. (2008). Zeolite-like metal-organic frameworks as platforms for applications: on metalloporphyrin-based catalysts, *J. Am. Chem. Soc.*, **130**, 12639–12641.
14. Ananthoji, R., Eubank, J. F., Nouar, F., Mouttaki, H., Eddaoudi, M., Harmon, J. P. (2011). Symbiosis of zeolite-like metal–organic frameworks (*rho*-ZMOF) and hydrogels: composites for controlled drug release, *J. Mater. Chem.*, **21**, 9587–9594.

15. An, J., Geib, S. J., Rosi, N. L. (2009). Cation-triggered drug release from a porous zinc-adeninate metal-organic framework, *J. Am. Chem. Soc.*, **131**, 8376–8377.

16. An, J., Rosi, N. L. (2010). Tuning MOF CO_2 adsorption properties via cation exchange, *J. Am. Chem. Soc.*, **132**, 5578–5579.

17. Li, T., Rosi, N. L. (2013). Screening and evaluating aminated cationic functional moieties for potential CO_2 capture applications using an anionc MOF scaffold, *Chem. Commun.*, **49**, 11385–11387.

18. An, J., Shade, C. M., Chengelis-Czegan, D. A., Petoud, S., Rosi, N. L. (2011). Zinc-adeninate metal-organic framework for aqueous encapsulation and sensitization of near-infrared and visible emitting lanthanide cations, *J. Am. Chem. Soc.*, **133**, 1220–1223.

19. An, J., Farha, O. K., Hupp, J. T., Pohl, E., Yeh, J. I., Rosi, N. L. (2012). Metal-adeninate vertices for the construction of an exceptionally porous metal-organic framework, *Nat. Commun.*, **3**, 604.

20. Lu, W. G., Jiang, L., Feng, X. L., Lu, T. B. (2009). Three-dimensional lanthanide anionic metal-organic frameworks with tunable luminescent properties induced by cation exchange, *Inorg. Chem.*, **48**, 6997–6999.

21. Plabst, M., McCusker, L. B., Bein, T. (2009). Exceptional ion-exchange selectivity in a flexible open framework lanthanum tetrakisphosphonate, *J. Am. Chem. Soc.*, **131**, 18112–18118.

22. Ortiz, G., Brandes, S., Rousselin, Y., Guilard, R. (2011). Selective CO_2 adsorption by a triazacyclononane-bridged microporous metal–organic framework, *Chem. Eur. J.*, **17**, 6689–6695.

23. Zhai, Q. G., Lin, Q., Wu, T., Wang, L., Zheng, S. T., Bu, X., Feng, P. (2012). High CO_2 and H_2 uptake in an anionic porous framework with amino-decorated polyhedral cages, *Chem. Mater.*, **24**, 2624–2626.

24. Zhao, X., Bu, X., Wu, T., Zheng, S. T., Wang, L., Feng, P. (2013). Selective anion exchange with nanogated isoreticular positive metal-organic frameworks, *Nat. Commun.*, **4**, 2344.

25. Masters, A. F., Maschmeyer, T. (2011). Zeolites: from curiosity to cornerstone, *Microporous Mesoporous Materials*, **142**, 423–438.

26. Hill, J. R., Subramanian, L., Maiti, A. (2005). *Molecular Modeling Techniques in Material Sciences* (Taylor & Francis, London).

27. Smit, B., Maesen, T. L. M. (2008). Molecular simulations of zeolites: adsorption, diffusion, and shape selectivity, *Chem. Rev.*, **108**, 4125–4184.

28. Combariza, A. F., Gomez, D. A., Sastre, G. (2013). Simulating the properties of small pore silica zeolites using interatomic potentials, *Chem. Soc. Rev.*, **42**, 114–127.

29. Yang, S. H., Lin, X., Blake, A. J., Thomas, K. M., Hubberstey, P., Champnes, N. R., Schroder, M. (2008). Enhancement of H_2 adsorption in Li^+-exchanged co-ordination framework materials, *Chem. Commun.*, **46**, 6108–6110.
30. Babarao, R., Jiang, J. W. (2009). Unprecedentedly high selective adsorption of gas mixtures in rho zeolite-like metal-organic framework: a molecular simulation study, *J. Am. Chem. Soc.*, **131**, 11417–11425.
31. Babarao, R., Jiang, J. W. (2011). Cation characterization and CO_2 capture in Li^+-exchanged metal-organic frameworks: from first-principles modeling to molecular simulation, *Ind. Eng. Chem. Res.*, **50**, 62–68.
32. Chen, Y. F., Jiang, J. W. (unpublished).
33. Toni, M. D., Coudert, F.-X., Paranthaman, S., Pullumbi, P., Boutin, A., Fuchs, A. H. (2012). Molecular simulation of a Zn-triazamacrocyle metal-organic frameworks family with extraframework anions, *J. Phys. Chem. C*, **116**, 2952–2959.
34. Düren, T., Bae, Y. S., Snurr, R. Q. (2009). Using molecular simulation to characterise metal-organic frameworks for adsorption applications, *Chem. Soc. Rev.*, **38**, 1203–1212.
35. Liu, D. H., Zhong, C. L. (2010). Understanding gas separation in metal-organic frameworks, *J. Mater. Chem.*, **20**, 10308–10318.
36. Jiang, J. W., Babarao, R., Hu, Z. Q. (2011). Molecular simulations for energy, environmental and pharmaceutical applications of nanoporous materials: from zeolites, metal-organic frameworks to protein crystals, *Chem. Soc. Rev.*, **40**, 3599–3612.
37. Jiang, J. W. (2012). Metal-organic frameworks for CO_2 capture: what are learned from molecular simulations, in *Coordination Polymers and Metal Organic Frameworks*, eds. Ortiz, O. L., Ramírez, L. D. (Nova Science).
38. Getman, R. B., Bae, Y. S., Wilmer, C. E., Snurr, R. Q. (2012). Review and analysis of molecular simulations of methane, hydrogen and acetylene storage in metal-organic frameworks, *Chem. Rev.*, **112**, 703–723.
39. Yang, Q. Y., Liu, D. H., Zhong, C. L., Li, J. R. (2013). Development of computational methodologies for metal-organic frameworks and their application in gas separations, *Chem. Rev.*, **113**, 8261–8323.
40. Belof, J. L., Stern, A. C., Eddaoudi, M., Space, B. (2007). On the mechanism of hydrogen storage in a metal-organic framework material, *J. Am. Chem. Soc.*, **129**, 15202–15210.
41. Dinca, M., Dailly, A., Liu, Y., Brown, C. M., Neumann, D. A., Long, J. R. (2006). Hydrogen storage in a microporous metal-organic framework

with exposed Mn^{2+} coordination sites, *J. Am. Chem. Soc.*, **128**, 16876–16883.

42. Georgiev, P. A., Albinati, A., Eckert, J. (2007). Room temperature isosteric heat of dihydrogen adsorption on Cu(I) cations in zeolite ZSM-5, *Chem. Phys. Lett.*, **449**, 182–185.
43. Georgiev, P. A., Albinati, A., Mojet, B. L., Ollivier, J., Eckert, J. (2007). Observation of exceptionally strong binding of molecular hydrogen in a porous material: formation of an η^2-H_2 complex in a Cu-exchanged ZSM-5 zeolite, *J. Am. Chem. Soc.*, **129**, 8086–8087.
44. Dinca, M., Han, W. S., Liu, Y., Dailly, A., Brown, C. M., Long, J. R. (2007). Observation of Cu^{2+}-H_2 interactions in a fully desolvated sodalite-type metal-organic framework, *Angew. Chem., Int. Ed.*, **46**, 1419–1422.
45. Jiang, J. W. (2009). Charged *soc* metal-organic framework for high-efficacy H_2 adsorption and syngas purification: atomistic simulation study, *AIChE J.*, **55**, 2422–2432.
46. Babarao, R., Eddaoudi, M., Jiang, J. W. (2010). Highly porous ionic *rht* metal-organic framework for H_2 and CO_2 storage and separation: a molecular simulation study, *Langmuir*, **26**, 11196–11203.
47. Pham, T., Forrest, K. A., Hogan, A., McLaughlin, K., Belof, J. L., Eckert, J., Space, B. (2014). Simulations of hydrogen sorption in *rht*-MOF-1: identifying the binding sites through explicit polarization and quantum rotation calculations, *J. Mater. Chem. A*, **2**, 2088–2100.
48. Plant, D. F., Maurin, G., Deroche, I., Gaberova, L., Llewellyn, P. (2006). CO_2 adsorption in alkali cation exchanged-Y, *Chem. Phys. Lett.*, **426**, 387.
49. Chen, Y. F., Nalaparaju, A., Eddaoudi, M., Jiang, J. W. (2012). CO_2 adsorption in mono-, di- and trivalent cation-exchanged metal-organic frameworks: a molecular simulation study, *Langmuir*, **28**, 3903–3910.
50. Ruthven, D. M., Farooq, S., Knaebel, K. S. (1994). *Pressure Swing Adsorption* (Wiley-VCH, New York).
51. Nalaparaju, A., Jiang, J. W. (unpublished).
52. He, P., Liu, H., Li, Y., Lei, Z., Huang, S., Wang, P., Tian, H. (2012). Enhanced selectivity and capacity of adsorption of CO_2 and CH_4 in zeolite-like metal-organic frameworks with different extra-framework cations: a molecular simulation study, *Mol. Simul.*, **38**, 72–83.
53. Demir, B., Ahunbay, M. G. (2013). CO_2/CH_4 separation in ion-exchanged zeolite-like metal organic frameworks with sodalite topology (*sod*-ZMOFs), *J. Phys. Chem. C*, **117**, 15647–15658.
54. Brandani, F., Ruthven, D. M. (2004). The effect of water on the adsorption of CO_2 and C_3H_8 on type X zeolites, *Ind. Eng. Chem. Res.*, **43**, 8339–8344.

55. Hutson, N. D., Zajic, S. C., Yang, R. T. (2000). Influence of residual water on the adsorption of atmospheric gases in Li-X zeolite: experiment and simulation, *Ind. Eng. Chem. Res.*, **39**, 1775–1780.
56. Moise, J. C., Bellat, J. P. (2005). Effect of preadsorbed water on the adsorption of *p*-xylene and *m*-xylene mixtures on BaX and BaY zeolites, *J. Phys. Chem. B*, **109**, 17239–17244.
57. Babarao, R., Jiang, J. W. (2009). Upgrade of natural gas in *rho* zeolite-like metal-organic framework and effect of water: a computational study, *Energy Environ. Sci.*, **2**, 1088–1093.
58. Chen, Y. F., Babarao, R., Sandler, S. I., Jiang, J. W. (2010). Metal organic framework MIL-101 for adsorption and effect of terminal water molecules: from quantum mechanics to molecular simulation, *Langmuir*, **26**, 8743–8750.
59. Chen, Y. F., Lee, J. Y., Babarao, R., Li, J., Jiang, J. W. (2010). A highly hydrophobic metal organic framework Zn(bdc)(ted)$_{0.5}$ for adsorption and separation of CH_3OH/H_2O and CO_2/CH_4: an integrated experimental and simulation study, *J. Phys. Chem. C*, **114**, 6602–6609.
60. Chen, Y. F., Jiang, J. W. (2010). A bio-metal-organic framework for highly selective CO_2 capture: a molecular simulation study, *ChemSusChem*, **3**, 982–988.
61. Shannon, M. A., Bohn, P. W., Elimelech, M., Georgiadis, J. G., Marinas, B. J., Mayes, A. M. (2008). Science and technology for water purification in the coming decades, *Nature*, **452**, 301–310.
62. Patrick, L. (2006). Lead toxicity, a review of the literature. Part 1: exposure, evaluation, and treatment, *Alter. Med. Rev.*, **11**, 2–22.
63. Nalaparaju, A., Jiang, J. W. (2012). Ion exchange in metal-organic framework for water purification: insight from molecular simulation, *J. Phys. Chem. C*, **116**, 6925–6931.
64. Ragauskas, A. J., Williams, C. K., Davison, B. H., Britovsek, G., Cairney, J., Eckert, C. A., Frederick, W. J., Hallett, J. P., Leak, D. J., Liotta, C. L., Mielenz, J. R., Murphy, R., Templer, R., Tschaplinski, T. (2006). The path forward for biofuels and biomaterials, *Science*, **311**, 484–488.
65. Nalaparaju, A., Babarao, R., Zhao, X. S., Jiang, J. W. (2009). Atomistic insight into adsorption, mobility, and vibration of water in ion-exchanged zeolite-like metal-organic frameworks, *ACS Nano*, **3**, 2563–2572.
66. Nalaparaju, A., Zhao, X. S., Jiang, J. W. (2010). Molecular understanding for the adsorption of water and alcohols in hydrophilic and hydrophobic zeolitic metal-organic frameworks, *J. Phys. Chem. C*, **114**, 11542–11550.

67. Nalaparaju, A., Zhao, X. S., Jiang, J. W. (2011). Biofuel purification by pervaporation and vapor permeation in metal-organic frameworks: a computational study, *Energy Environ. Sci.*, **4**, 2107–2116.
68. Langer, R., Tirrell, D. A. (2004). Designing materials for biology and medicine, *Nature*, **428**, 487–492.
69. Horcajada, P., Gref, R., Baati, T., Allan, P. K., Maurin, G., Couvreur, P., Ferey, G., Serre, C. (2012). Metal-organic frameworks in biomedicine, *Chem. Rev.*, **112**, 1232–1268.
70. Horcajada, P., Serre, C., Maurin, G., Ramsahye, N. A., Balas, F., Vallet-Regi, M., Sebban, M., Taulelle, F., Ferey, G. (2008). Flexible porous metal-organic frameworks for a controlled drug delivery, *J. Am. Chem. Soc.*, **130**, 6774–6780.
71. Babarao, R., Jiang, J. W. (2009). Unraveling the energetics and dynamics of ibuprofen in mesoporous metal-organic frameworks, *J. Phys. Chem. C*, **113**, 18287–18291.
72. Bernini, M. C., Fairen-Jimenez, D., Pasinetti, M., Ramirez-Pastor, A. J., Snurr, R. Q. (2014). Screening of bio-compatible metal–organic frameworks as potential drug carriers using Monte Carlo simulations, *J. Mater. Chem. B*, **2**, 766–774.

Chapter 12

Computational Modeling of Catalysis in Metal–Organic Frameworks

Qiuju Zhang and Liang Chen
Ningbo Institute of Materials Technology and Engineering,
Chinese Academy of Sciences, Ningbo 315201, People's Republic of China
chenliang@nimte.ac.cn

In this chapter, we first briefly review the experimental endeavors of generating active sites in MOFs for catalysis. After that, we concentrate on the recent progress of computational modeling on MOF-based catalysis. Finally, an outlook of future possible applications of MOFs for catalysis will be presented.

12.1 Introduction

It was initially proposed in the 1990s that metal–organic frameworks (MOFs) exhibit potential applications in catalysis, owing to their various constituted multifunctionalized organic linkers and the bound metal or metal clusters [1]. Although MOFs share some of the catalytically relevant features of other well-known porous materials, e.g., zeolites, mesoporous silicates or aluminosilicates, their catalytic properties are significantly different in several aspects including

Metal–Organic Frameworks: Materials Modeling towards Potential Engineering Applications
Edited by Jianwen Jiang

ISBN 978-981-4613-45-3 (Hardcover), 978-981-4613-46-0 (eBook)
www.panstanford.com

Table 12.1 Comparison of the structural, physical, and chemical properties of zeolites, mesoporous silica and alumina, and MOFs. From Ref. [2]. Reproduced by permission of the PCCP Owner Societies

	Zeolites	MOFs	Mesoporous silicates and aluminosilicates
Crystalline?	Yes	Yes	No
Homogeneous active sites?	Yes	Yes	No
Surface area	$<600\ m^2\ g^{-1}$	Up to $10{,}400\ m^2\ g^{-1}$[a]	$<2000\ m^2\ g^{-1}$
Cavity size	$\leq$ ca. 1 nm	Up to 4 nm	$\geq$ 2 nm
Diffusivity	Low	Low to high	High
Thermal stability	High	Low to medium[b]	Medium
Chemical stability	High	Variable	High
Chemical versatility	Low	High	Medium-low

[a] Maximum value published at the moment of this publication.
[b] Maximum value published at the moment of this publication: 540°C

much larger internal surface areas, uniform pore and cavity sizes. The comparison of the structural, physical and chemical properties of the three porous materials is summarized in Table 12.1 [2]. The pore size (up to 4 nm) of MOFs bridges the pore size gap between zeolites and mesoporous silicate materials. The abundant choice of structures makes it possible to generate an appropriate pore size and shape selectivity for particular catalytic reactions, while the small pore size of zeolites usually limits the catalytic transformation of large molecules [3]. Unfortunately, the catalyzed reaction temperature for MOFs could not be very high due to their low thermal stability, while zeolites with higher thermal and chemical stability could be used under relatively harsh conditions.

Despite the recent compelling developments of MOF-based catalysis, this area is still in an immature stage [2]. Actually, the experimental exploration is only starting to boom very recently, accompanied by even less theoretical investigations [4–8]. On the basis of the characteristics of the structures and methods of preparation, the several dozens of MOF-based catalysts reported to date fall into four categories: (1) those with active sites on the framework, (2) those with encapsulated active species, (3) those with incorporating active sites, and (4) those with active sites attached through postsynthetic modification (PSM). These various designed MOF-based catalysts exhibit many excellent catalytic properties, which have been successfully applied in biomimetic chemistry, photocatalysis, Lewis or Brønsted acid catalysis, base catalysis, enantioselective catalysis, C–C bond polymerization and gas catalysis [9].

In addition to experimental studies, the characteristics of extensive structures and regular porosity of MOFs provide possibilities for the theoretical design of catalytic active sites. For example, the carbonyl-ene reaction between HCHO@MOF-11 and propylene and hydroperoxide decomposition over copper and cobalt sites within MOFs were explored on the basis of quantum mechanics (QM) and molecular mechanics (MM) methods and density functional theory (DFT) methods, respectively [4, 8]. By means of the DFT method, a hybrid MOF W-Cu-BTC was designed to activate CO_2 molecules [10]. In addition, an efficient method of combining DFT with a genetic algorithm (GA) was employed recently to identify the most stable structures and evaluate the mobility of Au, Pd, and AuPd clusters in MOF-74 [11]. Although only a few computational modeling studies have been reported to date, the theoretical tool has already shown the power and capability to reveal the intrinsic catalytic mechanism on MOF-based materials. Thus, it is believed to be capable of predicting materials properties, guiding and complementing future experiments.

12.2 Experimentally Generating Active Sites in MOF-Based Catalysts

The strategies for building specific catalytic sites in MOFs generally include four categories

12.2.1 *Intrinsic Active Sites on the Framework*

The first strategy is to directly use metal nodes or organic linkers, the two basic components in frameworks, as catalytic active sites. Regarding the active sites on metal nodes, they can be further classified into intrinsic metal nodes and open metal nodes. Intrinsic metal nodes refer to active sites on the components of MOFs without

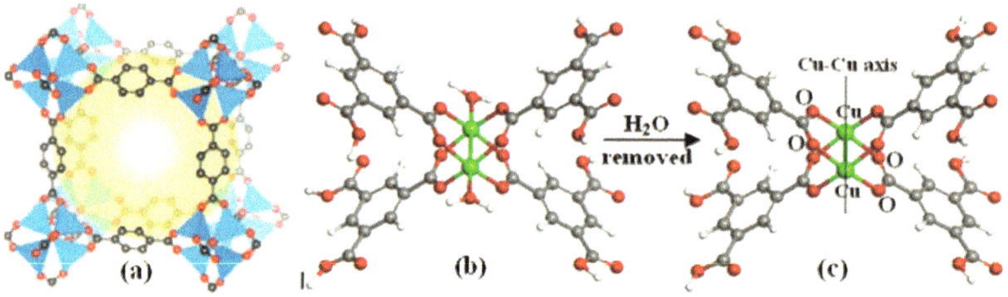

Figure 12.1 (a) The intrinsic metal cluster tetrahedral ZnO_4 (blue) in MOF-5 (b) the metal nodes Cu(II) in Cu-BTC, and (c) the open metal nodes Cu(II) in Cu-BTC after the removal of coordinated water molecules. Reprinted with permission from Ref. [13]. Copyright 2005 American Chemical Society. Reproduced from Ref. [15] by permission of Elsevier.

additional treatment. A good example, reported by Fujita et al. in 1994, is the 2D MOF $Cd(4,4'\text{-bpy})_2(NO_3)_2$,(bpy = bipyridine) with Cd(II), which could catalyze the cyanosilylation of aldehydes after aldehydes entered the pores as a Lewis acid [12]. Another example is the Zn_4O node in MOF-5 (Fig. 12.1a) [13], which is able to catalyze the Friedel–Crafts alkylation of both toluene and biphenyl. On the other hand, an open metal site is a type of active site obtained by additional treatment, that is removing the coordinated solvent to expose the metal nodes. Two typical MOFs Cu-BTC (HKUST-1) [14, 15] (Fig. 12.1b) and Matériaux de l'Institut Lavoisier (MIL)-101(Cr), with open metal sites have been widely investigated. Both unsaturated cations Cu(II) and Cr(III) were capable of catalyzing the cyanosilylation of carbonyl groups [16, 17]. In addition to metal nodes, organic linkers can also serve as active sites for catalysis. A typical example is NH_2-functionalized isoreticular metal–organic framework (IRMOF)-3 ($Zn_4O(atpa)_3$, atpa = 2-aminoterephthalate), which acted as an active basic catalyst for the Knoevenagel condensation of benzaldehyde with ethyl cyanoacetate [18].

12.2.2 *MOF-Encapsulated Active Species*

The second strategy is to encapsulate other active species into MOFs, involving metallic nanoparticles (NPs), molecular catalysts and polyoxyacids, etc. Here, encapsulation means that the active site

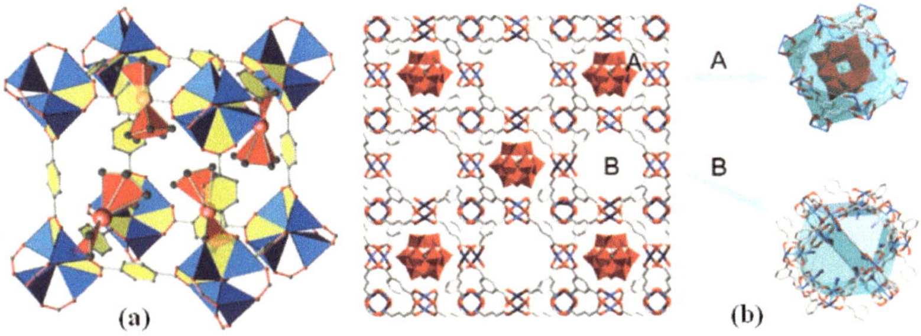

Figure 12.2 (a) MOF-5 cage (blue/yellow) with four incorporated $[(\eta^5\text{-}C_5H_5)Pd(\eta^3\text{-}C_3H_5)]$ precursors (red). (b) View of a (001) sheet with two kinds of pores, A and B, in **NENU-*n*** ($n = 1$–6). The Cu-BTC framework and Keggin polyanions are represented by wireframe and polyhedral models. Blue, red, and gray represent Cu, O, and C, respectively. From Ref. [22]. Copyright 2005 WILEY-VCH Verlag GmbH & Co. KGaA. Reprinted with permission from Ref. [27]. Copyright 2009 American Chemical Society.

is positioned within the pores of the MOF structure via noncovalent interactions. The framework is used as a porous support with a large surface area for the catalyst.

Transitional metal (TM) NPs are the most widely used active species, including Cu, Ru, Pd, Au, Ag, and Pt [19]. Recently, bimetallic NP@MOFs have also been successfully prepared by doping two types of TMs, such as RuPt and CuPd into MOFs [20, 21]. These MOF-capsulated NPs were generally prepared by the following four methods: (1) solvent-free gas-phase loading of Cu, Au and Pd NPs into MOF-5 to produce monometallic NP@MOF-5 [22], as shown in Fig. 12.2a; (2) solution impregnation of Ag, Au, Pd, and Cu particles into different MOF structures for example Pd@MIL-101(Cr) [21]; (3) incipient wetness impregnation mainly applied for preparing Pd@MOF-5 [23]; and (4) solid grinding specially developed for supporting nanosized Au clusters on different MOFs such as ZIF-8 [24, 25].

In addition, molecular catalysis such as polyoxometalates (POMs) ($H_nXM_{12}O_{40}$, X = Si, Ge, P, As; M = W, Mo], $[PW_{11}TiO_{40}]^{5-}$ and $[PW_{11}CoO_{39}]^{5-}$, have also been encapsulated into the large cavities of framework Cu-BTC and MOF-101 [26, 27]. The resulting MOF composites exhibited good catalytic activities in the oxidation of

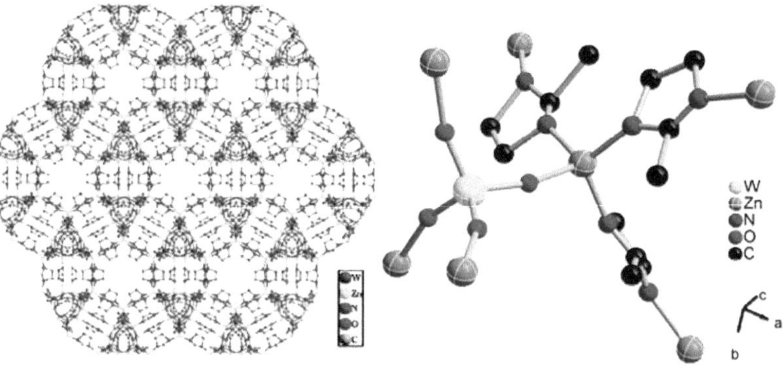

Figure 12.3 3D infinite coordination framework of (a) HZIF-1W viewed along the [111] direction and (b) coordination environments in the structure of HZIF-1W. From Ref. [28]. Copyright 2011 WILEY-VCH Verlag GmbH & Co. KGaA.

cyclohexane, enzymatic catalysis and the oxidation of α-pinene, respectively. Figure 12.2b displays a typical MOF-encapsulated composite, known as **NENU-*n***, or Northeast Normal University, $n =$ 1–6, in which POMs were encapsulated into Cu-BTC.

12.2.3 *Incorporating Active Sites*

The third strategy is known as incorporation, which refers to hybridizing other active species into MOFs as struts. By hybridizing a TO_4 ($T = W^{6+}$, Mo^{6+}) tetrahedron with a zinc-imidazolates unit, hybrid zeolitic imidazolate frameworks (HZIFs), HZIF-1W and HZIF-1Mo, were prepared by combining the structural features of zeolites and ZIFs, as shown in Fig. 12.3. The active sites WO_4 and MoO_4 act as heterogeneous catalysts for the selective oxidation of benzyl alcohol and the photocatalytic degradation of methyl orange [28]. Another example is the zinc-porphyrin-incorporated MOF (ZnPO MOF), which is a highly porous pillared paddlewheel MOF [29]. The ZnPO MOF exhibited activity for acyl transfer between N-acetylimidazole (NAI) and various pyridylcarbinols (PCs) and illustrated several interesting, catalysis-relevant structural and synthetic notions [30, 31].

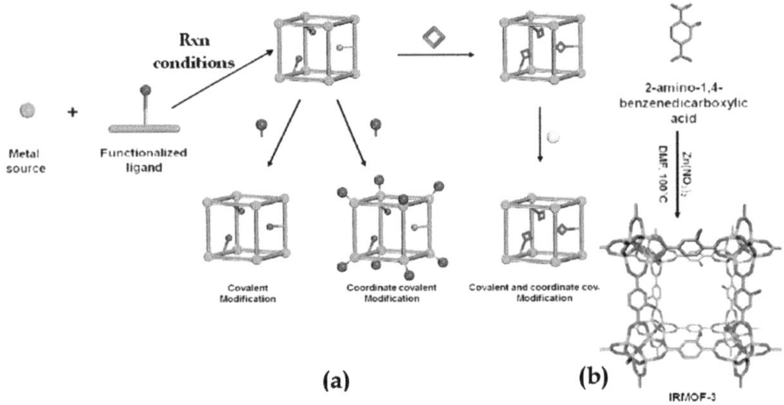

Figure 12.4 (a) Generic scheme for PSM of MOFs to form covalent, coordinate covalent, and a combination modifications. (b) Specific example of PSM on IRMOF-3 with benzoic anhydride. From Ref. [32]. Reproduced by permission of the Royal Society of Chemistry.

12.2.4 *Active Sites Attached through Postsynthetic Modification*

The last strategy is PSM, mainly using linkers or metal sites within MOFs as attachment points for catalysts, as shown in Fig. 12.4. Variable functional groups with catalytic activities, including metal complexes, amino groups, and carboxylates, can be attached to MOFs via noncovalent bonding, coordinative interactions or covalent bonding [32].

The amino group is the most widely used functionality in PSM. For example, MIL-101 was treated with a number of organic multifunctional amines such as ethylenediamine (ED), diethyelentriamine (DETA), and 3-aminopropyltrialkoxysilane (APS). All the three amino-functionalized MIL-101 (ED-MIL-101, DETA-MIL-101 and APS-MIL-101) showed higher activity for Knoevenagel condensation of benzaldehyde and ethyl cyanoacetate than unmodified MIL-101 [33]. Another amino group of 2-aminobenzenedicarboxylate (NH_2-BDC) was used for the modification of IRMOF-3 to produce IRMOF-3-AM1 [34, 35]. It was shown to be capable of reacting with different isocyanates. In addition, metal complexes, such as $M(CO)_3$ ($M = Cr$,

Mo, W) fragments were observed to bind to benzene rings of MOF-5 to generate classic arene organometllic "piano stool" complexes [36].

12.3 Computational Modeling of MOF-Based Catalysts

Compared to relatively comprehensive experimental investigations on MOF-based catalysts, theoretical modeling studies are scarce. Among this limited literature, the main focus is to study the catalytic properties of open metal sites and encapsulated alien active species. The commonly employed computational methods include first-principles DFT or Our-own-N-layer Integrated molecular Orbital + molecular Mechanics (ONIOM) methods within the framework of Vienna Ab initio Simulation Package (VASP) or Gaussian. By using the Gaussian program, the catalytic probabilities of Cu and Co cations in Cu MOFs and Co MOFs catalyzing hydoperoxide decomposition and the carbonyl-ene reaction catalyzed by the open Cu sites in MOF-11 have been studied [4, 8]. By utilizing VASP, the hybrid node W-Cu in W-Cu-BTC was designed to exhibit the Lewis base activity toward CO_2 [10]. In addition, MOF-encapsulated active sites were also theoretically studied, including Au, Pd clusters doped on MOF-74 [11] and Li^+-doped MOFs, for catalyzing formaldehyde trimerization [6]. Here, we present a relatively detailed description of calculation methods, modeling design and catalytic properties for the above-mentioned reactions.

12.3.1 *Catalyzed Reaction by Open Metal Nodes*

The first example is the carbonyl-ene reaction between formaldehyde (HCHO) and propylene ($CH_3CH{=}CH_2$) to produce 3-buten-1-ol catalyzed by the open Cu site in MOF-11 [8]. In the carbonyl-ene reaction, the catalysis normally requires a strongly electrophilic carbonyl to act as a Lewis acid. However, the disadvantage of a low boiling point (–19.5°C) and the tendency toward self-polymerization significantly prevents this reaction between HCHO and $CH_3CH{=}CH_2$. Porous-material MOFs are used as a support to encapsulate HCHO into its cavity to prevent aggregation, which obviously eliminates the possible problems concerning corrosion, handling, and toxic waste caused by the commonly used thermal or Lewis acid treatment. MOF-11 has an open metal site on the $Cu_2(CO_2)_4$ paddlewheel units and a spongy network of adamantine, which enables it to sever as the support to disperse HCHO molecules [37–40].

By using hybrid methods combining QM/MM methods [41–48] as well as the ONIOM method [44, 48] within a Gaussian 03 package, the carbonyl-ene reaction of HCH@MOF-11 and $CH_3CH{=}CH_2$ was investigated following three steps:

(1) The formation of HCHO@MOF-11 (Fig. 12.5a) via lone-pair electron interaction with an adsorption energy (E_{ads} = 12.34 kcal/mol).
(2) The carbonyl-ene reaction between HCHO@MOF-11 and CH_3 $CH{=}CH_2$ via a π interaction with a coadsorption energy of –19 kcal/mol (Fig. 12.5b). The following chemical reaction required an activation energy of 24.1 kcal/mol raised by transition state (TS) (Fig. 12.5c) to form the products (CH_3 $=CHCH_2CH_2OH$@MOF-11).
(3) The desorption of CH_2 $=CHCH_2CH_2OH$ from MOF-11, exothermically releasing 28.1 kcal/mol (Fig. 12.5d).

The corresponding energetic profile of the whole carbonyl-ene reaction of HCHO@MOF-11 and $CH_3CH{=}CH_2$ is plotted in Fig. 12.6. It can be clearly seen that the net activation energy of HCHO@MOF-11 is only 5.1 kcal/mol, which is much lower than that in the noncatalyzed reaction (31.3 kcal/mol). By removing all the linkers from MOF-11 to expose the naked Cu(I), the activation barrier would be further decreased. Overall, the calculated results suggested that MOF-11 might be a good support for use in catalysis. The reaction rate constants (K_r) and the equilibrium constants (K_{eq}) were calculated in the range of 300–600 K. The calculated K_r are quite small at low temperatures, while they increase with temperature. At 300–500 K, the forward reaction for 3-buten-1-ol formation is faster than the reverse reaction. At higher temperatures (550–600 K), this is reversed and K_{eq} decreases.

The second example of catalytic open metal nodes is the ethylbenzene hydroperoxide decomposition over Cu and Co sites within MOFs [4]. The QM study based on Gaussian program found that the steric effects around metal sites in the Co MOF and the

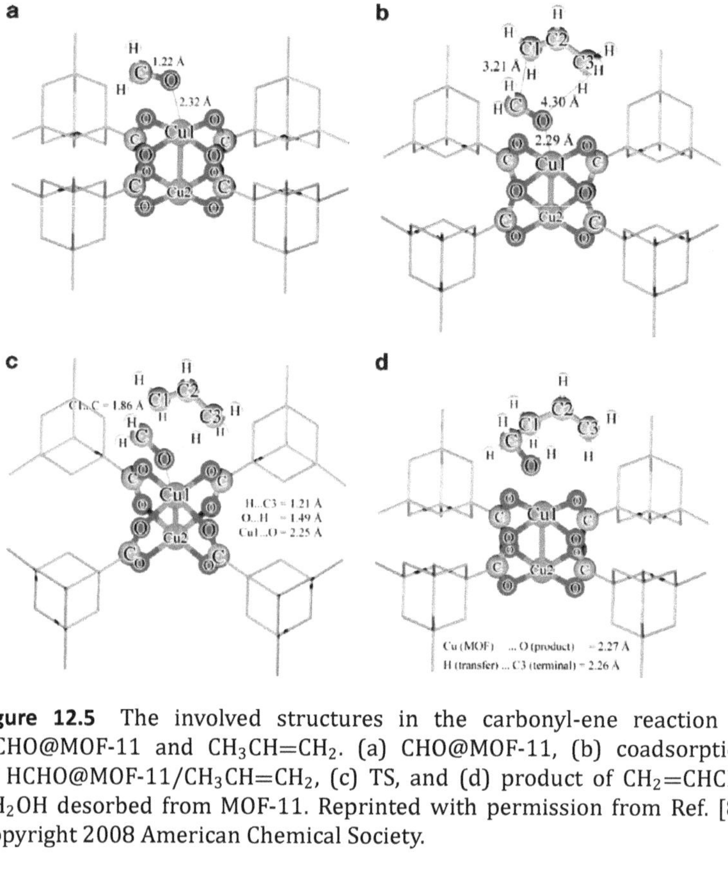

Figure 12.5 The involved structures in the carbonyl-ene reaction of HCHO@MOF-11 and $CH_3CH{=}CH_2$. (a) CHO@MOF-11, (b) coadsorption of HCHO@MOF-11/$CH_3CH{=}CH_2$, (c) TS, and (d) product of $CH_2{=}CHCH_2$ CH_2OH desorbed from MOF-11. Reprinted with permission from Ref. [8]. Copyright 2008 American Chemical Society.

Cu MOF limited their catalytic potential, while the surface atoms of MOFs might be catalytically active sites toward hydroperoxide decomposition. Although hydroperoxide could attach onto the Cu cation of Cu paddlewheel structure, it could not catalyze the O–O bond cleavage between the R group (O_{OR}) and the H atom (O_{OH}). The designed model involving unsaturated Cu clusters from the Cu MOF suggested that surface metal sites can reduce the activation barrier of hydroperoxide decomposition compared to liquid-phase decomposition.

Both Co(II) and (Cu) in these two frameworks are unsaturated, and the frameworks can serve as the support to prevent formation of the intermediate hydroperoxides. The Co MOF is also known as

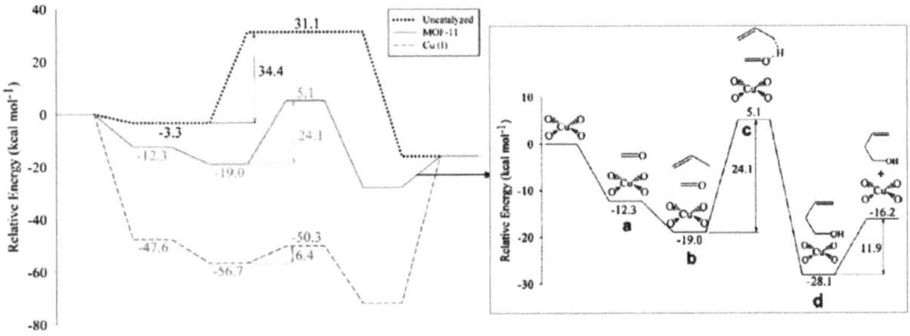

Figure 12.6 Comparison of the calculated energy profiles for the carbonyl-ene reaction of HCHO and $CH_3CH{=}CH_2$ in the MOF-11 system, the naked Cu(I) system, and the bare system. Inset: Corresponding structures along the carbonyl-ene reaction between HCHO and $CH_3CH{=}CH_2$ in MOF-11. The unit is kcal/mol. Reprinted with permission from Ref. [8]. Copyright 2008 American Chemical Society.

ZIF-9, in which each Co(II) cation is coordinated to the adjacent linkers in tetrahedral symmetry. [49]. The Cu MOF is composed of Cu(II) cations and 2-hydroxypyrimidinolate linkers in a 1:2 molar ratio. Cu(II) sites form a nearly square planar configuration with linkers [50].

By applying DFT calculations based on the B3LPY [51, 52] functional and cluster models of Co MOFs, Cu MOFs, and Cu paddlewheels around the active sites, various initial configurations of ethylbenzene hydroperoxide placed near metal sites Co(II) or Cu(II) were optimized. However, the Cu MOF is found to be energetically unfavorable to interact with the complex of the metal and ethylbenzene hydroperoxide. The hydroperoxide always left the metal site due to the large steric effect of the metal Co(II) and Cu(II). These results suggested that decomposition of tetralin hydroperoxide by metal ions inside the pores of Co MOFs and Cu MOFs is unlikely due to steric effects imposed by the linkers.

In contrast to Co MOFs and Cu MOFs, the metal site of Cu paddlewheel structure is not sterically blocked by its surrounding linkers, as shown in Fig. 12.7a,b, and hence leads to the formation of the hydroperoxide species. Since the point of spin crossing for the gas-phase reaction is estimated to be 36.0–36.7 kcal/mol higher in free energy than the stable ethylbenzene hydroperoxide,

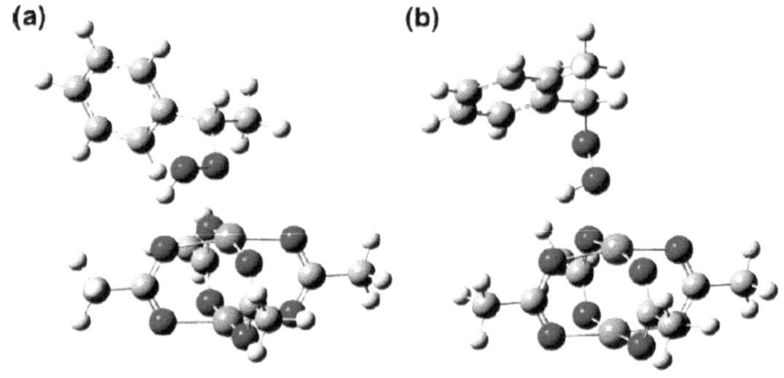

Figure 12.7 Two types of bound complexes of ethylbenzene hydroperoxide with the Cu paddlewheel structure. Reproduced from Ref. [4] by permission of Academic Press.

the paddlewheel does not appear to significantly catalyze the decomposition of hydroperoxides. The Mulliken charge analysis also indicated that no charge transfers between the Cu and O_{OR} or O_{OH} atoms.

The possible active sites of the external surface atoms of MOF crystals provide a model for a Cu site at the edge of the Cu MOF crystal. As shown in Fig. 12.8, the Cu atom in this model is coordinated to three organic linkers and a water molecule, which

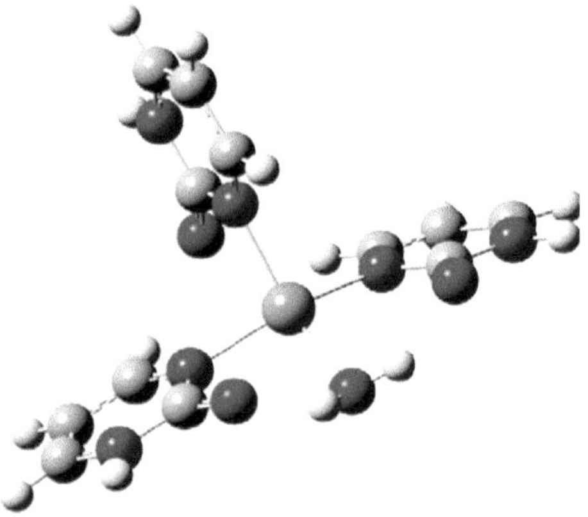

Figure 12.8 Model of a possible undercoordinated metal–ligand cluster on the exterior of a Cu MOF crystal. Reproduced from Ref. [4] by permission of Academic Press.

leads to undercoordinated Cu clusters. The barrier (25.3 kcal/mol) of tetralin hydroperoxide decomposition is significantly lower than that of the gas-phase reaction (36.0–36.7 kcal/mol). The Mulliken charge of the Cu atom in this complex is +0.65, indicating that the Cu atom has maintained a divalent oxidation state and has not been oxidized.

The last example is the acyl transfer reaction of 3-pyridylcarbinol (3-PC) and NAI catalyzed by the ZnPO MOF [53]. An acyl transfer reaction between 3-PC and NAI in the presence of a ZnPO MOF was observed to be enhanced 236-fold in the initial rate compared to the uncatalyzed reaction [5]. As presented in Fig. 12.9, the products were imidazole and 3-acetoxymethylpyridine (3-AMP). Acetonitrile (CH_3CN) was used as a solvent. Experiments with *ortho*- and *para*-isomers of PC showed very similar catalytic rates for 2-PC and 3-PC, and a two- to threefold lower rate for 4-PC. Thus, Shultz et al. [53] proposed that preconcentration was the major mechanism of catalysis because TS stabilization and favorable reactant alignment would be more sensitive to the difference in isomer structures.

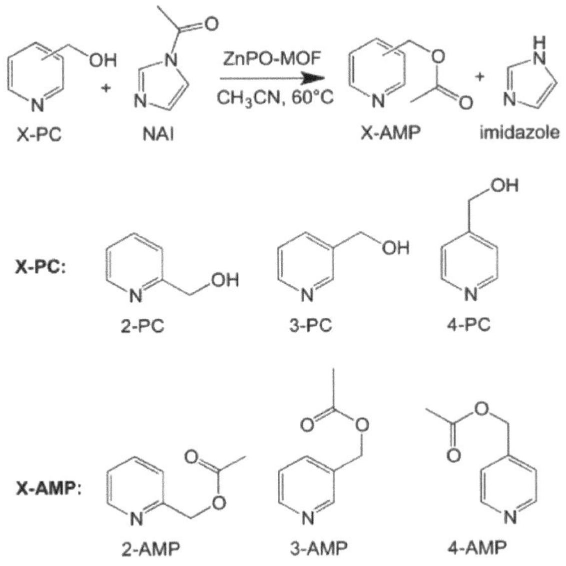

Figure 12.9 Acyl transfer reaction between PC and NAI. Reprinted with permission from Ref. [5]. Copyright 2012 American Chemical Society.

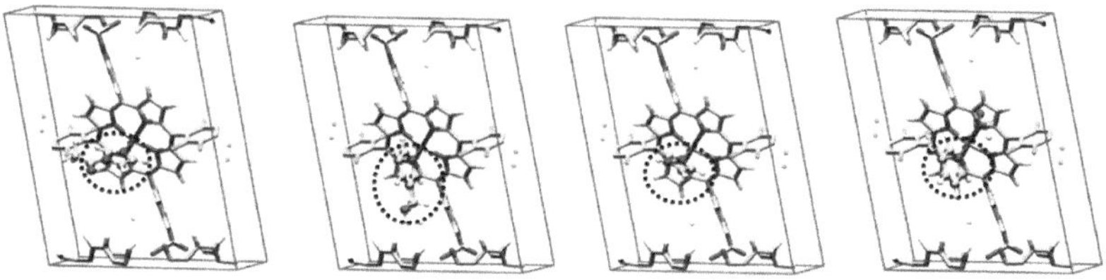

Figure 12.10 DFT-optimized geometries of ZnPO-MOF-bound reactants (a) NAI and (b) 3-PC and ZnPO-MOF-bound products (c) imidazole and (d) 3-AMP. Reprinted with permission from Ref. [5]. Copyright 2012 American Chemical Society.

A three-step approach was used to calculate the catalytic rate enhancement of the acyl transfer reaction due to preconcentration in the ZnPO MOF. The binding characteristics of all structures are calculated using the DFT method. On the basis of these results, the temperature-dependent equilibrium constants for reactant binding and product dissociation at the catalytic sites were computed. Finally, a kinetic model for heterogeneous MOF catalysis was proposed. The DFT-optimized geometries for NAI and 3-PC binding on the catalytic porphyrin-Zn site are presented in Fig. 12.10. Both of them prefer to bind on the porphyrin-Zn via their aromatic N atoms, while the solvent (CH_3CN) binds via the cyanide N atom. The binding properties listed in Table 12.2 indicate that the binding energies of NAI, 3-PC, and 4-PC are very similar, while the binding of 2-PC is compromised by steric repulsion between the prophyrin ring and the *ortho*-substituent on the pyridine. Similarly, the corresponding products of the catalyzing reaction including imidazole, 3-AMP, and 4-AMP show similar binding behavior to that of the reactants, while 2-AMP binds poorly due to unfavorable steric effects.

The following calculated reactant binding constants and product dissociation constants show that 3-PC has the highest binding constant, whereas 2-PC has the lowest binding constant due to steric repulsion. NAI, 3-PC, and 4-PC will effectively preconcentrate at the active sites of MOFs. The product dissociation constants indicate that 2-AMP dissociates much more readily than 3-AMP and 4-AMP. Imidazole has a low dissociation constant and consequently will be difficult to remove from the active sites of MOFs.

Table 12.2 Heterogeneous catalysis of the acyl transfer reaction between NAI and PC by 10 mol% of a crystalline ZnPO MOF[a] Reprinted with permission from Ref. [5]. Copyright 2012 American Chemical Society

			[X-Prophyrin] (mM)		[X-Porphyrin$_{MOF}$] (mM)		
	K_{NAI} (M^{-1})	K_{PC} (M^{-1})	X =NAI	X =PC	X =NAI	X =PC	Factor
2-PC	147	0.2	0.598	0.002	795	1	133
3-PC	147	1906	0.031	0.569	57	745	92
4-PC	147	76	0.333	0.266	526	272	118

[a][NAI]0 = 6 mM, $[PC]_0$ = 9 mM, $[PorphyrinMOF]_0$ = 0.802 M, and the reaction temperature is 60°C. Factor is preconcentration-based enhancement factor.

The catalyst-bound reactant concentrations inside MOFs are 2–3 orders of magnitude greater than the corresponding concentrations in homogeneous catalysis. The initial rate enhancement due to reactant preconcentration in a ZnPO MOF demonstrates that reactant preconcentration can independently increase the initial catalytic rate enhancement by up to 2 orders of magnitude. Preconcentration is the major component of the total rate enhancement in this system. In addition, the preconcentration factor is similar for all three isomers, just as the observed total rate enhancement is also similar for all the isomers.

12.3.2 *Knoevenagel Condensation by Framework Aniline-Like Groups*

Knoevenagel condensation is the nucleophilic addition of an active hydrogen compound to a carbonyl group, for example, an aldehyde or a ketone, followed by a dehydration reaction in which a molecule of water is eliminated (hence condensation). In this reaction, the catalyst is usually a basic amine and the product is often an α, β-conjugated enone, as shown in Fig. 12.11. Gascon et al. have studied the catalytic ability of the aniline-like amino groups of IRMOF-3 in Knoevenagel condensation in a recent experimental work, where the presence of the aromatic amino groups was found to have higher catalytic activity than that of the isolated aniline [54]. These findings indicated that the existence of local interactions within the IRMOF-3 framework was possibly able to increase the basicity of the amino groups and activate the reaction toward the condensation [7].

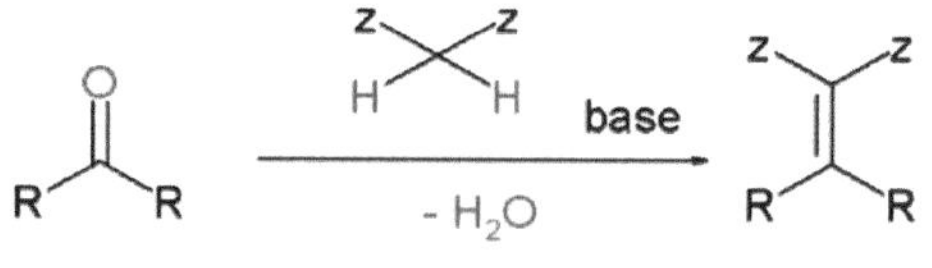

Figure 12.11 Schematic illustration of Knoevenagel condensation. From http://en.wikipedia.org/wiki/Knoevenagel_condensation.

By employing ONIOM calculations [41–48] within the Gaussian 03 program, the relative energetic and thermodynamic properties of Knoevenagel condensation on IRMOF-3 were studied, with the comparative analysis of these properties, aimed at characterizing and explaining the involvement of IRMOF-3 as a basic catalyst. Three structures with different model sizes were chosen, as shown in Fig. 12.12. The largest size of the IRMOF-3 model is b-1,4-BDC (Fig. 12.12a), with two tetrahedral ZnO units, saturated by benzoate groups The second size refers to f-1,4-BDC, and the smallest one is 2A-1,4-BDC by placing the amino groups of neighboring edges separated as far as possible.

The calculated basicity indicates that both intrinsic basicity (IB) and proton affinity (PA) increase from aniline to the f-2A-1,4-BDC

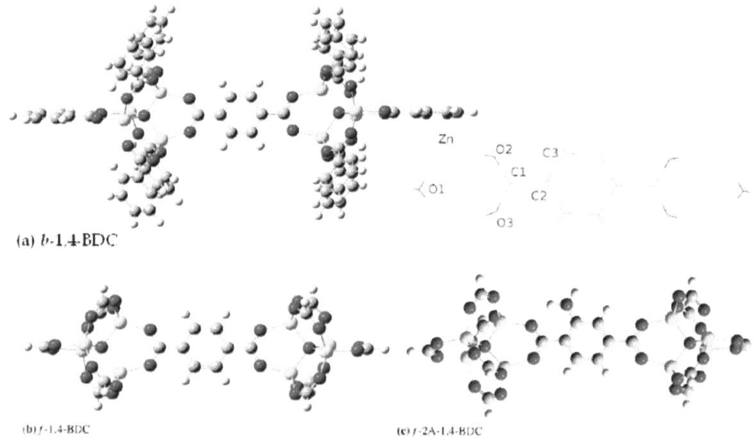

Figure 12.12 Three different model structures for calculation. Models (a) and (b) are for IRMOF-1, and model (c) is for IRMOF-3. From Ref. [7]. Reproduced by permission of the PCCP Owner Societies.

Table 12.3 BP86-calculated IB and PA parameters of aniline, DM2AT and f-2A-1,4-BDC derivatives, formed in the aminoalcohol and imine synthesis. From Ref. [7]. Reproduced by permission of the PCCP Owner Societies

	$R\text{-}NH_2 + C_6H_5CHO \rightarrow$ Aminoalcohol	
R-NH$_2$ derivative[a]	IB/kJ mol^{-1}	PA/kJ mol^{-1}
Aniline	925.3	955.6
DM2AT	926.2	956.4
f-2A-1,4-BDC	978.1	987.2
	Aminoalcohol $\rightarrow$ Imine + Water	
R-NH$_2$ derivative[a]	IB/kJ mol^{-1}	PA/kJ mol^{-1}
Aniline	954.8	982.6
DM2AT	970.1	1000.8
f-2A-1,4-BDC	1015.9	1030.1

[a]The reported derivatives are formed by the reaction of the R-NH$_2$ species (i.e., aniline, DM2AT, and f-2A-1,4-BDC) with benzaldehyde and by the following dehydration of the aminoalcohol.

fragment. The findings of Table 12.3 imply that the most basic derivatives are the imine derivatives. The basic strength of the aniline-like derivatives increases with the local molecular complexity, which agrees well with the results observed for the R-NH$_2$ starting species. Following the dehydration of the aminoalcohols obtained by the reaction of the R-NH$_2$ species with C_6H_5CHO molecules, the ethylcyanoacetate deprotonation process occurs, in which the imine derivatives act as the active species.

Figure 12.13 presents the energetics of the catalytic cycle and deactivation process of Knoevenagel condensation. In general, the behavior of the energetic values for different systems is quite regular, indicating lower differences for the smaller species, namely aniline and DM2AT. However, the energetics involved in the catalytic cycles are influenced by the local environment of the aromatic aminic groups, which implies that the stronger basicity of IRMOF-3 mainly depends on the formation of protonated IRMOF-3 acidic-conjugate derivatives. Along the catalytic cycles, the produced imine derivatives show the strongest basicity, irrespective of the starting phenylamine species. Imines obtained by IRMOF-3 fragments

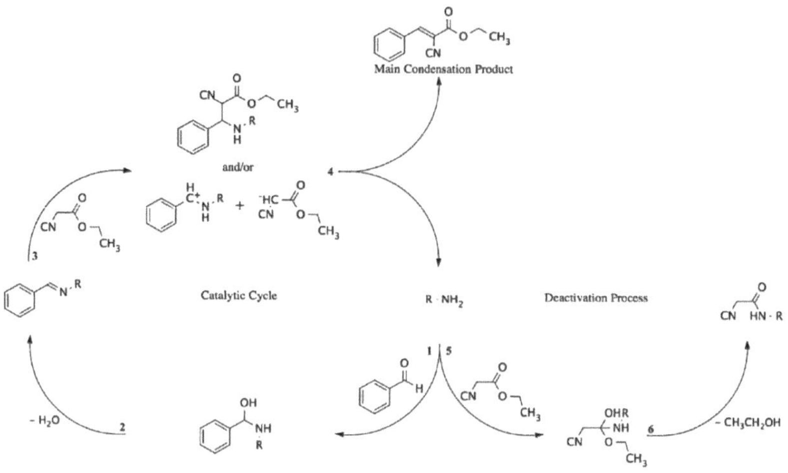

Figure 12.13 Knoevenagel condensation. The catalytic cycle, the deactivation process, and the main condensation product are explicitly pointed. From Ref. [7]. Reproduced by permission of the PCCP Owner Societies.

exhibits the strongest basicity because of the formed intramolecular O–H bonds originating from six-member rings.

12.3.3 *Incorporating W Ions into Cu-BTC to Activate CO_2*

Among the steps of recycling CO_2 as a source of carbon to produce useful chemicals [55], activation of CO_2 is a crucial and challenging step because it is an inert and thermodynamically stable compound. MOF-based catalysts have a remarkable advantage of coupling CO_2 capture and conversion [28]. To possess the probability of activating CO_2, the candidate MOF must satisfy the two criteria of unsaturated TM sites and their Lewis base characteristic. Tungsten (W) ions were selected to be partially incorporated into Cu-BTC to design W-Cu-BTC because of their relative cheapness and high activation toward activating CO_2 [56]. More importantly, it has been experimentally observed that Cr and Mo the same group (VIB) as W, were successfully used to synthesize Cr-BTC and Mo-BTC frameworks [57, 58].

By using a cluster model containing 82 atoms and GGA-PW91 [59–62] exchange correlation functional within VASP [63–65], the interaction strength of evacuating the H_2O molecule from the hybrid W ion in the W-Cu dimer was initially evaluated to be 0.62 eV. The W ion was estimated to be exposed by heating the frameworks to about 500 K, compared to $H_2O\cdots Mg$ (0.61 eV) in MOF-74 evacuated at 473 K [66]. The weak Lewis base exhibited by adsorbing CO was anticipated to be able to reduce CO_2. The optimized $CO_2\cdots$W-Cu-BTC adstructures are shown in Fig. 12.14. In the η_1^1(O) and η_2^1(O) models, CO_2 is linearly or angularly oriented on the W center through the O end, remaining in the original linear O=C=O structure and intact C=O bond lengths (~1.18 Å). The corresponding adsorption energies were calculated to be 0.22 eV and 0.24 eV, respectively. Two similar CO_2–W complexes η^1(C) and η^2(CO) bent the O–C–O angle from 180° to 136°–144°. In the η^2(CO) complex, CO_2 is obviously activated, evidenced by the bent O–C–O angle (136°) and two asymmetric C–O bonds in the lengths of 1.20 Å and 1.34 Å, which yields the highest E_{ads} of 1.34 eV. Unexpectedly, although the η^1(C) complex exhibit a significant bent angle (O–C–O) of 144°, its adsorption energy is much lower ($E_{ads} = 0.25$ eV).

The binding nature was analyzed by calculating the corresponding electron density difference (EDD) between the adstructures and separate components. An obvious charge transfer was observed from W to CO_2 in the activated complexes η^1(C), η^2(CO), and η^2(OO), evidenced by the overlapping of electron depletion around W and electron accumulation around C or O. The correspondingly enlarged electron density variations in the ranges of -0.4–0.5 e/Å^3, -0.45–0.7 e/Å^3, and -0.4–0.6 e/Å^3 indicated that W-Cu-BTC act as a π Lewis base.

By employing the B3LYP exchange correlation functional implemented in the Gaussian 09 package, the frontier molecular orbital theory of interactions between CO_2 and the open ions in Cu-BTC, W-BTC, and W-Cu-BTC were calculated. On the basis of the symmetry match principle, the lowest unoccupied molecular orbital (LUMO) of CO_2 could readily interact with W-5d$_{z2}$, the highest orbital molecular orbital (HOMO), in the hybrid W-Cu node by maximum overlapping. In contrast, the HOMO of the W ion in the completely W-substituted

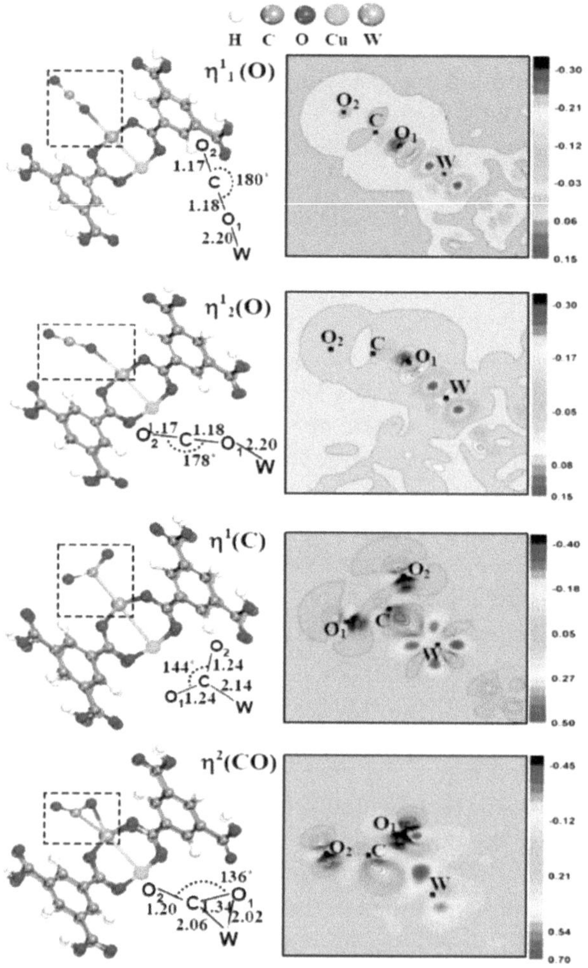

Figure 12.14 Structures and the corresponding electron density difference (EDD) of the five CO2–W–Cu–BTC complexes, with their detailed geometry parameters (distances in Å). From Ref. [10] Reproduced by permission of the Royal Society of Chemistry.

Cu-BTC (i.e., W-BTC) and the Cu ion in the primitive Cu-BTC are mainly distributed on the d_{xy} orbital, which limited their overlapping with the LUMO of CO_2. From this point, the asymmetric W-Cu dimer in the hybrid node is critical to create the Lewis base site due to

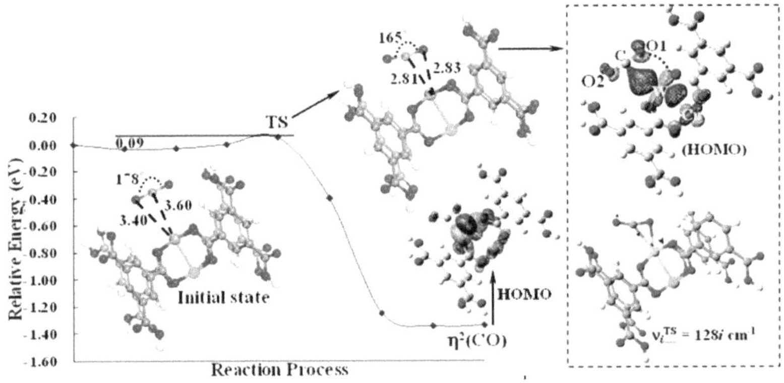

Figure 12.15 Reaction pathway of CO_2 activation on WCuBTC to yield the activated complex $\eta^{2(}$CO). The imaginary vibrational frequency mode and HOMO of TS are displayed in the dash-line box. The HOMO of $\eta^{2(}$CO) shows the disappearance of the $5dz^2$ character after the reaction (unit: eV). From Ref. [10] Reproduced by permission of the Royal Society of Chemistry.

the unequal electron negativity. The negligible activation barrier of 0.09 eV, as shown in Fig. 12.15, implied that the open W ion is kinetically favorable to activate CO_2 via forming two new C–W and O–W bonds.

The vibrational frequencies and core electron energies in the activated complexes were calculated to provide future IR and x-ray photoelectron spectroscopy (XPS) experiments for characterizing adstructures. Obviously, the appearance of the stretching mode of bent O=C=O at about 1810 cm^{-1} eliminates the O=C=O asymmetric stretching vibration. On the basis of Koopmans' theory, the IEs of C-1s in the activated η^1(C), η^2(CO), and η^2(OO) complexes were observed to down-shift by larger values of −1.97 eV, −2.14 eV, and −2.51 eV, respectively. For O-1s electrons in the complex η^2(CO), two distinct peaks down-shifted by −0.92 eV and −2.36 eV for O1-1s and O2-1s, respectively.

One of the important issues in the process of recycling CO_2 is the probability of releasing formed chemical products, because the W ion must be refreshed to accommodate new incoming CO_2. This could be achieved by hydrogenation of H atoms to form two newly chemicals, formic acid and methylene glycol. They will require the desorption energies of 0.54 eV and 0.88 eV, respectively, to leave the W-Cu dimer.

12.3.4 *Encapsulation of Active Sites*

Formaldehyde is difficult to preserve in its monomeric form, since it rapidly self-polymerizes into chain polymers. Lithium (Li)-decorated MOF-5 (Li-MOF-5) is expected to interact with formaldehyde to prevent this self-polymerization, which is similar to the Na cation in zeolites. By means of DFT calculations with the M06-L functional, trimerization of the encapsulated formaldehyde in Li-MOF-5 was found to take place in a single reaction step [6]. The simplest model consisting of two Zn_4O clusters connected by one 1,4-BDC linker was selected to mimic the active site of the MOF-5 linker. The Li cation coordinates symmetrically through a η^6-type interaction [67–69] to the hexagonal aromatic ring of the BDC linker with the average distance of C-Li of 2.39 Å (Fig. 12.16a) and $E_{\text{ads}} =$ –32.9 kcal/mol.

The different arrangement of the adsorbed formaldehydes is mainly due to the effect of the Li ion. Figure 12.16b shows three formaldehyde molecules coadsorbs on the Li ion through coordination of the lone-pair electrons on each formaldehyde

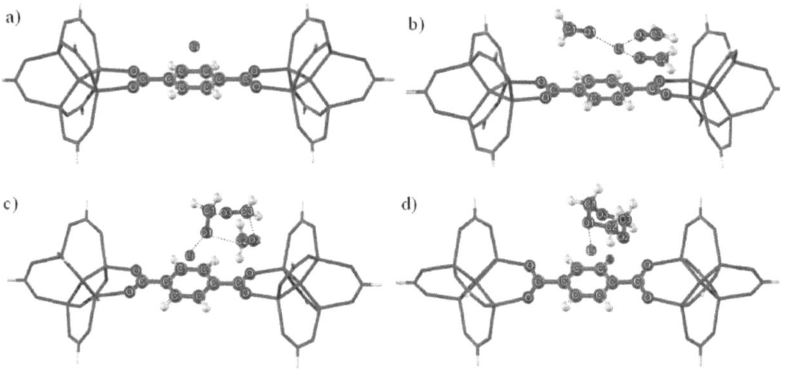

Figure 12.16 Optimized structures for (a) Li-MOF-5, (b) enclosed complex coordinated to Li for the formaldehyde trimerization in the Li-MOF-5 system, (c) TS, and (d) product. From Ref. [6]. Copyright 2012 WILEY-VCH Verlag GmbH & Co. KGaA.

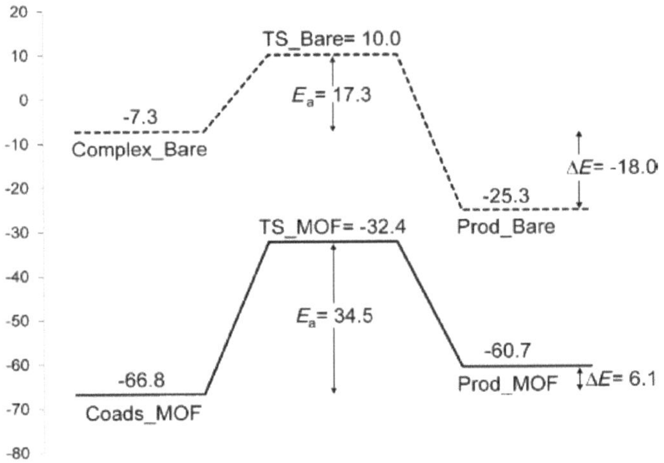

Figure 12.17 Complete energy profiles for formaldehyde trimerization on Li-MOF-5 and on the bare system, respectively. From Ref. [6]. Copyright 2012 WILEY-VCH Verlag GmbH & Co. KGaA.

oxygen atom. The stability of the adsorbed complex is additionally enhanced by the hydrogen bonding between the hydrogen atoms of two formaldehyde molecules and the oxygen atoms on the linker. Although the strong interaction between the oxygen atoms of trioxane and the Li ion results in an E_{ads} of -60.7 kcal/mol with respect to the reactants, the calculated activation energy (34.5 kcal/mol) of Li-MOF-5 is much higher than that of the bare system (17.3 kcal/mol), as shown in Fig. 12.17. In the bare system, the overall energy difference (ΔE) is exothermic by -18 kcal/mol and the overall reaction Gibbs energy (ΔG) is negative (-13.3 kcal/mol), indicating that trimerization on the bare system is a spontaneous reaction. In contrast, trimerization on Li-MOF-5 is endothermic, with an overall ΔG of 11 kcal/mol, indicating that such a reaction can hardly occur.

Metal clusters, which have a comparable size with the pore diameter, are another kind of possible encapsulated species to behave as active sites [70–72]. The widely used dopants within different MOFs are Au and Pd clusters. However, the complicated surface structure of MOFs makes it difficult to convincingly predict the most stable structures of adsorbed metal clusters, although it is

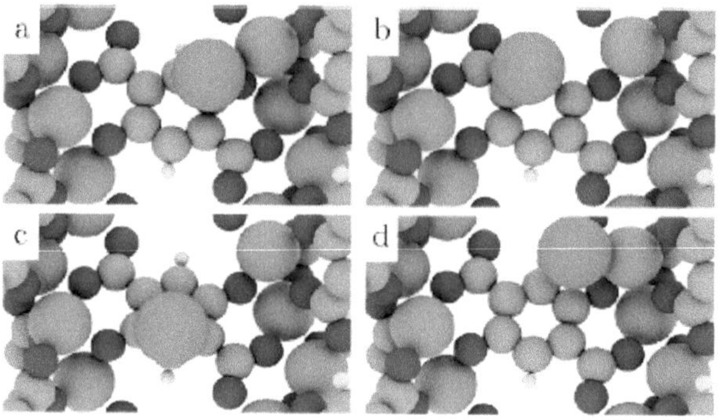

Figure 12.18 The four most stable Pd1 adsorption sites in the MOF-74 framework. Reprinted with permission from Ref. [11]. Copyright 2012 American Chemical Society.

critical to the development of these catalytic materials. An efficient method by combining DFT calculations with a genetic algorithm (GA) [73, 74] was employed to predict the most stable structures of Au, Pd, and AuPd clusters in the optimized primitive unit cell MOF-74 [11]. MOF-74 was chosen because the open metal site Zn is capable of binding metal clusters.

The four most stable configurations of single-metal-atom adsorption on MOF-74 for Pd are shown in Fig. 12.18. The E_{ads} for Pd-MOF-74 is in the range of 0.88–1.44 eV, while the E_{ads} of 0.24–0.66 eV for Au-MOF-74 is much weaker. It can be clearly observed that Pd is bound more strongly to MOF-74 than Au, due to its higher activity than Au. However, it is surprising that the open Zn ions do not offer significantly stronger binding sites than the aromatic ring. For both the Au and the Pd atom, the most favorable adsorption site includes the joint attachment of the metal atom to both Zn and C atoms. The second most favorable adsorption site of Au is coordinated only to a Zn atom. For the Pd atom, carbon is involved in all three most stable adsorption sites (Fig. 12.18a–d), while Zn is involved in only one adsorption structure. The bonding strength of the Au MOF is weaker than that of the Pd MOF, which can be understood as an unfavorable charge rearrangement in the framework. The density-derived electrostatic and chemical (DDEC) charges show that the Zn site donates more charge to Au than the adsorbed site not involving

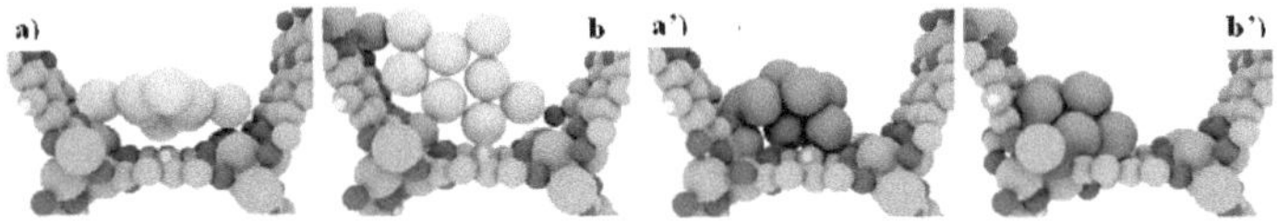

Figure 12.19 Side view of the most stable candidates found in a ring-centered run and a metal-centered run of the GA for (a, b) Au_8 and for (c, d) Pd_8. Reprinted with permission from Ref. [11]. Copyright 2012 American Chemical Society.

the Zn site. In contrast, C atoms gain charges from the Pd atom in all three most stable adsorption sites, indicating a covalent Pd–C bond.

By performing two separate runs of the GA, the most stable Au_8 and Pd_8 clusters are found to be more favorable on the metal center than the ring center, as displayed in Fig. 12.19b,d. Au_8 on the ring center (Fig. 12.19a) has a planar structure closely resembling the gas phase, whereas the structure in Fig. 12.19b is clearly collapsed. For Pd_8@MOF-74, the most stable cluster is the structure shown in Fig. 12.19d. The larger deformation makes one Zn atom translate more than 1.6 Å from being an open metal site in the neighboring pore to coordinating directly to the Pd cluster. The DDEC charges show that Pd_8 gains charge from the framework rather than donating charge to the framework for the Pd monomer. The binding strength of Pd_8 to the framework is almost three times stronger than Au_8, due to the 10-atom coordination of Pd_8 and only a 5-atom coordination of Au_8. The lower coordination number of Au_8 further illustrates that Au has a lower affinity toward the framework than Pd.

The most stable adsorption sites of Au_4Pd_4 clusters from each GA run are both centered on the aromatic ring with $E_b = -2.30$ eV and -2.09 eV, even though they were initially placed on the metal center. This is in contrast to the nonalloyed clusters, where the most stable sites are on top of the Zn atoms. The charge transfer in the adsorbed clusters is dominated by charge exchange between the Au and Pd atoms with a magnitude similar to the isolated cluster, which has a charge transfer of -0.50 eV from the Pd to the Au atoms.

The mobility of Au_8 and Pd_8 clusters along the pores of MOF-74 was explored with the climbing nudged elastic band (cNEB) method to locate the TS and evaluate the energy barrier of diffusion. The diffusion process of the most stable structure of Au_8@MOF-74

(Fig. 12.19a) was studied as an example to show how the cluster diffuses along pores. The low net diffusion barrier of no more than 0.5 eV at the TS indicates that the mobility of Au_8 clusters in MOF-74 could occur at low temperatures. In contrast, the Pd_8 cluster shows a higher net diffusion barrier of 1.59 eV for the structure in Fig. 12.19b', indicating that Pd_8 clusters are less mobile in MOF-74 than Au_8 clusters, as shown in Fig. 12.20. For both species, the ratio between the adsorption potential energy E_{ads} and the diffusion barrier E_b is ~2.1, implying that the mobility of adsorbed particles can be predicted from their adsorption energies. By substituting the metal-node Zn atom with a Mg atom in MOF-74, the isostructural Mg-MOF-74-supported Au and Pd clusters are found to more

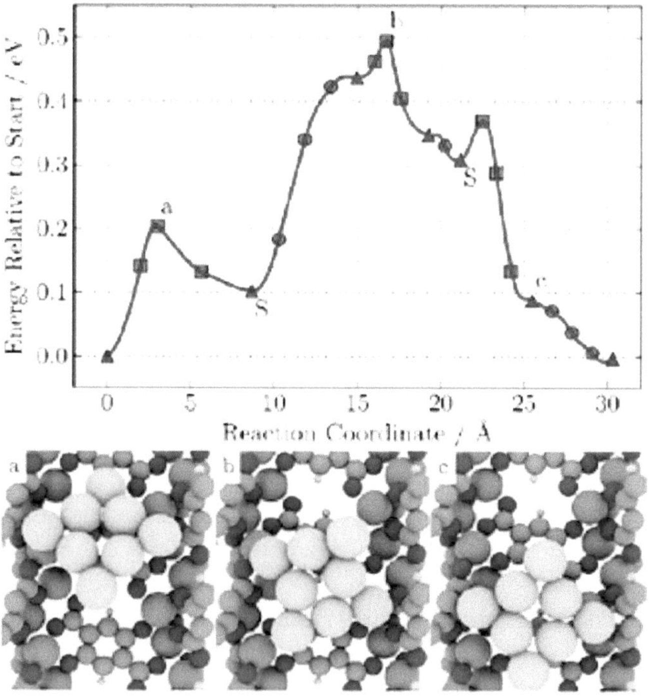

Figure 12.20 Diffusion paths of (a) Au_8 and (b) Pd_8 in the MOF-74 framework. The start and end points consist of the structure in Fig. 12.19a for Au_8 and Fig. 12.19b' for Pd_8, respectively. Reprinted with permission from Ref. [11]. Copyright 2012 American Chemical Society.

weakly bind with the three metal clusters. However, the pyridine-substituted MOF-74 will enhance the interaction, suggesting that N-based substitution of the framework may be an avenue to enhance the binding energy of adsorbed metal clusters in MOFs.

12.4 Applications and Outlook

The large amounts of synthesized MOF-based catalysts have been used in many catalyzed applications and are expected to be further exploited in broader areas. To date, they are mainly used toward catalyzing organic reactions, such as alcohol oxidation or Knoevenagel condensation. Au@MOFs showed higher catalytic activity than known-material-supported Au NPs in the aerobic oxidation of benzyl alcohol in a methanol solution. The second application is to use MOF-based catalysts as enantioselective catalysts, which was mostly studied by Lin et al. [75]. The homochiral MOFs with achiral catalytic sites, by incorporation of asymmetric catalysts and the PSM approach, can catalyze the transesterification between an aromatic ester and ethanol, olefin epoxidation reactions, and $ZnEt_2$ additions to aromatic aldehydes. Photocatalytic activity was also found for some MOF catalysts. Nanosized TiO_2 and ZnO can be hosted in MOF-5 to generate new types of quantum dot materials, and heterogeneous catalysts WO_4 and MoO_4 were incorporated into ZIFs for the selective oxidation of benzyl alcohol and the photocatalytic degradation of methyl orange [28]. Biomimetic catalysis could be suitable for some binuclear MOFs including Cu_2 paddlewheel units found in MOP-1, and hybrid flexible frameworks with linker rotation [76–78].

In addition, gasphase catalytic reactions involving CO and CO_2 have been recently realized. CO was oxidized in Pd-doped and PdO_2-doped MOF-5 [79] while the hybrid W-Cu-BTC has been theoretically designed to activate CO_2 molecules [10]. Although a number of MOF-based catalysts have been successfully synthesized and applied in various areas, we believe that this is just a start to attract the scientific community and will have great potential for developing many new highperformance catalyst materials.

References

1. Hoskins, B. F., Robson, R. (1990). Design and construction of a new class of scaffolding-like materials comprising infinite polymeric frameworks of 3D-linked molecular rods. A rappraisal of the $Zn(CN)_2$ and $Cd(CN)_2$sructures and the snthesis and structure of the diamond-related frameworks $[N(CH_3)_4]$ $[Cu^{I}Zn^{II}(CN)_4]$ and Cu^{I}[4,4',4'',4'''-tetracyanotetraphenylmethane] BF_4. $\chi C_6H_5NO_2$, *J. Am. Chem. Soc.*, **112**, 1546–1554.
2. Ranocchiari, M., van Bokhoven, J. A. (2011). Catalysis by metal-organic frameworks: Fundamentals and opportunities, *Phys. Chem. Chem. Phys.*, **13**, 6388–6396.
3. Cejka, J., Corma, A., Zones, S. (2010). *Zeolites and Catalysis: Synthesis, Reactions and Applications* (Wiley-VCH, Weinheim), 171–207.
4. Ryan, P., Konstantinov, I., Snurr, R. Q., Broadbelt, L. J. (2012). DFT investigation of hydroperoxide decomposition over copper and cobalt sites within metal-organic frameworks, *J. Catal.*, **286**, 95–102.
5. Roy, S., George, C. B., Ratner, M. A. (2012). Catalysis by a zinc-porphyrin-based metal-organic framework: from theory to computational design, *J. Phys. Chem. C*, **116**, 23494–23502.
6. Maihom, T., Choomwattana, S., Khongpracha, P., Probst, M., Limtrakul, J. (2012). Formaldehyde encapsulated in lithium-decorated metal-organic frameworks: a density functional theory study, *ChemPhysChem*, **13**, 245–249.
7. Cortese, R., Duca, D. (2011). A DFT study of IRMOF-3 catalysed knoevenagel condensation, *Phys. Chem. Chem. Phys.*, **13**, 15995–16004.
8. Choomwattana, S., Maihom, T., Khongpracha, P., Probst, M., Limtrakul, J. (2008). Structures and mechanisms of the carbonyl-ene reaction between MOF-11 encapsulated formaldehyde and propylene: an ONIOM study, *J. Phys. Chem. C*, **112**, 10855–10861.
9. Farrusseng, D., Aguado, S., Pinel, C. (2009). Metal-organic frameworks: opportunities for catalysis, *Angew. Chem., Int. Ed.*, **48**, 7502–7513.
10. Zhang, Q., Cao, L., Li, B., Chen, L. (2012). Catalyzed activation of CO_2 by a Lewis-base site in W-Cu-BTC hybrid metal organic frameworks, *Chem. Sci.*, **3**, 2708–2715.
11. Vilhelmsen, L. B., Walton, K. S., Sholl, D. S. (2012). Structure and mobility of metal clusters in MOFs: Au, Pd, and AuPd clusters in MOF-74, *J. Am. Chem. Soc.*, **134**, 12807–12816.

12. Fujita, M., Kwon, Y. J., Washizu, S., Ogura, K. (1994). Preparation, clathration ability, and catalysis of a 2-dimensinal square network material composed of cadmium(II) and 4,4'-bipyridine, *J. Am. Chem. Soc.*, **116**, 1151–1152.

13. Millward, A. R., Yaghi, O. M. (2005). Metal-organic frameworks with exceptionally high capacity for storage of carbon dioxide at room temperature, *J. Am. Chem. Soc.*, **127**, 17998–17999.

14. Chui, S. S. Y., Lo, S. M. F., Charmant, J. P. H., Orpen, A. G., Williams, I. D. (1999). A chemically functionalizable nanoporous material $[Cu_3(TMA)_2(H_2O)_3]_n$, *Science*, **283**, 1148–1150.

15. Zhou, C. G., Cao, L. J., Wei, S. H., Zhang, Q. J., Chen, L. (2011). A first principles study of gas adsorption on charged Cu-BTC, *Comput. Theor. Chem.*, **976**, 153–160.

16. Alaerts, L., Seguin, E., Poelman, H., Thibault-Starzyk, F., Jacobs, P. A., De Vos, D. E. (2006). Probing the lewis acidity and catalytic activity of the metal-organic framework $Cu_3(BTC)_2$ (BTC= benzene-1,3,5-tricarboxylate), *Chem. Eur. J.*, **12**, 7353–7363.

17. Henschel, A., Gedrich, K., Kraehnert, R., Kaskel, S. (2008). Catalytic properties of MIL-101, *Chem. Commun.*, 4192–4194.

18. Gascon, J., Aktay, U., Hernandez-Alonso, M. D., van Klink, G. P. M., Kapteijn, F. (2009). Amino-based metal-organic frameworks as stable, highly active basic catalysts, *J. Catal.*, **261**, 75–87.

19. Meilikhov, M., Yusenko, K., Esken, D., Turner, S., Van Tendeloo, G., Fischer, R. A. (2010). Metals@MOFs: loading MOFs with metal nanoparticles for hybrid functions, *Eur. J. Inorg. Chem.*, 3701–3714.

20. Schroeder, F., Henke, S., Zhang, X., Fischer, R. A. (2009). Simultaneous gas-phase loading of MOF-5 with two metal precursors: towards bimetallics@MOF, *Eur. J. Inorg. Chem.*, 3131–3140.

21. El-Shall, M. S., Abdelsayed, V., Khder, A. E. R. S., Hassan, H. M. A., El-Kaderi, H. M., Reich, T. E. (2009). Metallic and bimetallic nanocatalysts incorporated into highly porous coordination polymer MIL-101, *J. Mater. Chem.*, **19**, 7625–7631.

22. Hermes, S., Schröter, M. K., Schmid, R., Khodeir, L., Muhler, M., Tissler, A., Fischer, R. W., Fischer, R. A. (2005). Metal@MOF: loading of highly porous coordination polymers host lattices by metal organic chemical vapor deposition, *Angew. Chem., Int. Ed.*, **44**, 6237–6241.

23. Esken, D., Zhang, X., Lebedev, O. I., Schroeder, F., Fischer, R. A. (2009). Pd@MOF-5: Limitations of gas-phase infiltration and solution impregnation of $Zn_4O(BDC)_3$ (MOF-5) with metal-organic palladium

precursors for loading with Pd nanoparticles, *J. Mater. Chem.*, **19**, 1314–1319.

24. Ishida, T., Kawakita, N., Akita, T., Haruta, M. (2009). One-pot n-alkylation of primary amines to secondary amines by gold clusters supported on porous coordination polymers, *Gold Bull.*, **42**, 267–274.
25. Ishida, T., Nagaoka, M., Akita, T., Haruta, M. (2008). Deposition of gold clusters on porous coordination polymers by solid grinding and their catalytic activity in aerobic oxidation of alcohols, *Chem. Eur. J.*, **14**, 8456–8460.
26. Maksimchuk, N. V., Kovalenko, K. A., Arzumanov, S. S., Chesalov, Y. A., Melgunov, M. S., Stepanov, A. G., Fedin, V. P., Kholdeeva, O. A. (2010). Hybrid polyoxotungstate/MIL-101 materials: synthesis, characterization, and catalysis of H_2O_2-based alkene epoxidation, *Inorg. Chem.*, **49**, 2920–2930.
27. Sun, C. Y., Liu, S. X., Liang, D. D., Shao, K. Z., Ren, Y. H. and Su, Z. M. (2009). Highly stable crystalline catalysts based on a microporous metal-organic framework and polyoxometalates, *J. Am. Chem. Soc.*, **131**, 1883–1888.
28. Wang, F., Liu, Z. S., Yang, H., Tan, Y.-X., Zhang, J. (2011). Hybrid zeolitic imidazolate frameworks with catalytically active TO_4 building blocks, *Angew. Chem., Int. Ed.*, **50**, 450–453.
29. Ma, B. Q., Mulfort, K. L., Hupp, J. T. (2005). Microporous pillared paddle-wheel frameworks based on mixed-ligand coordination of zinc ions, *Inorg. Chem.*, **44**, 4912–4914.
30. Mackay, L. G., Wylie, R. S., Sanders, J. K. M. (1994). Catalytic acyl transfer by a cyclic porphyrin trimer—efficient turnover without product inhibition, *J. Am. Chem. Soc.*, **116**, 3141–3142.
31. Oliveri, C. G., Gianneschi, N. C., Nguyen, S. T., Mirkin, C. A., Stern, C. L., Wawrzak, Z., Pink, M. (2006). Supramolecular allosteric cofacial porphyrin complexes, *J. Am. Chem. Soc.*, **128**, 16286–16296.
32. Wang, Z., Cohen, S. M. (2009). Postsynthetic modification of metal-organic frameworks, *Chem. Soc. Rev.*, **38**, 1315–1329.
33. Hwang, Y. K., Hong, D. Y., Chang, J. S., Jhung, S. H., Seo, Y. K., Kim, J., Vimont, A., Daturi, M., Serre, C., Ferey, G. (2008). Amine grafting on coordinatively unsaturated metal centers of MOFs: consequences for catalysis and metal encapsulation, *Angew. Chem., Int. Ed.*, **47**, 4144–4148.
34. Eddaoudi, M., Kim, J., Rosi, N., Vodak, D., Wachter, J., O'Keeffe, M., Yaghi, O. M. (2002). Systematic design of pore size and functionality

in isoreticular MOFs and their application in methane storage, *Science*, **295**, 469–472.

35. Rowsell, J. L. C., Yaghi, O. M. (2006). Effects of functionalization, catenation, and variation of the metal oxide and organic linking units on the low-pressure hydrogen adsorption properties of metal-organic frameworks, *J. Am. Chem. Soc.*, **128**, 1304–1315.
36. Kaye, S. S., Long, J. R. (2008). Matrix isolation chemistry in a porous metal-organic framework: Photochemical substitutions of N_2 and H_2 in Zn_4O [(η^6-1,4-benzenedicarboxylate)$Cr(CO)_3$]$_3$, *J. Am. Chem. Soc.*, **130**, 806–807.
37. Thomas, J. M. (1999). Design, synthesis, and in situ characterization of new solid catalysts, *Angew. Chem., Int. Ed.*, **38**, 3589–3628.
38. Davis, M. E. (1999). New horizons for the use of porous materials as catalysts, *Stud. Surf. Sci. Catal.*, **121**, 23–32.
39. Holm, R. H., Kennepohl, P., Solomon, E. I. (1996). Structural and functional aspects of metal sites in biology, *Chem. Rev.*, **96**, 2239–2314.
40. Ramprasad, D., Pez, G. P., Toby, B. H., Markley, T. J., Pearlstein, R. M. (1995). Solid-state lithium cyanocobaltates with a high-capacity for reversible dioxygen binding: synthesis, reactivity, and structures, *J. Am. Chem. Soc.*, **117**, 10694–10701.
41. Limtrakul, J., Nanok, T., Jungsuttiwong, S., Khongpracha, P., Truong, T. N. (2001). Adsorption of unsaturated hydrocarbons on zeolites: the effects of the zeolite framework on adsorption properties of ethylene, *Chem. Phys. Lett.*, **349**, 161–166.
42. Khaliullin, R. Z., Bell, A. T., Kazansky, V. B. (2001). An experimental and density functional theory study of the interactions of CH_4 with H-ZSM-5, *J. Phys. Chem. A*, **105**, 10454–10461.
43. Limtrakul, J., Jungsuttiwong, S., Khongpracha, P. (2000). Adsorption of carbon monoxide on H-FAU and Li-FAU zeolites: an embedded cluster approach, *J. Mol. Struct.*, **525**, 153–162.
44. de Vries, A. H., Sherwood, P., Collins, S. J., Rigby, A. M., Rigutto, M., Kramer, G. J. (1999). Zeolite structure and reactivity by combined quantum-chemical-classical calculations, *J. Phys. Chem. B*, **103**, 6133–6141.
45. Sinclair, P. E., de Vries, A., Sherwood, P., Catlow, C. R. A., van Santen, R. A. (1998). Quantum-chemical studies of alkene chemisorption in chabazite: a comparison of cluster and embedded-cluster models, *J. Chem. Soc. Faraday Trans.*, **94**, 3401–3408.
46. Brandle, M., Sauer, J. (1998). Acidity differences between inorganic solids induced by their framework structure. A combined quantum

mechanics molecular mechanics ab initio study on zeolites, *J. Am. Chem. Soc.*, **120**, 1556–1570.

47. Greatbanks, S. P., Hillier, I. H., Burton, N. A., Sherwood, P. (1996). Adsorption of water and methanol on zeolite bronsted acid sites: an ab initio, embedded cluster study including electron correlation, *J. Chem. Phys.*, **105**, 3770–3776.
48. Svensson, M., Humbel, S., Froese, R. D. J., Matsubara, T., Sieber, S., Morokuma, K. (1996). ONIOM: a multilayered integrated MO+MM method for geometry optimizations and single point energy predictions. A test for Diels-Alder reactions and $Pt(P(t\text{-}Bu)_3)_2+H_2$ oxidative addition, *J. Phys. Chem. A*, **100**, 19357–19363.
49. Park, K. S., Ni, Z., Cote, A. P., Choi, J. Y., Huang, R., Uribe-Romo, F. J., Chae, H. K., O'Keeffe, M., Yaghi, O. M. (2006). Exceptional chemical and thermal stability of zeolitic imidazolate frameworks, *Proc. Natl. Acad. Sci. U. S. A.*, **103**, 10186–10191.
50. Tabares, L. C., Navarro, J. A. R., Salas, J. M. (2001). Cooperative guest inclusion by a zeolite analogue coordination polymer. Sorption behavior with gases and amine and group 1 metal salts, *J. Am. Chem. Soc.*, **123**, 383–387.
51. Lee, C. T., Yang, W. T., Parr, R. G. (1988). Development of the colle-salvetti correlation-energy formula into a functional of the electron-density *Phys. Rev. B*, **37**, 785–789.
52. Becke, A. D. (1988). Density-functional exchange-energy approximation with correct asymptotic-behavior, *Phys. Rev. A*, **38**, 3098–3100.
53. Shultz, A. M., Farha, O. K., Hupp, J. T., Nguyen, S. T. (2009). A catalytically active, permanently microporous MOF with metalloporphyrin struts, *J. Am. Chem. Soc.*, **131**, 4204–4295.
54. Gascon, J., Aktay, U., Hernandez-Alonso, M. D., Van Klink, G. P. M., Kapteijn, F. (2009). Amino-based metal-organic frameworks as stable, highly active basic catalysts, *J. Catal.*, **261**, 75–87.
55. Pan, M., Lin, X. M., Li, G. B. and Su, C. Y. (2011). Progress in the study of metal-organic materials applying naphthalene diimide (NDI) ligands, *Coord. Chem. Rev.*, **255**, 1921–1936.
56. Li, H.-J., Ho, J.-J. (2010). Density functional calculations on the hydrogenation of carbon dioxide on Fe(111) and W(111) surfaces, *J. Phys. Chem. C*, **114**, 1194–1200.
57. Murray, L. J., Dinca, M., Yano, J., Chavan, S., Bordiga, S., Brown, C. M., Long, J. R. (2010). Highly-selective and reversible O_2 binding in Cr_3(1,3,5-benzenetricarboxylate)$_2$, *J. Am. Chem. Soc.*, **132**, 7856–7863.

58. Kramer, M., Ulrich, S. B., Kaskel, S. (2006). Synthesis and properties of the metal-organic framework $Mo_3(BTC)_2$ TUDMOF-1), *J. Mater. Chem.*, **16**, 2245–2248.

59. Perdew, J. P., Yue, W. (1986). Accurate and simple density functional for the electronic exchange energy: generalized gradient approximation, *Phys. Rev. B*, **33**, 8800–8802.

60. Perdew, J. P. (1986). Density-functional approximation for the correlation-energy of the inhomogeneous electron gas, *Phys. Rev. B*, **33**, 8822–8824.

61. Lesar, R., Herschbach, D. R. (1983). Polarizability and quadrupole-moment of hydrogen molecule in a spheroidal box, *J. Phys. Chem.*, **87**, 5202–5206.

62. Langreth, D. C., Perdew, J. P. (1980). Theory of nonuniform electronic systems. Analysis of the gradient approximation and a generalization that works, *Phys. Rev. B*, **21**, 5469–5493.

63. Kresse, G., Furthmuller, J. (1996). Efficient iterative schemes for ab initio total-energy calculations using a plane-wave basis set, *Phys. Rev. B*, **54**, 11169–11186.

64. Kresse, G., Hafner, J. (1994). Norm-conserving and ultrasoft pseudopotentials for first-row and transition-elements, *J. Phys.: Condens. Matter.*, **6**, 8245–8257.

65. Kresse, G., Hafner, J. (1993). Ab-initio molecular-dynamics for open-shell transition-metals, *Phys. Rev. B*, **48**, 13115–13118.

66. Wu, H., Simmons, J. M., Srinivas, G., Zhou, W., Yildirim, T. (2010). Adsorption sites and binding nature of CO_2 in prototypical metal-organic frameworks: a combined neutron diffraction and first-principles study, *J. Phys. Chem. Lett.*, **1**, 1946–1951.

67. Mavrandonakis, A., Tylianakis, E., Stubos, A. K., Froudakis, G. E. (2008). Why li doping in MOFs enhances H_2 storage capacity? A multi-scale theoretical study, *J. Phys. Chem. C*, **112**, 7290–7294.

68. Han, S. S., Goddard, W. A., III (2007). Lithium-doped metal-organic frameworks for reversible H_2 storage at ambient temperature, *J. Am. Chem. Soc.*, **129**, 8422–8423.

69. Blomqvist, A., Araujo, C. M., Srepusharawoot, P., Ahuja, R. (2007). Li-decorated metal-organic framework 5: a route to achieving a suitable hydrogen storage medium, *Proc. Natl. Acad. Sci. U. S. A.*, **104**, 20173–20176.

70. Mueller, M., Turner, S., Lebedev, O. I., Wang, Y., van Tendeloo, G., Fischer, R. A. (2011). Au@ MOF-5 and Au/Mo_x@MOF-5 ($M = Zn, Ti; x = 1, 2$):

preparation and microstructural characterisation, *Eur. J. Inorg. Chem.*, 1876–1887.

71. Gu, X., Lu, Z.-H., Jiang, H.-L., Akita, T., Xu, Q. (2011). Synergistic catalysis of metal-organic framework-immobilized Au-Pd nanoparticles in dehydrogenation of formic acid for chemical hydrogen storage, *J. Am. Chem. Soc.*, **133**, 11822–11825.
72. Jiang, H. L., Liu, B., Akita, T., Haruta, M., Sakurai, H., Xu, Q. (2009). Au@ZIF-8: co oxidation over gold nanoparticles deposited to metal-organic framework, *J. Am. Chem. Soc.*, **131**, 11302–11303.
73. Vilhelmsen, L. B., Hammer, B. (2012). Systematic study of Au_6 to Au_{12} gold clusters on MgO(100) F centers using density-functional theory, *Phys. Rev. Lett.*, **108**, 126101–126101.
74. Deaven, D. M., Ho, K. M. (1995). Molecular-geometry optimization with a genetic algorithm, *Phys. Rev. Lett.*, **75**, 288–291.
75. Ma, L., Abney, C., Lin, W. (2009). Enantioselective catalysis with homochiral metal-organic frameworks, *Chem. Soc. Rev.*, **38**, 1248–1256.
76. Furukawa, H., Kim, J., Ockwig, N. W., O'Keeffe, M., Yaghi, O. M. (2008). Control of vertex geometry, structure dimensionality, functionality, and pore metrics in the reticular synthesis of crystalline metal-organic frameworks and polyhedra, *J. Am. Chem. Soc.*, **130**, 11650–11661.
77. Ni, Z., Yassar, A., Antoun, T., Yaghi, O. M. (2005). Porous metal-organic truncated octahedron constructed from paddle-wheel squares and terthiophene links, *J. Am. Chem. Soc.*, **127**, 12752–12753.
78. Eddaoudi, M., Kim, J., Wachter, J. B., Chae, H. K., O'Keeffe, M., Yaghi, O. M. (2001). Porous metal-organic polyhedra: 25 Å cuboctahedron constructed from 12 $Cu_2(CO_2)_4$ paddle-wheel building blocks, *J. Am. Chem. Soc.*, **123**, 4368–4369.
79. Ye, J.-Y., Liu, C.-J. (2011). $Cu_3(BTC)_2$: CO oxidation over MOF-based catalysts, *Chem. Commun.*, **47**, 2167–2169.

Chapter 13

Modeled Catalytic Properties of MOF-Based Compounds

Remedios Cortese and Dario Duca
Dipartimento di Fisica e Chimica, Università degli Studi di Palermo, Viale delle Scienze, Parco d'Orleans II, Ed. 17, I-90128 Palermo, Italy
dario.duca@unipa.it

13.1 Introduction

Metal–organic material (MOM) derivatives are organic-inorganic hybrid compounds characterized by typical arrangements of metal units and organic ligands, as shown in Fig. 13.1; these materials showing several periodicity degrees are exemplified by different groups of either discrete or polymeric structures. Although MOMs have existed for several decades [1–7], it was not until the early 1990s that they captured broad attention as it became evident that, typically, they are *facile* to prepare and easily adaptable to crystal engineering strategies [8–17].

In 1989 Hoskins and Robson, in fact, demonstrated that networks with predictable architectures could be engineered by exploiting interactions among metals and ligands containing multiple sites of

Metal–Organic Frameworks: Materials Modeling towards Potential Engineering Applications
Edited by Jianwen Jiang

ISBN 978-981-4613-45-3 (Hardcover), 978-981-4613-46-0 (eBook)
www.panstanford.com

Figure 13.1 Bulk structure of MOF-5, the first member of the IRMOF family [18]. Metallic centers are represented in yellow.

coordination [19, 20]. As a matter of fact, since the 1990s, crystal engineering and especially network-based material design emerged as a promising field of research. The elaborated approaches found their roots in the "node and spacer" conceptual scheme of Wells, which introduced the *idea* that inorganic crystals could be described as networks defined by metal ions (nodes) linked *via* bonds (spacer) to generate coordination polymers (CPs) [21, 22].

In the early 1990s, many 3D hybrid organic-inorganic lattices were produced. Despite that, most of the lattices above were not able to undergo guest exchange without incurring structure transformations [23]. Therefore, they could not be strictly considered porous compounds. This limitation influenced both their applications and their theoretical studies, and as a consequence, one of the main goals of crystal engineering research was the individuation of 3D MOMs able to survive guest exchanges. As a result, in the late 1990s Yaghi and Kitagawa reported the first examples of MOMs, which exhibited permanent porosity [24, 25]. These derivatives could be considered as second-generation MOMs for which Yaghi and Kitagawa coined the terms metal–organic framework (MOF) materials [26, 27] and porous coordination polymer (PCP) materials [28, 29], respectively.

To synthesize stable crystalline MOFs, rigid conjugated aromatic molecules are usually preferred, as organic components or linkers, over flexible species. Among the former the most widely used are polycarboxylate aromatic molecules, bipyridines, and polyazaheterocycles, as well as their derivatives [30]. Organic linkers determine chemical and structural properties of the resulting network. Usually, the properties above are specifically designed to provide the solid to be synthesized with peculiar characteristics. In fact, depending on the connector and on its coordination number and periodic chains typical arrays and nets can be obtained. The periodicity can be, thus, tuned according to the choice of the metallic centers and the framework structures, which in turn are controlled by the number of bonds involving the bridging linkers and by the coordination number of the employed metals. At variance with the use of naked (i.e., isolated) metal ions that of metal-complex connectors have the advantage both of offering control on the bond angles and in reducing the number of coordination sites.

Moreover, sites that are not functional to the synthesis or to the final employment of the synthesized materials can be blocked by macrocyclic and/or chelating ligands that are directly bound to given metal connectors and that specifically leave given sites free for the linkers [31]. In passing it has to be recalled that lack of control over the structure of solids—produced through traditional synthetic methods—could be actually related to structural changes, occurring on the reactants during the assembling processes. On the contrary, the design of robust extended networks can be realized by starting with well-defined and rigid molecular building blocks able to retain their structure throughout the construction processes. The concept of a secondary-building unit (SBU) is, for this, essential in order to classify and plan the directionality of MOFs [32].

SBUs are simple polyatomic structures having peculiar geometric characteristics. As an example, in Fig. 13.2 are reported two examples of SBUs based on polycarboxylate linkers. They are composed of inorganic clusters characterized by typical coordination spheres. The SBUs are linked to each other by the organic components of the material to form its framework. The reticulation of the SBUs involves self-assembly steps able to connect building units into molecular lattices by very strong bonds. Due to this, there is a direct

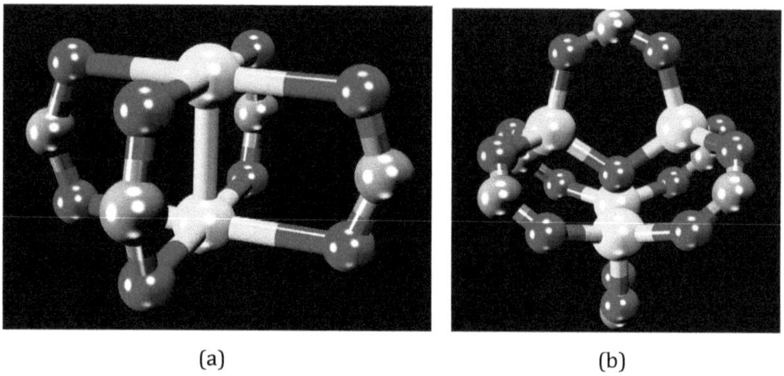

Figure 13.2 Common SBU arrangements characterizing carboxylate-based MOFs. (a) Square paddlewheel and (b) octahedral moieties.

connection between the starting molecular units and the shape of the resulting frameworks. A hierarchy of information is incorporated into the building blocks that make them sufficiently unique for the construction of a target network. Two essential elements of this approach are i) the knowledge of the possible networks that could be formed, depending on given building blocks, and ii) the determination of the geometric attributes that uniquely define the target network.

Catalysis was one of the first proposed applications for MOFs [33], which were actually thought as potential models to show contact points between homogeneous and heterogeneous catalysts. The MOFs activity was also compared to that of enzyme catalysts by Lee et al. [34]. Despite the outstanding textural properties and high metal contents of MOFs, however, their applications in catalysis have been up to now very limited. In fact, the possible use of MOF-based catalysis was demonstrated already nearly 15 years ago, but only recently extensive experimental research has been started. The lack of a well-established procedure for MOF design and synthesis—partly pointed out along the last years—certainly contributed to this delay.

In particular, MOFs were identified as promising heterogeneous catalysts, and significantly they may contain a large variety of organic moieties potentially distributed on differently sized surface areas that could undergo extended postsynthetic functionalizations without needing to face leaching phenomena. Possible applications of MOFs in the field of catalysis are pointed out in the following:

i) Heterogenization of well-defined homogeneous catalysts [35];
ii) Framework stabilization of otherwise short-lived catalysts [35];
iii) Framework encapsulation of molecular catalysts [36];
iv) Coupling of catalysis and chemical separation [33];
v) Postsynthetic incorporation of catalytic metal sites [37–39]; and
vi) Substrate size-selective catalysis transformation [40].

MOF materials are clearly characterized by three different elements: the metallic component, the organic ligand, and the cavity morphology. It is, therefore, possible to think of three different kinds of catalysts as suggested by Corma et al. [41].

The first group of catalysts includes materials with only one kind of metallic centers, which are involved at the same time as structural components and catalytic active sites. Other materials belonging to the same group of catalysts could contain two different types of metal: one is responsible for the catalytic activity, while the second has only a structural role. The active metal can be embodied in the framework either as single metal atoms or as metal clusters. One of the earliest report about MOM catalytic activity concerned a study of Fujita et al. [33] in 1994, dealing with the cyanosilylation of aldehydes by a 2D species, having the formula Cd(4,4'-*bpy*)$_2$(NO)$_3$: the active sites were Cd (II) ions showing Lewis acid catalysis. Other important examples reported in this field are those of Llabrés i Xamena et al. [42], which regard the catalytic activity of 2D square grids embedding single Pd (II) ions as nodes and 2-hydroxypyrimidinolates as linkers. Despite the initial coordinative saturation, the palladium centers in these materials proved to be capable of catalyzing alcohol oxidation, olefin hydrogenation, and Suzuki C–C coupling. Ravon et al. [43] reported catalysis by MOF-5 in Friedel–Crafts alkylations, ascribing the catalytic activity probably to the Zn_4O vertices. Up to now MOFs with metal catalytic sites were investigated in different reactions such as hydrogenation, isomerization, cyanosilylation, polimerization, oxidation, and enantioselective epoxidation. For a comprehensive list of reactions catalyzed by MOMs and MOFs see reviews [34, 41].

The second group of MOF-based catalysts has functional groups fixed in organic linkers that are able to catalyze given reactions. It is important that the linkers employed in building MOFs exhibit both the coordinative groups, ensuring structural stability, and the active groups, responsible for reactivity. The number of MOFs belonging to this ensemble with demonstrated catalytic activity is until now very limited. This occurrence is explained because metal atoms tend to coordinate the active organic groups, thus preventing the presence of free reactive moieties inside the framework. Seo et al. [44], in 2000, reported the first example describing the catalytic activity of one MOF containing ligands embedding reactive groups. This compound was synthesized involving Zn ions and chiral molecules containing carboxylic acids and pyridine moieties. The authors showed that the resulting solid was active in transterification processes because of the presence of uncoordinated pyridyl units. Gascon et al. [45] have studied the basic properties of isoreticular metal–organic framework (IRMOF)-3 and MIL-53 (NH_2). Both materials contained 2-amino-1,4-benzenedicarboxylate (2A-1,4-BDC) as organic linkers, which coordinated metal ions only through the carboxylic oxygen atoms, thus leaving free the amino groups. IRMOF-3 was found to be an active and stable catalyst in Knoevenagel condensation. Catalytic activity due to the linkers could also arise after postsynthetic modifications, transforming the linkers without changing the topology of the solid.

The third group consists of solids not specifically involved in catalysis. In fact, the porous system of the material provides just a physical environment (cage) in which i) catalysis occurs or ii) the catalytic centers are encapsulated. Müller et al. [46] studied the inclusion from the vapor phase of a series of 10 organometallic compounds inside MOF-5. Under *vacuum* and dry argon, it was observed that both the crystal structure and the organometallic species were unaltered after the inclusion. Once the organometallic compounds were incorporated, it was also possible to generate small metal nanoparticles in a weakly coordinating environment. These naked nanoparticles resulted highly reactive although extremely air sensitive.

13.2 Microporous Materials and Modeling

Microporous materials are characterized by crystalline frameworks having channels and cages that originate micropores of regular dimensions. The inner surface area, actually, delimits very large void volumes yielded by extremely narrow-pore-size distributions and embodying functional centers homogeneously dispersed in order to give very active microporous materials. These can be useful in ruling shape selectivity both of reactants and of products involved in given chemical processes. Because of confinement effects, the advantage of the shape caging inside microporous materials can turn into severe limitations when large bulky reagents are involved.

One of the most employed (and useful) comparisons performed in these years on porous solids, having catalytic properties, concerned MOF and zeolite derivatives. Both of them have a large surface area and uniform pore and cavity sizes; however, they differ, at least, on three important points: i) MOFs contain organic moieties, and hence they could be synthesized in a larger number of varieties than zeolites; ii) MOFs are not as stable as zeolites, which usually exhibit permanent microporosity also after solvent evaporation; and iii) MOFs, at variance with zeolites mainly built by tetrahedral blocks, are shaped from a wide variety of inorganic fragments connected by organic linkers having very different local topology.

This analogy has been used to design new generations of hybrid materials and their properties using strategies and concepts originating from the well-known zeolite field. In fact, it is generally accepted that the quantum mechanics (QM)-based methods are of general applicability. In particular, the variability and flexibility of MOF and zeolite frameworks do represent not a big deal. Nevertheless, according to Tafipolsky et al. [47] and Ramanan and Whittingham [48] one of the most important differences, occurring between MOFs and other porous materials, is the peculiar lattice flexibility of the latter, for example, shown by framework-breathing effects in response to external *stimuli* [49]. In this perspective atomistic simulations are expected to play a crucial role in the investigation and either in the development or in the decrement (depending on the required material functionality) of such effects.

13.3 Heterogeneous Catalysis: Methods and Models

In the science corner devoted to the modeling of heterogeneous catalytic processes, the different time and length scales involved determine an intrinsic impossibility to describe, by a single method, such a complexity. A hierarchy of methods has to be hence employed to fully elucidate a given catalytic process. These methods can be roughly categorized into QM methods, which describe electronic structures of involved systems, and molecular simulation (MS) methods, which furnish information on atomic structures and arrangements, either of isolated or of reacting species in given environments. The characterization of the catalytic reactivity requires the capability to describe formation and breaking of bonds during a specific reaction path. For this reason, QM-based methods appear to be more suitable [50]. Nevertheless MS approaches to study both catalyst structure and morphology as well as reaction kinetics would be more easily accomplished via atomistic-scale simulations, provided the appropriate interatomic and/or intermolecular potentials or the intrinsic kinetic data are available [51–55]. Furthermore, in the late 1990s the derivation of such data from ab initio calculations was suggested and applied [56, 57], thus allowing for a hierarchical approach to modeling.

Among all the possible computational methods, those based on density functional theory (DFT) [58] had a strong impact on modeling of heterogeneous catalytic systems [59]. Accuracy of DFT [60, 61] along with its applicability to large-sized structures makes it attractive for modeling heterogeneous catalytic systems. The information from DFT analysis concerns molecular properties, including structures, energetics, and distributions of electron density, just to mention a few. Different DFT models can be actually applied to mimic several characteristics of catalytic processes while they could be roughly separated into three groups: clusters, embedded clusters,

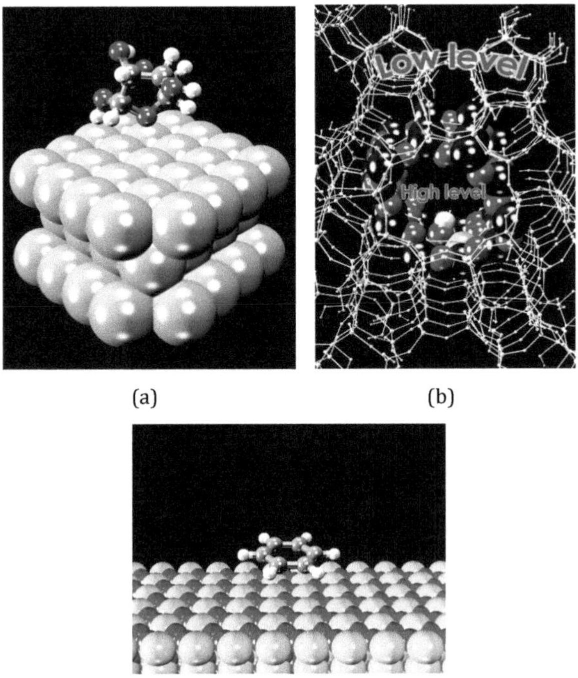

(a) (b)

(c)

Figure 13.3 Approaches used in DFT-based catalytic modeling. (a) Cluster approach: L-arabinose tautomer adsorbed on a ruthenium cluster [62]. (b) Embedded-cluster approach: Different level of theory employed in modeling reactivity inside ZSM-5 zeolite models [63]. (c) Periodic approach: TiO_2 surface model used to simulate adsorption processes [64].

and periodic systems. Application examples of the approaches above are given in Fig. 13.3.

Cluster models involve the use of small atom ensembles that can, in a way, be representative of the whole catalyst [65–67]. It is interesting to note that Civalleri et al. [68] used MOF-5 as a benchmark to calculate second derivatives of the atomic coordinates within the periodic CRYSTAL code [69], which use atomic-centered-basis functions. They compared periodic-structure and nonperiodic-cluster MOF-5 models using an identical level of theory. This led to the important conclusion that nonperiodic-cluster approaches can be also used to model periodic MOF-

5 derivatives. Beside reducing computational resources, the use of cluster models is generally useful for studying physisorption properties on preferential adsorption sites and the associated binding energies of guest molecules on different fragments of specific MOFs. Another advantage of using the cluster approach, related to low model sizes, is the possibility of using very high QM-level methods, such as MP2 or coupled cluster approaches, which yield accurate results, including correlation effects.

Embedded-cluster calculations are hybrid calculations usually employing both QM and molecular mechanics (MM)-based methods. QM/MM or QM/QM methods are built individuating one region of the model suitable for a high-level QM description and another region suitable for a lower-level description, viz. MM or also low-level QM methods [70]. The QM/MM approach found widespread use, in particular in modeling catalyzed reactions within zeolites.

Different embedding methods where covalent bonds cross the boundary between the QM and the MM region were proposed, and even models allowing molecules for dynamically moving into and out of the QM region were suggested. Of course, these approaches were used in several investigations of zeolite-catalyzed reactions [63, 71]. It is interesting to recall that most of the studies dealing with the analysis of the MOF catalytic cycle were performed employing embedded-cluster approaches. In this context, Choomwattana et al. [72], as an example, used a combination of a DFT–generalized gradient approximation (GGA) high-level method and the generic universal force field (UFF) to elucidate the mechanism of the carbonyl-ene reaction occurring between formaldehyde and propylene encapsulated in MOF-11. Braga et al. [73] employed a conformational analysis of protonated *cyclo*-[(S)-phenylalanyl-(S)-histidyl] and its complex with benzaldehyde within MOFs, while Cortese and Duca [74] studied the reaction pathway of the Knoevenagel condensation catalyzed by IRMOF-3.

A major limitation in the use of embedded approaches involving MM environments, according to Tafipolsky et al. [47], is the availability of accurate MM models. With regard to this, Tafipolsky et al. [75] developed a consistent force field parameterization using a genetic algorithm (GA) optimization method. On the basis of reference data computed at the DFT level for small nonperiodic model systems, they initially derived a parameter set for a $Zn_4O(O_2C)_6$ SBU, which was able to quantitatively reproduce the elastic constant of MOF-5, previously determined by periodic DFT. Then, they developed another parameter set for the copper paddlewheel $Cu_2(CO_2)_4$ SBU, which was validated by reproducing structure and dynamic properties of HKUST-1 [76]. According to the authors, this approach can be regarded as a multiscale simulation method with the force field systematically ruled by DFT calculations. Interestingly, the advanced force field could be used for different MOFs based on the same SBUs [77].

Periodic approaches are very often employed in the description of oxide and metallic surfaces, as well as bulk structures [78, 79]. However, even for the optimization of pretty small MOFs, substantial computational efforts are needed—in particular, for those periodic codes based on the representation of the wavefunctions by plane waves, for which the computing time scales with the volume of the unit cell. This, as might be expected, is very high in porous low-density materials, like MOFs. As a consequence, in the literature, few periodic dynamic studies are reported, most of them concerning first-principle molecular dynamics (MD) simulations of hydrogen adsorption on MOF-5 [80, 81].

The summarized ab initio approaches allow us to achieve information on macroscopic observables involved in a given catalytic cycle as the structural characteristics of minima and transition state (TS) species implied and the related activation energies and/or spectroscopic properties.

Modeling surface reactions implies full accounting of the electronic structure of the species involved in the processes. On the other hand, modeling other atomistic properties of microporous materials, such as adsorption sites, isotherms, and surface diffusion pathways, requires the simulation of larger systems and longer time scales. This can be important in characterizing a full catalytic cycle because (in addition to the reactive events at the active surface site) adsorption, desorption, and diffusion—also affected either by lateral interactions or by solvent and/or by-product effects—strongly influence the reaction kinetics [51, 53, 54, 56, 57, 82].

Atomistic simulations are directly or indirectly based on force fields and are often well suited for modeling weak interactions characterizing diffusion processes, since they are parameterized to handle such kind of properties [83]. Dispersion, repulsion, and electrostatic forces are typically accounted along with intramolecular forces (e.g., related to torsion energies for flexible adsorbates). The results depend not only on the parameter set (obtained by experimental or QM data) but also on the assignment of the atom types to the atoms taken into consideration in a particular chemical environment. MD simulations calculate time-dependent behaviors of specific molecular systems by integrating their classical equations of motion. In this case, the simulations generate information at the microscopic level related to atomic positions, velocities, and forces [84].

The conversion of microscopic information to macroscopic observable parameters like pressure, stress and strain tensors, energy, and heat capacity requires paradigms and strategies developed in the realm of statistical mechanics. Complete isotherms are typically computed using grand canonical Monte Carlo (GCMC) simulations [85, 86], a well-established technique, for example, in zeolite modeling [87, 88].

In GCMC simulations, volumes, temperatures, and chemical potentials are taken as fixed, while the number of molecules can change and the average number of adsorbed molecules, which corresponds to given gas-phase conditions, monitored. In such calculations the structure of the material framework is taken from the experiment and generally kept rigid. However, as pointed out previously, an important property of MOFs is their flexibility, and as a consequence, a number of fully flexible force fields need to be specifically developed [75] for them.

13.4 Adsorption Models

At the most basic level, catalysis modeling needs to encompass at least five elementary steps: adsorption, dissociation, activation, recombination/reaction, and desorption. It seems, hence, appropriate to give a general overview on the theoretical study reported in the literature dealing with gas adsorption into MOFs (the first basic event, which is present in any heterogeneous catalytic process). The concerning studies, even if not strictly aimed at modeling catalytic processes, actually represent an important background to understand the chemical reactivity regarding heterogeneous catalysis.

Most of the computational studies present in the literature dealing with MOFs concern gas adsorption, for example, H_2 [89–91] and CO_2 [92, 93]. The main attention was, however, on the possibility of using MOFs as hydrogen carriers [94]. With respect to this, the most interesting works can be divided into three broad sets: i) studies useful for predicting adsorption energies and favorite sites in the interaction with hydrogen [89], ii) studies aimed at determining the maximum amount of hydrogen that MOFs can store [95], and iii) studies devoted at evaluating hydrogen diffusion into MOFs [96]. The H_2 adsorption sites in MOFs are very often analyzed by QM calculations, either wavefunction- or DFT-based methods. Usually, hydrogen storage capacity is conversely predicted by employing GCMC simulations [97, 98], whereas kinetic properties of H_2 inside MOF frameworks are predicted by MD simulations.

Beyond hydrogen, larger molecules were studied inside MOFs using periodic QM calculations. Devi et al., for example, accurately reproduced the experimental crystal structures of the Zn 2,6-naphthalene host framework (isostructural with MOF-105) using periodic DFT-GGA energy minimizations and allowing both the atomic coordinates and the lattice parameters to change [99]. By calculating the binding energies of various aromatic hydrocarbons within the MOF frame, they were able to explain the selectivity to the uptake of p-xylene compared to that of other xylene isomers. Moreover, Hamel et al. [100] showed, on the basis of static geometry optimizations within the periodic DFT-GGA, that MOF-5 can be utilized as a rigid host matrix for the encapsulation of C_{60} and that in this way the density of states of the fullerene crystal can be fine-tuned. To our knowledge only few papers have been up to now published that concern modeling of catalytic properties, and on this subject we will go into more detail in the following section.

13.5 Modeling Catalysis within MOFs

To the best of our knowledge the first theoretical study on MOF applications in the catalyst field appeared in 2008, and until 2012 just seven papers came into view. A summary, pointing out models and methods employed in these investigations along with the obtained results, is given in the next subsections.

13.5.1 *Formaldehyde Catalytic Addition to Propylene on MOF-11*

In 2008 Choomwattana et al. [72] modeled 3-buten-1-ol production, starting from MOF-11 that embodied formaldehyde and propylene. This is an example of the carbonyl-ene reaction, the ene reaction being an indirect substituting addition of a compound with a double bond to an olefin with an allylic hydrogen. The model was built employing an embedded approach; thus, only a localized moiety into the framework was considered responsible for the electronic properties of the active sites. A QM/MM approach within an Our-own-N-layer Integrated molecular Orbital + molecular Mechanics (ONIOM) extrapolation scheme was considered.

The structure coordinates of MOF-11 were obtained by x-ray diffraction (XRD) data. The framework consisted of square-shaped $Cu_2(CO_2)_4$ paddlewheel units connected by 1,3,5,7-adamantane tetracarboxylate linkers with a PtS topology. One active paddlewheel unit was selected as the ONIOM model system and was treated by a B3LYP/6-31G(d,p) level of theory. The mesoporous framework, mostly characterized by van der Waals confining interactions, was the ONIOM real system accounted for by UFF. The MOF framework was kept frozen at the XRD crystallographic geometry, whereas parts of one paddlewheel active site as well as the adsorbed HCHO and $CH_3CH{=}CH_2$ molecules were fully optimized.

The results were used to predict kinetic parameters using TS theory. For the carbonyl-ene reaction, a concerted mechanism was proposed in which one C–C bond between the formaldehyde and the C1 propylene carbons is formed, the propylene proton being transferred. Assuming a reaction mechanism similar to that inferred in the presence of Na-faujasite (FAU), formaldehyde initially adsorbs over the paddlewheel active site of MOF-11 and then interacts with diffusing propylene via π interactions, followed by the chemical reaction to produce 3-buten-1-ol. The effects of the active site environment, consisting of organic linkers connected to four carboxylic groups (in turn connected to one Cu atom), were analyzed considering also the presence of Cu^+ ions. These, without the shielding effects of O atoms, caused larger electrostatic fields, leading to much lower process activation energies. The results obtained for MOF-11 were similar to those found for FAU zeolite catalysts [101], suggesting that MOF-11 derivatives could be good candidates of use in catalysis, having also an advantage with respect to the FAU ones due to their adjustable cavity sizes.

13.5.2 *Cyclo-[(S)-Phenylalanyl-(S)-Histidyl] Benzaldehyde-Complex Synthesis on MOFs*

Braga et al. [73] in 2008 showed that MOFs, like zeolite materials, could be used as confining systems for improving stereoselectivity in reactions with chiral inductors or chiral auxiliaries. *Cyclo*-[(S)-phenylalanyl-(S)-histidyl] is a cyclic dipeptide having excellent enantioselective catalytic activity in the hydrocyanation of aldehydes. In this reaction, it is protonated in the presence of optically active amines [102]. IRMOF-phen was proposed as the framework to encapsulate the dipeptide above to be complexed with benzaldehyde. Phenanthrene 2,7 dicarboxylate moieties were considered as linkers, substituted or not substituted by CH_3, Br, or even chiral groups.

Conformational analysis of the protonated *cyclo*-[(S)-phenylalanyl-(S)-histidyl], showed in Fig. 13.4, was performed placing inside the IRMOF-phen cavity either the protonated dipeptide (unbound form) or the protonated dipeptide bonded to a phenantrene ligand (bound form). The unbound form led to a chiral auxiliary species within the confined space of the IRMOF-phen cavities that could induce a pronounced chirality in asymmetric reactions,

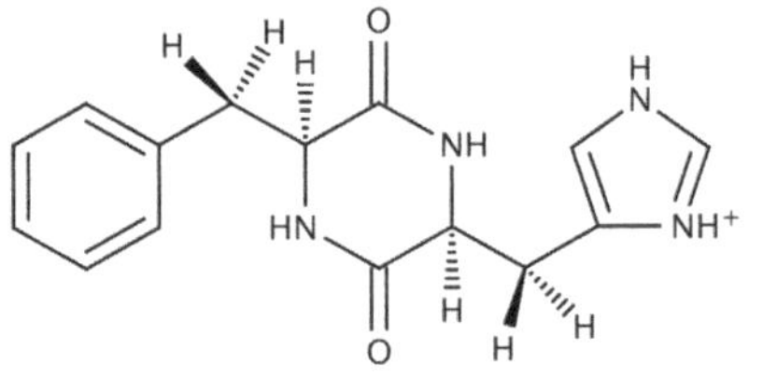

Figure 13.4 Protonated *cyclo*-[(S)-phenylalanyl-(S)-histidyl].

while the bound form showed a chiral cavity. Hydrogen bonding and a π-stacking complex was formed between the cyclic peptide and the benzaldehyde molecule within the IRMOF-phen cavity, suggesting that hydrocyanation of benzaldehyde could be highly enantioselective.

A systematic and stochastic conformational search was performed on the protonated *cyclo*-[(S)-phenylalanyl-(S)-histidyl]. These conformers were optimized within the cavity of the IRMOF-phen model, employing a cluster approach and freezing its coordinates. The protonated *cyclo*-[(S)-phenylalanyl-(S)-histidyl] was also bonded to position 4 of the phenantrene ring. The same analysis was performed for CH_3, Br, and other substituted IRMOF-phen derivatives. The IRMOF cavity seemed to induce significant changes in the conformational behavior of the protonated *cyclo*-[(S)-phenylalanyl-(S)-histidyl], especially in the unbound form. The confining space of the IRMOF cavity had significant effects upon the thermal conformational populations of the cyclic peptide when compared to its conformers in the gas phase. The cavity chemical environment influences these populations, and consequently, CH_3 groups in the phenantrene rings of IRMOF-phen lead to distinct thermally accessible conformers, while Br substituents just to one conformer, at room temperature.

Considering the chemical nature of the complexes devised by the benzaldehyde molecule and the protonated *cyclo*-[(S)-phenylalanyl-(S)-histidyl] species—either in the gas phase or embodied in the IRMOF-phen cavity—the authors used a QM/QM ONIOM approach, locally, employing the hybrid PBE1 functional of exchange and correlation that, as known, is able to improve the description of hydrogen bond formations. The low level was conversely outlined by the AM1 method.

Within the ONIOM approach above, it was calculated that the conformer complexes are stabilized by the formation of hydrogen bonds occurring between the hydrogen of the protonated hystidyl ring and the carbonyl oxygen of the benzaldehyde, as well as by π-stacking interactions taking place among the aromatic rings. The species advanced by the investigation discussed in this section were found to be compatible with the structure proposed for the TS characterizing the hydrocyanation of the benzaldehyde when enantioselectivity is observed. Therefore, the authors concluded that IRMOF cavities not only might change conformer populations but also could stabilize intermediate structures and complexes, properly designing cavity environments, hence improving the enantioselectivity of the catalyzed reactions.

13.5.3 *Knoevenagel Condensation between Benzaldehyde and Ethyl Cyanoacetate inside IRMOF-3*

Cortese and Duca [74] in 2011 modeled the Knoevenagel reaction between ethyl cyanoacetate and benzaldheyde catalyzed by IRMOF-3. Gascon et al. experimentally studying the same condensation [45], indeed, found a lower activity of aniline, used as the reaction catalyst, with respect of that of corresponding aniline-like groups present in the IRMOF-3 framework.

This occurrence was related to possible electronic and steric effects, influencing the chemistry of the aminoaromatic moieties framed inside the IRMOF-3 cavities. It was, as a consequence, assumed that a cluster treatment of the framework could be representative of local properties of the solid. Differently sized fragments were thus cut from the IRMOF-3 lattice and optimized at the DFT level, employing the BP86 functional of exchange and correlation. Zn atoms were described by the LANL2DZ basis set, while the 6-31G(d,p) basis set was utilized for the H, N, C, and O atoms. To take into account possible confinement effects produced by the lattice, a QM/QM approach within an ONIOM scheme was also applied to a large IRMOF-3 fragment, being the real system treated at PM3 level [74].

The calculation results showed that the experimentally found larger basicity of IRMOF-3, compared to simpler catalytic species, mostly depended on the formation of protonated IRMOF-3 acidic-conjugate derivatives, involving hydrogen bonds and originating quasi-planar six-term rings. Interestingly, it was advanced that the basicity of the simpler catalysts and of the corresponding embodied derivatives seems to not play a fundamental role in driving the different catalytic cycles involved.

Moreover, the energy trends, characterizing both the condensation catalytic cycles and the relative deactivation paths of IRMOF-

3 and aniline, clearly display that the increased activity of IRMOF-3 cannot be related to confinement effects and/or to lower energy differences, characterizing given steps involved in the catalytic processes. These findings hence suggested that other occurrences should drive the Knoevenagel processes.

In fact, one by-product formed in the cycles is water. This is able to poison the aminic site, resulting in its deactivation. Unlike the case of simpler phenylaminic species, this effect can be avoided in the IRMOF-3 applications because of its framework Zn centres that interact more strongly with the produced water molecules, so inhibiting their poisoning activity. The increased activity of IRMOF-3 could be, thus, explained by a lower deactivation of the catalyst caused by the produced water molecules.

13.5.4 *Electronic Effects of 5,5'-Substituents on the Enantioselectivity of (Salen)Mn Catalysts*

The study of Oxford et al. [103], performed in 2010, deals with (salen)Mn catalysts. The latter are well-known highly enantioselective catalysts in the epoxidation reactions of unfunctionalized olefins [104]. Because of the critical instability of these catalysts, however, much efforts were devoted in developing their heterogenized analogues [105]. However, both activity and selectivity of the catalysts usually suffer the consequent immobilization techniques. Referring to the inference that MOFs could conjugate the best features of homogeneous and heterogeneous catalysis, hence, to avoid the effects of the immobilization techniques, a MOF derivative was synthesized starting from biphenyldicarboxylate, zinc, and salen(Mn) catalyst (see Fig. 13.5) [35].

The obtained MOF derivative performed enantioselective epoxidation catalysis with enantiomeric excess values only slightly lower than those of the corresponding homogeneous catalyst. It was, however, unclear whether the coordination of the salem 5,5'-substituents to the Zn ions in the MOF or the steric constraints of the framework led to the enantioselectivity drop experimentally observed [106]. Consequent investigations of the electronic nature of the 5,5'-sustituents demonstrated that electron-donating groups were able to reduce the reactivity of the (salen)Mn=O intermedi-

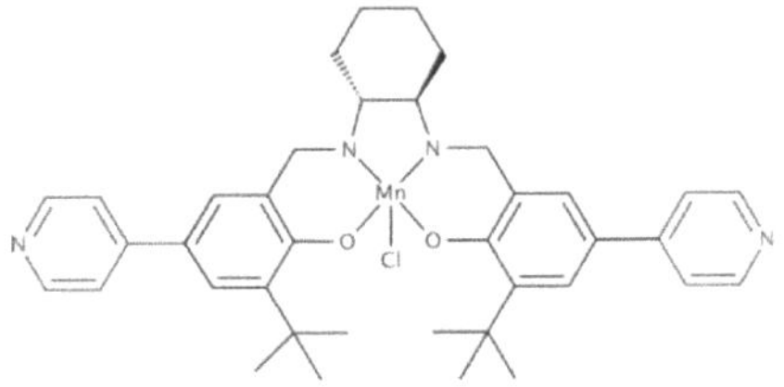

(a)

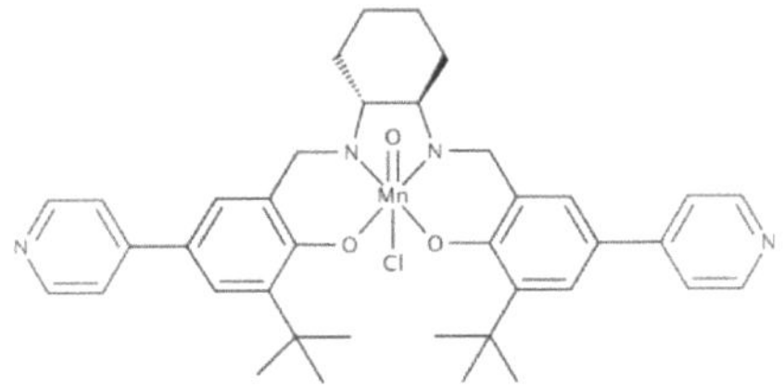

(b)

Figure 13.5 (a) (Salen)Mn catalyst used in the synthesis of MOF derivatives and its (b) oxo-form, which is considered responsible for the enantioselectivity of the catalyst.

ates, leading to larger enantioselectivity [106]. Hybrid QM/MM calculations were employed for studying these experimental findings.

Consequently, several (salen)Mn and salen(Mn)=O species were investigated using G03 [107] and TINKER [108] as QM and MM codes, respectively. The QM portion of the calculations was performed using the BP86 exchange and correlation functional with TZVP basis set centered on the Mn atom and SVP on all the other atoms. The MM portion was treated by a modified version of the DREIDING force field—DREIDING-SALEN, which was specifically developed [103]. At first, it was found that good agreement between QM/MM and full DFT results [109] could be reached with a larger model system, namely, including the phenyl rings of salen. For this purpose, several embedding methods were applied: mechanical embedding, straight electrostatic embedding, and redistributed charge and dipole methods, the latter being those better performing.

As previously mentioned, an understanding of the coordination effects of the 5-5'-substituents is a critical aspect in designing effective heterogeneous salen(Mn) catalysts. In particular, the coordination of pyridyl substituents to Zn paddlewheels was thoroughly analyzed for determining whether pyridyl coordination in MOF derivatives contributed to the lower enantioselectivity experimentally observed. Since two (salen)Mn/MOF crystallographical structures exist, two different cluster models were studied. It was concluded that the coordination of the pyridyl groups to the MOF Zn centers enhances the catalyst enantioselectivity. This, however, conflicts with the experimentally observed enantioselectivity decrease upon the MOF heterogenized pyridine substituted salen(Mn). Hence, it suggests that enantioselectivity drops in MOFs could originate by steric effects more significant than the electronic ones. In passing, it is here to be recalled that another weighty conclusion regarded the electron-withdrawing nature of the Zn ions in MOFs, which would have a negative, if any, influence on the enantioselectivity.

13.5.5 *Steric Effects on Enantioselective Epoxidation Catalyzed by (Salen)Mn in MOFs*

Oxford et al. [110] carried on with the previous study in 2011, dealing with steric effects on the enantioselectivity of heterogenized (salen)Mn catalysts on MOFs by classical atomistic modeling. In fact, the local environments surrounding the catalytic sites could inhibit some among the possible enantioselective pathways. Therefore, a molecular-level understanding of the role of the steric environments on the reagent molecules involved in a given process could help to elucidate the role of the support in asymmetric induction mechanisms [111]. Since it had been already concluded that steric effects could determine the enantioselectivity drops experimentally observed in the (salen)Mn/MOF catalysts [103], the authors of the conclusions above, because of the uniform distribution of the active sites present in those materials, inferred that they could offer a good case study for analyzing steric effects of heterogeneous environments on the mechanism of the asymmetric induction occurring on salen(Mn) catalysts.

A combination of hybrid MC simulations and classical optimizations was employed to determine the preferred diffusion paths of the 2-2 dimethyl-2H-chromene molecule to the active sites of either the homogeneous or both the two, crystallographically different (see preceding section), heterogenized (salen)Mn=O catalysts. In particular, potential energy surfaces for the rotation around the approach dihedral angle were built using constrained classical optimizations as implemented in RASPA [112].

The periodic MOF structure used in the constrained minimizations, performed by the DREIDING-SALEN force field [103] included $2\times2\times2$ unit cells. Global minima for the reactant–catalyst complexes were found at each angle of approach. Indeed, both the (salen)Mn/MOF catalysts showed the same pathway of approach either for *re* or *si* enantiofaces at variance with the homogeneous (salen)Mn catalyst for which different enantiofaces had different pathways. According to the authors, the approach direction changes might contribute to the enantioselectivity drop experimentally found, enabling the explanation of the R,R-enantiomer excess found for the homogeneous catalyst and suggesting that flexibility and interpenetration of the frameworks could play a significant role in controlling enantioselectivity.

13.5.6 *Effects of Lewis Acidity in Citronellal Cyclization Catalyzed by UiO-66*

In this work of 2012, Vermoortele et al. combined experimental and theoretical investigations to study the effects of the substituents toward the cyclization of citronellal catalyzed by UiO-66 derivatives [113]. The latter were obtained by singly including eight different 1,4-benzenedicarboxylate (BDC-X; X = H, NH_2, CH_3, OCH_3, F, Cl, Br, NO_2) species in the MOF structures. The reaction, summarized in Fig. 13.6, had been chosen because the selectivity to the products, namely, the isopulegol isomers, strongly depended on the Lewis acidity of the active sites. All the prepared materials resulted were catalytically active; however, the reaction rates increased with increasing electron-withdrawing properties of the groups on the linker. To rationalize the substituent electronic influence on the catalyst, molecular modeling was exerted on UiO-66 and UiO-66 (NO_2) derivatives. Ab initio calculations framed in a cluster approach were, conversely, performed to predict the reaction rates in the citronellal cyclization.

To identify the active sites into the citronellal cyclization four different clusters, built through fully hydrated UiO-66 unit

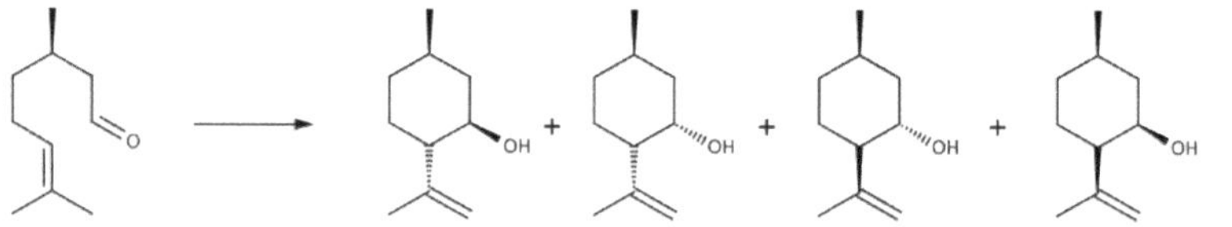

Figure 13.6 Cyclization of (+)-citronellal to isopulegol and its isomers.

cells, were optimized, making use of Vienna Ab initio Simulation Package (VASP) [114–117], by periodic DFT calculations: i) the fully hydrated cluster $[Zr_6O_4(OH)_4(HCOO)_{12}]$, ii) the fully dehydrated cluster $Zr_6O_6(HCOO)_{12}$, iii) a monodecarboxylated cluster $Zr_6O_5(OH)_3(HCOO)_{11}$, and a iv) a tridecarboxylated cluster $Zr_6O_7(OH)(HCOO)_9$.

TSs for the citronellal cyclization, producing isopulegol, were isolated on each of the four clusters using DFT geometry optimizations and the B3LYP functional of exchange and correlation. 4-31G* basis sets were used for all the atoms except Zr for which the LANL2DZ effective core potential and basis set were applied. Single-point energy refinements with the 6-31G+g(d) basis sets were also performed. Finally, the quasi-IRC approach was employed to get from TSs corresponding products' geometries.

Authors found that the only acceptable kinetic results were obtained for clusters mono- and tridecarboxylated. Therefore, they decided to select the monodecarboxylated one as the catalytic model. Following its application, it was advanced that nitrosubstitution in the catalyst could cause lowering of adsorption and of the reaction activation free energy. Interestingly, according to the authors, the substituents not only alter the Lewis acidic properties but also induce either stabilizing or destabilizing effects on the reactants, depending on their electronic properties.

13.5.7 *Epoxidation of Cyclohexene Catalyzed by Coordinatively Saturated Vanadium MIL-47*

In 2012 Leus et al. [118] theoretically and experimentally explored the properties of the coordinatively saturated vanadium Matriaux de l'Institut Lavoisier (MIL)-47 as a low-leaching heterogeneous catalyst in cyclohexene oxidation. MIL-47 has a 3D framework, where each V(IV) center is coordinated to four oxygen atoms (through four carboxylate groups and two oxygen atoms of the O–V–O axis), thus forming saturated octhaedral coordination nodes.

The approach selected was based on precedent studies modeling homogeneous $VO(acac)_2$ catalysts, for which authors found that, at first, vacant coordination sites should be created to activate the catalyst [119]. These structural defects in the periodic structure were hence simulated to analyze the heterogeneous reaction pathways, employing a cluster approach.

The model of the active sites included two vanadium atoms, bridged by two terephtalate linkers. Furthermore, into the vanadyl chains, only two V=O bonds were maintained while -OH cutoff groups were introduced in the terephtalate linkers. During some structure optimizations the outer carboxyl oxygen atoms were fixed at their crystallographic positions to prevent unphysical deformation of the cluster. Nevertheless, constrained and unconstrained models were employed to calculate catalytic properties of the coordinatively saturated vanadium MIL-47.

Two main epoxidation pathways were actually proposed and the energetics fully evaluated at the B3LYP/6-311+g(3df,2p)-D3//B3LYP/6-31G+g(d) level of theory. Constrained and unconstrained models predicted nearly the same reaction barriers and entropic contributions. The pathways were identified as direct and radical routes, respectively. They share the first step where the catalyst interacts with *tert*-butylhydroperoxide (TBHP) to produce a hydroperoxide of the catalyst. After this step, the hydroperoxide derivative either could evolve by a direct mechanism into an epoxide (being the catalyst regenerated by adding TBHP) or could produce a radical of the catalyst (being the V(IV) species oxidized to V(V) that further reacts with TBHP to give a complex, which can be epoxidated). The reaction pathways were found to be competitive, both forming a closed cycle of epoxidation and regeneration. Interestingly, in both cases the peroxide is of use in transforming inactive vanadium complexes into active species, showing vanadium in the oxidation states (IV) or (V).

Solvent effects were taken also into account because of the hydroxyl role in activating the catalysts (see below) and of the experimentally found larger preconditioning time found in the cyclohexene conversion when decane was used as the solvent. With respect to the former point, it was, indeed, observed that irrespective of the considered pathway, the vanadium catalyst needed one hydroxyl group to react with TBHP. Such a hydroxyl group could be easily introduced in a water-based medium, and thus water could help the formation of free coordination sites, enhancing the initial rate of cyclohexene conversion. Leus et al., however, stressed that water should be avoided as a pure solvent because it could accelerate undesired radical side reactions [118].

13.5.8 *Hydroperoxide Decomposition over Cu and Co Cations Embodied in MOFs*

In 2012 Ryan et al. [120] modeled the hydroperoxide decomposition within MOF embodying either copper or cobalt nodes (i.e., active sites) by means of DFT-based calculations using a cluster approach. Cluster models were extracted from MOF crystal structures, including active sites. The geometry optimization involved B3LYP as functional of exchange and correlation and as basis sets the TZVP for the metal centers and the 6-31G(d,p) for all the remaining atoms. Geometry optimizations were further followed by single-point energy calculations in which TZVP was maintained for the metallic centers, while the 6-311G(d,p) basis substituted the lower-level sets for all the other atoms.

Cobalt-containing MOF (Co MOF) derivatives, also known as zeolitic imidazolate framework (ZIF)-9, are composed by Co^{2+} cations interacting with anionic phenylimidazolate linkers and show sodalite-like structures. Copper-containing MOF (Cu MOF) derivatives are conversely composed of Cu^{2+} cations and 2-hydroxypyrimidinolate linkers in a 1:2 molar ratio. Copper sites are distributed in quasi-square planar configurations, which leave them coordinatively unsaturated and hence able to interact with guest molecules.

For the Co MOF systems, one cobalt atom and four surrounding linkers were included in the model. To lower the computational effort, the phenylimidazolate units were still replaced by imidazolate moieties. For the Cu MOF systems, one copper atom and four 2-hydroxypyrimidinolate linkers were instead modeled.

Although the linkers of both (Co and Cu) MOFs have anionic character and are interacting at both ends with metallic centers having cationic character, the cutoff atoms at the boundary moieties were in any case saturated by hydrogen. Additional calculations on Cu MOF models, showing copper paddlewheel nodes, were also carried out to deepen the role of the steric hindrance in the hydroperoxide decomposition on metal/MOFs.

The catalyzed hydroperoxide decomposition on MOFs should involve the Haber–Weiss cycle [121]. In this, the metallic cations undergo oxidation followed by reduction to decompose hydroperoxides into alkoxy and peroxy radical species, respectively. The proposed catalytic cycle starts through the formation of a bound species (occurring between a metallic site and the hydroperoxide), which is followed by the hydroperoxide O–O cleavage. After this, the alkoxy radical leaves, and another hydroperoxide species could enter to react and form a bound water molecule and a hydorperoxy radical. The catalytic cycle is complete when one water molecule desorbs.

The computational results suggest that the metallic sites within Co MOFs and Cu MOFs are limited with respect to their potential catalytic activity. In particular, steric effects around the metal cations might prevent direct interactions of the reactants with the metallic centers. As a consequence, authors suggested that surface catalysis could be dominant for these systems.

The deepening analysis concerning the copper paddlewheel conversely showed that, in this case, the linkers do not inhibit—due to steric reasons—guest molecules to complex copper atoms. Nevertheless, these clusters do not catalyze the O–O bond cleavage of hydroperoxide species. Instead, preliminary calculations involving undercoordinated metal sites (on MOF external surfaces) showed that they could have catalytic behavior.

References

1. Baur, R., Schwarzenbach, G. (1960). Neue einschlussverbindungen vom typus des nickelcyanid-ammoniak-benzols, *Helv. Chim. Acta*, **43**, 842–847.

2. Billard, R.D., Wilkinson, G. (1963). Triethylenetetramine complexes of cobalt(III) and rhodium(III), *J. Chem. Soc.*, 3193–3200.
3. Griffin, W.P. (1962). Cyanide complexes of the transition metals, *Q. Rev. Chem. Soc.*, **16**, 188–207.
4. Iwamoto, T., Nakano, T., Morita, M., Miyoshi, T., Miyamoto, T., Sasaki, Y. (1968). The Hofmann-type clathrate: $M(NH_3)2M'(CN)_4$ *Inorg. Chim. Acta*, **2**, 313–316.
5. Krishnamurty, K.V., Harris, G.M., Sastri, V.S. (1970). Chemistry of the metal carbonato complexes, *Chem. Rev.*, **70**, 171–197.
6. Rayner, J.H., Powell, H.M. (1952). Structure of molecular compounds. Part X. Crystal structure of the compound of benzene with an ammonia-nickel cyanide complex, *J. Chem. Soc.*, 319–328.
7. Walton, R.A. (1965). The reactions of metal halides with alkyl cyanides, *Q. Rev., Chem. Soc.*, **19**, 126–143.
8. Biradha, K., Sarkar, M., Rajput, L. (2006). Crystal engineering of coordination polymers using 4,4'-bipyridine as a bond between transition metal atoms, *Chem. Commun.*, **40**, 4169–4179.
9. Cheetham, A.K., Rao, C.N.R., Feller, R.K. (2006). Structural diversity and chemical trends in hybrid inorganic-organic framework materials, *Chem. Commun.*, **46**, 4780–4795.
10. James, S.L. (2003). Metal-organic frameworks, *Chem. Soc. Rev.*, **32**, 276–288.
11. Moulton, B., Zaworotko, M.J. (2001). From molecules to crystal engineering: supramolecular isomerism and polymorphism in network solids, *Chem. Rev.*, **101**, 1629–1658.
12. Robin, A.Y., Fromm, K.M. (2006). Coordination polymer networks with O- and N-donors: what they are, why and how they are made, *Coord. Chem. Rev.*, **250**, 2127–2157.
13. Robson, R. (2008). Design and its limitations in the construction of bi- and poly-nuclear coordination complexes and coordination polymers (aka MOFs): a personal view, *Dalton Trans.*, **38**, 5113–5131.
14. Rosi, N.L., Eddaoudi, M., Kim, J., O'Keeffe, M., Yaghi, O.M. (2002). Advances in the chemistry of metal-organic frameworks, *CrystEngComm.*, **4**, 401–404.
15. Rosseinsky, M.J. (2004). Recent developments in metal-organic framework chemistry: design, discovery, permanent porosity and flexibility, *Microporous Mesoporous Mater.*, **73**, 15–30.

16. Shimizu, G.K.H. (2005). Assembly of metal ions and ligands with adaptable coordinative tendencies as a route to functional metal-organic solids, *J. Solid State Chem.*, **178**, 2519–2526.

17. Zaworotko, M.J. (2007). Molecules to crystals, crystals to molecules ... and back again? *Cryst. Growth Des.*, **7**, 4–9.

18. Eddaoudi, M., Kim, J., Rosi, N., Vodak, D., Watcher, J., O'Keeffe, M., Yaghi, O.M. (2002). Systematic design of pore size and functionality in isoreticular MOFs and their application in methane storage, *Science* **295**, 469–472.

19. Hoskins, B.F., Robson, R. (1989). Infinite polymeric frameworks consisting of three dimensionally linked rod-like segments, *J. Am. Chem. Soc.*, **111**, 5962–5964.

20. Hoskins, B.F., Robson, R. (1990). Design and construction of a new class of scaffolding-like materials comprising infinite polymeric frameworks of 3D-linked molecular rods. A reappraisal of the zinc cyanide and cadmium cyanide structures and the synthesis and structure of the diamond-related frameworks $[N(CH_3)_4][CuIZnII(CN)_4]$ and CuI[4,4',4",4"'-tetracyanotetraphenylmethane]$BF_4.xC_6H_5NO$, *J. Am. Chem. Soc.*, **112**, 1546–1554.

21. Wells, A.F. (1954a). The geometrical basis of crystal chemistry. Part 1, *Acta Crystallogr.*, **7**, 535–544.

22. Wells, A.F. (1954b). The geometrical basis of crystal chemistry. Part 2, *Acta Crystallogr.*, **7**, 545–554.

23. Gardner, G.B., Venkataraman, D., Moore, J.S., Lee, S. (1995). Spontaneous assembly of a hinged coordination network, *Nature*, **374**, 792–795.

24. Kondo, M., Yoshitomi, T., Seki, K., Matsuzaka, H., Kitagawa, S. (1997). Three-dimensional framework with channeling cavities for small molecules: [M(4,4'-bpy)3(NO3)4x(H2O)]n (M = Co, Ni, Zn), *Angew. Chem., Int. Ed. Engl.*, **36**, 1725–1727.

25. Yaghi, O.M., Li, G., Li, H. (1995a). Selective binding and removal of guests in a microporous metal-organic framework, *Nature*, **378**, 703–706.

26. Yaghi, O.M., Li, H. (1995). Hydrothermal synthesis of a metal-organic framework containing large rectangular channels, *J. Am. Chem. Soc.*, **117**, 10401–10402.

27. Yaghi, O.M., Richardson, D.A., Li, G., Davis, E., Groy, T.L. (1995b). Open-framework solids with diamond-like structures prepared from

clusters and metal-organic building blocks, *Mater. Res. Soc. Symp. Proc.*, **371**, 15–19.

28. Noro, S., Kitagawa, S., Kondo, M., Seki, K. (2000). A new, methane adsorbent, porous coordination polymer [CuSiF6(4,4'-bipyridine)2]n, *Angew. Chem., Int. Ed.*, **39**, 2082–2084.
29. Shimomura, S., Horike, S., Kitagawa, S. (2007). Chemistry and application of porous coordination polymers, *Stud. Surf. Sci. Catal.*, **170B**, 1983–1990.
30. Kitagawa, S., Kitaura, R., Noro, S. (2004). Functional porous coordination polymers, *Angew. Chem., Int. Ed.*, **43**, 2334–2375.
31. Furukawa, H., Kim, J., Ockwig, N.W., O'Keeffe, M., Yaghi, O.M. (2008). Control of vertex geometry, structure dimensionality, functionality, and pore metrics in the reticular synthesis of crystalline metal-organic frameworks and polyhedra, *J. Am. Chem. Soc.*, **130**, 11650–11661.
32. Tranchemontagne, D.J., Mendoza-Cortés, J.L., O'Keeffe, M., Yaghi, O.M. (2009). Secondary building units, nets and bonding in the chemistry of metal-organic frameworks, *Chem. Soc. Rev.*, **38**, 1257–1283.
33. Fujita, M., Kwon, Y.J., Washizu, S., Ogura, K. (1994). Preparation, clathration ability, and catalysis of a two-dimensional square network material composed of cadmium(II) and 4,4'-bipyridine, *J. Am. Chem. Soc.*, **116**, 1151–1152.
34. Lee, J.Y., Farha, O.K., Roberts, J., Scheidt, K.A., Nguyen, S.T., Hupp, J.T. (2009). Metal-organic framework materials as catalysts, *Chem. Soc. Rev.*, **38**, 1450–1459.
35. Cho, S.H., Ma, B., Nguyen, S.T., Hupp, J.T., Albrecht-Schmitt, T.E. (2006). A metal-organic framework material that functions as an enantioselective catalyst for olefin epoxidation, *Chem. Commun.*, 2563–2565.
36. Alkordi, M.H., Liu, Y., Larsen, R.W., Eubank, J.F., Eddaoudi, M. (2008). Zeolite-like metal-organic frameworks as platforms for applications on metalloporphyrin-based catalysts, *J. Am. Chem. Soc.*, **130**, 12639–12641.
37. Chui, S.S., Lo, S.M., Charmant, J.P., Orpen, A.G., Williams, I.D. (1999). A chemically functionalizable nanoporous material $[Cu_3(TMA)_2(H_2O)_3]_n$, *Science*, **283**, 1148–1150.
38. Farha, O.K., Mulfort, K.L., Hupp, J.T. (2008). An example of node-based post assembly elaboration of a hydrogen-sorbing, metal-organic framework material, *Inorg. Chem.*, **47**, 10223–10225.

39. Wu, C.D., Hu, A., Zhang, L., Lin, W. (2005). A homochiral porous metal-organic framework for highly enantioselective heterogeneous asymmetric catalysis, *J. Am. Chem. Soc.*, **127**, 8940–8941.
40. Horike, S., Dinca, M., Tamaki, K., Long, J.R. (2008). Size-selective lewis acid catalysis in a microporous metal-organic framework with exposed Mn^{2+} coordination sites, *J. Am. Chem. Soc.*, **130**, 5954–5855.
41. Corma, A., García, H. and Llabrés i Xamena, F.X. (2010). Engineering metal organic frameworks for heterogeneous catalysis, *Chem. Rev.*, **110**, 4606–4655.
42. Llabrés i Xamena, F.X., Abad, A., Corma, A., García, H. (2007). MOFs as catalysts: activity, reusability and shape-selectivity of a Pd-containing MOF, *J. Catal.*, **250**, 294–298.
43. Ravon, U., Domine, M.E., Gaudillere, C., Desmartin-Chomel, A., Farrusseng, D. (2008). MOFs as acid catalysts with shape selectivity properties, *New J. Chem.*, **32**, 937–940.
44. Seo, J.S., Whang, D., Lee, H., Jun, S.I., Oh, J., Jeon, Y.J., Kim, K. (2000). A homochiral metal–organic porous material for enantioselective separation and catalysis, *Nature*, **404**, 982–986.
45. Gascon, J., Aktay, U., Hernandez-Alonso, M.D., van Klink, G.P.M., Kapteijn, F. (2009). Amino-based metal-organic frameworks as stable, highly active basic catalysts, *J. Catal.*, **261**, 75–87.
46. Müller, M., Hermes, S., Kähler, K., vanden Berg, M.W.E., Muhler, M., Fischer, R.A. (2008). Loading of MOF-5 with Cu and ZnO nanoparticles by gas–phase infiltration with organometallic precursor: properties of Cu/ZnO@MOF as catalyst for methanol synthesis, *Chem. Mater.*, **20**, 4576–4587.
47. Tafipolsky, M., Amirjalayer, S., Schmid, R. (2010a). Atomistic theoretical models for nanoporous hybrid materials, *Microporous Mesoporous Mater.*, **129**, 304–318.
48. Ramanan, A., Whittingham, M.S. (2006). How molecules turn into solids: the case of self-assembled metal-organic frameworks, *Cryst. Growth Des.*, **6**, 2419–2421.
49. Ferey, G., Serre, C. (2009). Large breathing effects in three–dimensional porous hybrid matter: facts, analyses, rules and consequences, *Chem. Soc. Rev.*, **38**, 1380–1399.
50. Neyman, K.M., Illas, F. (2005). Theoretical aspects of heterogeneous catalysis: applications of density functional methods, *Catal. Today*, **105**, 2–16.

51. Duca, D., Baranyai, P., Vidoczy, T. (1998). A Monte Carlo model for the hydrogenation of ethylene on Pt catalysts, *J. Comput. Chem.*, **19**, 396–403.
52. Duca, D., Barone, G., Giuffrida, S., Varga, relax Zs. (2007). IDEA: interface dynamics and energetics algorithm, *J. Comput. Chem.*, **28**, 2483–2499.
53. Duca, D., Barone, G., Varga, Z. (2001). Hydrogenation of acetylene-ethylene mixtures on Pd catalysts: computational study on the surface mechanism and on the influence of the carbonaceous deposits, *Catal. Lett.*, **72**, 17–23.
54. Duca, D., Botar, L., Vidoczy, T. (1996). Monte Carlo simulations of ethylene hydrogenation on Pt catalysts, *J. Catal.*, **162**, 260–267.
55. Mei, D., Sheth, P.A., Neurock, M., Smith, C.M. (2006). First-principles-based kinetic Monte Carlo simulation of the selective hydrogenation of acetylene over Pd(111), *J. Catal.*, **242**, 1–15.
56. Duca, D., LaManna, G., Russo, M.R. (1999). Computational studies on surface reaction mechanism: ethylene hydrogenation on platinum catalysts, *Phys. Chem. Chem. Phys.*, **1**, 1375–1382.
57. Duca, D., LaManna, G., Varga, Z., Vidoczy, T. (2000). Hydrogenation of acetylene-ethylene mixtures on Pd catalysts: study of the surface mechanism by computational approaches. Metal dispersion and activity of the catalyst, *Theor. Chem. Acc.*, **104**, 302–311.
58. Khon, W., Sham, L.J. (1965). Self-consistent equations including exchange and correlation effects, *Phys. Rev. A*, **140**, 1133–1138.
59. Nørskov, J.K., Bligaard, T., Rossmeisl, J. Christensen, C.H. (2009). Towards the computational design of solid catalysts, *Nat Chem.*, **1**, 37–46.
60. Hohenberg, P., Kohn, W. (1964). Inhomogeneous electron gas, *Phys. Rev. B* **136**, B864–B871.
61. Kohn, W., Sham, L.S. (1965). Self-consistent equations including exchange and correlation effects, *Phys. Rev.*, **140**, A1133–A1138.
62. Cortese, R., Duca, D., Sifontes-Herrera, V.A., Murzin, D.Y. (2012). L-arabinose conformers adsorption on ruthenium surfaces: a DFT study, *J. Phys. Chem. C*, **116**, 14908–14916.
63. Ferrante, F., Rubino, T., Duca, D. (2011). Butene isomerization and double-bond migration on the H-ZSM-5 outer surface: a density functional theory study, *J. Phys. Chem. C*, **115**, 14862–14868.
64. Arena, F., Ferrante, F., Spadaro, L., Prestianni, A., Raneri, A., Duca, D. (2011). Factors controlling the energy of nitrogen monolayer coverage

on high surface area catalyst oxide carriers, *J. Phys. Chem. C*, **115**, 24728–24733.

65. Armata, N., Baldissin, G., Barone, G., Cortese, R., D'Anna, V., Ferrante, F., Giuffrida, S., LiManni, G., Prestianni, A., Rubino, T., Duca, D. (2009). Structural and kinetic DFT characterization of materials to rationalize catalytic performance, *Top. Catal.*, **52**, 444–455.
66. Barone, G., Casella, G., Giuffrida, S., Duca, D. (2007). H-ZSM-5 modified zeolite: quantum chemical models of acidic sites, *J. Phys. Chem. C*, **111**, 13033–13043.
67. Czekaj, I., Wambach, J., Kröcher, O. (2009). Modelling catalyst surfaces using DFT cluster calculations, *Int. J. Mol. Sci.*, **10**, 4310–4329.
68. Civalleri, B., Napoli, F., Noel, Y., Roetti, C., Dovesi, R. (2006). Ab-initio prediction of materials properties with CRYSTAL: MOF-5 as a case study, *Cryst. Eng. Com.*, **8**, 364–371.
69. Dovesi, R., Saunders, V.R., Roetti, C., Orlando, R., Zicovich-Wilson, C.M., Pascale, F., Civalleri, B., Doll, K., Harrison, N.M., Bush, I.J., D'Arco, P., Llunell, M. (2009). CRYSTAL09 User's Manual., University of Torino Torino.
70. Svensson, M., Humbel, S., Froese, R. D.J., Matsubara, T., Sieber, S., Morokuma, K. (1996). ONIOM: a multilayered integrated MO + MM method for geometry optimizations and single point energy predictions. A test for Diels-Alder reactions and Pt(P(t-Bu)3)2 + H2 oxidative addition, *J. Phys. Chem.*, **100**, 19357–19363.
71. Svelle, C., Tuma, S., Rozanska, X., Kerber, T., Sauer, J. (2009). Quantum chemical modeling of zeolite-catalyzed methylation reactions: toward chemical accuracy for barriers, *J. Am. Chem. Soc.*, **131**, 816–825.
72. Choomwattana, S., Maihom, T., Khongpracha, P., Probst, M., Limtrakul, J. (2008). Structures and mechanisms of the carbonyl-ene reaction between MOF-11 encapsulated formaldehyde and propylene: an ONIOM study, *J. Phys. Chem.*, **112**, 10855–10861.
73. Braga, C.F., Longo, R.L. (2008). Conformational analysis of protonated cyclo-[(S)-phenylalanyl-(S)-histidyl] and its complex with benzaldehyde within metal-organic frameworks (MOFs), *J. Braz. Chem. Soc.*, **19**, 321–329.
74. Cortese, R., Duca, D. (2011). A DFT study of IRMOF-3 catalyzed Knoevenagel condensation, *Phys. Chem. Chem. Phys.*, **13**, 15995–16004.
75. Tafipolsky, M., Schmid, R. (2009). Systematic first principles parameterization of force fields for metal-organic frameworks using a genetic algorithm approach, *J. Phys. Chem. B*, **113**, 1341–1352.

76. Tafipolsky, M., Amirjalayes, S., Schmid, R. (2010b). First-principles-derived force field for copper paddle-wheel-based metal-organic frameworks, *J. Phys. Chem. C*, **114**, 14402–14409.
77. Amirjalayes, S., Tafipolsky, M., Schmid, R. (2011). Exploring network topologies of copper paddle-wheel-based metal organic frameworks with a first principles derived force field, *J. Phys. Chem. C*, **115**, 15133–15139.
78. Calatayuda, M. (2010). Ethylene glycol interaction on alkaline earth oxides: a periodic DFT study, *Catal. Today*, **152**, 88–92.
79. Loffreda, D. (2006). Theoretical insight of adsorption thermodynamics of multifunctional molecules on metal surfaces, *Surf. Science*, **600**, 2103–2112.
80. Mueller, T., Ceder, G. (2005). A density functional theory study of hydrogen adsorption in MOF-5, *J. Phys. Chem. B*, **109**, 17974–17983.
81. Mulder, F.M., Dingemans, T.J., Wagemaker, M., Kearley, G.J. (2005). Modelling of hydrogen adsorption in the metal organic framework MOF-5, *Chem. Phys.*, **317**, 113–118.
82. Barone, G., Duca, D. (2002). Hydrogenation of 2,4-dinitrotoluene on Pd/C catalysts: computational study on the influence of the steric hindrance of the surface species and of the metal dispersion on the reaction mechanism, *J. Catal.*, **211**, 296–307.
83. Maitland, G.C., Rigby, M., Smith, E.B., Wakeham, W.A. (1981). *Intermolecular Forces: Their Origin and Determination* (Clarendon Press).
84. Leach, A.R. (2001). *Molecular Modeling: Principles and Applications* (Pearson Education EMA).
85. Woods, G.B., Rowlinson, J.S. (1989). Computer simulations of fluids in zeolites X and Y, *J. Chem. Soc, Faraday Trans. 2*, **85**, 765–781.
86. Soto, J.L., Myers, A.L. (1981). Monte Carlo studies of adsorption in molecular sieves, *Mol. Phys.*, **42**, 971–983.
87. Demontis, P., Suffritti, G.B. (1997). Structure and dynamics of zeolites investigated by molecular dynamics, *Chem. Rev.*, **97**, 2845–2878.
88. Smit, B., Maesen, T.L.M. (2008). Molecular simulations of zeolites: adsorption, diffusion, and shape selectivity, *Chem. Rev.*, **108**, 4125–4184.
89. Han, S.S., Mendoza-Cortes, J.L., Goddard III, W.A. (2009). Recent advances on simulation and theory of hydrogen storage in metal-organic frameworks and covalent organic frameworks, *Chem. Soc. Rev.*, **38**, 1460–1476.

90. Sagara, T., Klassen, J., Ganz, E. (2004). Computational study of hydrogen binding by metal-organic framework-5, *J. Chem. Phys.*, **121**, 12543–12548.

91. Sillar, K., Hofmann, A., Sauer, J. (2009). Ab initio study of hydrogen adsorption in MOF-5, *J. Am. Chem. Soc.*, **131**, 4134–4150.

92. Yang, Q., Zhong, C., Chen, J.F. (2008). Computational study of CO2 storage in metal-organic frameworks, *J. Phys. Chem. C*, **112**, 1562–1569.

93. Zheng, C., Liu, D., Yang, Q., Zhong, C., Mi, J. (2009). Computational study on the influences of framework charges on CO2 uptake in metal-organic frameworks, *Ind. Eng. Chem. Res.*, **48**, 10479–10484.

94. Keskin, S., Liu, J., Rankin, R.B., Johnson, J.K., Sholl, D.S. (2009). Progress, opportunities, and challenges for applying atomically detailed modeling to molecular adsorption and transport in metal-organic framework materials, *Ind. Eng. Chem. Res.*, **48**, 2355–2371.

95. Garberoglio, G., Skoulidas, A.I., Johnson, J.K. (2005). Adsorption of gases in metal organic materials: comparison of simulations and experiments, *J. Phys. Chem. B*, **109**, 13094–13103.

96. Yang, Q., Zhong, C. (2005). Molecular simulation of adsorption and diffusion of hydrogen in metal-organic frameworks, *J. Phys. Chem. B*, **109**, 11862–11864.

97. Allen, M.P., Tildesley, D.J. (1987). *Computer Simulation of Liquids* (Oxford University Press).

98. Frenkel, D., Smit, B. (2002). *Understanding Molecular Simulation: From Algorithms to Applications* (Academic Press).

99. Devi, R.N., Edgar, M., Gonzalez, J., Slawin, D.P., Tunstall, D.P., Grewal, P., Cox, P.A., Wright, P.A. (2004). Structural studies and computer simulation of the inclusion of aromatic hydrocarbons in a zinc 2,6-naphthalene dicarboxylate framework compound, *J. Phys. Chem. B*, **108**, 535–543.

100. Hamel, S., V., T. and Côté, M. (2005). Fullerene in a metal-organic matrix: design of the electronic structure, *Phys. Rev. Lett.*, **95**, 146403–146407.

101. Sangthong, W., Probst, M., Limtrakul, J. (2005). Computational study of the carbonyl-ene reaction of encapsulated formaldehyde in Na-FAU zeolite, *J. Mol. Struct.*, **748**, 119–127.

102. Lowry, T.H., Richardson, K.S. (1987). *Mechanism and Theory in Organic Chemistry* (Harper Collins, New York).

103. Oxford, G.A.E., Snurr, R.Q., Broadbelt, L.J. (2010). Hybrid quantum mechanics/molecular mechanics investigation of (salen)Mn for use in metal-organic frameworks, *Ind. Eng. Chem.*, **49**, 10965–10973.

104. Yoon, T.P., Jacobsen, E.N. (2003). Privileged chiral catalysts, *Science* **299**, 1691–1693.

105. Baleizão, C., Garcia, H. (2006). Chiral salen complexes: an overview to recoverable and reusable homogeneous and heterogeneous catalysts, *Chem. Rev.*, **106**, 3987–4043.

106. Jacobsen, E.N., Zhang, W., Guler, M.L. (1991). Electronic tuning of asymmetric catalysts, *J. Am. Chem. Soc.*, **113**, 6703–6704.

107. Frisch, M.J., Trucks, G.W., Schlegel, H.B., Scuseria, G.E., Robb, M.A., Cheeseman, J.R., MontgomeryJr., J.A., Vreven, T., Kudin, K.N., Burant, J.C., Millam, J.M., Iyengar, S.S., Tomasi, J., Barone, V., Mennucci, B., Cossi, M., Scalmani, G., Rega, N., Petersson, G.A., Nakatsuji, H., Hada, M., Ehara, M., Toyota, K., Fukuda, R., Hasegawa, J., Ishida, M., Nakajima, T., Honda, Y., Kitao, O., Nakai, H., Klene, M., Li, X., Knox, J.E., Hratchian, H.P., Cross, J.B., Bakken, V., Adamo, C., Jaramillo, J., Gomperts, R., Stratmann, R.E., Yazyev, O., Austin, A.J., Cammi, R., Pomelli, C., Ochterski, J.W., Ayala, P.Y., Morokuma, K., Voth, G.A., Salvador, P., Dannenberg, J.J., Zakrzewski, V.G., Dapprich, S., Daniels, A.D., Strain, M.C., Farkas, O., Malick, D.K., Rabuck, A.D., Raghavachari, K., Foresman, J.B., Ortiz, J.V., Cui, Q., Baboul, A.G., Clifford, S., Cioslowski, J., Stefanov, B.B., Liu, G., Liashenko, A., Piskorz, P., Komáromi, I., Martin, R.L., Fox, D.J., Keith, T., Al-Laham, M.A., Peng, C.Y., Nanayakkara, A., Challacombe, M., Gill, P. M.W., Johnson, B., Chen, W., Wong, M.W., Gonzalez, C., Pople, J.A. (2005). Gaussian 03, Revision D.02, Gaussian, Inc., Wallingford CT.

108. Ponder, J.P. (2004). TINKER 4.2, Software Tools for Molecular Design, Washington University St. Louis MO.

109. Cavallo, L. Jacobsen, H. (2003). Electronic effects in (salen)Mn-based epoxidation catalyst, *J. Org. Chem.*, **68**, 6202–6207.

110. Oxford, G.A.O., Dubbeldam, D., Broadbelt, L.J., Snurr, R.Q. (2011). Elucidating steric effects on enantioselective epoxidation catalyed by (salen)Mn in metal-organic frameworks, *J. Mol. Cat. A: Chem.*, **334**, 89–97.

111. Malek, K., Jansen, A.P.J., Li, C., van Santen, R.A. (2007). Enantioselectivity of immobilized Mn-salen complexes: a computational study, *J. Catal.*, **246**, 127–135.

112. Dubbeldam, D., Calero, S., Ellis, D.E., Snurr, R.Q. (2008). RASPA 1.0, Northwestern University Evanston.

113. Vermoortele, F., Vandichel, M., Van deVoorde, B., Ameloot, R., Waroquier, M., VanSpeybroeck, V., DeVos, D.E. (2012). Electronic effects of linker substitution on Lewis acid catalysis with metal-organic frameworks, *Angew. Chem., Int. Ed. Engl.*, **51**, 4887–4890.

114. Kresse, G., Furthmüller, J. (1996a). Efficiency of ab-initio total energy calculations for metals and semiconductors using a plane-wave basis set, *Comput. Mat. Sci.*, **6**, 15–19.

115. Kresse, G., Furthmüller, J. (1996b). Efficient iterative schemes for ab initio total-energy calculations using a plane-wave basis set, *Phys. Rev. B*, **54**, 11169–11175.

116. Kresse, G., Hafner, J. (1993). Ab initio molecular dynamics for liquid metals, *Phys. Rev. B*, **47**, 558–562.

117. Kresse, G., Hafner, J. (1994). Ab initio molecular-dynamics simulation of the liquid-metal-amorphous-semiconductor transition in germanium, *Phys. Rev. B*, **49**, 14251–14255.

118. Leus, K., Vandichel, M., Ying-Ya, L., Mulylaert, I., Musschoot, J., Pyl, S., Vrienlinck, H., Callens, F., Marin, G.B., Detavernier, C., Wiper, P.V., Khmyak, Y.Z., Waroquier, M., VanSpeybroeck, V., Van DerVoort, P. (2012). The coordinatively saturated vanadium MIL-47 as low leaching heterogeneous catalyst in the oxidation of cyclohexene, *J. Catal.*, **285**, 196–207.

119. Leus, K., Mulylaert, I., Vandichel, M., Marin, G.B., Waroquier, M., VanSpeybroeck, V., Van DerVoort, P. (2010). The remarkable catalytic activity of the saturated metal organic framework V-MIL-47 in the cyclohexene oxidation, *Chem. Commun.*, **46**, 5085–5087.

120. Ryan, P., Konstantinov, I., Snurr, R.Q. J., B.L. (2012). DFT investigation of hydorperoxide decomposition over copper and cobalt sites within metal-organic frameworks, *J. Catal.*, **286**, 95–102.

121. Llabrés i Xamena, F.X., Casanova, O., Tailleur, R.G., García, H., Corma, A. (2008). Metal organic frameworks (MOFs) as catalysts: a combination of Cu^{2+} and Co^{2+} MOFs as an efficient catalyst for tetralin oxidation, *J. Catal.*, **255**, 220–227.

Index